[redacted]

# FREE VIBRATIONS
# OF
# CIRCULAR CYLINDRICAL SHELLS

# FREE VIBRATIONS OF CIRCULAR CYLINDRICAL SHELLS

*by*

ANTHONY E. ARMENÀKAS, Ph.D.
*Polytechnic Institute of Brooklyn, Brooklyn, New York*

DENOS C. GAZIS, Ph.D.
*International Business Machines, Yorktown Heights, New York*

GEORGE HERRMANN, D.Sc.
*Northwestern University, Evanston, Illinois*



THE QUEEN'S AWARD TO INDUSTRY 1966

PERGAMON PRESS
OXFORD · LONDON · EDINBURGH · NEW YORK
TORONTO · SYDNEY · PARIS · BRAUNSCHWEIG

Pergamon Press Ltd., Headington Hill Hall, Oxford
4 & 5 Fitzroy Square, London W. 1
Pergamon Press (Scotland) Ltd., 2 & 3 Teviot Place, Edinburgh 1
Pergamon Press Inc., 44-01 21st Street, Long Island City, New York 11101
Pergamon of Canada Ltd., 207 Queen's Quay West, Toronto 1
Pergamon Press (Aust.) Pty. Ltd., 19a Boundary Street, Rushcutters Bay, N.S.W. 2011 Australia
Pergamon Press S.A.R.L., 24 rue des Ecoles, Paris 5$^e$
Vieweg & Sohn GmbH, Burgplatz 1, Braunschweig


First edition 1969
Library of Congress Catalog Card No. 66–25530

*Printed in Great Britain by J. W. Arrowsmith Ltd., Bristol*
08 011732 5

# CONTENTS

## PART I



## PART II

# PART I

# PREFACE

DUE to more demanding requirements imposed upon thin-walled structures subjected to dynamic loads, there has been an increasing activity during recent years in the field of vibrations of shells. Moreover, interest in the area of wave propagation in elastic media has grown considerably as a result of novel technological applications, such as delay lines and frequency control devices.

This book is intended principally for those engaged in the aforementioned fields. The text includes a self-contained treatment of the problem of propagation of plane harmonic waves along a hollow circular cylinder, within the framework of the three-dimensional theory of elasticity. The tables present frequencies of free vibration of such cylinders for a wide range of axial and circumferential wavelengths and of shell dimensions. In addition, graphs of frequency spectra and associated mode shapes are included.

Most of the numerical results presented in this book were obtained in the course of research sponsored by the Air Force Office of Scientific Research under grant AF-AFOSR-100-64 to Northwestern University. The authors wish to acknowledge the interest in this work expressed by Dr. J. Pomerantz of the Mechanics Division of AFOSR.

The numerical results were obtained on an IBM 7094 computer. The machine program is appended for the convenience of any user who may wish to carry out computations in a range not covered by the tables given here.

The authors would like to express their appreciation to Mr. Conrad Hipkins of the Institute for Space Studies of the National Aeronautics and Space Administration in New York, N.Y.; to Mr. Daniel Hegarty of the Cooper Union for the Advancement of Science and Art, New York, N.Y.; and to Mr. H. Sheen of the IBM Thomas J. Watson Research Center for their competent assistance in the preparation of the tables.

A.E.A.
D.C.G.
G.H.

# PARTIAL LIST OF SYMBOLS

$z, \theta, r$ = cylindrical coordinates

$R$ = radius of the middle surface of the shell

$a$ = inner radius of shell

$b$ = outer radius of shell

$H$ = thickness of shell

$u_z, u_\theta, u_r$ = axial, tangential, and radial components of displacement, respectively

$e_{ij}$ = components of strain

$\sigma_{ij}$ = components of stress

$n$ = number of circumferential waves

$\xi$ = wave number in axial direction

$L$ = axial half wavelength

$\omega$ = circular frequency

$\omega_s$ = lowest frequency of the first thickness-shear mode of an infinite plate of thickness $H$

$\Omega=\omega/\omega_s$ = normalized frequency

$v_1$ = velocity of propagation of dilatational waves

$v_2$ = velocity of propagation of distortional waves

$\rho$ = density

$\lambda, \mu$ = Lamé's constants

$\nu$ = Poisson's ratio

1.

# INTRODUCTION

Elastic energy is transmitted through an infinite isotropic medium by means of two types of waves, dilatational and distortional. Each type propagates with its own speed which depends only on the elastic constants and the density of the medium. When either one of the two types impinges on a free boundary, it can be shown that, in general, waves of both types are generated. As a result, waves in bounded elastic media are generally mixtures of dilatation and distortion propagating with a speed which is no longer constant but depends on the wavelength of the disturbance. Knowledge of this dispersive character of waves in bounded elastic solids is necessary for the study of wave propagation in such media.

The solution of problems of wave propagation in bounded elastic media, on the basis of the three-dimensional theory of elasticity is, generally, difficult. Such solutions have been obtained only for infinite trains of harmonic waves traveling along infinitely long plates [1, 2], semi-infinite media [3], and bodies bounded by circular or elliptical cylindrical surfaces [4–12]. These solutions yield transcendental relationships between frequency and wavelength from which an infinite number of dispersion curves are obtained. The evaluation of these curves has been simplified by the use of modern high-speed computers. However, this is the extent to which three-dimensional elasticity has produced solutions of vibration problems. If an additional boundary is introduced, as in a finite plate or cylinder, a superposition of an infinite number of plate or cylinder modes is generally required in order to satisfy the additional boundary conditions. The complexity of such problems, as well as those associated with the treatment of transients has led to the establishment of various approximate plate and shell theories. These theories involve a finite number of modes and result in much simpler frequency equations. Furthermore, they need not be limited to circular cylinders, which are the only ones extensively treated on the basis of the theory of elasticity. In order to ascertain the validity of a shell theory, an acceptable test, short of an experiment, is a comparison of its results with those of the corresponding three-dimensional elastic solution, when available [13].

This volume contains tables of natural frequencies and graphs of representative mode shapes of harmonic elastic waves propagating in an infinitely long isotropic hollow cylinder. They have been obtained on the basis of the linear, three-dimensional theory of elasticity by solving the associated eigenvalue problem for stress-free cylindrical surfaces. Their accuracy, therefore, is limited solely by the assumptions inherent to this theory.

The tables may be used directly in obtaining satisfactory approximations to the frequency of standing waves propagated in shells of finite length. As is well known, the modal characteristics of a wave traveling in an infinite circular cylindrical shell are equivalent to those of a standing wave in a finite cylinder of length equal to an integer multiple of the half wavelength and restrained in such a manner that, at the ends, the tangential and radial components of the displacement and the axial component of the stress vanish. For other end conditions, the results obtained on the basis of the theory of elasticity offer a reasonably satisfactory approximation. This is particularly true for cylinders whose end restraints are such that the component of the axial displacement is not restricted [14], and their length is large compared to the maximum dimension of their cross-section.

The tables, moreover, offer a check on the validity, and an estimate of the range of applicability of various shell theories which have been proposed in the past or may be proposed in the future.

In the sequel, a theoretical introduction is presented followed by a description of the tables and an explanation of their use.

# FREQUENCY EQUATION

THE following derivation is based on the treatment of reference 9. The interested reader may find references to other related works in the general bibliography given in references 15, 16 and 17.

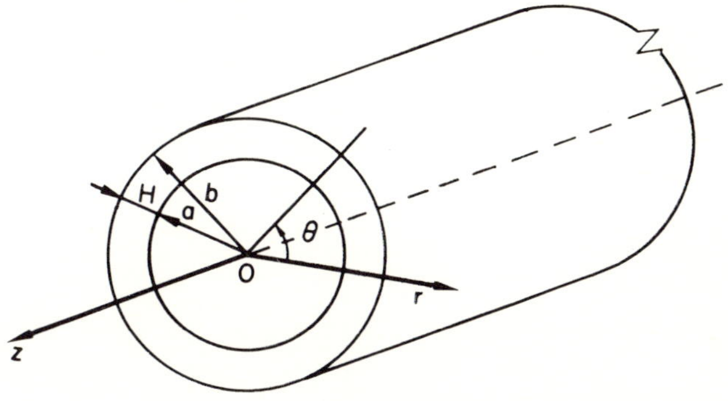


FIG. 1. Reference coordinates and dimensions.

Consider the isotropic elastic cylinder shown in Fig. 1 and extending from $-\infty$ to $+\infty$ in the $z$-direction. The equations of motion are, in invariant form,

$$\mu\nabla^2\mathbf{u}+(\lambda+\mu)\nabla\nabla\cdot\mathbf{u} = \rho(\partial^2\mathbf{u}/\partial t^2) \quad (1)$$

where $\mathbf{u}$ is the displacement vector, $\rho$ is the density, $\lambda$ and $\mu$ are Lamé's constants, and $\nabla^2$ is the three-dimensional Laplace operator. The eigenvalue problem consists of finding solutions $\mathbf{u}$ which satisfy equations (1) and the boundary conditions

$$\sigma_{rr} = \sigma_{rz} = \sigma_{r\theta} = 0 \qquad \text{at} \qquad r = a \quad \text{and} \quad r = b \quad (2)$$

This problem may be solved conveniently by using the Lamé potentials [2] $\phi$ and $\mathbf{H}$ defined by

$$\mathbf{u} = \nabla\phi+\nabla\times\mathbf{H} \quad (3)$$

with

$$\nabla\cdot\mathbf{H} = F(\mathbf{r}, t) \quad (4)$$

In equation (4), $F$ is a function of the coordinate vector $\mathbf{r}$ and the time, which can be chosen arbitrarily due to the *gauge invariance* [18] of the transformation of equation (3). The displacement equations of motion (1) are satisfied if the potentials $\phi$ and $\mathbf{H}$ satisfy the wave equations

$$\begin{aligned} v_1^2\nabla^2\phi &= \partial^2\phi/\partial t^2 \\ v_2^2\nabla^2\mathbf{H} &= \partial^2\mathbf{H}/\partial t^2 \end{aligned} \quad (5)$$

where $v_1$ and $v_2$ are the velocity of propagation in an infinite medium of dilatational and distortional waves, respectively, and are given by

$$\begin{aligned} v_1^2 &= (\lambda+2\mu)/\rho \\ v_2^2 &= \mu/\rho \end{aligned} \quad (6)$$

Assuming solutions of the wave equations (5) in the form

$$\begin{aligned} \phi &= f(r)\cos n\theta \cos(\omega t+\xi z) \\ H_r &= g_r(r)\sin n\theta \sin(\omega t+\xi z) \\ H_\theta &= g_\theta(r)\cos n\theta \sin(\omega t+\xi z) \\ H_z &= g_3(r)\sin n\theta \cos(\omega t+\xi z) \end{aligned} \tag{7}$$

and using the differential operator

$$\mathscr{B}_{n,x} = \left[\frac{\partial^2}{\partial x^2}+\frac{1}{x}\frac{\partial}{\partial x}-\left(\frac{n^2}{x^2}-1\right)\right] \tag{8}$$

we obtain

$$\begin{aligned} \mathscr{B}_{n,\alpha r}[f] &= 0 \\ \mathscr{B}_{n,\beta r}[g_3] &= 0 \\ \mathscr{B}_{n+1,\beta r}[g_r-g_\theta] &= 0 \\ \mathscr{B}_{n-1,\beta r}[g_r+g_\theta] &= 0 \end{aligned} \tag{9}$$

where

$$\alpha^2 = \omega^2/v_1^2-\xi^2 \qquad \beta^2 = \omega^2/v_2^2-\xi^2 \tag{10}$$

The general solution of the Bessel-type equations (9) is

$$\begin{aligned} f &= AZ_n(\alpha_1 r)+BW_n(\alpha_1 r) \\ g_3 &= A_3 Z_n(\beta_1 r)+B_3 W_n(\beta_1 r) \\ 2g_1 = (g_r-g_\theta) &= 2A_1 Z_{n+1}(\beta_1 r)+2B_1 W_{n+1}(\beta_1 r) \\ 2g_2 = (g_r+g_\theta) &= 2A_2 Z_{n-1}(\beta_1 r)+2B_2 W_{n-1}(\beta_1 r) \end{aligned} \tag{11}$$

where $Z_n$ and $W_n$ denote the Bessel functions $J_n$ and $Y_n$, or the modified Bessel functions, $I_n$ and $K_n$ of the arguments $\alpha_1 r = |\alpha r|$ and $\beta_1 r = |\beta r|$ depending on whether $\alpha$ and $\beta$, as determined by equations (10), are real or imaginary. The proper selections of Bessel functions to be used appear in Table 1. It may be shown that any one of the three potentials $g_i (i = 1, 2, 3)$ can be set equal to zero without any loss of generality of the solution.

TABLE 1

BESSEL FUNCTIONS USED AT DIFFERENT INTERVALS OF THE FREQUENCY

| Interval | Functions used |
|---|---|
| $v_1\xi < \omega$ | $J(\alpha r)$, $Y(\alpha r)$, $J(\beta r)$, $Y(\beta r)$ |
| $v_2\xi < \omega < v_1\xi$ | $I(\alpha_1 r)$, $K(\alpha_1 r)$, $J(\beta r)$, $Y(\beta r)$ |
| $\omega < v_2\xi$ | $I(\alpha_1 r)$, $K(\alpha_1 r)$, $I(\beta_1 r)$, $K(\beta_1 r)$ |

After setting $g_r = -g_\theta = g_1$, the displacements, strains and stresses are expressed in terms of the functions $f$, $g_1$, and $g_3$, by substituting equations (3), (7), and (11) into the well-known displacement–strain and stress–strain relationships of linear elasticity theory. The displacements are given by

$$\begin{aligned} u_r &= [f'+(n/r)g_3+\xi g_1]\cos n\theta \cos(\omega t+\xi z) \\ u_\theta &= [-(n/r)f+\xi g_1-g_3']\sin n\theta \cos(\omega t+\xi z) \\ u_z &= [-\xi f-g_1'-(n+1)(g_1/r)]\cos n\theta \sin(\omega t+\xi z) \end{aligned} \tag{12}$$

and the stresses by

$$\sigma_{rr} = \left\{ -\lambda(a^2+\xi^2)f + 2\mu\left[ f'' + \frac{n}{r}\left( g_3' - \frac{g_3}{r} \right) + \xi g_1' \right] \right\} \cos n\theta \cos(\omega t + \xi z)$$

$$\sigma_{r\theta} = \mu\left\{ -\frac{2n}{r}\left( f' - \frac{f}{r} \right) - (2g_3'' + \beta^2 g_3) - \xi\left( \frac{n+1}{r} g_1 - g_1' \right) \right\} \sin n\theta \cos(\omega t + \xi z) \tag{13}$$

$$\sigma_{rz} = \mu\left\{ -2\xi f' - \frac{n}{r} g_1' + \left[ -\frac{n(n+1)}{r^2} + (\beta^2 - \xi^2) \right] g_1 - \frac{n\xi}{r} g_3 \right\} \cos n\theta \sin(\omega t + \xi z)$$

where primes denote differentiation with respect to $r$. Entering equations (11) in the boundary conditions (2), one finds the homogeneous system of linear equations

$$c_{ij} X_j = 0 \qquad (i, j = 1, 6) \tag{14}$$

where $X_j$ denote the vector $X_j = \{A, B, A_1, B_1, A_3, B_3\}$ and $c_{ij}$ the matrix whose first three rows are given by

$$\begin{aligned}
c_{11} &= [2n(n-1) - (\beta^2 - \xi^2)a^2] Z_n(\alpha_1 a) + 2\lambda_1 \alpha_1 a Z_{n+1}(\alpha_1 a) \\
c_{12} &= 2\xi\beta_1 a^2 Z_n(\beta_1 a) - 2\xi a(n+1) Z_{n+1}(\beta_1 a) \\
c_{13} &= -2n(n-1) Z_n(\beta_1 a) + 2\lambda_2 n\beta_1 a Z_{n+1}(\beta_1 a) \\
c_{14} &= [2n(n-1) - (\beta^2 - \xi^2)a^2] W_n(\alpha_1 a) + 2\alpha_1 a W_{n+1}(\alpha_1 a) \\
c_{15} &= 2\lambda_2 \xi\beta_1 a^2 W_n(\beta_1 a) - 2(n+1)\xi a W_{n+1}(\beta_1 a) \\
c_{16} &= -2n(n-1) W_n(\beta_1 a) + 2n\beta_1 a W_{n+1}(\beta_1 a) \\
c_{21} &= 2n(n-1) Z_n(\alpha_1 a) - 2\lambda_1 n\alpha_1 a Z_{n+1}(\alpha_1 a) \\
c_{22} &= -\xi\beta_1 a^2 Z_n(\beta_1 a) + 2\xi a(n+1) Z_{n+1}(\beta_1 a) \\
c_{23} &= -[2n(n-1) - \beta^2 a^2] Z_n(\beta_1 a) - 2\lambda_2 \beta_1 a Z_{n+1}(\beta_1 a) \\
c_{24} &= 2n(n-1) W_n(\alpha_1 a) - 2n\alpha_1 a W_{n+1}(\alpha_1 a) \\
c_{25} &= -\lambda_2 \xi\beta_1 a^2 W_n(\beta_1 a) + 2\xi a(n+1) W_{n+1}(\beta_1 a) \\
c_{26} &= -[2n(n-1) - \beta^2 a^2] W_n(\beta_1 a) - 2\beta_1 a W_{n+1}(\beta_1 a) \\
c_{31} &= -2n\xi a Z_n(\alpha_1 a) + 2\lambda_1 \xi\alpha_1 a^2 Z_{n+1}(\alpha_1 a) \\
c_{32} &= -n\beta_1 a Z_n(\beta_1 a) + (\beta^2 - \xi^2) a^2 Z_{n+1}(\beta_1 a) \\
c_{33} &= n\xi a Z_n(\beta_1 a) \\
c_{34} &= -2n\xi a W_n(\alpha_1 a) + 2\xi\alpha_1 a^2 W_{n+1}(\alpha_1 a) \\
c_{35} &= -\lambda_2 n\beta_1 a W_n(\beta_1 a) + (\beta^2 - \xi^2) a^2 W_{n+1}(\beta_1 a) \\
c_{36} &= n\xi a W_n(\beta_1 a)
\end{aligned} \tag{15}$$

The remaining three rows are obtained from the first three by substitution of $b$ for $a$. In equations (15) $\lambda_1$ and $\lambda_2$ are parameters which are introduced in order to account for the differences in the recursion and differentiation formulas between the different kinds of Bessel functions. The value of $\lambda_j$ is 1 when $J$ and $Y$ functions are used and $-1$ when $I$ and $K$ functions are used.

A nontrivial solution of equation (14) requires the vanishing of the determinant $|c_{ij}|$. Hence,

$$|c_{ij}| = 0 \qquad (i, j \neq 1, \ldots, 6) \tag{16}$$

is the frequency equation. For given dimensions and elastic constants, equation (16) is an implicit transcendental function of $H/L$ and $\omega/\omega_s$ where $L$ is the wavelength and

$$\omega_s = \pi v_2/H \tag{17}$$

is the lowest simple thickness-shear frequency of an infinite plate of thickness $H$ and the same elastic constants as those of the cylinder. The amplitudes $X_j$ are proportional to the cofactors of any row of the matrix $c_{ij}$ which satisfies the frequency equation (16). Given these amplitudes, the dependence of the displacements $u_r$, $u_\theta$, and $u_z$ on the radius $r$ can be computed using equations (3), (7), and (11). This was done in this volume for a few of the computed natural modes, in order to show the variation of the mode-shape with a variation of the geometric parameters of the problem.

The frequency equation (16) degenerates into simpler forms for a number of special cases, as follows:

1. *Motion independent of z.* For infinitely long wavelengths, i.e. for $\xi = 0$, equation (16) breaks into the product of subdeterminants

$$D_1 D_2 = 0 \tag{18}$$

where

$$D_1 = \begin{vmatrix} c_{11} & c_{13} & c_{14} & c_{16} \\ c_{21} & c_{23} & c_{24} & c_{26} \\ c_{41} & c_{43} & c_{44} & c_{46} \\ c_{51} & c_{53} & c_{54} & c_{56} \end{vmatrix} \quad \text{and} \quad D_2 = \begin{vmatrix} c_{32} & c_{35} \\ c_{62} & c_{65} \end{vmatrix} \tag{19}$$

and the terms $c_{ij}$ are given by equations (15) with $\xi = 0$. Equation (18) is satisfied if either $D_1$ or $D_2$ is equal to zero. The case $D_1 = 0$ corresponds to *plane-strain vibrations* [8], in which the displacement component $u_z$ is equal to zero. The case $D_2 = 0$ corresponds to *longitudinal* (or axial) *shear* vibrations, i.e. motion involving only longitudinal displacements $u_z$. These two types of motion are uncoupled when the motion is independent of the longitudinal coordinate $z$, but they are coupled for a nonzero wave-number $\xi$.

2. *Motion independent of $\theta$* (axially symmetric). In this case the frequency equation (16) degenerates into

$$D_3 D_4 = 0 \tag{20}$$

where

$$D_3 = \begin{vmatrix} c_{11} & c_{12} & c_{14} & c_{15} \\ c_{31} & c_{32} & c_{34} & c_{35} \\ c_{41} & c_{42} & c_{44} & c_{45} \\ c_{61} & c_{62} & c_{64} & c_{65} \end{vmatrix} \qquad D_4 = \begin{vmatrix} c_{23} & c_{26} \\ c_{53} & c_{56} \end{vmatrix} \tag{21}$$

and the terms $c_{ij}$ are given by equation (16) with $n = 0$. Equation $D_3 = 0$ corresponds to *longitudinal waves*, involving only displacements $u_r$ and $u_z$ independent of $\theta$. Equation $D_4 = 0$ corresponds to *torsional waves*, involving nodal cylinders. The lowest torsional mode which entails the rotation of each transverse section of the cylinder as a whole about its center is not adequately described by the Bessel equations (9). This mode corresponds to $\beta^2 = 0$ and, therefore,

$$\omega = \xi v_2 \tag{22}$$

When both $n$ and $\xi$ vanish, the frequency equation for plane strain degenerates to

$$D_1 = D_5 D_6 = 0$$

where

$$D_5 = \begin{vmatrix} c_{11} & c_{14} \\ c_{41} & c_{44} \end{vmatrix} \qquad D_6 = \begin{vmatrix} c_{23} & c_{26} \\ c_{53} & c_{56} \end{vmatrix} \tag{23}$$

Equations $D_5 = 0$ and $D_6 = 0$ correspond to radial and shear modes, respectively. A more detailed discussion of all the degenerate cases is given in references 8 and 9.

The aforementioned special cases are characterized by the fact that the eigenmodes of the hollow cylinder are separated into uncoupled families of modes with restricted displacement fields. For example, when $\xi = 0$, the motion in the plane $(r, z)$ is uncoupled from the motion in the longitudinal direction, $z$. When $n = 0$ the transverse shear motion, along the $\theta$-direction, is uncoupled from the motion in the plane $(r, z)$. When both $n$ and $\xi$ are equal to zero, the eigenmodes are separated into three uncoupled families of modes each associated with a displacement in the direction $r$, $\theta$, or $z$. In this latter case, the determinant $D_6$ becomes identical to $D_4$, implying that the frequencies of the plane strain, axisymmetric shear modes are the cut-off frequencies ($\xi = 0$) of the torsional waves. The frequency of the $j$th torsional modes corresponding to a wave number $\xi \neq 0$ is given by

$$\omega_j = [\omega_{0j}^2 + \xi^2 v_2^2]^{\frac{1}{2}} \tag{24}$$

where $\omega_{0j}$ is the $j^{th}$ cut-off frequency.

These three uncoupled families of modes are analogous to those of uncoupled thickness modes in an infinite isotropic plate, namely, the thickness-stretch modes and the thickness-shear modes in two mutually perpendicular directions. An exception is the lowest mode associated with motion in the $r$-direction, generally referred to as the "breathing mode". This mode corresponds to a unidirectional radial displacement across the thickness of the cylinder. In the case of a plate, the counterpart of this mode is the rigid displacement normal to the plane of the plate. When both $\xi$ and $n$ are different from zero, the motion in the $r$- and $z$-directions is coupled. The coupling of all the eigenmodes is reflected in the absence of intersections amongst the various dispersion curves.

# NUMERICAL COMPUTATIONS

THE six first roots $\omega/\omega_s$ of the frequency equation (16) were computed for values of $H/L$ in the interval $0 \leqq H/L \leqq 1.0$, and for a wide range of the geometric parameters. The numerical computations were done on a 7094 IBM computer using the following "Interval halving" iteration routine.

For a fixed value of $H/L$, the determinant (16) is a function of $\Omega = \omega/\omega_s$ alone. It is evaluated starting from an initial value $\Omega_i$ and proceeding at discrete intervals $\Delta\Omega$ up to a prescribed upper bound $\Omega_u$. A change of sign of the determinant across a $\Delta\Omega$ interval indicates a root in that interval. The interval is then halved and the direction of scanning is reversed. The process is repeated until a root is obtained within a prescribed accuracy. Additional roots are obtained by re-entering successively into the iteration subroutine with a new initial value of $\Omega$ just above the last computed root.

This subroutine is particularly effective for finding a long sequence of roots of a sine-like function. However, if the scanning interval is so large that two roots may exist between two successive evaluation points, no change of sign of the function will be observed, and consequently the two roots will be missed.

The general equation (16) was used in the absence of any special symmetry, while the equations $D_j = 0$ $(j = 1, 4)$ given in (19) and (21) were used for special degenerate cases.

The amplitudes $u_r$, $u_\theta$, and $u_z$ were computed at twenty points across the thickness of the cylinder, as explained in the preceding section. They were normalized with respect to the largest of the sixty values of the displacement components computed for each mode.

The following range of parameters was used in the computation:

Poisson's ratio: $\nu = 0.30$.

Ratio of thickness to mean radius: $0.01 \leqq H/R \leqq 2$ (the case $H/R = 2$ corresponds to a solid cylinder).

Ratio of thickness to half-wavelength: $0 \leqq H/L \leqq 1.00$.

Number of waves around the circumference: $n = 0, 1, 2, 3, 4$.

The accuracy of the computations, i.e. the ratio of the possible error to the value of the root, is $10^{-5}$.

The first six eigenvalues were computed for all values of $n$ except $n = 0$, the axisymmetric case. In this case six roots were computed for the family of longitudinal modes, $D_3 = 0$ in equation (21). Moreover, the frequency of the first torsional mode may be obtained from relation (22), and the frequencies of the two higher torsional modes can be readily established by introducing their cut-off frequencies tabulated on page 17, into equation (24). Nine eigenmodes are given for the axisymmetric motion, inasmuch as this case has a greater range of applicability since the axisymmetric modes are more easily excited both in the laboratory and in practice.

# DISCUSSION OF THE RESULTS

For a few of the computed modes, the results of the numerical computations are illustrated in the plots of the dispersion curves (pages 169–180) and the mode-shapes (pages 181–201). On the basis of these plots, the following general remarks may be made:

It is convenient to identify the various curves by observing the nature of the corresponding modes for very long wavelengths. For each value of the wavelength, the lowest three of the six modes computed are associated with displacements that do not change sign across the thickness of the cylinder. The displacements of the upper three modes involve a change of sign and can be termed, essentially, thickness modes. The three lowest modes are those contained in shell theories such as the bending theory in which the radial component of displacement is assumed constant across the thickness, whereas the tangential and axial components are assumed to vary linearly with the thickness coordinate.

For $n \geqq 1$, the lowest mode is the flexural mode, associated with large radial displacements; the second higher mode is an axial shear mode, associated with large $z$-displacements; the third is a "breathing mode", with large tangential displacements; the fourth and fifth are thickness-shear modes associated with motions in the $z$- and $\theta$-directions, respectively. Finally, the sixth mode is a thickness-stretch mode with predominantly radial displacements.

For $n = 0$, the lowest mode is the torsional one followed by the lowest extensional mode and the breathing mode. Then come a circumferential shear (torsional) mode, an axial shear mode, and a thickness-stretch mode, all three characterized by one cylindrical nodal surface near the middle of the thickness. In the case of $n = 0$, two higher thickness modes are also evaluated, a torsional and an axial shear mode, involving two nodal surfaces across the thickness. In addition, a higher axial shear mode is given which involves three nodal surfaces.

Dispersion is observed for every mode with the exception of the lowest torsional mode which is identical to that described by the elementary torsion theory.* There are two dispersion parameters in the case of a hollow cylinder, its radius, $R$, and its thickness, $H$. The influence of these parameters on dispersion will be followed in the example of the lowest dispersion curve of the upper figure of page 171, which contains the dispersion curves for $n = 1$, and $H/R = 0.1$, i.e. a relatively thin-walled cylinder. When the wavelength, $2L$, is very large in comparison with $R$, the lowest mode is essentially a bending mode in which the entire cross-section moves almost uniformly in a direction normal to the axis of the cylinder. This mode is fairly adequately described by ordinary beam theory, which predicts no dispersion. Dispersion first becomes noticeable when the half-wavelength becomes comparable to the radius, i.e. for $H/L \approx 0.1$, $R/L \approx 1.0$. Thereafter, for increasing values of $H/L$ the motion of the cylinder is predominantly a shearing motion across its diameter. When $L$ approaches values comparable to $H$, the displacements first approximate those of a plate of the same thickness, $H$, and then, as $H/L \to \infty$, they tend to become surface wave displacements, confined near the two surfaces of the cylinder.

An inspection of the dispersion curves for relatively thin-walled cylinders reveals that these curves are almost independent of $H/R$ and $n$, with the exception of a small region near the origin. This is understandable if it is remembered that the eigenmodes associated with points everywhere except in this small region near the origin can be described as thickness vibration modes. As a

*See reference 9, p. 572.

consequence, the controlling factor in the nature of dispersion is the ratio $H/L$. Variation of $n$ changes the effective wavelength, $2\pi R/n$, in the circumferential direction. However, the ratio of this wavelength to the wall thickness remains small for thin-walled cylinders, and, hence, the effect of the variation in $n$ considered in the computations is negligible. It should be observed that the ratio $H/R$ does affect both the values of the cut-off frequencies, i.e. the frequencies for $H/L = 0$, and the nature of the dispersion curves in certain limited regions. This can be seen if one goes to the limit $H/R \to 0$ when the dispersion curves become those of an infinite plate. In this case the cut-off frequencies tend to the Lamb thickness-vibration frequencies [2], and the eigenmodes are separated into two uncoupled families. The first of these is the family of the Rayleigh–Lamb modes [1, 2] corresponding to motions in the sagittal plane, i.e. the plane defined by the normal to the plane of the plate and the direction of propagation. The other is the family of face-shear modes associated with motion in the plane of the plate normal to the direction of propagation. The dispersion curves corresponding to these two families are either tangent or intersecting at various points. The corresponding dispersion curves of a thin shell have the typical configuration of "weak coupling" (Fig. 2) of the plate modes, namely, they come close

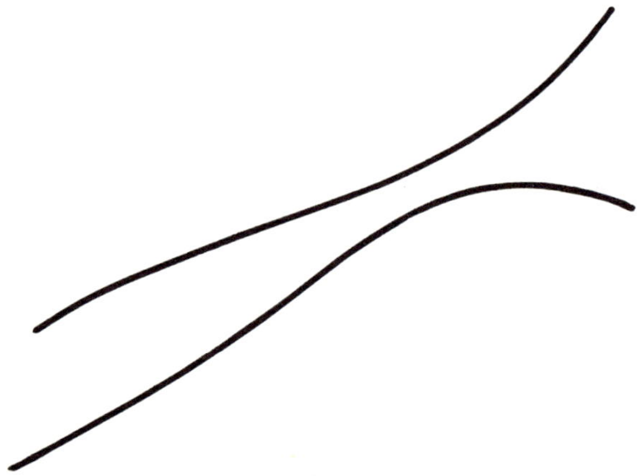

FIG. 2. "Weak coupling" of modes.

together near the region of intersection or contact of the dispersion curves of the plate. The degree of coupling, and consequently the degree of separation of the dispersion curves of the cylinder, is affected by the ratio $H/R$.

The rate at which energy is transported in a medium is proportional to the group velocity, i.e. the slope of the dispersion curves. From a physical consideration, therefore, the group velocity is positive. The dispersion curve of the sixth mode exhibits a minimum and, hence, to the right of this minimum the slope is positive, and to the left it is negative. This can be justified by noting that the frequency equation (16) will yield, for negative values of $H/L$, dispersion curves which are reflections of those for positive $H/L$ about the ordinate axis. When the group velocity of a portion of a dispersion curve is negative, its reflected image corresponds to a positive group velocity and a negative phase velocity.

# EXPLANATION OF THE FREQUENCY TABLES AND GRAPHS OF THE MODE SHAPES

ON every page of the tables, for a specified value of $n$ and $H/R$, the normalized frequencies of the first three or six modes are tabulated for a number of values of $H/L$, ranging from infinite wavelength ($H/L = 0$) to $H/L = 1.0$. Frequencies of the first six modes are tabulated for $H/R = 0.01$ and for $H/R$ between 0.1 and 2.0 at intervals of 0.1. However, for small values of $H/R$, the frequencies of the first three modes vary considerably with $H/R$ and, hence, they are tabulated for some intermediate values of $H/R$. The torsional modes are excluded from the tables inasmuch as the first mode is readily obtainable from (22), and the higher torsional modes may be obtained on the basis of equation (24), using the values of cut-off frequencies given in the first page of the tables.

The dispersion curves are plotted on pages 169 to 180 for $H/R = 0.01$, 0.1, 1.0, and 2.0, and for all values of $n$. The mode shapes of the first six modes are illustrated on pages 181 to 201. For axisymmetric motion ($n = 0$) the torsional modes are plotted separately on page 181. On each page, the normalized displacement components, $u_r$, $u_\theta$, $u_z$, are plotted as functions of the radial coordinate for specified values of $H/R$, $H/L$, and $n$. Mode shapes are plotted for $n = 0$, 1, 2, 3, 4, for $H/R = 0.1$ and 1.0, and for $H/L = 0.1$ and 1.0.

# BIBLIOGRAPHY


1. LORD RAYLEIGH, "On the Free Vibrations of an Infinite Plate of Homogeneous Isotropic Elastic Matter", *Proc. London Math. Soc.* **20**, 225–34 (1889).

2. H. LAMB, "On Waves in an Elastic Plate", *Proc. Royal Soc. London* **A93**, 114–28 (1917).

3. LORD RAYLEIGH, "On Waves Propagated Along the Plane Surface of an Elastic Solid", *Proc. London Math. Soc.* **17**, 4–11 (1887).

4. L. POCHHAMMER, "Über die Fortpflanzungsgeschwindigkeiten Kleiner Schwingungen in Einem Unbegrenzten Isotropen Kreiscylinder", *Z. Math.* **81**, 324–36 (1876).

5. C. CHREE, "The Equations of an Isotropic Elastic Solid in Polar and Cylindrical Coordinates, Their Solutions and Applications", *Trans. Cambridge Phil. Soc.* **14**, 250 (1889).

6. J. GHOSH, "Longitudinal Vibrations of a Hollow Cylinder", *Bull. Calcutta Math. Soc.* **14**, 31–40 (1923–24).

7. G. HERRMANN and I. MIRSKY, "Three-Dimensional and Shell Theory Analysis of Axially-Symmetric Motions of Cylinders", *J. Appl. Mech.* **23**, 4, 563–8 (1956).

8. D. C. GAZIS, "Exact Analysis of the Plane-Strain Vibrations of Thick-Walled Hollow Cylinders", *J. Accoust. Soc. Am.* **30**, 786–94 (1958).

9. D. C. GAZIS, "Three-Dimensional Investigation of the Propagation of Waves of Hollow Circular Cylinders—I. Analytical Foundation, II. Numerical Results", *J. Acoust. Soc. Am.* **31**, 568–78 (1959).

10. J. E. GREENSPON, "Flexural Vibrations of a Thick-Walled Hollow Cylinder", *Proc. Third U.S. National Congress Applied Mechanics* (1958), pp. 163–73.

11. P. K. WONG, J. MIKLOWITZ, and R. A. SCOTT, "Propagation of Harmonic Flexural Waves in an Infinite Elastic Rod of Elliptical Cross Section", J. Acoust. Soc. Am. **40**, 393–98 (1966).

12. A. E. ARMENÀKAS, "Propagation of Harmonic Waves in Composite Circular Cylindrical Shells 1: Theoretical Investigation" A.I.A.A.J., Vol. 5 No. 4 April 1967 pp. 740–744.

13. A. E. ARMENÀKAS, "On the Accuracy of Some Dynamic Shell Theories" J. Eng. Mech. Div. ASCE Vol. 93 No. EM5 October 1966 pp. 95–109.

14. K. FORSBERG, "Influence of Boundary Conditions on the Modal Characteristics of Thin Cylindrical Shells", *A.I.A.A. J.*, Vol. 2, No. 12, pp. 2150–57, Dec. 1964.

15. C. G. GROSS and K. FORSBERG, "Vibrations of Thin Shells—A Partially Annotated Bibliography". SB–63–43, Lockheed Missiles and Space Company (April 1963).

16. W. A. NASH, "Bibliography on Shells and Shell-Like Structures", David W. Taylor Model Basin, Report No. 863 (Nov. 1954); also Supplement, University of Florida, Dept. of Eng. Mechanics (June 1957).

17. N. H. ABRAMSON, H. J. PLASS, and E. A. RIPPERGER, "Stress Wave Propagation in Rods and Beams", *Advances in Applied Mechanics* (Academic Press, New York, 1958), Vol. V, pp. 111–94.

18. See, for example, P. M. Morse and H. Feshbach, *Methods of Theoretical Physics* (McGraw-Hill Book Company, New York, 1953), Part 1, p. 297.

# PART II

CUT-OFF FREQUENCIES OF TORSIONAL MODES

| $H/R$ | $\Omega_1$ | $\Omega_2$ |
|---|---|---|
| 0.010 | 1.00002 | 2.00001 |
| 0.020 | 1.00008 | 2.00004 |
| 0.030 | 1.00017 | 2.00009 |
| 0.040 | 1.00030 | 2.00015 |
| 0.050 | 1.00048 | 2.00024 |
| 0.060 | 1.00068 | 2.00034 |
| 0.070 | 1.00093 | 2.00047 |
| 0.080 | 1.00122 | 2.00061 |
| 0.090 | 1.00154 | 2.00077 |
| 0.100 | 1.00190 | 2.00095 |
| 0.120 | 1.00274 | 2.00137 |
| 0.140 | 1.00372 | 2.00187 |
| 0.160 | 1.00486 | 2.00244 |
| 0.180 | 1.00616 | 2.00310 |
| 0.200 | 1.00760 | 2.00383 |
| 0.250 | 1.01188 | 2.00601 |
| 0.300 | 1.01710 | 2.00870 |
| 0.350 | 1.02328 | 2.01191 |
| 0.400 | 1.03041 | 2.01567 |
| 0.450 | 1.03849 | 2.02000 |
| 0.500 | 1.04752 | 2.02491 |
| 0.600 | 1.06842 | 2.03666 |
| 0.700 | 1.09310 | 2.05121 |
| 0.800 | 1.12152 | 2.06894 |
| 0.900 | 1.15360 | 2.09034 |
| 1.000 | 1.18920 | 2.11601 |
| 1.100 | 1.22812 | 2.14666 |
| 1.200 | 1.27002 | 2.18306 |
| 1.300 | 1.31445 | 2.22601 |
| 1.400 | 1.36077 | 2.27616 |
| 1.500 | 1.40822 | 2.33372 |
| 1.600 | 1.45593 | 2.39808 |
| 1.700 | 1.50305 | 2.46753 |
| 1.800 | 1.54885 | 2.53937 |
| 1.900 | 1.59281 | 2.61066 |
| 2.000 | 1.63472 | 2.67929 |

$H/R = 0.010 \qquad n = 0.$

| $H/L$ | $\Omega_1$ | $\Omega_2$ | $\Omega_3$ | $\Omega_4$ | $\Omega_5$ | $\Omega_6$ |
|---|---|---|---|---|---|---|
| 0. | 0. | 0.00538 | 1.00001 | 1.87084 | 2.00000 | 3.00002 |
| 0.001 | 0.00160 | 0.00541 | 1.00001 | 1.87084 | 2.00001 | 3.00000 |
| 0.005 | 0.00499 | 0.00869 | 1.00005 | 1.87078 | 2.00009 | 3.00001 |
| 0.010 | 0.00511 | 0.01699 | 1.00018 | 1.87059 | 2.00036 | 3.00002 |
| 0.020 | 0.00516 | 0.03384 | 1.00070 | 1.86981 | 2.00135 | 3.00000 |
| 0.030 | 0.00531 | 0.05073 | 1.00156 | 1.86855 | 2.00299 | 2.99998 |
| 0.040 | 0.00568 | 0.06762 | 1.00275 | 1.86685 | 2.00525 | 2.99995 |
| 0.050 | 0.00639 | 0.08452 | 1.00429 | 1.86474 | 2.00806 | 2.99990 |
| 0.060 | 0.00749 | 0.10140 | 1.00617 | 1.86227 | 2.01137 | 2.99985 |
| 0.070 | 0.00903 | 0.11829 | 1.00838 | 1.85949 | 2.01515 | 2.99980 |
| 0.080 | 0.01097 | 0.13517 | 1.01091 | 1.85646 | 2.01935 | 2.99973 |
| 0.090 | 0.01327 | 0.15204 | 1.01377 | 1.85321 | 2.02392 | 2.99966 |
| 0.100 | 0.01590 | 0.16891 | 1.01696 | 1.84980 | 2.02880 | 2.99958 |
| 0.120 | 0.02213 | 0.20262 | 1.02425 | 1.84258 | 2.03946 | 2.99941 |
| 0.140 | 0.02951 | 0.23629 | 1.03275 | 1.83504 | 2.05104 | 2.99922 |
| 0.160 | 0.03795 | 0.26992 | 1.04240 | 1.82736 | 2.06338 | 2.99902 |
| 0.180 | 0.04738 | 0.30350 | 1.05315 | 1.81967 | 2.07639 | 2.99878 |
| 0.200 | 0.05773 | 0.33701 | 1.06495 | 1.81206 | 2.08993 | 2.99856 |
| 0.220 | 0.06892 | 0.37046 | 1.07772 | 1.80462 | 2.10395 | 2.99832 |
| 0.240 | 0.08091 | 0.40383 | 1.09141 | 1.79739 | 2.11841 | 2.99810 |
| 0.260 | 0.09364 | 0.43711 | 1.10597 | 1.79043 | 2.13324 | 2.99790 |
| 0.280 | 0.10704 | 0.47029 | 1.12133 | 1.78378 | 2.14843 | 2.99768 |
| 0.300 | 0.12106 | 0.50338 | 1.13743 | 1.77747 | 2.16393 | 2.99753 |
| 0.320 | 0.13566 | 0.53634 | 1.15424 | 1.77154 | 2.17974 | 2.99741 |
| 0.340 | 0.15079 | 0.56917 | 1.17170 | 1.76600 | 2.19583 | 2.99731 |
| 0.360 | 0.16640 | 0.60185 | 1.18974 | 1.76089 | 2.21218 | 2.99729 |
| 0.380 | 0.18244 | 0.63438 | 1.20835 | 1.75623 | 2.22879 | 2.99731 |
| 0.400 | 0.19890 | 0.66674 | 1.22747 | 1.75205 | 2.24562 | 2.99741 |
| 0.420 | 0.21571 | 0.69890 | 1.24706 | 1.74836 | 2.26270 | 2.99761 |
| 0.440 | 0.23287 | 0.73085 | 1.26710 | 1.74520 | 2.27997 | 2.99788 |
| 0.460 | 0.25032 | 0.76257 | 1.28753 | 1.74259 | 2.29746 | 2.99824 |
| 0.480 | 0.26806 | 0.79403 | 1.30833 | 1.74056 | 2.31515 | 2.99873 |
| 0.500 | 0.28605 | 0.82522 | 1.32948 | 1.73913 | 2.33303 | 2.99930 |
| 0.520 | 0.30427 | 0.85610 | 1.35093 | 1.73834 | 2.35111 | 3.00001 |
| 0.540 | 0.32269 | 0.88665 | 1.37267 | 1.73821 | 2.36934 | 3.00086 |
| 0.560 | 0.34130 | 0.91683 | 1.39468 | 1.73879 | 2.38775 | 3.00185 |
| 0.580 | 0.36008 | 0.94661 | 1.41692 | 1.74011 | 2.40635 | 3.00299 |
| 0.600 | 0.37902 | 0.97595 | 1.43937 | 1.74220 | 2.42508 | 3.00430 |
| 0.620 | 0.39810 | 1.00480 | 1.46202 | 1.74510 | 2.44398 | 3.00577 |
| 0.640 | 0.41730 | 1.03315 | 1.48484 | 1.74886 | 2.46302 | 3.00741 |
| 0.660 | 0.43662 | 1.06092 | 1.50782 | 1.75353 | 2.48220 | 3.00924 |
| 0.680 | 0.45604 | 1.08808 | 1.53094 | 1.75913 | 2.50151 | 3.01127 |
| 0.700 | 0.47555 | 1.11458 | 1.55418 | 1.76574 | 2.52097 | 3.01350 |
| 0.720 | 0.49515 | 1.14037 | 1.57752 | 1.77340 | 2.54055 | 3.01594 |
| 0.740 | 0.51482 | 1.16543 | 1.60095 | 1.78212 | 2.56025 | 3.01862 |
| 0.760 | 0.53456 | 1.18968 | 1.62447 | 1.79197 | 2.58008 | 3.02150 |
| 0.780 | 0.55435 | 1.21311 | 1.64803 | 1.80298 | 2.60000 | 3.02463 |
| 0.800 | 0.57420 | 1.23567 | 1.67164 | 1.81517 | 2.62004 | 3.02802 |
| 0.820 | 0.59410 | 1.25735 | 1.69528 | 1.82858 | 2.64019 | 3.03166 |
| 0.840 | 0.61403 | 1.27814 | 1.71894 | 1.84318 | 2.66042 | 3.03554 |
| 0.860 | 0.63401 | 1.29803 | 1.74260 | 1.85900 | 2.68079 | 3.03972 |
| 0.880 | 0.65401 | 1.31703 | 1.76625 | 1.87603 | 2.70122 | 3.04416 |
| 0.900 | 0.67405 | 1.33515 | 1.78988 | 1.89423 | 2.72175 | 3.04890 |
| 0.920 | 0.69411 | 1.35245 | 1.81348 | 1.91357 | 2.74236 | 3.05395 |
| 0.940 | 0.71418 | 1.36896 | 1.83703 | 1.93399 | 2.76304 | 3.05927 |
| 0.960 | 0.73428 | 1.38472 | 1.86051 | 1.95545 | 2.78381 | 3.06494 |
| 0.980 | 0.75439 | 1.39979 | 1.88394 | 1.97789 | 2.80464 | 3.07089 |
| 1.000 | 0.77452 | 1.41423 | 1.90726 | 2.00124 | 2.82556 | 3.07719 |

$H/R = 0.100 \qquad n = 0.$

| $H/L$ | $\Omega_1$ | $\Omega_2$ | $\Omega_3$ | $\Omega_4$ | $\Omega_5$ | $\Omega_6$ |
|---|---|---|---|---|---|---|
| 0. | 0. | 0.05387 | 1.00038 | 1.87168 | 2.00019 | 3.00014 |
| 0.001 | 0.00161 | 0.05387 | 1.00039 | 1.87168 | 2.00021 | 3.00014 |
| 0.005 | 0.00805 | 0.05393 | 1.00043 | 1.87162 | 2.00029 | 3.00014 |
| 0.010 | 0.01605 | 0.05412 | 1.00056 | 1.87142 | 2.00054 | 3.00015 |
| 0.020 | 0.03141 | 0.05528 | 1.00108 | 1.87065 | 2.00154 | 3.00012 |
| 0.030 | 0.04358 | 0.05977 | 1.00194 | 1.86939 | 2.00320 | 3.00010 |
| 0.040 | 0.04858 | 0.07150 | 1.00314 | 1.86767 | 2.00546 | 3.00007 |
| 0.050 | 0.05003 | 0.08686 | 1.00468 | 1.86555 | 2.00828 | 3.00002 |
| 0.060 | 0.05068 | 0.10310 | 1.00655 | 1.86306 | 2.01160 | 2.99998 |
| 0.070 | 0.05115 | 0.11964 | 1.00876 | 1.86028 | 2.01540 | 2.99993 |
| 0.080 | 0.05162 | 0.13630 | 1.01130 | 1.85723 | 2.01960 | 2.99988 |
| 0.090 | 0.05219 | 0.15302 | 1.01416 | 1.85397 | 2.02419 | 2.99980 |
| 0.100 | 0.05292 | 0.16977 | 1.01735 | 1.85054 | 2.02910 | 2.99973 |
| 0.120 | 0.05503 | 0.20332 | 1.02464 | 1.84330 | 2.03975 | 2.99956 |
| 0.140 | 0.05824 | 0.23689 | 1.03315 | 1.83575 | 2.05134 | 2.99937 |
| 0.160 | 0.06284 | 0.27045 | 1.04282 | 1.82805 | 2.06370 | 2.99915 |
| 0.180 | 0.06879 | 0.30397 | 1.05358 | 1.82034 | 2.07671 | 2.99893 |
| 0.200 | 0.07614 | 0.33744 | 1.06538 | 1.81272 | 2.09028 | 2.99868 |
| 0.220 | 0.08480 | 0.37085 | 1.07816 | 1.80527 | 2.10431 | 2.99846 |
| 0.240 | 0.09468 | 0.40420 | 1.09186 | 1.79803 | 2.11878 | 2.99824 |
| 0.260 | 0.10564 | 0.43746 | 1.10642 | 1.79106 | 2.13362 | 2.99802 |
| 0.280 | 0.11757 | 0.47063 | 1.12179 | 1.78441 | 2.14880 | 2.99783 |
| 0.300 | 0.13037 | 0.50369 | 1.13790 | 1.77809 | 2.16432 | 2.99766 |
| 0.320 | 0.14394 | 0.53664 | 1.15472 | 1.77214 | 2.18013 | 2.99753 |
| 0.340 | 0.15819 | 0.56947 | 1.17218 | 1.76660 | 2.19622 | 2.99746 |
| 0.360 | 0.17306 | 0.60215 | 1.19024 | 1.76148 | 2.21258 | 2.99744 |
| 0.380 | 0.18848 | 0.63467 | 1.20886 | 1.75681 | 2.22918 | 2.99746 |
| 0.400 | 0.20438 | 0.66702 | 1.22798 | 1.75261 | 2.24601 | 2.99756 |
| 0.420 | 0.22073 | 0.69918 | 1.24758 | 1.74893 | 2.26309 | 2.99775 |
| 0.440 | 0.23747 | 0.73113 | 1.26761 | 1.74576 | 2.28036 | 2.99802 |
| 0.460 | 0.25457 | 0.76285 | 1.28805 | 1.74313 | 2.29786 | 2.99839 |
| 0.480 | 0.27198 | 0.79431 | 1.30886 | 1.74109 | 2.31554 | 2.99888 |
| 0.500 | 0.28969 | 0.82550 | 1.33001 | 1.73966 | 2.33342 | 2.99946 |
| 0.520 | 0.30766 | 0.85639 | 1.35147 | 1.73885 | 2.35148 | 3.00017 |
| 0.540 | 0.32586 | 0.88694 | 1.37322 | 1.73871 | 2.36972 | 3.00103 |
| 0.560 | 0.34427 | 0.91713 | 1.39522 | 1.73928 | 2.38814 | 3.00200 |
| 0.580 | 0.36287 | 0.94691 | 1.41746 | 1.74058 | 2.40672 | 3.00315 |
| 0.600 | 0.38165 | 0.97626 | 1.43993 | 1.74265 | 2.42545 | 3.00444 |
| 0.620 | 0.40058 | 1.00513 | 1.46257 | 1.74554 | 2.44435 | 3.00593 |
| 0.640 | 0.41965 | 1.03348 | 1.48540 | 1.74928 | 2.46340 | 3.00757 |
| 0.660 | 0.43884 | 1.06127 | 1.50838 | 1.75394 | 2.48256 | 3.00940 |
| 0.680 | 0.45815 | 1.08844 | 1.53150 | 1.75953 | 2.50188 | 3.01145 |
| 0.700 | 0.47756 | 1.11496 | 1.55474 | 1.76612 | 2.52134 | 3.01367 |
| 0.720 | 0.49706 | 1.14077 | 1.57808 | 1.77375 | 2.54092 | 3.01611 |
| 0.740 | 0.51665 | 1.16583 | 1.60152 | 1.78245 | 2.56062 | 3.01877 |
| 0.760 | 0.53630 | 1.19010 | 1.62503 | 1.79229 | 2.58042 | 3.02167 |
| 0.780 | 0.55602 | 1.21355 | 1.64859 | 1.80327 | 2.60037 | 3.02480 |
| 0.800 | 0.57580 | 1.23613 | 1.67220 | 1.81546 | 2.62041 | 3.02819 |
| 0.820 | 0.59563 | 1.25782 | 1.69584 | 1.82883 | 2.64055 | 3.03183 |
| 0.840 | 0.61551 | 1.27863 | 1.71949 | 1.84343 | 2.66079 | 3.03571 |
| 0.860 | 0.63543 | 1.29854 | 1.74314 | 1.85923 | 2.68113 | 3.03989 |
| 0.880 | 0.65538 | 1.31755 | 1.76678 | 1.87625 | 2.70156 | 3.04433 |
| 0.900 | 0.67537 | 1.33571 | 1.79041 | 1.89443 | 2.72209 | 3.04907 |
| 0.920 | 0.69538 | 1.35302 | 1.81400 | 1.91376 | 2.74270 | 3.05410 |
| 0.940 | 0.71542 | 1.36955 | 1.83753 | 1.93417 | 2.76338 | 3.05944 |
| 0.960 | 0.73547 | 1.38531 | 1.86100 | 1.95565 | 2.78413 | 3.06511 |
| 0.980 | 0.75555 | 1.40040 | 1.88441 | 1.97807 | 2.80498 | 3.07106 |
| 1.000 | 0.75630 | 1.41484 | 1.90773 | 2.00141 | 2.82588 | 3.07736 |

$H/R = 0.120$ $n = 0.$ $H/R = 0.140$

| $H/L$ | $\Omega_1$ | $\Omega_2$ | $\Omega_3$ | $\Omega_1$ | $\Omega_2$ | $\Omega_3$ | $H/L$ |
|---|---|---|---|---|---|---|---|
| 0. | 0. | 0.06467 | 1.00055 | 0. | 0.07549 | 1.00075 | 0. |
| 0.001 | 0.00161 | 0.06467 | 1.00055 | 0.00161 | 0.07550 | 1.00075 | 0.001 |
| 0.005 | 0.00806 | 0.06472 | 1.00059 | 0.00806 | 0.07567 | 1.00079 | 0.005 |
| 0.010 | 0.01607 | 0.06488 | 1.00072 | 0.01609 | 0.07567 | 1.00092 | 0.010 |
| 0.020 | 0.03174 | 0.06570 | 1.00124 | 0.03190 | 0.07630 | 1.00144 | 0.020 |
| 0.030 | 0.04587 | 0.06816 | 1.00210 | 0.04684 | 0.07792 | 1.00230 | 0.030 |
| 0.040 | 0.05520 | 0.07552 | 1.00330 | 0.05916 | 0.08225 | 1.00350 | 0.040 |
| 0.050 | 0.05876 | 0.08872 | 1.00484 | 0.06610 | 0.09202 | 1.00504 | 0.050 |
| 0.060 | 0.06009 | 0.10423 | 1.00672 | 0.06889 | 0.10601 | 1.00692 | 0.060 |
| 0.070 | 0.06080 | 0.12044 | 1.00893 | 0.07017 | 0.12160 | 1.00913 | 0.070 |
| 0.080 | 0.06136 | 0.13692 | 1.01147 | 0.07095 | 0.13778 | 1.01167 | 0.080 |
| 0.090 | 0.06192 | 0.15353 | 1.01433 | 0.07160 | 0.15421 | 1.01454 | 0.090 |
| 0.100 | 0.06258 | 0.17021 | 1.07021 | 0.07226 | 0.17078 | 1.01772 | 0.100 |
| 0.120 | 0.06441 | 0.20367 | 1.02482 | 0.07391 | 0.20410 | 1.02503 | 0.120 |
| 0.140 | 0.06717 | 0.23718 | 1.03333 | 0.07634 | 0.23753 | 1.03355 | 0.140 |
| 0.160 | 0.07114 | 0.27069 | 1.04300 | 0.07980 | 0.27100 | 1.04322 | 0.160 |
| 0.180 | 0.07638 | 0.30419 | 1.05377 | 0.08449 | 0.30445 | 1.05399 | 0.180 |
| 0.200 | 0.08302 | 0.33764 | 1.06557 | 0.09043 | 0.33788 | 1.06579 | 0.200 |
| 0.220 | 0.09097 | 0.37104 | 1.07835 | 0.09777 | 0.37126 | 1.07859 | 0.220 |
| 0.240 | 0.10019 | 0.40437 | 1.09206 | 0.10634 | 0.40457 | 1.09229 | 0.240 |
| 0.260 | 0.11055 | 0.43762 | 1.10662 | 0.11610 | 0.43781 | 1.10687 | 0.260 |
| 0.280 | 0.12196 | 0.47078 | 1.12199 | 0.12696 | 0.47096 | 1.12224 | 0.280 |
| 0.300 | 0.13430 | 0.50384 | 1.13811 | 0.13881 | 0.50401 | 1.13837 | 0.300 |
| 0.320 | 0.14747 | 0.53678 | 1.15493 | 0.15154 | 0.53695 | 1.15519 | 0.320 |
| 0.340 | 0.16138 | 0.56960 | 1.17240 | 0.16507 | 0.56976 | 1.17265 | 0.340 |
| 0.360 | 0.17595 | 0.60228 | 1.19046 | 0.17930 | 0.60244 | 1.19072 | 0.360 |
| 0.380 | 0.19110 | 0.63480 | 1.20908 | 0.19416 | 0.63495 | 1.20934 | 0.380 |
| 0.400 | 0.20678 | 0.66715 | 1.22821 | 0.20958 | 0.66730 | 1.22847 | 0.400 |
| 0.420 | 0.22292 | 0.69930 | 1.24781 | 0.22550 | 0.69946 | 1.24808 | 0.420 |
| 0.440 | 0.23949 | 0.73125 | 1.26784 | 0.24186 | 0.73140 | 1.26813 | 0.440 |
| 0.460 | 0.25643 | 0.76298 | 1.28829 | 0.25862 | 0.76313 | 1.28857 | 0.460 |
| 0.480 | 0.27371 | 0.79444 | 1.30909 | 0.27575 | 0.79459 | 1.30928 | 0.480 |
| 0.500 | 0.29130 | 0.82564 | 1.33026 | 0.29319 | 0.82579 | 1.33053 | 0.500 |
| 0.520 | 0.30916 | 0.85652 | 1.35171 | 0.31092 | 0.85667 | 1.35200 | 0.520 |
| 0.540 | 0.32726 | 0.88707 | 1.37345 | 0.32891 | 0.88723 | 1.37374 | 0.540 |
| 0.560 | 0.34559 | 0.91726 | 1.39547 | 0.34714 | 0.91742 | 1.39576 | 0.560 |
| 0.580 | 0.36411 | 0.94705 | 1.41771 | 0.36556 | 0.94721 | 1.41800 | 0.580 |
| 0.600 | 0.38281 | 0.97640 | 1.44017 | 0.38418 | 0.97657 | 1.44046 | 0.600 |
| 0.620 | 0.40168 | 1.00528 | 1.46282 | 0.40297 | 1.00544 | 1.46311 | 0.620 |
| 0.640 | 0.42069 | 1.03363 | 1.48565 | 0.42192 | 1.03380 | 1.48595 | 0.640 |
| 0.660 | 0.43983 | 1.06142 | 1.50863 | 0.44099 | 1.06160 | 1.50893 | 0.660 |
| 0.680 | 0.45909 | 1.08860 | 1.53176 | 0.46019 | 1.08879 | 1.53205 | 0.680 |
| 0.700 | 0.47845 | 1.11512 | 1.55500 | 0.47950 | 1.11532 | 1.55529 | 0.700 |
| 0.720 | 0.49791 | 1.14094 | 1.47835 | 0.49891 | 1.14115 | 1.57864 | 0.720 |
| 0.740 | 0.51746 | 1.16601 | 1.60177 | 0.51841 | 1.16623 | 1.60207 | 0.740 |
| 0.760 | 0.53708 | 1.19029 | 1.62528 | 0.53799 | 1.19051 | 1.62558 | 0.760 |
| 0.780 | 0.55676 | 1.21374 | 1.64885 | 0.55764 | 1.21398 | 1.64914 | 0.780 |
| 0.800 | 0.57651 | 1.23633 | 1.67245 | 0.57735 | 1.23658 | 1.67275 | 0.800 |
| 0.820 | 0.59632 | 1.25804 | 1.69608 | 0.59712 | 1.25830 | 1.69638 | 0.820 |
| 0.840 | 0.61617 | 1.27886 | 1.71973 | 0.61695 | 1.27912 | 1.72003 | 0.840 |
| 0.860 | 0.63606 | 1.29877 | 1.74339 | 0.63681 | 1.29905 | 1.74368 | 0.860 |
| 0.880 | 0.65599 | 1.31780 | 1.76703 | 0.65672 | 1.31809 | 1.76731 | 0.880 |
| 0.900 | 0.67596 | 1.33596 | 1.79064 | 0.67666 | 1.33625 | 1.79092 | 0.900 |
| 0.920 | 0.69595 | 1.35328 | 1.81422 | 0.69662 | 1.35358 | 1.81449 | 0.920 |
| 0.940 | 0.71597 | 1.36981 | 1.83775 | 0.71661 | 1.37012 | 1.83802 | 0.940 |
| 0.960 | 0.73601 | 1.38558 | 1.86122 | 0.73663 | 1.38590 | 1.86148 | 0.960 |
| 0.980 | 0.75606 | 1.40068 | 1.88462 | 0.75667 | 1.40099 | 1.88487 | 0.980 |
| 1.000 | 0.77613 | 1.41512 | 1.90794 | 0.77672 | 1.41545 | 1.90818 | 1.000 |

$H/R = 0.160$ $n = 0.$ $H/R = 0.180$

| $H/L$ | $\Omega_1$ | $\Omega_2$ | $\Omega_3$ | $\Omega_1$ | $\Omega_2$ | $\Omega_3$ | $H/L$ |
|---|---|---|---|---|---|---|---|
| 0. | 0. | 0.08634 | 1.00098 | 0. | 0.09720 | 1.00124 | 0. |
| 0.001 | 0.00161 | 0.08634 | 1.00098 | 0.00161 | 0.09720 | 1.00124 | 0.001 |
| 0.005 | 0.00806 | 0.08637 | 1.00102 | 0.00806 | 0.09724 | 1.00128 | 0.005 |
| 0.010 | 0.01610 | 0.08649 | 1.00115 | 0.01610 | 0.09733 | 1.00141 | 0.010 |
| 0.020 | 0.03199 | 0.08701 | 1.00166 | 0.03205 | 0.09777 | 1.00193 | 0.020 |
| 0.030 | 0.04732 | 0.08820 | 1.00253 | 0.04760 | 0.09872 | 1.00279 | 0.030 |
| 0.040 | 0.06116 | 0.09097 | 1.00373 | 0.06221 | 0.10068 | 1.00399 | 0.040 |
| 0.050 | 0.07138 | 0.09742 | 1.00527 | 0.07460 | 0.10492 | 1.00554 | 0.050 |
| 0.060 | 0.07666 | 0.10888 | 1.00716 | 0.08285 | 0.11337 | 1.00742 | 0.060 |
| 0.070 | 0.07902 | 0.12333 | 1.00937 | 0.08705 | 0.12593 | 1.00963 | 0.070 |
| 0.080 | 0.08027 | 0.13898 | 1.01191 | 0.08916 | 0.14067 | 1.01218 | 0.080 |
| 0.090 | 0.08113 | 0.15513 | 1.01477 | 0.09041 | 0.15635 | 1.01504 | 0.090 |
| 0.100 | 0.08186 | 0.17152 | 1.01796 | 0.09134 | 0.17248 | 1.01823 | 0.100 |
| 0.120 | 0.08347 | 0.20464 | 1.02528 | 0.09302 | 0.20531 | 1.02555 | 0.120 |
| 0.140 | 0.08566 | 0.23797 | 1.03380 | 0.09507 | 0.23849 | 1.03408 | 0.140 |
| 0.160 | 0.08875 | 0.27136 | 1.04347 | 0.09787 | 0.27180 | 1.04376 | 0.160 |
| 0.180 | 0.09294 | 0.30477 | 1.05424 | 0.10166 | 0.30515 | 1.05454 | 0.180 |
| 0.200 | 0.09836 | 0.33816 | 1.06606 | 0.10659 | 0.33850 | 1.06636 | 0.200 |
| 0.220 | 0.10506 | 0.37152 | 1.07886 | 0.11274 | 0.37182 | 1.07916 | 0.220 |
| 0.240 | 0.11299 | 0.40481 | 1.09257 | 0.12014 | 0.40509 | 1.09288 | 0.240 |
| 0.260 | 0.12220 | 0.43803 | 1.10714 | 0.12874 | 0.43829 | 1.10746 | 0.260 |
| 0.280 | 0.13250 | 0.47117 | 1.12253 | 0.13849 | 0.47142 | 1.12285 | 0.280 |
| 0.300 | 0.14383 | 0.50421 | 1.13865 | 0.14935 | 0.50445 | 1.13898 | 0.300 |
| 0.320 | 0.15612 | 0.53714 | 1.15548 | 0.16114 | 0.53737 | 1.15581 | 0.320 |
| 0.340 | 0.16923 | 0.56995 | 1.17295 | 0.17383 | 0.57017 | 1.17329 | 0.340 |
| 0.360 | 0.18309 | 0.60262 | 1.19102 | 0.18731 | 0.60283 | 1.19136 | 0.360 |
| 0.380 | 0.19763 | 0.63513 | 1.20965 | 0.20150 | 0.63534 | 1.20999 | 0.380 |
| 0.400 | 0.21276 | 0.66747 | 1.22878 | 0.21632 | 0.66768 | 1.22913 | 0.400 |
| 0.420 | 0.22843 | 0.69963 | 1.24839 | 0.23171 | 0.69983 | 1.24875 | 0.420 |
| 0.440 | 0.24457 | 0.73158 | 1.26844 | 0.24761 | 0.73178 | 1.26880 | 0.440 |
| 0.460 | 0.26113 | 0.76330 | 1.28889 | 0.26395 | 0.76350 | 1.28925 | 0.460 |
| 0.480 | 0.27808 | 0.79477 | 1.30970 | 0.28070 | 0.79497 | 1.31007 | 0.480 |
| 0.500 | 0.29536 | 0.82596 | 1.33086 | 0.29780 | 0.82616 | 1.33123 | 0.500 |
| 0.520 | 0.31295 | 0.85685 | 1.35233 | 0.31523 | 0.85705 | 1.35270 | 0.520 |
| 0.540 | 0.33081 | 0.88741 | 1.37407 | 0.33295 | 0.88762 | 1.37445 | 0.540 |
| 0.560 | 0.34891 | 0.91761 | 1.39609 | 0.35093 | 0.91781 | 1.39647 | 0.560 |
| 0.580 | 0.36724 | 0.94740 | 1.41833 | 0.36914 | 0.94761 | 1.41871 | 0.580 |
| 0.600 | 0.38576 | 0.97676 | 1.44080 | 0.38755 | 0.97698 | 1.44118 | 0.600 |
| 0.620 | 0.40447 | 1.00565 | 1.46345 | 0.40616 | 1.00587 | 1.46384 | 0.620 |
| 0.640 | 0.42333 | 1.03401 | 1.48629 | 0.42493 | 1.03425 | 1.48668 | 0.640 |
| 0.660 | 0.44233 | 1.06182 | 1.50927 | 0.44385 | 1.06206 | 1.50966 | 0.660 |
| 0.680 | 0.46147 | 1.08901 | 1.53240 | 0.46291 | 1.08926 | 1.53278 | 0.680 |
| 0.700 | 0.48072 | 1.11555 | 1.55564 | 0.48209 | 1.11581 | 1.55603 | 0.700 |
| 0.720 | 0.50007 | 1.14138 | 1.57899 | 0.50138 | 1.14166 | 1.57938 | 0.720 |
| 0.740 | 0.51952 | 1.16648 | 1.60242 | 0.52077 | 1.16676 | 1.60281 | 0.740 |
| 0.760 | 0.53905 | 1.19077 | 1.62592 | 0.54025 | 1.19107 | 1.62632 | 0.760 |
| 0.780 | 0.55865 | 1.21424 | 1.64949 | 0.55980 | 1.21455 | 1.64988 | 0.780 |
| 0.800 | 0.57833 | 1.23686 | 1.67309 | 0.57943 | 1.23718 | 1.67348 | 0.800 |
| 0.820 | 0.59806 | 1.25860 | 1.69672 | 0.59911 | 1.25892 | 1.69711 | 0.820 |
| 0.840 | 0.61784 | 1.27943 | 1.72037 | 0.61886 | 1.27977 | 1.72075 | 0.840 |
| 0.860 | 0.63767 | 1.29936 | 1.74400 | 0.63865 | 1.29972 | 1.74438 | 0.860 |
| 0.880 | 0.65754 | 1.31841 | 1.76764 | 0.65848 | 1.31878 | 1.76802 | 0.880 |
| 0.900 | 0.67745 | 1.33659 | 1.79125 | 0.67836 | 1.33697 | 1.79160 | 0.900 |
| 0.920 | 0.69739 | 1.35393 | 1.81481 | 0.69827 | 1.35432 | 1.81517 | 0.920 |
| 0.940 | 0.71736 | 1.37047 | 1.83832 | 0.71821 | 1.37087 | 1.83868 | 0.940 |
| 0.960 | 0.73736 | 1.38626 | 1.86178 | 0.73817 | 1.38669 | 1.86212 | 0.960 |
| 0.980 | 0.75736 | 1.40137 | 1.88516 | 0.75816 | 1.40179 | 1.88549 | 0.980 |
| 1.000 | 0.77739 | 1.41584 | 1.90846 | 0.77816 | 1.41627 | 1.90878 | 1.000 |

$H/R = 0.200 \qquad n = 0.$

| $H/L$ | $\Omega_1$ | $\Omega_2$ | $\Omega_3$ | $\Omega_4$ | $\Omega_5$ | $\Omega_6$ |
|---|---|---|---|---|---|---|
| 0. | 0. | 0.10810 | 1.00153 | 1.87427 | 2.00077 | 3.00052 |
| 0.010 | 0.01611 | 0.10821 | 1.00170 | 1.87401 | 2.00111 | 3.00051 |
| 0.020 | 0.03209 | 0.10859 | 1.00222 | 1.87322 | 2.00213 | 3.00051 |
| 0.030 | 0.04778 | 0.10937 | 1.00308 | 1.87193 | 2.00381 | 3.00049 |
| 0.040 | 0.06282 | 0.11087 | 1.00429 | 1.87018 | 2.00609 | 3.00046 |
| 0.050 | 0.07645 | 0.11383 | 1.00583 | 1.86803 | 2.00894 | 3.00041 |
| 0.060 | 0.08720 | 0.11975 | 1.00771 | 1.86552 | 2.01231 | 3.00037 |
| 0.070 | 0.09386 | 0.12981 | 1.00993 | 1.86269 | 2.01614 | 3.00032 |
| 0.080 | 0.09737 | 0.14309 | 1.01248 | 1.85961 | 2.02038 | 3.00024 |
| 0.090 | 0.09931 | 0.15802 | 1.01535 | 1.85632 | 2.02499 | 3.00020 |
| 0.100 | 0.10059 | 0.17373 | 1.01854 | 1.85286 | 2.02993 | 3.00010 |
| 0.120 | 0.10250 | 0.20615 | 1.02586 | 1.84554 | 2.04064 | 2.99993 |
| 0.140 | 0.10451 | 0.23912 | 1.03439 | 1.83793 | 2.05229 | 2.99976 |
| 0.160 | 0.10710 | 0.27231 | 1.04408 | 1.83019 | 2.06471 | 2.99954 |
| 0.180 | 0.11057 | 0.30559 | 1.05487 | 1.82244 | 2.07775 | 2.99932 |
| 0.200 | 0.11508 | 0.33888 | 1.06669 | 1.81477 | 2.09134 | 2.99910 |
| 0.220 | 0.12075 | 0.37216 | 1.07950 | 1.80729 | 2.10540 | 2.99885 |
| 0.240 | 0.12762 | 0.40541 | 1.09323 | 1.80001 | 2.11988 | 2.99863 |
| 0.260 | 0.13570 | 0.43859 | 1.10781 | 1.79301 | 2.13475 | 2.99844 |
| 0.280 | 0.14494 | 0.47169 | 1.12321 | 1.78632 | 2.14995 | 2.99824 |
| 0.300 | 0.15524 | 0.50471 | 1.13935 | 1.77998 | 2.16548 | 2.99807 |
| 0.320 | 0.16658 | 0.53762 | 1.15619 | 1.77400 | 2.18129 | 2.99795 |
| 0.340 | 0.17885 | 0.57041 | 1.17367 | 1.76843 | 2.19739 | 2.99788 |
| 0.360 | 0.19192 | 0.60306 | 1.19175 | 1.76329 | 2.21375 | 2.99785 |
| 0.380 | 0.20574 | 0.63557 | 1.21038 | 1.75859 | 2.23035 | 2.99790 |
| 0.400 | 0.22024 | 0.66791 | 1.22953 | 1.75438 | 2.24719 | 2.99800 |
| 0.420 | 0.23534 | 0.70006 | 1.24915 | 1.75065 | 2.26427 | 2.99819 |
| 0.440 | 0.25097 | 0.73200 | 1.26920 | 1.74746 | 2.28155 | 2.99846 |
| 0.460 | 0.26707 | 0.76372 | 1.28965 | 1.74481 | 2.29904 | 2.99883 |
| 0.480 | 0.28360 | 0.79519 | 1.31048 | 1.74274 | 2.31673 | 2.99932 |
| 0.500 | 0.30052 | 0.82639 | 1.33164 | 1.74127 | 2.33461 | 2.99990 |
| 0.520 | 0.31777 | 0.85728 | 1.35311 | 1.74042 | 2.35267 | 3.00063 |
| 0.540 | 0.33533 | 0.88785 | 1.37488 | 1.74025 | 2.37091 | 3.00149 |
| 0.560 | 0.35317 | 0.91805 | 1.39689 | 1.74078 | 2.38933 | 3.00249 |
| 0.580 | 0.37124 | 0.94786 | 1.41914 | 1.74203 | 2.40790 | 3.00361 |
| 0.600 | 0.38954 | 0.97723 | 1.44161 | 1.74406 | 2.42663 | 3.00493 |
| 0.620 | 0.40804 | 1.00613 | 1.46427 | 1.74691 | 2.44552 | 3.00640 |
| 0.640 | 0.42671 | 1.03451 | 1.48711 | 1.75060 | 2.46455 | 3.00806 |
| 0.660 | 0.44555 | 1.06233 | 1.51010 | 1.75520 | 2.48372 | 3.00989 |
| 0.680 | 0.46452 | 1.08954 | 1.53322 | 1.76074 | 2.50305 | 3.01193 |
| 0.700 | 0.48363 | 1.11610 | 1.55648 | 1.76727 | 2.52249 | 3.01415 |
| 0.720 | 0.50285 | 1.14196 | 1.57982 | 1.77484 | 2.54207 | 3.01662 |
| 0.740 | 0.52217 | 1.16708 | 1.60325 | 1.78349 | 2.56174 | 3.01928 |
| 0.760 | 0.54158 | 1.19140 | 1.62676 | 1.79326 | 2.58157 | 3.02219 |
| 0.780 | 0.56108 | 1.21490 | 1.65031 | 1.80419 | 2.60149 | 3.02531 |
| 0.800 | 0.58066 | 1.23753 | 1.67392 | 1.81630 | 2.62151 | 3.02871 |
| 0.820 | 0.60029 | 1.25929 | 1.69545 | 1.82963 | 2.64165 | 3.03234 |
| 0.840 | 0.61999 | 1.28016 | 1.72118 | 1.84418 | 2.66189 | 3.03623 |
| 0.860 | 0.63975 | 1.30013 | 1.74481 | 1.85994 | 2.68220 | 3.04040 |
| 0.880 | 0.65954 | 1.31920 | 1.76844 | 1.87691 | 2.70264 | 3.04484 |
| 0.900 | 0.67938 | 1.33741 | 1.79202 | 1.89507 | 2.72314 | 3.04960 |
| 0.920 | 0.69925 | 1.35477 | 1.81557 | 1.91438 | 2.74375 | 3.05463 |
| 0.940 | 0.71916 | 1.37134 | 1.83907 | 1.93477 | 2.76443 | 3.05998 |
| 0.960 | 0.73909 | 1.38715 | 1.86250 | 1.95621 | 2.78518 | 3.06562 |
| 0.980 | 0.75905 | 1.40227 | 1.88587 | 1.97864 | 2.80601 | 3.07160 |
| 1.000 | 0.77902 | 1.41676 | 1.90915 | 2.00199 | 2.82690 | 3.07790 |

$H/R = 0.250 \qquad n = 0.$

| $H/L$ | $\Omega_1$ | $\Omega_2$ | $\Omega_3$ |
|---|---|---|---|
| 0. | 0. | 0.13547 | 1.00239 |
| 0.001 | 0.00161 | 0.13547 | 1.00239 |
| 0.005 | 0.00806 | 0.13549 | 1.00243 |
| 0.010 | 0.01611 | 0.13555 | 1.00256 |
| 0.020 | 0.03215 | 0.13583 | 1.00308 |
| 0.030 | 0.04802 | 0.13637 | 1.00395 |
| 0.040 | 0.06356 | 0.13729 | 1.00516 |
| 0.050 | 0.07852 | 0.13886 | 1.00670 |
| 0.060 | 0.09237 | 0.14158 | 1.00859 |
| 0.070 | 0.10421 | 0.14637 | 1.01081 |
| 0.080 | 0.11298 | 0.15428 | 1.01336 |
| 0.090 | 0.11855 | 0.16547 | 1.01624 |
| 0.100 | 0.12189 | 0.17896 | 1.01944 |
| 0.120 | 0.12556 | 0.20927 | 1.02678 |
| 0.140 | 0.12797 | 0.24133 | 1.03533 |
| 0.160 | 0.13039 | 0.27403 | 1.04502 |
| 0.180 | 0.13333 | 0.30701 | 1.05583 |
| 0.200 | 0.13708 | 0.34012 | 1.06768 |
| 0.220 | 0.14179 | 0.37326 | 1.08051 |
| 0.240 | 0.14757 | 0.40640 | 1.09426 |
| 0.260 | 0.15447 | 0.43950 | 1.10887 |
| 0.280 | 0.16249 | 0.47255 | 1.12427 |
| 0.300 | 0.17162 | 0.50552 | 1.14044 |
| 0.320 | 0.18179 | 0.53839 | 1.15729 |
| 0.340 | 0.19294 | 0.57115 | 1.17480 |
| 0.360 | 0.20499 | 0.60378 | 1.19289 |
| 0.380 | 0.21786 | 0.63627 | 1.21154 |
| 0.400 | 0.23148 | 0.66859 | 1.23070 |
| 0.420 | 0.24577 | 0.70074 | 1.25034 |
| 0.440 | 0.26068 | 0.73267 | 1.27041 |
| 0.460 | 0.27614 | 0.76439 | 1.29087 |
| 0.480 | 0.29207 | 0.79587 | 1.31170 |
| 0.500 | 0.30844 | 0.82706 | 1.33288 |
| 0.520 | 0.32520 | 0.85797 | 1.35436 |
| 0.540 | 0.34231 | 0.88854 | 1.37613 |
| 0.560 | 0.35974 | 0.91876 | 1.39816 |
| 0.580 | 0.37744 | 0.94858 | 1.42042 |
| 0.600 | 0.39540 | 0.97797 | 1.44289 |
| 0.620 | 0.41359 | 1.00689 | 1.46556 |
| 0.640 | 0.43197 | 1.03529 | 1.48841 |
| 0.660 | 0.45054 | 1.06314 | 1.51140 |
| 0.680 | 0.46928 | 1.09038 | 1.53453 |
| 0.700 | 0.48816 | 1.11697 | 1.55779 |
| 0.720 | 0.50717 | 1.14287 | 1.51813 |
| 0.740 | 0.52631 | 1.16802 | 1.60458 |
| 0.760 | 0.54555 | 1.19239 | 1.62807 |
| 0.780 | 0.58430 | 1.23861 | 1.65163 |
| 0.800 | 0.58430 | 1.23861 | 1.67523 |
| 0.820 | 0.60379 | 1.26041 | 1.69885 |
| 0.840 | 0.62336 | 1.28132 | 1.72246 |
| 0.860 | 0.64299 | 1.30133 | 1.74608 |
| 0.880 | 0.66267 | 1.32044 | 1.76968 |
| 0.900 | 0.68239 | 1.33868 | 1.79325 |
| 0.920 | 0.70216 | 1.35610 | 1.81678 |
| 0.940 | 0.72197 | 1.37270 | 1.84024 |
| 0.960 | 0.74181 | 1.38855 | 1.86365 |
| 0.980 | 0.76169 | 1.40370 | 1.88698 |
| 1.000 | 0.78158 | 1.41822 | 1.91023 |

$H/R = 0.300 \qquad n = 0.$

| $H/L$ | $\Omega_1$ | $\Omega_2$ | $\Omega_3$ | $\Omega_4$ | $\Omega_5$ | $\Omega_6$ |
|---|---|---|---|---|---|---|
| 0. | 0. | 0.16306 | 1.00344 | 1.87871 | 2.00174 | 3.00118 |
| 0.001 | 0.00161 | 0.16307 | 1.00345 | 1.87870 | 2.00175 | 3.00117 |
| 0.005 | 0.00806 | 0.16308 | 1.00349 | 1.87864 | 2.00184 | 3.00117 |
| 0.010 | 0.01612 | 0.16313 | 1.00362 | 1.87843 | 2.00209 | 3.00117 |
| 0.020 | 0.03218 | 0.16335 | 1.00414 | 1.87762 | 2.00314 | 3.00115 |
| 0.030 | 0.04814 | 0.16375 | 1.00501 | 1.87630 | 2.00484 | 3.00115 |
| 0.040 | 0.06389 | 0.16441 | 1.00622 | 1.87451 | 2.00718 | 3.00110 |
| 0.050 | 0.07932 | 0.16543 | 1.00777 | 1.87229 | 2.01009 | 3.00107 |
| 0.060 | 0.09422 | 0.16704 | 1.00967 | 1.86972 | 2.01351 | 3.00103 |
| 0.070 | 0.10820 | 0.16960 | 1.01189 | 1.86683 | 2.01741 | 3.00098 |
| 0.080 | 0.12067 | 0.17372 | 1.01445 | 1.86369 | 2.02171 | 3.00090 |
| 0.090 | 0.13087 | 0.18060 | 1.01734 | 1.86032 | 2.02638 | 3.00085 |
| 0.100 | 0.13829 | 0.18946 | 1.02055 | 1.85681 | 2.03137 | 3.00076 |
| 0.120 | 0.14654 | 0.21489 | 1.02790 | 1.84938 | 2.04219 | 3.00061 |
| 0.140 | 0.15060 | 0.24491 | 1.03647 | 1.84167 | 2.05393 | 3.00041 |
| 0.160 | 0.15351 | 0.27664 | 1.04620 | 1.83383 | 2.06643 | 3.00020 |
| 0.180 | 0.15637 | 0.30908 | 1.05703 | 1.82600 | 2.07953 | 3.00000 |
| 0.200 | 0.15970 | 0.34184 | 1.06890 | 1.81827 | 2.09318 | 2.99976 |
| 0.220 | 0.16376 | 0.37476 | 1.08176 | 1.81072 | 2.10729 | 2.99954 |
| 0.240 | 0.16872 | 0.40773 | 1.09553 | 1.80339 | 2.12180 | 2.99932 |
| 0.260 | 0.17468 | 0.44072 | 1.11016 | 1.79634 | 2.13670 | 2.99912 |
| 0.280 | 0.18168 | 0.47367 | 1.12560 | 1.78960 | 2.15193 | 2.99895 |
| 0.300 | 0.18974 | 0.50657 | 1.14179 | 1.78321 | 2.16747 | 2.99878 |
| 0.320 | 0.19884 | 0.53939 | 1.15867 | 1.77720 | 2.18330 | 2.99868 |
| 0.340 | 0.20893 | 0.57211 | 1.17619 | 1.77157 | 2.19942 | 2.99861 |
| 0.360 | 0.21997 | 0.60470 | 1.19431 | 1.76639 | 2.21579 | 2.99858 |
| 0.380 | 0.23188 | 0.63717 | 1.21297 | 1.76166 | 2.23240 | 2.99863 |
| 0.400 | 0.24459 | 0.66947 | 1.23216 | 1.75739 | 2.24924 | 2.99875 |
| 0.420 | 0.25804 | 0.70160 | 1.25181 | 1.75362 | 2.26632 | 2.99895 |
| 0.440 | 0.27216 | 0.73352 | 1.27188 | 1.75038 | 2.28360 | 2.99922 |
| 0.460 | 0.28688 | 0.76524 | 1.29237 | 1.74767 | 2.30109 | 2.99961 |
| 0.480 | 0.30214 | 0.79671 | 1.31322 | 1.74554 | 2.31878 | 3.00010 |
| 0.500 | 0.31790 | 0.82791 | 1.33441 | 1.74402 | 2.33665 | 3.00068 |
| 0.520 | 0.33411 | 0.85882 | 1.35590 | 1.74312 | 2.35471 | 3.00142 |
| 0.540 | 0.35070 | 0.88941 | 1.37768 | 1.74289 | 2.37294 | 3.00227 |
| 0.560 | 0.36766 | 0.91964 | 1.39972 | 1.74335 | 2.39136 | 3.00327 |
| 0.580 | 0.38293 | 0.94948 | 1.42199 | 1.74453 | 2.40992 | 3.00442 |
| 0.600 | 0.40249 | 0.97889 | 1.44448 | 1.74649 | 2.42865 | 3.00574 |
| 0.620 | 0.42032 | 1.00783 | 1.46715 | 1.74925 | 2.44752 | 3.00723 |
| 0.640 | 0.43836 | 1.03627 | 1.49001 | 1.75287 | 2.46654 | 3.00889 |
| 0.660 | 0.45662 | 1.06415 | 1.51302 | 1.75737 | 2.48572 | 3.01072 |
| 0.680 | 0.47507 | 1.09143 | 1.53615 | 1.76282 | 2.50503 | 3.01276 |
| 0.700 | 0.49368 | 1.11806 | 1.55941 | 1.76925 | 2.52444 | 3.01501 |
| 0.720 | 0.51245 | 1.14400 | 1.58276 | 1.77672 | 2.54402 | 3.01745 |
| 0.740 | 0.53135 | 1.16919 | 1.60620 | 1.78527 | 2.56370 | 3.02014 |
| 0.760 | 0.55038 | 1.19361 | 1.62971 | 1.79494 | 2.58347 | 3.02304 |
| 0.780 | 0.56952 | 1.21720 | 1.65327 | 1.80576 | 2.60339 | 3.02617 |
| 0.800 | 0.58875 | 1.23994 | 1.67685 | 1.81778 | 2.62341 | 3.02956 |
| 0.820 | 0.60807 | 1.26180 | 1.70045 | 1.83100 | 2.64353 | 3.03320 |
| 0.840 | 0.62748 | 1.28276 | 1.72407 | 1.84546 | 2.66375 | 3.03710 |
| 0.860 | 0.64696 | 1.30282 | 1.74767 | 1.86114 | 2.68406 | 3.04128 |
| 0.880 | 0.66649 | 1.32199 | 1.77125 | 1.87804 | 2.70447 | 3.04575 |
| 0.900 | 0.68608 | 1.34029 | 1.79479 | 1.89615 | 2.72498 | 3.05048 |
| 0.920 | 0.70573 | 1.35774 | 1.81828 | 1.91540 | 2.74555 | 3.05551 |
| 0.940 | 0.72542 | 1.37439 | 1.84170 | 1.93578 | 2.76621 | 3.06086 |
| 0.960 | 0.74515 | 1.39028 | 1.86507 | 1.95720 | 2.78694 | 3.06652 |
| 0.980 | 0.76492 | 1.40547 | 1.88836 | 1.97960 | 2.80776 | 3.07250 |
| 1.000 | 0.78471 | 1.42002 | 1.91156 | 2.00294 | 2.82864 | 3.07880 |

$H/R = 0.350$ $n = 0.$

| $H/L$ | $\Omega_1$ | $\Omega_2$ | $\Omega_3$ |
|---|---|---|---|
| 0. | 0. | 0.19094 | 1.00470 |
| 0.001 | 0.00161 | 0.19094 | 1.00470 |
| 0.005 | 0.00806 | 0.19096 | 1.00475 |
| 0.010 | 0.01618 | 0.19100 | 1.00488 |
| 0.020 | 0.03220 | 0.19117 | 1.00540 |
| 0.030 | 0.04820 | 0.19148 | 1.00627 |
| 0.040 | 0.06407 | 0.19198 | 1.00748 |
| 0.050 | 0.07973 | 0.19272 | 1.00904 |
| 0.060 | 0.09507 | 0.19382 | 1.01094 |
| 0.070 | 0.10991 | 0.19544 | 1.01318 |
| 0.080 | 0.12398 | 0.19788 | 1.01575 |
| 0.090 | 0.13686 | 0.20155 | 1.01864 |
| 0.100 | 0.14799 | 0.20703 | 1.02186 |
| 0.120 | 0.16341 | 0.22503 | 1.02924 |
| 0.140 | 0.17132 | 0.25090 | 1.03784 |
| 0.160 | 0.17578 | 0.28067 | 1.04760 |
| 0.180 | 0.17916 | 0.31208 | 1.05846 |
| 0.200 | 0.18247 | 0.34425 | 1.07036 |
| 0.220 | 0.18620 | 0.37678 | 1.08324 |
| 0.240 | 0.19060 | 0.40950 | 1.09705 |
| 0.260 | 0.19584 | 0.44230 | 1.11171 |
| 0.280 | 0.20202 | 0.47512 | 1.12718 |
| 0.300 | 0.20917 | 0.50791 | 1.14339 |
| 0.320 | 0.21732 | 0.54065 | 1.16030 |
| 0.340 | 0.22645 | 0.57330 | 1.17785 |
| 0.360 | 0.23652 | 0.60585 | 1.19599 |
| 0.380 | 0.24749 | 0.63827 | 1.21468 |
| 0.400 | 0.25930 | 0.67055 | 1.23389 |
| 0.420 | 0.27190 | 0.70265 | 1.25356 |
| 0.440 | 0.28520 | 0.73457 | 1.27366 |
| 0.460 | 0.29916 | 0.76627 | 1.29416 |
| 0.480 | 0.31372 | 0.79774 | 1.31503 |
| 0.500 | 0.32883 | 0.82894 | 1.33624 |
| 0.520 | 0.34442 | 0.85986 | 1.35774 |
| 0.540 | 0.36045 | 0.89046 | 1.37954 |
| 0.560 | 0.37688 | 0.92071 | 1.40159 |
| 0.580 | 0.39367 | 0.95057 | 1.42387 |
| 0.600 | 0.41078 | 0.98000 | 1.44638 |
| 0.620 | 0.42818 | 1.00898 | 1.46907 |
| 0.640 | 0.44585 | 1.03744 | 1.49193 |
| 0.660 | 0.46376 | 1.06537 | 1.51494 |
| 0.680 | 0.48188 | 1.09269 | 1.53809 |
| 0.700 | 0.50018 | 1.11937 | 1.56135 |
| 0.720 | 0.51869 | 1.14536 | 1.58472 |
| 0.740 | 0.53730 | 1.17061 | 1.60816 |
| 0.760 | 0.55608 | 1.19508 | 1.63166 |
| 0.780 | 0.57499 | 1.21873 | 1.65521 |
| 0.800 | 0.59401 | 1.24154 | 1.67880 |
| 0.820 | 0.61314 | 1.26345 | 1.70240 |
| 0.840 | 0.63236 | 1.28448 | 1.72599 |
| 0.860 | 0.65166 | 1.30461 | 1.74958 |
| 0.880 | 0.67102 | 1.32385 | 1.77312 |
| 0.900 | 0.69046 | 1.34220 | 1.79663 |
| 0.920 | 0.70996 | 1.35971 | 1.82008 |
| 0.940 | 0.72951 | 1.37642 | 1.84347 |
| 0.960 | 0.74911 | 1.39236 | 1.86678 |
| 0.980 | 0.76875 | 1.40760 | 1.89003 |
| 1.000 | 0.78843 | 1.42219 | 1.91319 |

$H/R = 0.400 \qquad n = 0.$

| $H/L$ | $\Omega_1$ | $\Omega_2$ | $\Omega_3$ | $\Omega_4$ | $\Omega_5$ | $\Omega_6$ |
|---|---|---|---|---|---|---|
| 0. | 0. | 0.21915 | 1.00616 | 1.88521 | 2.00315 | 3.00212 |
| 0.001 | 0.00161 | 0.21915 | 1.00616 | 1.88521 | 2.00316 | 3.00212 |
| 0.005 | 0.00806 | 0.21916 | 1.00620 | 1.88514 | 2.00323 | 3.00212 |
| 0.010 | 0.01612 | 0.21919 | 1.00634 | 1.88492 | 2.00351 | 3.00212 |
| 0.020 | 0.03221 | 0.21933 | 1.00686 | 1.88407 | 2.00460 | 3.00210 |
| 0.030 | 0.04824 | 0.21958 | 1.00774 | 1.88269 | 2.00637 | 3.00207 |
| 0.040 | 0.06418 | 0.21996 | 1.00895 | 1.88082 | 2.00877 | 3.00205 |
| 0.050 | 0.07996 | 0.22052 | 1.01052 | 1.87852 | 2.01176 | 3.00200 |
| 0.060 | 0.09553 | 0.22133 | 1.01243 | 1.87585 | 2.01528 | 3.00198 |
| 0.070 | 0.11079 | 0.22247 | 1.01467 | 1.87287 | 2.01926 | 3.00193 |
| 0.080 | 0.12560 | 0.22410 | 1.01725 | 1.86961 | 2.02366 | 3.00186 |
| 0.090 | 0.13973 | 0.22644 | 1.02016 | 1.86617 | 2.02842 | 3.00178 |
| 0.100 | 0.15289 | 0.22980 | 1.02339 | 1.86255 | 2.03350 | 3.00171 |
| 0.120 | 0.17462 | 0.24126 | 1.03080 | 1.85496 | 2.04447 | 3.00156 |
| 0.140 | 0.18849 | 0.26089 | 1.03943 | 1.84710 | 2.05635 | 3.00137 |
| 0.160 | 0.19656 | 0.28701 | 1.04922 | 1.83913 | 2.06895 | 3.00117 |
| 0.180 | 0.20110 | 0.31653 | 1.06012 | 1.83119 | 2.08216 | 3.00095 |
| 0.200 | 0.20495 | 0.34765 | 1.07205 | 1.82336 | 2.09588 | 3.00073 |
| 0.220 | 0.20871 | 0.37954 | 1.08498 | 1.81571 | 2.11005 | 3.00054 |
| 0.240 | 0.21284 | 0.41184 | 1.09882 | 1.80831 | 2.12462 | 3.00032 |
| 0.260 | 0.21760 | 0.44435 | 1.11352 | 1.80118 | 2.13956 | 3.00012 |
| 0.280 | 0.22315 | 0.47696 | 1.12902 | 1.79437 | 2.15482 | 2.99995 |
| 0.300 | 0.22956 | 0.50960 | 1.14527 | 1.78184 | 2.18625 | 2.99980 |
| 0.320 | 0.23690 | 0.54221 | 1.16221 | 1.78184 | 2.18652 | 2.99971 |
| 0.340 | 0.25417 | 0.57478 | 1.17979 | 1.77615 | 2.20238 | 2.99963 |
| 0.360 | 0.25436 | 0.60725 | 1.19796 | 1.77091 | 2.21877 | 2.99963 |
| 0.380 | 0.26446 | 0.63962 | 1.21668 | 1.76611 | 2.23539 | 2.99968 |
| 0.400 | 0.27540 | 0.67185 | 1.23591 | 1.76178 | 2.25224 | 2.99980 |
| 0.420 | 0.28715 | 0.70393 | 1.25561 | 1.75794 | 2.26931 | 3.00002 |
| 0.440 | 0.29964 | 0.73582 | 1.27573 | 1.75463 | 2.28661 | 3.00032 |
| 0.460 | 0.31284 | 0.76751 | 1.29625 | 1.75186 | 2.30410 | 3.00071 |
| 0.480 | 0.32667 | 0.79897 | 1.31714 | 1.74966 | 2.32178 | 3.00120 |
| 0.500 | 0.34108 | 0.83018 | 1.33836 | 1.74805 | 2.33966 | 3.00181 |
| 0.520 | 0.35603 | 0.86110 | 1.35990 | 1.74707 | 2.35771 | 3.00254 |
| 0.540 | 0.37146 | 0.89171 | 1.38171 | 1.74675 | 2.37593 | 3.00342 |
| 0.560 | 0.38733 | 0.92198 | 1.40378 | 1.74711 | 2.39433 | 3.00442 |
| 0.580 | 0.40360 | 0.95186 | 1.42608 | 1.74819 | 2.41288 | 3.00559 |
| 0.600 | 0.42022 | 0.98133 | 1.44860 | 1.75005 | 2.43159 | 3.00691 |
| 0.620 | 0.43717 | 1.01033 | 1.47130 | 1.75269 | 2.45047 | 3.00840 |
| 0.640 | 0.45442 | 1.03884 | 1.49418 | 1.75617 | 2.46948 | 3.01006 |
| 0.660 | 0.47193 | 1.06680 | 1.51721 | 1.76055 | 2.48862 | 3.01191 |
| 0.680 | 0.48969 | 1.09417 | 1.54037 | 1.76586 | 2.50791 | 3.01396 |
| 0.700 | 0.50765 | 1.12090 | 1.56363 | 1.77215 | 2.52732 | 3.01621 |
| 0.720 | 0.52582 | 1.14696 | 1.58701 | 1.77947 | 2.54687 | 3.01867 |
| 0.740 | 0.54416 | 1.17227 | 1.61046 | 1.78787 | 2.56653 | 3.02133 |
| 0.760 | 0.56266 | 1.19681 | 1.63397 | 1.79737 | 2.58630 | 3.02426 |
| 0.780 | 0.58131 | 1.22053 | 1.65753 | 1.80804 | 2.60620 | 3.02741 |
| 0.800 | 0.60009 | 1.24341 | 1.68110 | 1.81991 | 2.62620 | 3.03081 |
| 0.820 | 0.61898 | 1.26541 | 1.70468 | 1.83300 | 2.64629 | 3.03444 |
| 0.840 | 0.63798 | 1.28651 | 1.72828 | 1.84731 | 2.66650 | 3.03835 |
| 0.860 | 0.65708 | 1.30672 | 1.75182 | 1.86288 | 2.68679 | 3.04252 |
| 0.880 | 0.67626 | 1.32602 | 1.77535 | 1.87967 | 2.70718 | 3.04699 |
| 0.900 | 0.69552 | 1.34446 | 1.79881 | 1.89768 | 2.72764 | 3.05175 |
| 0.920 | 0.71485 | 1.36203 | 1.82222 | 1.91688 | 2.74819 | 3.05678 |
| 0.940 | 0.73425 | 1.37880 | 1.84556 | 1.93719 | 2.76885 | 3.06213 |
| 0.960 | 0.75369 | 1.39481 | 1.86883 | 1.95859 | 2.78955 | 3.06779 |
| 0.980 | 0.77319 | 1.41010 | 1.89201 | 1.98098 | 2.81033 | 3.07380 |
| 1.000 | 0.79274 | 1.42474 | 1.91512 | 2.00431 | 2.83118 | 3.08010 |

$H/R = 0.450$ $n = 0.$

| $H/L$ | $\Omega_1$ | $\Omega_2$ | $\Omega_3$ |
|---|---|---|---|
| 0. | 0. | 0.24773 | 1.00782 |
| 0.001 | 0.00161 | 0.24773 | 1.00783 |
| 0.005 | 0.00806 | 0.24773 | 1.00786 |
| 0.010 | 0.01612 | 0.24776 | 1.00800 |
| 0.020 | 0.03222 | 0.24787 | 1.00853 |
| 0.030 | 0.04827 | 0.24807 | 1.00941 |
| 0.040 | 0.06425 | 0.24837 | 1.01063 |
| 0.050 | 0.08011 | 0.24880 | 1.01221 |
| 0.060 | 0.09582 | 0.24942 | 1.01412 |
| 0.070 | 0.11131 | 0.25027 | 1.01638 |
| 0.080 | 0.12651 | 0.25144 | 1.01897 |
| 0.090 | 0.14128 | 0.25306 | 1.02189 |
| 0.100 | 0.15546 | 0.25530 | 1.02513 |
| 0.120 | 0.18105 | 0.26271 | 1.03258 |
| 0.140 | 0.20081 | 0.27619 | 1.04125 |
| 0.160 | 0.21361 | 0.29697 | 1.05108 |
| 0.180 | 0.22134 | 0.32323 | 1.06202 |
| 0.200 | 0.22658 | 0.35252 | 1.07400 |
| 0.220 | 0.23090 | 0.38334 | 1.08696 |
| 0.240 | 0.23511 | 0.41496 | 1.10084 |
| 0.260 | 0.23965 | 0.44702 | 1.11559 |
| 0.280 | 0.24477 | 0.47931 | 1.13114 |
| 0.300 | 0.25062 | 0.51171 | 1.14742 |
| 0.320 | 0.25730 | 0.54416 | 1.16440 |
| 0.340 | 0.26482 | 0.57659 | 1.18201 |
| 0.360 | 0.27325 | 0.60897 | 1.20023 |
| 0.380 | 0.28253 | 0.64126 | 1.21898 |
| 0.400 | 0.29267 | 0.67343 | 1.23824 |
| 0.420 | 0.30361 | 0.70546 | 1.25797 |
| 0.440 | 0.31532 | 0.73732 | 1.27812 |
| 0.460 | 0.32775 | 0.76899 | 1.29867 |
| 0.480 | 0.34085 | 0.80043 | 1.31958 |
| 0.500 | 0.35457 | 0.83163 | 1.34083 |
| 0.520 | 0.36886 | 0.86256 | 1.36238 |
| 0.540 | 0.38366 | 0.89318 | 1.38421 |
| 0.560 | 0.39894 | 0.92346 | 1.40630 |
| 0.580 | 0.41466 | 0.95337 | 1.42863 |
| 0.600 | 0.43077 | 0.98286 | 1.45116 |
| 0.620 | 0.44724 | 1.01191 | 1.47389 |
| 0.640 | 0.46403 | 1.04046 | 1.49677 |
| 0.660 | 0.48112 | 1.06847 | 1.51981 |
| 0.680 | 0.49847 | 1.09590 | 1.54300 |
| 0.700 | 0.51607 | 1.12269 | 1.56628 |
| 0.720 | 0.53389 | 1.14881 | 1.58966 |
| 0.740 | 0.55191 | 1.17420 | 1.61312 |
| 0.760 | 0.57010 | 1.19882 | 1.63664 |
| 0.780 | 0.58846 | 1.22262 | 1.66018 |
| 0.800 | 0.60697 | 1.24558 | 1.66018 |
| 0.820 | 0.62561 | 1.26768 | 1.70735 |
| 0.840 | 0.64437 | 1.28886 | 1.73092 |
| 0.860 | 0.66325 | 1.30916 | 1.75445 |
| 0.880 | 0.68222 | 1.32855 | 1.77794 |
| 0.900 | 0.70128 | 1.34707 | 1.80137 |
| 0.920 | 0.72042 | 1.36473 | 1.82473 |
| 0.940 | 0.73964 | 1.38157 | 1.84800 |
| 0.960 | 0.75892 | 1.39766 | 1.87120 |
| 0.980 | 0.77826 | 1.41300 | 1.89432 |
| 1.000 | 0.79765 | 1.42771 | 1.91735 |

$H/R = 0.500 \qquad n = 0.$

| $H/L$ | $\Omega_1$ | $\Omega_2$ | $\Omega_3$ | $\Omega_4$ | $\Omega_5$ | $\Omega_6$ |
|---|---|---|---|---|---|---|
| 0. | 0. | 0.27673 | 1.00969 | 1.89408 | 2.00501 | 3.00336 |
| 0.010 | 0.01612 | 0.27676 | 1.00987 | 1.89377 | 2.00539 | 3.00336 |
| 0.050 | 0.08021 | 0.27757 | 1.01410 | 1.88698 | 2.01402 | 3.00326 |
| 0.100 | 0.15695 | 0.28235 | 1.02710 | 1.87032 | 2.03642 | 3.00297 |
| 0.120 | 0.18468 | 0.28742 | 1.03459 | 1.86250 | 2.04761 | 3.00282 |
| 0.140 | 0.20859 | 0.29652 | 1.04330 | 1.85443 | 2.05967 | 3.00264 |
| 0.160 | 0.22674 | 0.31166 | 1.05318 | 1.84629 | 2.07242 | 3.00245 |
| 0.180 | 0.23881 | 0.33326 | 1.06416 | 1.83820 | 2.08575 | 3.00224 |
| 0.200 | 0.24661 | 0.35958 | 1.07620 | 1.83023 | 2.09958 | 3.00204 |
| 0.220 | 0.25224 | 0.38863 | 1.08921 | 1.82247 | 2.11384 | 3.00183 |
| 0.240 | 0.25701 | 0.41916 | 1.10315 | 1.81495 | 2.12848 | 3.00163 |
| 0.260 | 0.26167 | 0.45051 | 1.11794 | 1.80773 | 2.14347 | 3.00145 |
| 0.280 | 0.26662 | 0.48232 | 1.13353 | 1.80083 | 2.15879 | 3.00129 |
| 0.300 | 0.27211 | 0.51438 | 1.14987 | 1.79429 | 2.17440 | 3.00116 |
| 0.320 | 0.27828 | 0.54657 | 1.16689 | 1.78812 | 2.19029 | 3.00107 |
| 0.340 | 0.28521 | 0.57881 | 1.18455 | 1.78236 | 2.20644 | 3.00101 |
| 0.360 | 0.29296 | 0.61105 | 1.20279 | 1.77703 | 2.22284 | 3.00102 |
| 0.380 | 0.30153 | 0.64322 | 1.22159 | 1.77215 | 2.23947 | 3.00108 |
| 0.400 | 0.31092 | 0.67531 | 1.24088 | 1.76774 | 2.25634 | 3.00123 |
| 0.420 | 0.32112 | 0.70728 | 1.26064 | 1.76382 | 2.27342 | 3.00144 |
| 0.440 | 0.33208 | 0.73909 | 1.28083 | 1.76041 | 2.29070 | 3.00174 |
| 0.460 | 0.34378 | 0.77072 | 1.30141 | 1.75755 | 2.30819 | 3.00215 |
| 0.480 | 0.35616 | 0.80215 | 1.32235 | 1.75525 | 2.32587 | 3.00266 |
| 0.500 | 0.36919 | 0.83334 | 1.34362 | 1.75354 | 2.34374 | 3.00328 |
| 0.520 | 0.38281 | 0.86426 | 1.36520 | 1.75244 | 2.36178 | 3.00404 |
| 0.540 | 0.39698 | 0.89488 | 1.38706 | 1.75200 | 2.38000 | 3.00492 |
| 0.560 | 0.41166 | 0.92518 | 1.40917 | 1.75223 | 2.39838 | 3.00594 |
| 0.580 | 0.42681 | 0.95511 | 1.43152 | 1.75318 | 2.41692 | 3.00710 |
| 0.600 | 0.44238 | 0.98464 | 1.45407 | 1.75488 | 2.43562 | 3.00844 |
| 0.620 | 0.45834 | 1.01372 | 1.47682 | 1.75737 | 2.45446 | 3.00994 |
| 0.640 | 0.47465 | 1.04232 | 1.49973 | 1.76070 | 2.47346 | 3.01162 |
| 0.660 | 0.49129 | 1.07038 | 1.52279 | 1.76490 | 2.49258 | 3.01347 |
| 0.680 | 0.50822 | 1.09787 | 1.54599 | 1.77002 | 2.51185 | 3.01553 |
| 0.700 | 0.52542 | 1.12474 | 1.56929 | 1.77611 | 2.53124 | 3.01779 |
| 0.720 | 0.54286 | 1.15093 | 1.59269 | 1.78323 | 2.55076 | 3.02026 |
| 0.740 | 0.56053 | 1.17640 | 1.61616 | 1.79141 | 2.57039 | 3.02297 |
| 0.760 | 0.57839 | 1.20111 | 1.63969 | 1.80071 | 2.59015 | 3.02587 |
| 0.780 | 0.59644 | 1.22501 | 1.66325 | 1.81116 | 2.61001 | 3.02904 |
| 0.800 | 0.61465 | 1.24807 | 1.68683 | 1.82281 | 2.62998 | 3.03244 |
| 0.820 | 0.63302 | 1.27025 | 1.71041 | 1.83569 | 2.65005 | 3.03608 |
| 0.840 | 0.65152 | 1.29155 | 1.73396 | 1.84982 | 2.67023 | 3.04000 |
| 0.860 | 0.67015 | 1.31194 | 1.75747 | 1.86521 | 2.69049 | 3.04420 |
| 0.880 | 0.68889 | 1.33144 | 1.78092 | 1.88185 | 2.71084 | 3.04866 |
| 0.900 | 0.70773 | 1.35006 | 1.80430 | 1.89974 | 2.73129 | 3.05342 |
| 0.920 | 0.72666 | 1.36781 | 1.82760 | 1.91883 | 2.75181 | 3.05848 |
| 0.940 | 0.74568 | 1.38475 | 1.85082 | 1.93908 | 2.77241 | 3.06383 |
| 0.960 | 0.76477 | 1.40090 | 1.87394 | 1.96042 | 2.79309 | 3.06951 |
| 0.980 | 0.78394 | 1.41634 | 1.89698 | 1.98280 | 2.81384 | 3.07548 |
| 1.000 | 0.80317 | 1.43110 | 1.91993 | 2.00612 | 2.83465 | 3.08180 |

$H/R = 0.600 \qquad n = 0.$

| $H/L$ | $\Omega_1$ | $\Omega_2$ | $\Omega_3$ | $\Omega_4$ | $\Omega_5$ | $\Omega_6$ |
|---|---|---|---|---|---|---|
| 0. | 0. | 0.33621 | 1.01408 | 1.90578 | 2.00739 | 3.00499 |
| 0.010 | 0.01612 | 0.33622 | 1.01426 | 1.90544 | 2.00780 | 3.00497 |
| 0.050 | 0.08034 | 0.33667 | 1.01855 | 1.89807 | 2.01700 | 3.00488 |
| 0.100 | 0.15854 | 0.33935 | 1.03174 | 1.88045 | 2.04033 | 3.00460 |
| 0.120 | 0.18828 | 0.34205 | 1.03932 | 1.87231 | 2.05181 | 3.00445 |
| 0.140 | 0.21626 | 0.34664 | 1.04814 | 1.86398 | 2.06412 | 3.00428 |
| 0.160 | 0.24142 | 0.35426 | 1.05814 | 1.85561 | 2.07707 | 3.00410 |
| 0.180 | 0.26239 | 0.36635 | 1.06925 | 1.84732 | 2.09057 | 3.00391 |
| 0.200 | 0.27829 | 0.38389 | 1.08141 | 1.83918 | 2.10453 | 3.00371 |
| 0.220 | 0.28957 | 0.40648 | 1.09455 | 1.83127 | 2.11889 | 3.00352 |
| 0.240 | 0.29770 | 0.43274 | 1.10861 | 1.82362 | 2.13362 | 3.00333 |
| 0.260 | 0.30412 | 0.46132 | 1.12353 | 1.81627 | 2.14869 | 3.00316 |
| 0.280 | 0.30979 | 0.49127 | 1.13924 | 1.80926 | 2.16406 | 3.00301 |
| 0.300 | 0.31533 | 0.52206 | 1.15569 | 1.80261 | 2.17971 | 3.00290 |
| 0.320 | 0.32109 | 0.55335 | 1.17282 | 1.79634 | 2.19564 | 3.00282 |
| 0.340 | 0.32731 | 0.58492 | 1.19058 | 1.79048 | 2.21182 | 3.00278 |
| 0.360 | 0.33411 | 0.61665 | 1.20893 | 1.78504 | 2.22824 | 3.00281 |
| 0.380 | 0.34158 | 0.64845 | 1.22782 | 1.78006 | 2.24489 | 3.00289 |
| 0.400 | 0.34976 | 0.68025 | 1.24720 | 1.77554 | 2.26176 | 3.00304 |
| 0.420 | 0.35867 | 0.71199 | 1.26704 | 1.77151 | 2.27885 | 3.00328 |
| 0.440 | 0.36830 | 0.74363 | 1.28731 | 1.76800 | 2.29613 | 3.00360 |
| 0.460 | 0.37866 | 0.77514 | 1.30796 | 1.76502 | 2.31362 | 3.00402 |
| 0.480 | 0.38970 | 0.80648 | 1.32898 | 1.76259 | 2.33129 | 3.00455 |
| 0.500 | 0.40141 | 0.83761 | 1.35032 | 1.76075 | 2.34914 | 3.00518 |
| 0.520 | 0.41374 | 0.86851 | 1.37196 | 1.75951 | 2.36717 | 3.00596 |
| 0.540 | 0.42666 | 0.89913 | 1.39388 | 1.75892 | 2.38537 | 3.00684 |
| 0.560 | 0.44013 | 0.92944 | 1.41606 | 1.75899 | 2.40374 | 3.00789 |
| 0.580 | 0.45412 | 0.95941 | 1.43846 | 1.75977 | 2.42226 | 3.00908 |
| 0.600 | 0.46858 | 0.98900 | 1.46107 | 1.76128 | 2.44093 | 3.01042 |
| 0.620 | 0.48348 | 1.01817 | 1.48387 | 1.76357 | 2.45976 | 3.01193 |
| 0.640 | 0.49879 | 1.04687 | 1.50684 | 1.76668 | 2.47872 | 3.01363 |
| 0.660 | 0.51447 | 1.07506 | 1.52995 | 1.77065 | 2.49783 | 3.01551 |
| 0.680 | 0.53049 | 1.10269 | 1.55319 | 1.77554 | 2.51706 | 3.01758 |
| 0.700 | 0.54683 | 1.12971 | 1.57655 | 1.78137 | 2.53642 | 3.01984 |
| 0.720 | 0.56347 | 1.15608 | 1.59999 | 1.78822 | 2.55591 | 3.02232 |
| 0.740 | 0.58036 | 1.18175 | 1.62351 | 1.79612 | 2.57551 | 3.02502 |
| 0.760 | 0.59750 | 1.20666 | 1.64707 | 1.80512 | 2.59524 | 3.02795 |
| 0.780 | 0.61486 | 1.23079 | 1.67066 | 1.81528 | 2.61506 | 3.03113 |
| 0.800 | 0.63242 | 1.25408 | 1.69426 | 1.82664 | 2.63500 | 3.03453 |
| 0.820 | 0.65017 | 1.27650 | 1.71784 | 1.83923 | 2.65503 | 3.03819 |
| 0.840 | 0.66809 | 1.29805 | 1.74138 | 1.85309 | 2.67517 | 3.04213 |
| 0.860 | 0.68617 | 1.31869 | 1.76485 | 1.86823 | 2.69539 | 3.04633 |
| 0.880 | 0.70439 | 1.33843 | 1.78824 | 1.88466 | 2.71571 | 3.05079 |
| 0.900 | 0.72272 | 1.35728 | 1.81153 | 1.90236 | 2.73611 | 3.05556 |
| 0.920 | 0.74120 | 1.37527 | 1.83470 | 1.92131 | 2.75659 | 3.06064 |
| 0.940 | 0.75977 | 1.39242 | 1.85775 | 1.94146 | 2.77715 | 3.06598 |
| 0.960 | 0.77844 | 1.40879 | 1.88069 | 1.96275 | 2.79778 | 3.07166 |
| 0.980 | 0.79720 | 1.42441 | 1.90354 | 1.98509 | 2.81848 | 3.07765 |
| 1.000 | 0.81605 | 1.43935 | 1.92629 | 2.00841 | 2.83925 | 3.08398 |

$H/R = 0.700 \qquad n = 0.$

| $H/L$ | $\Omega_1$ | $\Omega_2$ | $\Omega_3$ | $\Omega_4$ | $\Omega_5$ | $\Omega_6$ |
|---|---|---|---|---|---|---|
| 0. | 0. | 0.39798 | 1.01937 | 1.92097 | 2.01034 | 3.00699 |
| 0.010 | 0.01612 | 0.39798 | 1.01955 | 1.92058 | 2.01080 | 3.00699 |
| 0.050 | 0.08041 | 0.39812 | 1.02393 | 1.91229 | 2.02090 | 3.00690 |
| 0.100 | 0.15933 | 0.39944 | 1.03737 | 1.89333 | 2.04553 | 3.00664 |
| 0.120 | 0.18994 | 0.40092 | 1.04509 | 1.88478 | 2.05740 | 3.00650 |
| 0.140 | 0.21956 | 0.40345 | 1.05405 | 1.87611 | 2.07002 | 3.00634 |
| 0.160 | 0.24771 | 0.40760 | 1.06421 | 1.86745 | 2.08323 | 3.00617 |
| 0.180 | 0.27365 | 0.41418 | 1.07548 | 1.85891 | 2.09693 | 3.00599 |
| 0.200 | 0.29644 | 0.42418 | 1.08780 | 1.85057 | 2.11105 | 3.00581 |
| 0.220 | 0.31524 | 0.43855 | 1.10111 | 1.84247 | 2.12555 | 3.00563 |
| 0.240 | 0.32984 | 0.45756 | 1.11534 | 1.83466 | 2.14038 | 3.00546 |
| 0.260 | 0.34090 | 0.48061 | 1.13041 | 1.82717 | 2.15553 | 3.00531 |
| 0.280 | 0.34952 | 0.50667 | 1.14628 | 1.82002 | 2.17097 | 3.00518 |
| 0.300 | 0.35673 | 0.53474 | 1.16288 | 1.81324 | 2.18668 | 3.00508 |
| 0.320 | 0.36329 | 0.56411 | 1.18015 | 1.80685 | 2.20264 | 3.00502 |
| 0.340 | 0.36969 | 0.59432 | 1.19805 | 1.80087 | 2.21885 | 3.00500 |
| 0.360 | 0.37625 | 0.62505 | 1.21652 | 1.79531 | 2.23530 | 3.00504 |
| 0.380 | 0.38318 | 0.65609 | 1.23553 | 1.79021 | 2.25196 | 3.00515 |
| 0.400 | 0.39061 | 0.68732 | 1.25503 | 1.78557 | 2.26884 | 3.00532 |
| 0.420 | 0.39861 | 0.71863 | 1.27498 | 1.78141 | 2.28592 | 3.00558 |
| 0.440 | 0.40723 | 0.74994 | 1.29534 | 1.77776 | 2.30321 | 3.00592 |
| 0.460 | 0.41640 | 0.78119 | 1.31609 | 1.77464 | 2.32068 | 3.00636 |
| 0.480 | 0.42639 | 0.81234 | 1.33720 | 1.77206 | 2.33834 | 3.00691 |
| 0.500 | 0.43693 | 0.84334 | 1.35863 | 1.77006 | 2.35618 | 3.00758 |
| 0.520 | 0.44809 | 0.87415 | 1.38035 | 1.76866 | 2.37419 | 3.00836 |
| 0.540 | 0.45984 | 0.90472 | 1.40235 | 1.76788 | 2.39237 | 3.00928 |
| 0.560 | 0.47216 | 0.93502 | 1.42461 | 1.76775 | 2.41071 | 3.01033 |
| 0.580 | 0.48501 | 0.96501 | 1.44708 | 1.76832 | 2.42920 | 3.01153 |
| 0.600 | 0.49837 | 0.99465 | 1.46977 | 1.76960 | 2.44785 | 3.01289 |
| 0.620 | 0.51221 | 1.02389 | 1.49265 | 1.77165 | 2.46664 | 3.01443 |
| 0.640 | 0.52649 | 1.05270 | 1.51569 | 1.77449 | 2.48557 | 3.01613 |
| 0.660 | 0.54119 | 1.08102 | 1.53888 | 1.77817 | 2.50464 | 3.01802 |
| 0.680 | 0.55626 | 1.10881 | 1.56220 | 1.78275 | 2.52384 | 3.02012 |
| 0.700 | 0.57170 | 1.13602 | 1.58562 | 1.78826 | 2.54316 | 3.02240 |
| 0.720 | 0.58746 | 1.16259 | 1.60914 | 1.79475 | 2.56261 | 3.02488 |
| 0.740 | 0.60353 | 1.18849 | 1.63272 | 1.80229 | 2.58217 | 3.02760 |
| 0.760 | 0.61988 | 1.21366 | 1.65637 | 1.81091 | 2.60185 | 3.03055 |
| 0.780 | 0.63648 | 1.23805 | 1.68002 | 1.82068 | 2.62163 | 3.03372 |
| 0.800 | 0.65333 | 1.26163 | 1.70368 | 1.83164 | 2.64152 | 3.03714 |
| 0.820 | 0.67040 | 1.28436 | 1.72730 | 1.84384 | 2.66151 | 3.04082 |
| 0.840 | 0.68767 | 1.30621 | 1.75086 | 1.85731 | 2.68159 | 3.04477 |
| 0.860 | 0.70512 | 1.32716 | 1.77432 | 1.87210 | 2.70177 | 3.04896 |
| 0.880 | 0.72275 | 1.34722 | 1.79766 | 1.88821 | 2.72203 | 3.05345 |
| 0.900 | 0.74054 | 1.36638 | 1.82086 | 1.90565 | 2.74238 | 3.05822 |
| 0.920 | 0.75847 | 1.38467 | 1.84389 | 1.92440 | 2.76281 | 3.06329 |
| 0.940 | 0.77653 | 1.40210 | 1.86676 | 1.94441 | 2.78331 | 3.06867 |
| 0.960 | 0.79472 | 1.41874 | 1.88947 | 1.96561 | 2.80388 | 3.07434 |
| 0.980 | 0.81301 | 1.43461 | 1.91205 | 1.98792 | 2.82453 | 3.08035 |
| 1.000 | 0.83142 | 1.44978 | 1.93453 | 2.01123 | 2.84524 | 3.08668 |

$H/R = 0.800 \qquad n = 0.$

| $H/L$ | $\Omega_1$ | $\Omega_2$ | $\Omega_3$ | $\Omega_4$ | $\Omega_5$ | $\Omega_6$ |
|---|---|---|---|---|---|---|
| 0. | 0. | 0.46243 | 1.02562 | 1.94954 | 2.01397 | 3.00950 |
| 0.010 | 0.01612 | 0.46242 | 1.02580 | 1.94006 | 2.01451 | 3.00949 |
| 0.050 | 0.08046 | 0.46227 | 1.03030 | 1.93029 | 2.02609 | 3.00941 |
| 0.100 | 0.15980 | 0.46250 | 1.04407 | 1.90943 | 2.05255 | 3.00917 |
| 0.120 | 0.19086 | 0.46315 | 1.05197 | 1.90036 | 2.06492 | 3.00904 |
| 0.140 | 0.22128 | 0.46445 | 1.06113 | 1.89127 | 2.07793 | 3.00889 |
| 0.160 | 0.25080 | 0.46674 | 1.07149 | 1.88227 | 2.09144 | 3.00873 |
| 0.180 | 0.27904 | 0.47046 | 1.08298 | 1.87344 | 2.10537 | 3.00857 |
| 0.200 | 0.30549 | 0.47621 | 1.09552 | 1.86486 | 2.11968 | 3.00840 |
| 0.220 | 0.32950 | 0.48471 | 1.10905 | 1.85656 | 2.13433 | 3.00824 |
| 0.240 | 0.35038 | 0.49671 | 1.12348 | 1.84856 | 2.14929 | 3.00809 |
| 0.260 | 0.36772 | 0.51268 | 1.13877 | 1.84090 | 2.16453 | 3.00796 |
| 0.280 | 0.38166 | 0.53256 | 1.15483 | 1.83360 | 2.18004 | 3.00785 |
| 0.300 | 0.39280 | 0.55577 | 1.17162 | 1.82668 | 2.19580 | 3.00778 |
| 0.320 | 0.40200 | 0.58152 | 1.18907 | 1.82015 | 2.21180 | 3.00774 |
| 0.340 | 0.41002 | 0.60907 | 1.20714 | 1.81403 | 2.22804 | 3.00775 |
| 0.360 | 0.41744 | 0.63784 | 1.22577 | 1.80834 | 2.24450 | 3.00782 |
| 0.380 | 0.42468 | 0.66744 | 1.24493 | 1.80310 | 2.26117 | 3.00794 |
| 0.400 | 0.43202 | 0.69758 | 1.26457 | 1.79832 | 2.27805 | 3.00814 |
| 0.420 | 0.43965 | 0.72806 | 1.28465 | 1.79402 | 2.29513 | 3.00842 |
| 0.440 | 0.44769 | 0.75874 | 1.30515 | 1.79022 | 2.31240 | 3.00879 |
| 0.460 | 0.45622 | 0.78951 | 1.32601 | 1.78694 | 2.32986 | 3.00925 |
| 0.480 | 0.46527 | 0.82028 | 1.34723 | 1.78419 | 2.34750 | 3.00982 |
| 0.500 | 0.47489 | 0.85101 | 1.36877 | 1.78200 | 2.36531 | 3.01051 |
| 0.520 | 0.48507 | 0.88161 | 1.39060 | 1.78040 | 2.38329 | 3.01131 |
| 0.540 | 0.49581 | 0.91205 | 1.41270 | 1.77941 | 2.40144 | 3.01225 |
| 0.560 | 0.50710 | 0.94226 | 1.43505 | 1.77905 | 2.41974 | 3.01333 |
| 0.580 | 0.51892 | 0.97222 | 1.45763 | 1.77936 | 2.43820 | 3.01455 |
| 0.600 | 0.53125 | 1.00187 | 1.48041 | 1.78037 | 2.45680 | 3.01594 |
| 0.620 | 0.54407 | 1,03117 | 1.50338 | 1.78212 | 2.47555 | 3.01749 |
| 0.640 | 0.55736 | 1.06007 | 1.52652 | 1.78464 | 2.49444 | 3.01922 |
| 0.660 | 0.57108 | 1.08852 | 1.54982 | 1.78797 | 2.51346 | 3.02113 |
| 0.680 | 0.58521 | 1.11648 | 1.57324 | 1.79217 | 2.53261 | 3.02323 |
| 0.700 | 0.59972 | 1.14389 | 1.59678 | 1.79727 | 2.55188 | 3.02555 |
| 0.720 | 0.61460 | 1.17070 | 1.62041 | 1.80333 | 2.57127 | 3.02804 |
| 0.740 | 0.62981 | 1.19686 | 1.64412 | 1.81039 | 2.59078 | 3.03078 |
| 0.760 | 0.64534 | 1.22233 | 1.66787 | 1.81853 | 2.61040 | 3.03373 |
| 0.780 | 0.66115 | 1.24704 | 1.69165 | 1.82777 | 2.63013 | 3.03692 |
| 0.800 | 0.67724 | 1.27097 | 1.71542 | 1.83820 | 2.64996 | 3.04037 |
| 0.820 | 0.69358 | 1.29406 | 1.73915 | 1.84986 | 2.66988 | 3.04404 |
| 0.840 | 0.71015 | 1.31629 | 1.76280 | 1.86280 | 2.68990 | 3.04799 |
| 0.860 | 0.72693 | 1.33763 | 1.78633 | 1.87708 | 2.71002 | 3.05221 |
| 0.880 | 0.74392 | 1.35808 | 1.80969 | 1.89272 | 2.73021 | 3.05672 |
| 0.900 | 0.76109 | 1.37763 | 1.83283 | 1.90976 | 2.75049 | 3.06148 |
| 0.920 | 0.77843 | 1.39630 | 1.85573 | 1.92820 | 2.77085 | 3.06656 |
| 0.940 | 0.79593 | 1.41410 | 1.87839 | 1.94799 | 2.79129 | 3.07194 |
| 0.960 | 0.81358 | 1.43109 | 1.90083 | 1.96907 | 2.81179 | 3.07763 |
| 0.980 | 0.83136 | 1.44729 | 1.92307 | 1.99134 | 2.83236 | 3.08365 |
| 1.000 | 0.84927 | 1.46276 | 1.94518 | 2.01467 | 2.85300 | 3.08998 |

$H/R = 0.900$ $\quad$ $n = 0.$

| $H/L$ | $\Omega_1$ | $\Omega_2$ | $\Omega_3$ | $\Omega_4$ | $\Omega_5$ | $\Omega_6$ |
|---|---|---|---|---|---|---|
| 0. | 0. | 0.52989 | 1.03289 | 1.96577 | 2.01838 | 3.01259 |
| 0.010 | 0.01612 | 0.52987 | 1.03308 | 1.96510 | 2.01911 | 3.01257 |
| 0.050 | 0.08049 | 0.52942 | 1.03774 | 1.95262 | 2.03338 | 3.01249 |
| 0.100 | 0.16009 | 0.52862 | 1.05196 | 1.92919 | 2.06237 | 3.01228 |
| 0.120 | 0.19142 | 0.52859 | 1.06010 | 1.91950 | 2.07533 | 3.01216 |
| 0.140 | 0.22229 | 0.52897 | 1.06952 | 1.90992 | 2.08877 | 3.01203 |
| 0.160 | 0.25255 | 0.53001 | 1.08016 | 1.90055 | 2.10261 | 3.01189 |
| 0.180 | 0.28197 | 0.53198 | 1.09193 | 1.89142 | 2.11679 | 3.01174 |
| 0.200 | 0.31027 | 0.53526 | 1.10475 | 1.88258 | 2.13130 | 3.01160 |
| 0.220 | 0.33707 | 0.54030 | 1.11856 | 1.87407 | 2.14609 | 3.01146 |
| 0.240 | 0.36189 | 0.54765 | 1.13327 | 1.86588 | 2.16116 | 3.01134 |
| 0.260 | 0.38423 | 0.55786 | 1.14882 | 1.85806 | 2.17648 | 3.01123 |
| 0.280 | 0.40368 | 0.57142 | 1.16513 | 1.85060 | 2.19205 | 3.01115 |
| 0.300 | 0.42011 | 0.58849 | 1.18216 | 1.84354 | 2.20785 | 3.01110 |
| 0.320 | 0.43378 | 0.60888 | 1.19984 | 1.83687 | 2.22388 | 3.01109 |
| 0.340 | 0.44520 | 0.63207 | 1.21811 | 1.83062 | 2.24013 | 3.01113 |
| 0.360 | 0.45502 | 0.65746 | 1.23695 | 1.82479 | 2.25660 | 3.01122 |
| 0.380 | 0.46383 | 0.68447 | 1.25629 | 1.81941 | 2.27326 | 3.01138 |
| 0.400 | 0.47209 | 0.71263 | 1.27610 | 1.81448 | 2.29013 | 3.01161 |
| 0.420 | 0.48015 | 0.74161 | 1.29634 | 1.81003 | 2.30719 | 3.01192 |
| 0.440 | 0.48825 | 0.77113 | 1.31698 | 1.80607 | 2.32443 | 3.01231 |
| 0.460 | 0.49657 | 0.80101 | 1.33800 | 1.80261 | 2.34185 | 3.01281 |
| 0.480 | 0.50521 | 0.83110 | 1.35935 | 1.79968 | 2.35946 | 3.01341 |
| 0.500 | 0.51425 | 0.86129 | 1.38101 | 1.79729 | 2.37723 | 3.01412 |
| 0.520 | 0.52374 | 0.89150 | 1.40297 | 1.79547 | 2.39516 | 3.01495 |
| 0.540 | 0.53371 | 0.92164 | 1.42519 | 1.79424 | 2.41326 | 3.01592 |
| 0.560 | 0.54417 | 0.95166 | 1.44766 | 1.79362 | 2.43151 | 3.01702 |
| 0.580 | 0.55513 | 0.98149 | 1.47035 | 1.79365 | 2.44991 | 3.01827 |
| 0.600 | 0.56656 | 1.01108 | 1.49325 | 1.79434 | 2.46846 | 3.01968 |
| 0.620 | 0.57848 | 1.04038 | 1.51634 | 1.79575 | 2.48714 | 3.02127 |
| 0.640 | 0.59085 | 1.06933 | 1.53961 | 1.79789 | 2.50597 | 3.02301 |
| 0.660 | 0.60366 | 1.09789 | 1.56303 | 1.80082 | 2.52492 | 3.02494 |
| 0.680 | 0.61689 | 1.12600 | 1.58659 | 1.80456 | 2.54400 | 3.02707 |
| 0.700 | 0.63052 | 1.15362 | 1.61027 | 1.80917 | 2.56321 | 3.02939 |
| 0.720 | 0.64453 | 1.18068 | 1.63406 | 1.81470 | 2.58253 | 3.03191 |
| 0.740 | 0.65889 | 1.20714 | 1.65793 | 1.82119 | 2.60197 | 3.03466 |
| 0.760 | 0.67359 | 1.23293 | 1.68187 | 1.82870 | 2.62151 | 3.03763 |
| 0.780 | 0.68861 | 1.25803 | 1.70585 | 1.83729 | 2.64116 | 3.04084 |
| 0.800 | 0.70392 | 1.28235 | 1.72983 | 1.84701 | 2.66092 | 3.04428 |
| 0.820 | 0.71950 | 1.30587 | 1.75378 | 1.85793 | 2.68076 | 3.04798 |
| 0.840 | 0.73535 | 1.32856 | 1.77764 | 1.87011 | 2.70070 | 3.05194 |
| 0.860 | 0.75143 | 1.35038 | 1.80137 | 1.88364 | 2.72073 | 3.05617 |
| 0.880 | 0.76774 | 1.37131 | 1.82488 | 1.89857 | 2.74085 | 3.06067 |
| 0.900 | 0.78426 | 1.39135 | 1.84811 | 1.91497 | 2.76104 | 3.06546 |
| 0.920 | 0.80098 | 1.41050 | 1.87099 | 1.93288 | 2.78131 | 3.07055 |
| 0.940 | 0.81787 | 1.42878 | 1.89348 | 1.95232 | 2.80166 | 3.07593 |
| 0.960 | 0.83494 | 1.44622 | 1.91559 | 1.97321 | 2.82207 | 3.08163 |
| 0.980 | 0.85216 | 1.46285 | 1.93739 | 1.99543 | 2.84255 | 3.08766 |
| 1.000 | 0.86953 | 1.47872 | 1.95897 | 2.01883 | 2.86309 | 3.09400 |

$H/R = 1.000 \qquad n = 0.$

| $H/L$ | $\Omega_1$ | $\Omega_2$ | $\Omega_3$ | $\Omega_4$ | $\Omega_5$ | $\Omega_6$ |
|---|---|---|---|---|---|---|
| 0. | 0. | 0.60059 | 1.04127 | 1.99847 | 2.02371 | 3.01636 |
| 0.010 | 0.01612 | 0.60056 | 1.04147 | 1.99711 | 2.02514 | 3.01636 |
| 0.050 | 0.08051 | 0.59977 | 1.04635 | 1.97904 | 2.04498 | 3.01630 |
| 0.100 | 0.16029 | 0.59789 | 1.06120 | 1.95264 | 2.07688 | 3.01611 |
| 0.120 | 0.19180 | 0.59720 | 1.06966 | 1.94237 | 2.09038 | 3.01601 |
| 0.140 | 0.22295 | 0.59676 | 1.07943 | 1.93238 | 2.10919 | 3.01589 |
| 0.160 | 0.25364 | 0.59677 | 1.09043 | 1.92268 | 2.11830 | 3.01577 |
| 0.180 | 0.28374 | 0.59744 | 1.10258 | 1.91331 | 2.13268 | 3.01565 |
| 0.200 | 0.31307 | 0.59903 | 1.11577 | 1.90427 | 2.14732 | 3.01554 |
| 0.220 | 0.34139 | 0.60185 | 1.12994 | 1.89559 | 2.16220 | 3.01543 |
| 0.240 | 0.36841 | 0.60626 | 1.14500 | 1.88726 | 2.17733 | 3.01533 |
| 0.260 | 0.39380 | 0.61267 | 1.16087 | 1.87931 | 2.19269 | 3.01525 |
| 0.280 | 0.41716 | 0.62153 | 1.17750 | 1.87174 | 2.20827 | 3.01520 |
| 0.300 | 0.43816 | 0.63321 | 1.19482 | 1.86456 | 2.22408 | 3.01519 |
| 0.320 | 0.45659 | 0.64798 | 1.21277 | 1.85778 | 2.24010 | 3.01521 |
| 0.340 | 0.47249 | 0.66582 | 1.23130 | 1.85142 | 2.25632 | 3.01529 |
| 0.360 | 0.48612 | 0.68649 | 1.25036 | 1.84548 | 2.27275 | 3.01542 |
| 0.380 | 0.49792 | 0.70958 | 1.26992 | 1.83998 | 2.28937 | 3.01561 |
| 0.400 | 0.50838 | 0.73458 | 1.28993 | 1.83493 | 2.30619 | 3.01588 |
| 0.420 | 0.51796 | 0.76106 | 1.31036 | 1.83034 | 2.32319 | 3.01622 |
| 0.440 | 0.52703 | 0.78862 | 1.33118 | 1.82623 | 2.34038 | 3.01666 |
| 0.460 | 0.53587 | 0.81697 | 1.35235 | 1.82261 | 2.35774 | 3.01718 |
| 0.480 | 0.54471 | 0.84587 | 1.37385 | 1.81950 | 2.37527 | 3.01783 |
| 0.500 | 0.55369 | 0.87513 | 1.39566 | 1.81691 | 2.39297 | 3.01856 |
| 0.520 | 0.56291 | 0.90461 | 1.41775 | 1.81487 | 2.41083 | 3.01943 |
| 0.540 | 0.57247 | 0.93420 | 1.44010 | 1.81330 | 2.42885 | 3.02043 |
| 0.560 | 0.58239 | 0.96381 | 1.46270 | 1.81251 | 2.44702 | 3.02156 |
| 0.580 | 0.59272 | 0.99334 | 1.48552 | 1.81224 | 2.46534 | 3.02285 |
| 0.600 | 0.60347 | 1.02274 | 1.50855 | 1.81261 | 2.48380 | 3.02428 |
| 0.620 | 0.61465 | 1.05194 | 1.53177 | 1.81364 | 2.50240 | 3.02589 |
| 0.640 | 0.62625 | 1.08087 | 1.55518 | 1.81538 | 2.52113 | 3.02766 |
| 0.660 | 0.63827 | 1.10948 | 1.57874 | 1.81786 | 2.54000 | 3.02962 |
| 0.680 | 0.65070 | 1.13772 | 1.60246 | 1.82110 | 2.55899 | 3.03176 |
| 0.700 | 0.66352 | 1.16551 | 1.62631 | 1.82516 | 2.57810 | 3.03410 |
| 0.720 | 0.67672 | 1.19282 | 1.65029 | 1.83008 | 2.59732 | 3.03665 |
| 0.740 | 0.69028 | 1.21958 | 1.67437 | 1.83590 | 2.61666 | 3.03942 |
| 0.760 | 0.70419 | 1.24574 | 1.69854 | 1.84267 | 2.63611 | 3.04241 |
| 0.780 | 0.71843 | 1.27124 | 1.72279 | 1.85044 | 2.65566 | 3.04563 |
| 0.800 | 0.73297 | 1.29603 | 1.74708 | 1.85928 | 2.67531 | 3.04910 |
| 0.820 | 0.74781 | 1.32005 | 1.77138 | 1.86923 | 2.69505 | 3.05281 |
| 0.840 | 0.76293 | 1.34328 | 1.79565 | 1.88038 | 2.71489 | 3.05677 |
| 0.860 | 0.77831 | 1.36567 | 1.81982 | 1.89280 | 2.73481 | 3.06102 |
| 0.880 | 0.79393 | 1.38719 | 1.84379 | 1.90660 | 2.75481 | 3.06553 |
| 0.900 | 0.80978 | 1.40784 | 1.86743 | 1.92191 | 2.77489 | 3.07031 |
| 0.920 | 0.82585 | 1.42759 | 1.89059 | 1.93887 | 2.79505 | 3.07540 |
| 0.940 | 0.84211 | 1.44647 | 1.91310 | 1.95760 | 2.81528 | 3.08079 |
| 0.960 | 0.85857 | 1.46449 | 1.93493 | 1.97814 | 2.83557 | 3.08649 |
| 0.980 | 0.87520 | 1.48168 | 1.95615 | 2.00032 | 2.85593 | 3.09251 |
| 1.000 | 0.89200 | 1.49809 | 1.97696 | 2.02388 | 2.87635 | 3.09887 |

$H/R = 1.100$ $n = 0.$

| $H/L$ | $\Omega_1$ | $\Omega_2$ | $\Omega_3$ | $\Omega_4$ | $\Omega_5$ | $\Omega_6$ |
|---|---|---|---|---|---|---|
| 0. | 0. | 0.67458 | 1.05086 | 2.03019 | 2.04119 | 3.02105 |
| 0.010 | 0.01612 | 0.67453 | 1.05107 | 2.02756 | 2.04391 | 3.02105 |
| 0.050 | 0.08053 | 0.67331 | 1.05627 | 2.00590 | 2.06729 | 3.02099 |
| 0.100 | 0.16043 | 0.67017 | 1.07200 | 1.97876 | 2.09987 | 3.02084 |
| 0.120 | 0.19206 | 0.66876 | 1.08090 | 1.96840 | 2.11342 | 3.02076 |
| 0.140 | 0.22340 | 0.66749 | 1.09115 | 1.95836 | 2.12724 | 3.02067 |
| 0.160 | 0.25438 | 0.66652 | 1.10264 | 1.94865 | 2.14131 | 3.02058 |
| 0.180 | 0.28491 | 0.66604 | 1.11527 | 1.93927 | 2.15564 | 3.02048 |
| 0.200 | 0.31485 | 0.66627 | 1.12894 | 1.93024 | 2.17021 | 3.02040 |
| 0.220 | 0.34407 | 0.66743 | 1.14357 | 1.92155 | 2.18501 | 3.02032 |
| 0.240 | 0.37238 | 0.66978 | 1.15905 | 1.91323 | 2.20005 | 3.02026 |
| 0.260 | 0.39955 | 0.67362 | 1.17534 | 1.90528 | 2.21532 | 3.02023 |
| 0.280 | 0.42532 | 0.67927 | 1.19234 | 1.89771 | 2.23080 | 3.02022 |
| 0.300 | 0.44940 | 0.68706 | 1.21000 | 1.89052 | 2.24650 | 3.02024 |
| 0.320 | 0.47151 | 0.69732 | 1.22827 | 1.88373 | 2.26241 | 3.02031 |
| 0.340 | 0.49146 | 0.71026 | 1.24709 | 1.87734 | 2.27852 | 3.02043 |
| 0.360 | 0.50918 | 0.72599 | 1.26642 | 1.87137 | 2.29483 | 3.02060 |
| 0.380 | 0.52477 | 0.74441 | 1.28621 | 1.86582 | 2.31134 | 3.02084 |
| 0.400 | 0.53850 | 0.76528 | 1.30644 | 1.86071 | 2.32803 | 3.02115 |
| 0.420 | 0.55071 | 0.78823 | 1.32707 | 1.85605 | 2.34492 | 3.02154 |
| 0.440 | 0.56179 | 0.81288 | 1.34806 | 1.85185 | 2.36198 | 3.02202 |
| 0.460 | 0.57210 | 0.83886 | 1.36940 | 1.84812 | 2.37921 | 3.02259 |
| 0.480 | 0.58193 | 0.86586 | 1.39105 | 1.84488 | 2.39662 | 3.02328 |
| 0.500 | 0.59152 | 0.89361 | 1.41300 | 1.84214 | 2.41419 | 3.02406 |
| 0.520 | 0.60105 | 0.92189 | 1.43522 | 1.83993 | 2.43192 | 3.02496 |
| 0.540 | 0.61067 | 0.95054 | 1.45770 | 1.83825 | 2.44981 | 3.02498 |
| 0.560 | 0.62047 | 0.97942 | 1.48041 | 1.83713 | 2.46785 | 3.02717 |
| 0.580 | 0.63052 | 1.00840 | 1.50335 | 1.83660 | 2.48604 | 3.02849 |
| 0.600 | 0.64088 | 1.03740 | 1.52649 | 1.83666 | 2.50437 | 3.02998 |
| 0.620 | 0.65157 | 1.06633 | 1.54983 | 1.83736 | 2.52284 | 3.03160 |
| 0.640 | 0.66261 | 1.09510 | 1.57335 | 1.83871 | 2.54144 | 3.03340 |
| 0.660 | 0.67402 | 1.12366 | 1.59704 | 1.84075 | 2.56017 | 3.03539 |
| 0.680 | 0.68580 | 1.15194 | 1.62089 | 1.84351 | 2.57903 | 3.03756 |
| 0.700 | 0.69794 | 1.17986 | 1.64489 | 1.84702 | 2.59800 | 3.03993 |
| 0.720 | 0.71044 | 1.20738 | 1.66903 | 1.85133 | 2.61709 | 3.04250 |
| 0.740 | 0.72330 | 1.23443 | 1.69330 | 1.85645 | 2.63629 | 3.04528 |
| 0.760 | 0.73649 | 1.26095 | 1.71770 | 1.86245 | 2.65560 | 3.04830 |
| 0.780 | 0.75001 | 1.28688 | 1.74222 | 1.86936 | 2.67501 | 3.05152 |
| 0.800 | 0.76384 | 1.31217 | 1.76684 | 1.87722 | 2.69452 | 3.05499 |
| 0.820 | 0.77797 | 1.33676 | 1.79155 | 1.88607 | 2.71413 | 3.05873 |
| 0.840 | 0.79239 | 1.36060 | 1.81634 | 1.89597 | 2.73382 | 3.06269 |
| 0.860 | 0.80708 | 1.38366 | 1.84116 | 1.90697 | 2.75360 | 3.06693 |
| 0.880 | 0.82202 | 1.40588 | 1.86597 | 1.91914 | 2.77346 | 3.07145 |
| 0.900 | 0.83721 | 1.42725 | 1.89064 | 1.93260 | 2.79339 | 3.07625 |
| 0.920 | 0.85263 | 1.44776 | 1.91495 | 1.94758 | 2.81340 | 3.08133 |
| 0.940 | 0.86826 | 1.46739 | 1.93845 | 1.96452 | 2.83348 | 3.08673 |
| 0.960 | 0.88410 | 1.48615 | 1.96051 | 1.98401 | 2.85362 | 3.09242 |
| 0.980 | 0.90013 | 1.50408 | 1.98096 | 2.00618 | 2.87382 | 3.09844 |
| 1.000 | 0.91635 | 1.52119 | 2.00430 | 2.03036 | 2.89408 | 3.10479 |

$H/R = 1.200 \qquad n = 0.$

| $H/L$ | $\Omega_1$ | $\Omega_2$ | $\Omega_3$ | $\Omega_4$ | $\Omega_5$ | $\Omega_6$ |
|---|---|---|---|---|---|---|
| 0. | 0. | 0.754154 | 1.06176 | 2.03804 | 2.09753 | 3.02688 |
| 0.010 | 0.01612 | 0.75146 | 1.06200 | 2.03748 | 2.09815 | 3.02688 |
| 0.050 | 0.08054 | 0.74967 | 1.06767 | 2.02657 | 2.11079 | 3.02684 |
| 0.100 | 0.16054 | 0.74492 | 1.08466 | 2.00482 | 2.13793 | 3.02673 |
| 0.120 | 0.19225 | 0.74266 | 1.09419 | 1.99562 | 2.15029 | 3.02667 |
| 0.140 | 0.22373 | 0.74044 | 1.10509 | 1.98647 | 2.16317 | 3.02661 |
| 0.160 | 0.25491 | 0.73843 | 1.11722 | 1.97747 | 2.17650 | 3.02656 |
| 0.180 | 0.28572 | 0.73682 | 1.13048 | 1.96867 | 2.19019 | 3.02650 |
| 0.200 | 0.31607 | 0.73578 | 1.14475 | 1.96012 | 2.20421 | 3.02646 |
| 0.220 | 0.34586 | 0.73552 | 1.15994 | 1.95186 | 2.21854 | 3.02642 |
| 0.240 | 0.37497 | 0.73624 | 1.17595 | 1.94389 | 2.23315 | 3.02641 |
| 0.260 | 0.40324 | 0.73815 | 1.19270 | 1.93625 | 2.24802 | 3.02640 |
| 0.280 | 0.43048 | 0.74150 | 1.21013 | 1.92895 | 2.26314 | 3.02647 |
| 0.300 | 0.45650 | 0.74654 | 1.22818 | 1.92199 | 2.27850 | 3.02655 |
| 0.320 | 0.48107 | 0.75353 | 1.24679 | 1.91541 | 2.29409 | 3.02667 |
| 0.340 | 0.50397 | 0.76271 | 1.26592 | 1.90920 | 2.30990 | 3.02684 |
| 0.360 | 0.52503 | 0.77429 | 1.28552 | 1.90338 | 2.32593 | 3.02707 |
| 0.380 | 0.54416 | 0.78836 | 1.30555 | 1.89796 | 2.34216 | 3.02737 |
| 0.400 | 0.56137 | 0.80493 | 1.32599 | 1.89295 | 2.35859 | 3.02774 |
| 0.420 | 0.57679 | 0.82386 | 1.34679 | 1.88837 | 2.37522 | 3.02818 |
| 0.440 | 0.59066 | 0.84490 | 1.36795 | 1.88422 | 2.39203 | 3.02871 |
| 0.460 | 0.60327 | 0.86777 | 1.38942 | 1.88052 | 2.40902 | 3.02934 |
| 0.480 | 0.61492 | 0.89213 | 1.41119 | 1.87729 | 2.42619 | 3.03007 |
| 0.500 | 0.62588 | 0.91770 | 1.43324 | 1.87453 | 2.44353 | 3.03091 |
| 0.520 | 0.63641 | 0.94420 | 1.45555 | 1.87226 | 2.46104 | 3.03187 |
| 0.540 | 0.64668 | 0.97142 | 1.47810 | 1.87051 | 2.47871 | 3.03296 |
| 0.560 | 0.65687 | 0.99914 | 1.50088 | 1.86928 | 2.49653 | 3.03418 |
| 0.580 | 0.66709 | 1.02722 | 1.52387 | 1.86860 | 2.51451 | 3.03554 |
| 0.600 | 0.67744 | 1.05552 | 1.54706 | 1.86849 | 2.53262 | 3.03706 |
| 0.620 | 0.68797 | 1.08393 | 1.57043 | 1.86897 | 2.55088 | 3.03873 |
| 0.640 | 0.69875 | 1.11235 | 1.59299 | 1.87006 | 2.56928 | 3.04057 |
| 0.660 | 0.70979 | 1.14069 | 1.61771 | 1.87180 | 2.58780 | 3.04259 |
| 0.680 | 0.72113 | 1.16887 | 1.64159 | 1.87420 | 2.60646 | 3.04479 |
| 0.700 | 0.73278 | 1.19682 | 1.66562 | 1.87731 | 2.62523 | 3.04718 |
| 0.720 | 0.74474 | 1.22447 | 1.68979 | 1.88114 | 2.64412 | 3.04978 |
| 0.740 | 0.75702 | 1.25175 | 1.71410 | 1.88574 | 2.66313 | 3.05260 |
| 0.760 | 0.76962 | 1.27860 | 1.73854 | 1.89114 | 2.68224 | 3.05560 |
| 0.780 | 0.78252 | 1.30496 | 1.76312 | 1.89738 | 2.70145 | 3.05885 |
| 0.800 | 0.79572 | 1.33076 | 1.78782 | 1.90448 | 2.72076 | 3.06234 |
| 0.820 | 0.80922 | 1.35594 | 1.81264 | 1.91248 | 2.74017 | 3.06607 |
| 0.840 | 0.82299 | 1.38046 | 1.83759 | 1.92142 | 2.75967 | 3.07006 |
| 0.860 | 0.83704 | 1.40425 | 1.86268 | 1.93132 | 2.77925 | 3.07430 |
| 0.880 | 0.85135 | 1.42728 | 1.88789 | 1.94219 | 2.79891 | 3.07880 |
| 0.900 | 0.86591 | 1.44950 | 1.91323 | 1.95405 | 2.81865 | 3.08359 |
| 0.920 | 0.88070 | 1.47090 | 1.93871 | 1.96691 | 2.83846 | 3.08867 |
| 0.940 | 0.89572 | 1.49144 | 1.96432 | 1.98078 | 2.85833 | 3.09405 |
| 0.960 | 0.91096 | 1.51114 | 1.98980 | 1.99586 | 2.87827 | 3.09973 |
| 0.980 | 0.92639 | 1.52999 | 2.01051 | 2.01682 | 2.89827 | 3.10574 |
| 1.000 | 0.94292 | 1.54801 | 2.02730 | 2.04272 | 2.91832 | 3.11207 |

$H/R = 1.300$   $n = 0.$

| $H/L$ | $\Omega_1$ | $\Omega_2$ | $\Omega_3$ | $\Omega_4$ | $\Omega_5$ | $\Omega_6$ |
|---|---|---|---|---|---|---|
| 0. | 0. | 0.83062 | 1.07415 | 2.04763 | 2.17242 | 3.03422 |
| 0.010 | 0.01612 | 0.83051 | 1.07442 | 2.04739 | 2.17272 | 3.03421 |
| 0.050 | 0.08055 | 0.82789 | 1.08082 | 2.04200 | 2.17982 | 3.03420 |
| 0.100 | 0.16062 | 0.82095 | 1.09968 | 2.02812 | 2.19905 | 3.03415 |
| 0.120 | 0.19240 | 0.81761 | 1.11008 | 2.02136 | 2.20895 | 3.03412 |
| 0.140 | 0.22397 | 0.81425 | 1.12184 | 2.01427 | 2.21975 | 3.03410 |
| 0.160 | 0.25530 | 0.81106 | 1.13482 | 2.00702 | 2.23130 | 3.03409 |
| 0.180 | 0.28631 | 0.80823 | 1.14886 | 1.99972 | 2.24346 | 3.03408 |
| 0.200 | 0.31694 | 0.80594 | 1.16385 | 1.99246 | 2.25615 | 3.03409 |
| 0.220 | 0.34712 | 0.80435 | 1.17968 | 1.98532 | 2.26931 | 3.03412 |
| 0.240 | 0.37676 | 0.80364 | 1.19627 | 1.97834 | 2.28288 | 3.03417 |
| 0.260 | 0.40574 | 0.80398 | 1.21354 | 1.97158 | 2.29682 | 3.03425 |
| 0.280 | 0.43393 | 0.80556 | 1.23141 | 1.96506 | 2.31110 | 3.03436 |
| 0.300 | 0.46119 | 0.80855 | 1.24984 | 1.95881 | 2.32569 | 3.03451 |
| 0.320 | 0.48736 | 0.81316 | 1.26878 | 1.95285 | 2.34057 | 3.03470 |
| 0.340 | 0.51224 | 0.81960 | 1.28818 | 1.94721 | 2.35573 | 3.03495 |
| 0.360 | 0.53569 | 0.82804 | 1.30801 | 1.94191 | 2.37115 | 3.03525 |
| 0.380 | 0.55754 | 0.83865 | 1.32824 | 1.93696 | 2.38681 | 3.03563 |
| 0.400 | 0.57770 | 0.85152 | 1.34882 | 1.93238 | 2.40271 | 3.03607 |
| 0.420 | 0.59615 | 0.86669 | 1.36975 | 1.92818 | 2.41883 | 3.03660 |
| 0.440 | 0.61296 | 0.88407 | 1.39099 | 1.92439 | 2.43516 | 3.03720 |
| 0.460 | 0.62827 | 0.90351 | 1.41252 | 1.92100 | 2.45169 | 3.03791 |
| 0.480 | 0.64229 | 0.92479 | 1.43432 | 1.91805 | 2.46843 | 3.03870 |
| 0.500 | 0.65524 | 0.94766 | 1.45638 | 1.91554 | 2.48535 | 3.03961 |
| 0.520 | 0.66737 | 0.97186 | 1.47867 | 1.91349 | 2.50245 | 3.04064 |
| 0.540 | 0.67890 | 0.99713 | 1.50119 | 1.91193 | 2.51973 | 3.04179 |
| 0.560 | 0.69003 | 1.02326 | 1.52391 | 1.91085 | 2.53717 | 3.04307 |
| 0.580 | 0.70091 | 1.05003 | 1.54683 | 1.91030 | 2.55478 | 3.04449 |
| 0.600 | 0.71169 | 1.07728 | 1.56992 | 1.91028 | 2.57255 | 3.04606 |
| 0.620 | 0.72246 | 1.10488 | 1.59318 | 1.91082 | 2.59046 | 3.04778 |
| 0.640 | 0.73330 | 1.13268 | 1.61660 | 1.91194 | 2.60853 | 3.04967 |
| 0.660 | 0.74429 | 1.16057 | 1.64016 | 1.91367 | 2.62672 | 3.05173 |
| 0.680 | 0.75546 | 1.18848 | 1.66385 | 1.91603 | 2.64506 | 3.05397 |
| 0.700 | 0.76685 | 1.21629 | 1.68766 | 1.91907 | 2.66352 | 3.05639 |
| 0.720 | 0.77848 | 1.24394 | 1.71159 | 1.92280 | 2.68211 | 3.05902 |
| 0.740 | 0.79036 | 1.27135 | 1.73561 | 1.92728 | 2.70081 | 3.06184 |
| 0.760 | 0.80251 | 1.29845 | 1.75971 | 1.93253 | 2.71963 | 3.06488 |
| 0.780 | 0.81494 | 1.32517 | 1.78389 | 1.93860 | 2.73855 | 3.06814 |
| 0.800 | 0.82763 | 1.35144 | 1.80812 | 1.94554 | 2.75758 | 3.07163 |
| 0.820 | 0.84060 | 1.37720 | 1.83239 | 1.95340 | 2.77670 | 3.07536 |
| 0.840 | 0.85383 | 1.40240 | 1.85666 | 1.96224 | 2.79592 | 3.07934 |
| 0.860 | 0.86732 | 1.42697 | 1.88091 | 1.97212 | 2.81522 | 3.08357 |
| 0.880 | 0.88106 | 1.45086 | 1.90507 | 1.98313 | 2.83461 | 3.08807 |
| 0.900 | 0.89504 | 1.47402 | 1.92906 | 1.99537 | 2.85407 | 3.09284 |
| 0.920 | 0.90926 | 1.49642 | 1.95276 | 2.00898 | 2.87361 | 3.09791 |
| 0.940 | 0.92371 | 1.51803 | 1.97600 | 2.02414 | 2.89321 | 3.10326 |
| 0.960 | 0.93837 | 1.53883 | 1.99856 | 2.04106 | 2.91288 | 3.10891 |
| 0.980 | 0.95324 | 1.55880 | 2.02027 | 2.05990 | 2.93260 | 3.11486 |
| 1.000 | 0.96831 | 1.57796 | 2.04112 | 2.08065 | 2.95238 | 3.12115 |

$H/R = 1.400$ $n = 0.$

| $H/L$ | $\Omega_1$ | $\Omega_2$ | $\Omega_3$ | $\Omega_4$ | $\Omega_5$ | $\Omega_6$ |
|---|---|---|---|---|---|---|
| 0. | 0. | 0.91027 | 1.08816 | 2.05940 | 2.27219 | 3.04358 |
| 0.010 | 0.01612 | 0.91010 | 1.08849 | 2.05929 | 2.27237 | 3.04358 |
| 0.050 | 0.08055 | 0.90615 | 1.09612 | 2.05656 | 2.27681 | 3.04359 |
| 0.100 | 0.16068 | 0.89605 | 1.11781 | 2.04877 | 2.28996 | 3.04362 |
| 0.120 | 0.19251 | 0.89132 | 1.12939 | 2.04462 | 2.29725 | 3.04365 |
| 0.140 | 0.22416 | 0.88659 | 1.14227 | 2.04007 | 2.30552 | 3.04368 |
| 0.160 | 0.25559 | 0.88207 | 1.15626 | 2.03521 | 2.31466 | 3.04373 |
| 0.180 | 0.28675 | 0.87795 | 1.17120 | 2.03014 | 2.32459 | 3.04379 |
| 0.200 | 0.31759 | 0.87440 | 1.18697 | 2.02495 | 2.33522 | 3.04388 |
| 0.220 | 0.34805 | 0.87157 | 1.20347 | 2.01971 | 2.34647 | 3.04398 |
| 0.240 | 0.37806 | 0.86959 | 1.22061 | 2.01449 | 2.35829 | 3.04412 |
| 0.260 | 0.40752 | 0.86861 | 1.23834 | 2.00934 | 2.37062 | 3.04429 |
| 0.280 | 0.43636 | 0.86876 | 1.25660 | 2.00430 | 2.38340 | 3.04450 |
| 0.300 | 0.46445 | 0.87018 | 1.27533 | 1.99943 | 2.39662 | 3.04475 |
| 0.320 | 0.49167 | 0.87303 | 1.29450 | 1.99475 | 2.41022 | 3.04504 |
| 0.340 | 0.51789 | 0.87745 | 1.31407 | 1.99029 | 2.42419 | 3.04539 |
| 0.360 | 0.54296 | 0.88361 | 1.33402 | 1.98609 | 2.43850 | 3.04581 |
| 0.380 | 0.56675 | 0.89163 | 1.35431 | 1.98217 | 2.45312 | 3.04628 |
| 0.400 | 0.58913 | 0.90165 | 1.37492 | 1.97855 | 2.46804 | 3.04683 |
| 0.420 | 0.61002 | 0.91375 | 1.39583 | 1.97526 | 2.48324 | 3.04746 |
| 0.440 | 0.62939 | 0.92795 | 1.41702 | 1.97231 | 2.49871 | 3.04818 |
| 0.460 | 0.64725 | 0.94422 | 1.43847 | 1.96972 | 2.51442 | 3.04898 |
| 0.480 | 0.66371 | 0.96245 | 1.46015 | 1.96752 | 2.53038 | 3.04989 |
| 0.500 | 0.67890 | 0.98248 | 1.48206 | 1.96573 | 2.54656 | 3.05090 |
| 0.520 | 0.69300 | 1.00411 | 1.50418 | 1.96435 | 2.56296 | 3.05203 |
| 0.540 | 0.70621 | 1.02712 | 1.52648 | 1.96342 | 2.57956 | 3.05327 |
| 0.560 | 0.71872 | 1.05129 | 1.54897 | 1.96296 | 2.59636 | 3.05466 |
| 0.580 | 0.73071 | 1.07641 | 1.57161 | 1.96298 | 2.61336 | 3.05615 |
| 0.600 | 0.74233 | 1.10229 | 1.59440 | 1.96351 | 2.63053 | 3.05780 |
| 0.620 | 0.75373 | 1.12876 | 1.61732 | 1.96457 | 2.64787 | 3.05961 |
| 0.640 | 0.76500 | 1.15567 | 1.64036 | 1.96619 | 2.66538 | 3.06156 |
| 0.660 | 0.77625 | 1.18288 | 1.66350 | 1.96840 | 2.68305 | 3.06368 |
| 0.680 | 0.78754 | 1.21028 | 1.68672 | 1.97122 | 2.70087 | 3.06597 |
| 0.700 | 0.79892 | 1.23776 | 1.71001 | 1.97471 | 2.71884 | 3.06845 |
| 0.720 | 0.81045 | 1.26524 | 1.73335 | 1.97888 | 2.73694 | 3.07111 |
| 0.740 | 0.82215 | 1.29263 | 1.75672 | 1.98379 | 2.75518 | 3.07397 |
| 0.760 | 0.83405 | 1.31984 | 1.78010 | 1.98947 | 2.77354 | 3.07703 |
| 0.780 | 0.84616 | 1.34680 | 1.80346 | 1.99598 | 2.79202 | 3.08031 |
| 0.800 | 0.85849 | 1.37345 | 1.82678 | 2.00337 | 2.81062 | 3.08381 |
| 0.820 | 0.87106 | 1.39972 | 1.85003 | 2.01171 | 2.82932 | 3.08753 |
| 0.840 | 0.88386 | 1.42553 | 1.87318 | 2.02104 | 2.84813 | 3.09150 |
| 0.860 | 0.89690 | 1.45084 | 1.89620 | 2.03144 | 2.86703 | 3.09571 |
| 0.880 | 0.91016 | 1.47558 | 1.91904 | 2.04297 | 2.88603 | 3.10017 |
| 0.900 | 0.92365 | 1.49970 | 1.94167 | 2.05569 | 2.90511 | 3110491 |
| 0.920 | 0.93737 | 1.52316 | 1.96406 | 2.06965 | 2.92427 | 3.10992 |
| 0.940 | 0.95131 | 1.54590 | 1.98620 | 2.08488 | 2.94350 | 3.11520 |
| 0.960 | 0.96546 | 1.56791 | 2.00809 | 2.10139 | 2.96281 | 3.12080 |
| 0.980 | 0.97981 | 1.58915 | 2.02975 | 2.11916 | 2.98217 | 3.12667 |
| 1.000 | 0.99437 | 1.60961 | 2.05123 | 2.13813 | 3.00159 | 3.13287 |

$H/R = 1.500 \qquad n = 0.$

| $H/L$ | $\Omega_1$ | $\Omega_2$ | $\Omega_3$ | $\Omega_4$ | $\Omega_5$ | $\Omega_6$ |
|---|---|---|---|---|---|---|
| 0. | 0. | 0.98812 | 1.10401 | 2.07400 | 2.40375 | 3.05576 |
| 0.010 | 0.00161 | 0.98783 | 1.10445 | 2.07394 | 2.40386 | 3.05576 |
| 0.050 | 0.08056 | 0.98152 | 1.11434 | 2.07266 | 2.40688 | 3.05581 |
| 0.100 | 0.16074 | 0.96687 | 1.14023 | 2.06883 | 2.41611 | 3.05596 |
| 0.120 | 0.19260 | 0.96045 | 1.15328 | 2.06672 | 2.42139 | 3.05606 |
| 0.140 | 0.22431 | 0.95420 | 1.16742 | 2.06434 | 2.42753 | 3.05617 |
| 0.160 | 0.25582 | 0.94831 | 1.18245 | 2.06173 | 2.43446 | 3.05631 |
| 0.180 | 0.28709 | 0.94297 | 1.19824 | 2.05895 | 2.44215 | 3.05648 |
| 0.200 | 0.31809 | 0.93829 | 1.21468 | 2.05604 | 2.45055 | 3.05668 |
| 0.220 | 0.34875 | 0.93439 | 1.23170 | 2.05305 | 2.45962 | 3.05691 |
| 0.240 | 0.37903 | 0.93138 | 1.24925 | 2.05002 | 2.46931 | 3.05718 |
| 0.260 | 0.40885 | 0.92935 | 1.26726 | 2.04699 | 2.47957 | 3.05749 |
| 0.280 | 0.43813 | 0.92841 | 1.28571 | 2.04401 | 2.49039 | 3.05784 |
| 0.300 | 0.46681 | 0.92867 | 1.30455 | 2.04111 | 2.50170 | 3.05825 |
| 0.320 | 0.49476 | 0.93022 | 1.32377 | 2.03833 | 2.51350 | 3.05871 |
| 0.340 | 0.52190 | 0.93320 | 1.34332 | 2.03570 | 2.52574 | 3.05922 |
| 0.360 | 0.54809 | 0.93770 | 1.36319 | 2.03324 | 2.53840 | 3.05980 |
| 0.380 | 0.57324 | 0.94387 | 1.38336 | 2.03100 | 2.55145 | 3.06046 |
| 0.400 | 0.59721 | 0.95179 | 1.40381 | 2.02899 | 2.56488 | 3.06118 |
| 0.420 | 0.61992 | 0.96158 | 1.42451 | 2.02723 | 2.57865 | 3.06199 |
| 0.440 | 0.64128 | 0.97328 | 1.44546 | 2.02577 | 2.59276 | 3.06287 |
| 0.460 | 0.66127 | 0.98694 | 1.46663 | 2.02462 | 2.60718 | 3.06386 |
| 0.480 | 0.67989 | 1.00251 | 1.48800 | 2.02379 | 2.62190 | 3.06494 |
| 0.500 | 0.69720 | 1.01994 | 1.50956 | 2.02333 | 2.63690 | 3.06612 |
| 0.520 | 0.71330 | 1.03908 | 1.53130 | 2.02324 | 2.65217 | 3.06742 |
| 0.540 | 0.72833 | 1.05979 | 1.55319 | 2.02356 | 2.66769 | 3.06883 |
| 0.560 | 0.74244 | 1.08189 | 1.57522 | 2.02430 | 2.68346 | 3.07036 |
| 0.580 | 0.75580 | 1.10517 | 1.59738 | 2.02550 | 2.69946 | 3.07202 |
| 0.600 | 0.76857 | 1.12947 | 1.61965 | 2.02717 | 2.71568 | 3.07383 |
| 0.620 | 0.78089 | 1.15459 | 1.64201 | 2.02935 | 2.73211 | 3.07577 |
| 0.640 | 0.79289 | 1.18038 | 1.66444 | 2.03206 | 2.74873 | 3.07786 |
| 0.660 | 0.80467 | 1.20668 | 1.68693 | 2.03533 | 2.76556 | 3.08012 |
| 0.680 | 0.81635 | 1.23337 | 1.70946 | 2.03921 | 2.78256 | 3.08252 |
| 0.700 | 0.82798 | 1.26032 | 1.73200 | 2.04371 | 2.79974 | 3.08511 |
| 0.720 | 0.83963 | 1.28744 | 1.75455 | 2.04888 | 2.81708 | 3.08787 |
| 0.740 | 0.85135 | 1.31462 | 1.77708 | 2.05476 | 2.83458 | 3.09082 |
| 0.760 | 0.86318 | 1.34178 | 1.79956 | 2.06138 | 2.85223 | 3.09397 |
| 0.780 | 0.87515 | 1.36883 | 1.82199 | 2.06880 | 2.87003 | 3.09731 |
| 0.800 | 0.88728 | 1.39570 | 1.84434 | 2.07704 | 2.88795 | 3.10087 |
| 0.820 | 0.89959 | 1.42233 | 1.86659 | 2.08616 | 2.90601 | 3.10465 |
| 0.840 | 0.91209 | 1.44864 | 1.88873 | 2.09620 | 2.92419 | 3.10864 |
| 0.860 | 0.92479 | 1.47457 | 1.91076 | 2.10719 | 2.94248 | 3.11286 |
| 0.880 | 0.93769 | 1.50006 | 1.93266 | 2.11916 | 2.96088 | 3.11734 |
| 0.900 | 0.95079 | 1.52505 | 1.95444 | 2.13213 | 2.97938 | 3.12206 |
| 0.920 | 0.96410 | 1.54949 | 1.97613 | 2.14612 | 2.99798 | 3.12705 |
| 0.940 | 0.97760 | 1.57333 | 1.99773 | 2.16111 | 3.01666 | 3.13230 |
| 0.960 | 0.99131 | 1.59654 | 2.01929 | 2.17710 | 3.03542 | 3.13783 |
| 0.980 | 1.00522 | 1.61906 | 2.04084 | 2.19405 | 3.05425 | 3.14365 |
| 1.000 | 1.01931 | 1.64088 | 2.06243 | 2.21192 | 3.07315 | 3.14978 |

$H/R = 1.600$    $n = 0.$

| $H/L$ | $\Omega_1$ | $\Omega_2$ | $\Omega_3$ | $\Omega_4$ | $\Omega_5$ | $\Omega_6$ |
|---|---|---|---|---|---|---|
| 0. | 0. | 1.06130 | 1.12193 | 2.09228 | 2.57094 | 3.07197 |
| 0.010 | 0.00161 | 1.06072 | 1.12265 | 2.09225 | 2.57103 | 3.07196 |
| 0.050 | 0.08056 | 1.04967 | 1.13715 | 2.09181 | 2.57323 | 3.07207 |
| 0.100 | 0.16078 | 1.02922 | 1.16847 | 2.09050 | 2.58003 | 3.07243 |
| 0.120 | 0.19268 | 1.02121 | 1.18289 | 2.08977 | 2.58399 | 3.07264 |
| 0.140 | 0.22443 | 1.01367 | 1.19803 | 2.08895 | 2.58862 | 3.07290 |
| 0.160 | 0.25600 | 1.00674 | 1.21379 | 2.08806 | 2.59291 | 3.07320 |
| 0.180 | 0.28737 | 1.0050 | 1.23007 | 2.08711 | 2.59985 | 3.07354 |
| 0.200 | 0.31848 | 0.99504 | 1.24685 | 2.08613 | 2.60640 | 3.07394 |
| 0.220 | 0.34930 | 0.99042 | 1.26406 | 2.08514 | 2.61356 | 3.07439 |
| 0.240 | 0.37978 | 0.98672 | 1.28168 | 2.08417 | 2.62130 | 3.07489 |
| 0.260 | 0.40986 | 0.98400 | 1.29968 | 2.08324 | 2.62959 | 3.07545 |
| 0.280 | 0.43948 | 0.98233 | 1.31804 | 2.08237 | 2.63842 | 3.07607 |
| 0.300 | 0.46857 | 0.98180 | 1.33673 | 2.08158 | 2.64776 | 3.07675 |
| 0.320 | 0.49705 | 0.98247 | 1.35574 | 2.08091 | 2.65759 | 3.07750 |
| 0.340 | 0.52484 | 0.98445 | 1.37504 | 2.08037 | 2.66789 | 3.07833 |
| 0.360 | 0.55184 | 0.98781 | 1.39461 | 2.07999 | 2.67864 | 3.07922 |
| 0.380 | 0.57794 | 0.99267 | 1.41445 | 2.07979 | 2.68982 | 3.08020 |
| 0.400 | 0.60306 | 0.99910 | 1.43452 | 2.07980 | 2.70141 | 3.08127 |
| 0.420 | 0.62709 | 1.00721 | 1.45482 | 2.08003 | 2.71340 | 3.08243 |
| 0.440 | 0.64996 | 1.01706 | 1.47534 | 2.08052 | 2.72575 | 3.08369 |
| 0.460 | 0.67159 | 1.02870 | 1.49604 | 2.08129 | 2.73846 | 3.08504 |
| 0.480 | 0.69195 | 1.04216 | 1.51692 | 2.08235 | 2.75152 | 3.08649 |
| 0.500 | 0.71106 | 1.05740 | 1.53797 | 2.08373 | 2.76490 | 3.08805 |
| 0.520 | 0.72894 | 1.07428 | 1.55915 | 2.08545 | 2.77858 | 3.08974 |
| 0.540 | 0.74569 | 1.09300 | 1.58047 | 2.08755 | 2.79256 | 3.09155 |
| 0.560 | 0.76140 | 1.11311 | 1.60190 | 2.09004 | 2.80683 | 3.09349 |
| 0.580 | 0.77620 | 1.13459 | 1.62343 | 2.09294 | 2.82135 | 3.09557 |
| 0.600 | 0.79023 | 1.15724 | 1.64504 | 2.09629 | 2.83613 | 3.09779 |
| 0.620 | 0.80364 | 1.18092 | 1.66671 | 2.10011 | 2.85115 | 3.10017 |
| 0.640 | 0.81654 | 1.20547 | 1.68843 | 2.10443 | 2.86640 | 3.10270 |
| 0.660 | 0.82906 | 1.23072 | 1.71018 | 2.10927 | 2.88185 | 3.10540 |
| 0.680 | 0.84131 | 1.25654 | 1.73194 | 2.11468 | 2.89751 | 3.10829 |
| 0.700 | 0.85337 | 1.28280 | 1.75370 | 2.12066 | 2.91335 | 3.11136 |
| 0.720 | 0.86533 | 1.30939 | 1.77544 | 2.12726 | 2.92935 | 3.11463 |
| 0.740 | 0.87725 | 1.33620 | 1.79714 | 2.13451 | 2.94551 | 3.11812 |
| 0.760 | 0.88918 | 1.36313 | 1.81880 | 2.14244 | 2.96180 | 3.12184 |
| 0.780 | 0.90116 | 1.39010 | 1.84041 | 2.15107 | 2.97820 | 3.12581 |
| 0.800 | 0.91324 | 1.41703 | 1.86195 | 2.16044 | 2.99469 | 3.13006 |
| 0.820 | 0.92542 | 1.44385 | 1.88341 | 2.17058 | 3.01124 | 3.13461 |
| 0.840 | 0.93775 | 1.47048 | 1.90482 | 2.18150 | 3.02780 | 3.13951 |
| 0.860 | 0.95022 | 1.49685 | 1.92615 | 2.19322 | 3.04432 | 3.14478 |
| 0.880 | 0.96286 | 1.52291 | 1.94744 | 2.20576 | 3.06075 | 3.15052 |
| 0.900 | 0.97567 | 1.54860 | 1.96869 | 2.21912 | 3.07696 | 3.15684 |
| 0.920 | 0.98866 | 1.57386 | 1.98994 | 2.23330 | 3.09284 | 3.16383 |
| 0.940 | 1.00182 | 1.59864 | 2.01120 | 2.24830 | 3.10821 | 3.17170 |
| 0.960 | 1.01517 | 1.62289 | 2.03253 | 2.26410 | 3.12281 | 3.18070 |
| 0.980 | 1.02870 | 1.64657 | 2.05394 | 2.28067 | 3.13643 | 3.19104 |
| 1.000 | 1.04241 | 1.66965 | 2.07549 | 2.29798 | 3.14890 | 3.20289 |

$H/R = 1.700 \qquad n = 0.$

| $H/L$ | $\Omega_1$ | $\Omega_2$ | $\Omega_3$ | $\Omega_4$ | $\Omega_5$ | $\Omega_6$ |
|---|---|---|---|---|---|---|
| 0. | 0. | 1.12702 | 1.14217 | 2.11544 | 2.76611 | 3.09410 |
| 0.010 | 0.01612 | 1.12491 | 1.14441 | 2.11544 | 2.76617 | 3.09411 |
| 0.050 | 0.08057 | 1.10475 | 1.16791 | 2.11551 | 2.76786 | 3.09435 |
| 0.100 | 0.16081 | 1.08014 | 1.20301 | 2.11576 | 2.77310 | 3.09511 |
| 0.120 | 0.19274 | 1.07136 | 1.21796 | 2.11592 | 2.77615 | 3.09556 |
| 0.140 | 0.22453 | 1.06327 | 1.23337 | 2.11613 | 2.77974 | 3.09610 |
| 0.160 | 0.25615 | 1.05592 | 1.24921 | 2.11640 | 2.78387 | 3.09672 |
| 0.180 | 0.28759 | 1.04934 | 1.26545 | 2.11674 | 2.78851 | 3.09744 |
| 0.200 | 0.31879 | 1.04357 | 1.28209 | 2.11715 | 2.79366 | 3.09825 |
| 0.220 | 0.34974 | 1.03865 | 1.29909 | 2.11764 | 2.79932 | 3.09915 |
| 0.240 | 0.38037 | 1.03463 | 1.31644 | 2.11824 | 2.80546 | 3.10015 |
| 0.260 | 0.41065 | 1.03156 | 1.33413 | 2.11895 | 2.81207 | 3.10125 |
| 0.280 | 0.44052 | 1.02950 | 1.35213 | 2.11978 | 2.81914 | 3.10247 |
| 0.300 | 0.46993 | 1.02849 | 1.37044 | 2.12076 | 2.82666 | 3.10381 |
| 0.320 | 0.49880 | 1.02862 | 1.38903 | 2.12189 | 2.83461 | 3.10523 |
| 0.340 | 0.52707 | 1.02994 | 1.40788 | 2.12320 | 2.84298 | 3.10680 |
| 0.360 | 0.55465 | 1.03254 | 1.42699 | 2.12469 | 2.85174 | 3.10849 |
| 0.380 | 0.58146 | 1.03649 | 1.44634 | 2.12638 | 2.86089 | 3.11032 |
| 0.400 | 0.60741 | 1.04187 | 1.46591 | 2.12830 | 2.87040 | 3.11230 |
| 0.420 | 0.63242 | 1.04877 | 1.48568 | 2.13045 | 2.88025 | 3.11443 |
| 0.440 | 0.65639 | 1.05726 | 1.50565 | 2.13286 | 2.89044 | 3.11674 |
| 0.460 | 0.67927 | 1.06739 | 1.52579 | 2.13555 | 2.90093 | 3.11919 |
| 0.480 | 0.70099 | 1.07921 | 1.54610 | 2.13853 | 2.91170 | 3.12185 |
| 0.500 | 0.72153 | 1.09273 | 1.56655 | 2.14182 | 2.92273 | 3.12471 |
| 0.520 | 0.74090 | 1.10793 | 1.58712 | 2.14545 | 2.93400 | 3.12781 |
| 0.540 | 0.75912 | 1.12476 | 1.60782 | 2.14942 | 2.94548 | 3.13112 |
| 0.560 | 0.77627 | 1.14314 | 1.62861 | 2.15378 | 2.95714 | 3.13473 |
| 0.580 | 0.79242 | 1.16294 | 1.64949 | 2.15852 | 2.96894 | 3.13859 |
| 0.600 | 0.80769 | 1.18406 | 1.67043 | 2.16369 | 2.98085 | 3.14279 |
| 0.620 | 0.82220 | 1.20633 | 1.69143 | 2.16929 | 2.99282 | 3.14736 |
| 0.640 | 0.83606 | 1.22961 | 1.71247 | 2.17536 | 3.00481 | 3.15230 |
| 0.660 | 0.84940 | 1.25376 | 1.73353 | 2.18192 | 3.01678 | 3.15769 |
| 0.680 | 0.86233 | 1.27864 | 1.75460 | 2.18898 | 3.02865 | 3.16359 |
| 0.700 | 0.87494 | 1.30411 | 1.77567 | 2.19658 | 3.04038 | 3.17001 |
| 0.720 | 0.88732 | 1.33006 | 1.70673 | 2.20473 | 3.05190 | 3.17705 |
| 0.740 | 0.89955 | 1.35637 | 1.81777 | 2.21346 | 3.06315 | 3.18474 |
| 0.760 | 0.91170 | 1.38294 | 1.83878 | 2.22279 | 3.07408 | 3.19315 |
| 0.780 | 0.92381 | 1.40968 | 1.85976 | 2.23274 | 3.08464 | 3.20230 |
| 0.800 | 0.93594 | 1.43651 | 1.88070 | 2.24333 | 3.09482 | 3.21222 |
| 0.820 | 0.94811 | 1.46334 | 1.90161 | 2.25458 | 3.10460 | 3.22291 |
| 0.840 | 0.96036 | 1.49011 | 1.92250 | 2.26649 | 3.11399 | 3.23435 |
| 0.860 | 0.97271 | 1.51675 | 1.94337 | 2.27908 | 3.12304 | 3.24652 |
| 0.880 | 0.98519 | 1.54318 | 1.96424 | 2.29235 | 3.13178 | 3.25936 |
| 0.900 | 0.99779 | 1.56936 | 1.98514 | 2.30631 | 3.14026 | 3.27283 |
| 0.920 | 1.01055 | 1.59522 | 2.00608 | 2.32095 | 3.14856 | 3.28684 |
| 0.940 | 1.02345 | 1.62072 | 2.02709 | 2.33627 | 3.15672 | 3.30135 |
| 0.960 | 1.03651 | 1.64580 | 2.04821 | 2.35224 | 3.16479 | 3.31633 |
| 0.980 | 1.04974 | 1.67041 | 2.06947 | 2.36887 | 3.17284 | 3.33166 |
| 1.000 | 1.06313 | 1.69453 | 2.09090 | 2.38612 | 3.18090 | 3.34736 |

$H/R = 1.800 \qquad n = 0.$

| $H/L$ | $\Omega_1$ | $\Omega_2$ | $\Omega_3$ | $\Omega_4$ | $\Omega_5$ | $\Omega_6$ |
|---|---|---|---|---|---|---|
| 0. | 0. | 1.16502 | 1.18335 | 2.14514 | 2.96084 | 3.12536 |
| 0.010 | 0.01612 | 1.16330 | 1.18520 | 2.14515 | 2.96088 | 3.12537 |
| 0.050 | 0.08057 | 1.14456 | 1.20717 | 2.14552 | 2.96206 | 3.12595 |
| 0.100 | 0.16084 | 1.12068 | 1.24115 | 2.14670 | 2.96572 | 3.12776 |
| 0.120 | 0.19278 | 1.11210 | 1.25568 | 2.14741 | 2.96785 | 3.12883 |
| 0.140 | 0.22461 | 1.10417 | 1.27067 | 2.14825 | 2.97034 | 3.13012 |
| 0.160 | 0.25628 | 1.09693 | 1.28608 | 2.14924 | 2.97319 | 3.13162 |
| 0.180 | 0.28777 | 1.09042 | 1.30189 | 2.15037 | 2.97639 | 3.13334 |
| 0.200 | 0.31905 | 1.08468 | 1.31808 | 2.15166 | 2.97992 | 3.13529 |
| 0.220 | 0.35009 | 1.07974 | 1.33463 | 2.15310 | 2.98376 | 3.13749 |
| 0.240 | 0.38085 | 1.07565 | 1.35153 | 2.15472 | 2.98790 | 3.13995 |
| 0.260 | 0.41128 | 1.07244 | 1.36875 | 2.15652 | 2.99233 | 3.14267 |
| 0.280 | 0.44135 | 1.07018 | 1.38628 | 2.15849 | 2.99702 | 3.14568 |
| 0.300 | 0.47100 | 1.06891 | 1.40410 | 2.16067 | 3.00195 | 3.14899 |
| 0.320 | 0.50017 | 1.06868 | 1.42220 | 2.16305 | 3.00709 | 3.15261 |
| 0.340 | 0.52880 | 1.06956 | 1.44056 | 2.16564 | 3.01243 | 3.15657 |
| 0.360 | 0.55682 | 1.07162 | 1.45917 | 2.16846 | 3.01794 | 3.16089 |
| 0.380 | 0.58415 | 1.07492 | 1.47800 | 2.17152 | 3.02358 | 3.16558 |
| 0.400 | 0.61073 | 1.07953 | 1.49706 | 2.17483 | 3.02934 | 3.17066 |
| 0.420 | 0.63646 | 1.08554 | 1.51631 | 2.17841 | 3.03519 | 3.17617 |
| 0.440 | 0.66127 | 1.09300 | 1.53575 | 2.18226 | 3.04110 | 3.18210 |
| 0.460 | 0.68509 | 1.10198 | 1.55535 | 2.18640 | 3.04705 | 3.18849 |
| 0.480 | 0.70785 | 1.11253 | 1.57512 | 2.19084 | 3.05302 | 3.19535 |
| 0.500 | 0.72952 | 1.12469 | 1.59502 | 2.19561 | 3.05899 | 3.20267 |
| 0.520 | 0.75007 | 1.13845 | 1.61505 | 2.20071 | 3.06495 | 3.21049 |
| 0.540 | 0.76051 | 1.15380 | 1.63520 | 2.20616 | 3.07089 | 3.21878 |
| 0.560 | 0.78787 | 1.17069 | 1.65544 | 2.21198 | 3.07680 | 3.22757 |
| 0.580 | 0.80520 | 1.18905 | 1.67576 | 2.21819 | 3.08267 | 3.23680 |
| 0.600 | 0.82158 | 1.20877 | 1.69615 | 2.22479 | 3.08851 | 3.24660 |
| 0.620 | 0.83711 | 1.22973 | 1.71660 | 2.23182 | 3.09433 | 3.25682 |
| 0.640 | 0.85190 | 1.25182 | 1.73710 | 2.23928 | 3.10013 | 3.26750 |
| 0.660 | 0.86605 | 1.27489 | 1.75763 | 2.24720 | 3.10591 | 3.27863 |
| 0.680 | 0.87967 | 1.29882 | 1.77817 | 2.25559 | 3.11169 | 3.29018 |
| 0.700 | 0.89286 | 1.32347 | 1.79873 | 2.26447 | 3.11749 | 3.30213 |
| 0.720 | 0.90572 | 1.34872 | 1.81930 | 2.27385 | 3.12331 | 3.31448 |
| 0.740 | 0.91833 | 1.37446 | 1.83986 | 2.28376 | 3.12916 | 3.32720 |
| 0.760 | 0.93076 | 1.40058 | 1.86042 | 2.29421 | 3.13507 | 3.34028 |
| 0.780 | 0.94307 | 1.42699 | 1.88096 | 2.30521 | 3.14103 | 3.35370 |
| 0.800 | 0.95532 | 1.45360 | 1.90151 | 2.31677 | 3.14707 | 3.36744 |
| 0.820 | 0.96755 | 1.48032 | 1.92204 | 2.32891 | 3.15320 | 3.38148 |
| 0.840 | 0.97980 | 1.50708 | 1.94259 | 2.34164 | 3.15943 | 3.39581 |
| 0.860 | 0.99211 | 1.53382 | 1.96315 | 2.35495 | 3.16577 | 3.41042 |
| 0.880 | 1.00448 | 1.56045 | 1.98374 | 2.36885 | 3.17223 | 3.42528 |
| 0.900 | 1.01696 | 1.58693 | 2.00438 | 2.38335 | 3.17883 | 3.44039 |
| 0.920 | 1.02955 | 1.61320 | 2.02509 | 2.39844 | 3.18557 | 3.45572 |
| 0.940 | 1.04226 | 1.63920 | 2.04590 | 2.41411 | 3.19246 | 3.47128 |
| 0.960 | 1.05511 | 1.66487 | 2.06683 | 2.43046 | 3.19952 | 3.48703 |
| 0.980 | 1.06809 | 1.69018 | 2.08791 | 2.44717 | 3.20676 | 3.50297 |
| 1.000 | 1.08123 | 1.71508 | 2.10918 | 2.46453 | 3.21418 | 3.51909 |

$H/R = 1.900 \qquad n = 0.$

| $H/L$ | $\Omega_1$ | $\Omega_2$ | $\Omega_3$ | $\Omega_4$ | $\Omega_5$ | $\Omega_6$ |
|---|---|---|---|---|---|---|
| 0. | 0. | 1.19078 | 1.22956 | 2.18355 | 3.11847 | 3.17090 |
| 0.010 | 0.01612 | 1.18997 | 1.23050 | 2.18357 | 3.11846 | 3.17098 |
| 0.050 | 0.08058 | 1.17653 | 1.24704 | 2.18409 | 3.11823 | 3.17285 |
| 0.100 | 0.16086 | 1.15497 | 1.27838 | 2.18575 | 3.11777 | 3.17854 |
| 0.120 | 0.19283 | 1.14685 | 1.29226 | 2.18673 | 3.11760 | 3.18177 |
| 0.140 | 0.22467 | 1.13924 | 1.30669 | 2.18790 | 3.11749 | 3.18549 |
| 0.160 | 0.25638 | 1.13223 | 1.32159 | 2.18926 | 3.11746 | 3.18969 |
| 0.180 | 0.28792 | 1.12588 | 1.33694 | 2.19080 | 3.11752 | 3.19434 |
| 0.200 | 0.31926 | 1.12022 | 1.35268 | 2.19254 | 3.11770 | 3.19942 |
| 0.220 | 0.35038 | 1.11530 | 1.36881 | 2.19449 | 3.11800 | 3.20491 |
| 0.240 | 0.38124 | 1.11118 | 1.38528 | 2.19664 | 3.11844 | 3.21081 |
| 0.260 | 0.41180 | 1.10788 | 1.40209 | 2.19901 | 3.11902 | 3.21710 |
| 0.280 | 0.44202 | 1.10546 | 1.41921 | 2.20159 | 3.11976 | 3.22377 |
| 0.300 | 0.47186 | 1.10396 | 1.43663 | 2.20440 | 3.12065 | 3.23082 |
| 0.320 | 0.50126 | 1.10344 | 1.45432 | 2.20745 | 3.12170 | 3.23821 |
| 0.340 | 0.53018 | 1.10396 | 1.47228 | 2.21074 | 3.12291 | 3.24598 |
| 0.360 | 0.55853 | 1.10557 | 1.49048 | 2.21428 | 3.12429 | 3.25409 |
| 0.380 | 0.58627 | 1.10833 | 1.50892 | 2.21807 | 3.12583 | 3.26256 |
| 0.400 | 0.61332 | 1.11232 | 1.52757 | 2.22214 | 3.12753 | 3.27136 |
| 0.420 | 0.63960 | 1.11760 | 1.54642 | 2.22648 | 3.12940 | 3.28051 |
| 0.440 | 0.66505 | 1.12423 | 1.56546 | 2.23112 | 3.13144 | 3.28998 |
| 0.460 | 0.68959 | 1.13229 | 1.58467 | 2.23605 | 3.13364 | 3.29978 |
| 0.480 | 0.71316 | 1.14181 | 1.60404 | 2.24130 | 3.13601 | 3.30988 |
| 0.500 | 0.73570 | 1.15285 | 1.62355 | 2.24687 | 3.13854 | 3.32032 |
| 0.520 | 0.75719 | 1.16543 | 1.64319 | 2.25277 | 3.14124 | 3.33106 |
| 0.540 | 0.77760 | 1.17955 | 1.66295 | 2.25903 | 3.14411 | 3.34211 |
| 0.560 | 0.79695 | 1.19518 | 1.68281 | 2.26564 | 3.14715 | 3.35345 |
| 0.580 | 0.81526 | 1.21228 | 1.70276 | 2.27264 | 3.15035 | 3.36510 |
| 0.600 | 0.83259 | 1.23078 | 1.72278 | 2.28002 | 3.15373 | 3.37701 |
| 0.620 | 0.84902 | 1.25057 | 1.74287 | 2.28780 | 3.15727 | 3.38921 |
| 0.640 | 0.86464 | 1.27156 | 1.76302 | 2.29600 | 3.16098 | 3.40170 |
| 0.660 | 0.87953 | 1.29363 | 1.78321 | 2.30463 | 3.16487 | 3.41444 |
| 0.680 | 0.89381 | 1.31664 | 1.80343 | 2.31370 | 3.16893 | 3.42746 |
| 0.700 | 0.90756 | 1.34049 | 1.82367 | 2.32324 | 3.17316 | 3.44072 |
| 0.720 | 0.92089 | 1.36504 | 1.84394 | 2.33324 | 3.17757 | 3.45424 |
| 0.740 | 0.93388 | 1.39018 | 1.86421 | 2.34373 | 3.18216 | 3.46800 |
| 0.760 | 0.94661 | 1.41581 | 1.88450 | 2.35472 | 3.18692 | 3.48200 |
| 0.780 | 0.95915 | 1.44183 | 1.90480 | 2.36621 | 3.19187 | 3.49623 |
| 0.800 | 0.97156 | 1.46815 | 1.92510 | 2.37822 | 3.19700 | 3.51067 |
| 0.820 | 0.98389 | 1.49467 | 1.94542 | 2.39076 | 3.20232 | 3.52535 |
| 0.840 | 0.99619 | 1.52133 | 1.96575 | 2.40382 | 3.20783 | 3.54021 |
| 0.860 | 1.00849 | 1.54804 | 1.98612 | 2.41742 | 3.21354 | 3.55529 |
| 0.880 | 1.02082 | 1.57475 | 2.00652 | 2.43156 | 3.21944 | 3.57055 |
| 0.900 | 1.03321 | 1.60139 | 2.02699 | 2.44624 | 3.22553 | 3.58601 |
| 0.920 | 1.04569 | 1.62790 | 2.04752 | 2.46146 | 3.23184 | 3.60164 |
| 0.940 | 1.05826 | 1.65423 | 2.06816 | 2.47720 | 3.23834 | 3.61744 |
| 0.960 | 1.07094 | 1.68031 | 2.08891 | 2.49348 | 3.24506 | 3.63340 |
| 0.980 | 1.08374 | 1.70612 | 2.10981 | 2.51027 | 3.25200 | 3.64951 |
| 1.000 | 1.09668 | 1.73159 | 2.13088 | 2.52757 | 3.25915 | 3.66578 |

$H/R = 2.000 \qquad n = 0.$

| $H/L$ | $\Omega_1$ | $\Omega_2$ | $\Omega_3$ | $\Omega_4$ | $\Omega_5$ | $\Omega_6$ |
|---|---|---|---|---|---|---|
| 0. | 0. | 1.21967 | 1.26589 | 2.23313 | 3.22399 | 3.23832 |
| 0.010 | 0.01612 | 1.21900 | 1.26669 | 2.23315 | 3.22379 | 3.23861 |
| 0.050 | 0.08058 | 1.20706 | 1.28166 | 2.23372 | 3.22000 | 3.24402 |
| 0.100 | 0.16088 | 1.18643 | 1.31180 | 2.23549 | 3.21395 | 3.25516 |
| 0.120 | 0.19286 | 1.17847 | 1.32535 | 2.23654 | 3.21173 | 3.26032 |
| 0.140 | 0.22473 | 1.17096 | 1.33950 | 2.23778 | 3.20972 | 3.26584 |
| 0.160 | 0.25647 | 1.16399 | 1.35414 | 2.23922 | 3.20793 | 3.27168 |
| 0.180 | 0.28804 | 1.15761 | 1.36924 | 2.24087 | 3.20636 | 3.27783 |
| 0.200 | 0.31944 | 1.15190 | 1.38476 | 2.24272 | 3.20501 | 3.28429 |
| 0.220 | 0.35062 | 1.14688 | 1.40065 | 2.24478 | 3.20389 | 3.29105 |
| 0.240 | 0.38156 | 1.14261 | 1.41691 | 2.24705 | 3.20300 | 3.29810 |
| 0.260 | 0.41223 | 1.13913 | 1.43350 | 2.24955 | 3.20233 | 3.30546 |
| 0.280 | 0.44258 | 1.13648 | 1.45040 | 2.25227 | 3.20188 | 3.31311 |
| 0.300 | 0.47257 | 1.13470 | 1.46760 | 2.25522 | 3.20165 | 3.32106 |
| 0.320 | 0.50216 | 1.13385 | 1.48508 | 2.25841 | 3.20165 | 3.32931 |
| 0.340 | 0.53129 | 1.13398 | 1.50283 | 2.26184 | 3.20186 | 3.33784 |
| 0.360 | 0.55991 | 1.13515 | 1.52082 | 2.26553 | 3.20229 | 3.34667 |
| 0.380 | 0.58796 | 1.13740 | 1.53905 | 2.26948 | 3.20293 | 3.35578 |
| 0.400 | 0.61538 | 1.14081 | 1.55749 | 2.27369 | 3.20378 | 3.36518 |
| 0.420 | 0.64209 | 1.14543 | 1.57614 | 2.27818 | 3.20484 | 3.37487 |
| 0.440 | 0.66803 | 1.15133 | 1.59497 | 2.28296 | 3.20611 | 3.38484 |
| 0.460 | 0.69313 | 1.15858 | 1.61398 | 2.28803 | 3.20758 | 3.39509 |
| 0.480 | 0.71732 | 1.16722 | 1.63316 | 2.29341 | 3.20925 | 3.40562 |
| 0.500 | 0.74055 | 1.17731 | 1.65248 | 2.29910 | 3.21113 | 3.41643 |
| 0.520 | 0.76277 | 1.18887 | 1.67193 | 2.30512 | 3.21320 | 3.42750 |
| 0.540 | 0.78396 | 1.20193 | 1.69151 | 2.31148 | 3.21548 | 3.43885 |
| 0.560 | 0.80411 | 1.21648 | 1.71119 | 2.31818 | 3.21795 | 3.45045 |
| 0.580 | 0.82322 | 1.23248 | 1.73097 | 2.32525 | 3.22062 | 3.46231 |
| 0.600 | 0.84135 | 1.24990 | 1.75084 | 2.33269 | 3.22349 | 3.47443 |
| 0.620 | 0.85854 | 1.26865 | 1.77078 | 2.34052 | 3.22654 | 3.48681 |
| 0.640 | 0.87486 | 1.28865 | 1.79078 | 2.34874 | 3.22980 | 3.49943 |
| 0.660 | 0.89040 | 1.30979 | 1.81083 | 2.35737 | 3.23325 | 3.51230 |
| 0.680 | 0.90526 | 1.33196 | 1.83092 | 2.36643 | 3.23689 | 3.52543 |
| 0.700 | 0.91952 | 1.35504 | 1.85105 | 2.37592 | 3.24073 | 3.53877 |
| 0.720 | 0.93328 | 1.37891 | 1.87120 | 2.38586 | 3.24477 | 3.55234 |
| 0.740 | 0.94663 | 1.40347 | 1.89137 | 2.39626 | 3.24900 | 3.56613 |
| 0.760 | 0.95965 | 1.42861 | 1.91156 | 2.40712 | 3.25342 | 3.58015 |
| 0.780 | 0.97241 | 1.45421 | 1.93176 | 2.41847 | 3.25804 | 3.59439 |
| 0.800 | 0.98498 | 1.48020 | 1.95198 | 2.43030 | 3.26286 | 3.60883 |
| 0.820 | 0.99742 | 1.50648 | 1.97221 | 2.44263 | 3.26788 | 3.62348 |
| 0.840 | 1.00978 | 1.53298 | 1.99246 | 2.45546 | 3.27310 | 3.63832 |
| 0.860 | 1.02210 | 1.55961 | 2.01274 | 2.46880 | 3.27852 | 3.65336 |
| 0.880 | 1.03441 | 1.58631 | 2.03305 | 2.48266 | 3.28415 | 3.66857 |
| 0.900 | 1.04675 | 1.61301 | 2.05341 | 2.49702 | 3.28998 | 3.68395 |
| 0.920 | 1.05914 | 1.63966 | 2.07383 | 2.51190 | 3.29602 | 3.69953 |
| 0.940 | 1.07161 | 1.66619 | 2.09433 | 2.52729 | 3.30228 | 3.71525 |
| 0.960 | 1.08416 | 1.69256 | 2.11493 | 2.54319 | 3.30875 | 3.73115 |
| 0.980 | 1.09682 | 1.71871 | 2.13566 | 2.55958 | 3.31543 | 3.74719 |
| 1.000 | 1.10959 | 1.74460 | 2.15652 | 2.57647 | 3.32234 | 3.76336 |

$H/R = 0.010 \qquad n = 1.$

| $H/L$ | $\Omega_1$ | $\Omega_2$ | $\Omega_3$ | $\Omega_4$ | $\Omega_5$ | $\Omega_6$ |
|---|---|---|---|---|---|---|
| 0. | 0. | 0.00318 | 0.00761 | 1.00001 | 1.00004 | 1.87081 |
| 0.001 | 0.000318 | 0.00355 | 0.00770 | 1.00002 | 1.00004 | 1.87082 |
| 0.005 | 0.00308 | 0.00671 | 0.01048 | 1.00004 | 1.00007 | 1.87075 |
| 0.010 | 0.00455 | 0.01069 | 0.01786 | 1.00007 | 1.00019 | 1.87056 |
| 0.020 | 0.00503 | 0.02028 | 0.03427 | 1.00023 | 1.00071 | 1.86979 |
| 0.030 | 0.00525 | 0.03018 | 0.05102 | 1.00048 | 1.00157 | 1.86853 |
| 0.040 | 0.00566 | 0.04013 | 0.06784 | 1.00083 | 1.00277 | 1.86683 |
| 0.050 | 0.00638 | 0.05010 | 0.08469 | 1.00128 | 1.00431 | 1.86471 |
| 0.060 | 0.00749 | 0.06009 | 0.10155 | 1.00183 | 1.00619 | 1.86225 |
| 0.070 | 0.00903 | 0.07007 | 0.11841 | 1.00248 | 1.00839 | 1.85947 |
| 0.080 | 0.01098 | 0.08006 | 0.13528 | 1.00323 | 1.01093 | 1.85644 |
| 0.090 | 0.01327 | 0.09006 | 0.15214 | 1.00407 | 1.01379 | 1.85320 |
| 0.100 | 0.01592 | 0.10005 | 0.16900 | 1.00502 | 1.01697 | 1.84978 |
| 0.120 | 0.02215 | 0.12004 | 0.20269 | 1.00720 | 1.02426 | 1.84257 |
| 0.140 | 0.02953 | 0.14004 | 0.23635 | 1.00978 | 1.03277 | 1.83502 |
| 0.160 | 0.03797 | 0.16003 | 0.26997 | 1.01275 | 1.04242 | 1.82734 |
| 0.180 | 0.04739 | 0.18003 | 0.30354 | 1.01610 | 1.05317 | 1.81965 |
| 0.200 | 0.05774 | 0.20003 | 0.33705 | 1.01984 | 1.06497 | 1.81205 |
| 0.220 | 0.06894 | 0.22002 | 0.37050 | 1.02395 | 1.07774 | 1.80460 |
| 0.240 | 0.08093 | 0.24002 | 0.40386 | 1.02843 | 1.09143 | 1.79739 |
| 0.260 | 0.09365 | 0.26002 | 0.43714 | 1.03327 | 1.10598 | 1.79043 |
| 0.280 | 0.10705 | 0.28002 | 0.47033 | 1.03849 | 1.12134 | 1.78378 |
| 0.300 | 0.12108 | 0.30002 | 0.50340 | 1.04406 | 1.13745 | 1.77747 |
| 0.320 | 0.13567 | 0.32002 | 0.53636 | 1.04998 | 1.15425 | 1.77153 |
| 0.340 | 0.15080 | 0.34002 | 0.56919 | 1.05625 | 1.17170 | 1.76600 |
| 0.360 | 0.16641 | 0.36002 | 0.60188 | 1.06286 | 1.18976 | 1.76089 |
| 0.380 | 0.18245 | 0.38002 | 0.63441 | 1.06979 | 1.20836 | 1.75624 |
| 0.400 | 0.19891 | 0.40001 | 0.66676 | 1.07706 | 1.22748 | 1.75204 |
| 0.420 | 0.21572 | 0.42001 | 0.69892 | 1.08465 | 1.24708 | 1.74837 |
| 0.440 | 0.23288 | 0.44001 | 0.73086 | 1.09255 | 1.26711 | 1.74520 |
| 0.460 | 0.25033 | 0.46001 | 0.76258 | 1.10075 | 1.28754 | 1.74260 |
| 0.480 | 0.26807 | 0.48001 | 0.79405 | 1.10926 | 1.30834 | 1.74055 |
| 0.500 | 0.28606 | 0.50001 | 0.82523 | 1.11806 | 1.32949 | 1.73913 |
| 0.520 | 0.30427 | 0.52001 | 0.85611 | 1.12715 | 1.35094 | 1.73834 |
| 0.540 | 0.32270 | 0.54001 | 0.88666 | 1.13651 | 1.37268 | 1.73821 |
| 0.560 | 0.34131 | 0.56001 | 0.91684 | 1.14615 | 1.39468 | 1.73880 |
| 0.580 | 0.36009 | 0.58001 | 0.94662 | 1.15605 | 1.41693 | 1.74011 |
| 0.600 | 0.37903 | 0.60001 | 0.97596 | 1.16622 | 1.43937 | 1.74219 |
| 0.620 | 0.39811 | 0.62001 | 1.00482 | 1.17663 | 1.46203 | 1.74510 |
| 0.640 | 0.41731 | 0.64001 | 1.03316 | 1.18729 | 1.48485 | 1.74886 |
| 0.660 | 0.43663 | 0.66001 | 1.06093 | 1.19819 | 1.50783 | 1.75352 |
| 0.680 | 0.45605 | 0.68001 | 1.08809 | 1.20932 | 1.53095 | 1.75914 |
| 0.700 | 0.47556 | 0.70001 | 1.11459 | 1.22068 | 1.55419 | 1.76575 |
| 0.720 | 0.49516 | 0.72001 | 1.14039 | 1.23226 | 1.57753 | 1.77340 |
| 0.740 | 0.51482 | 0.74001 | 1.16543 | 1.24405 | 1.60097 | 1.78212 |
| 0.760 | 0.53456 | 0.76001 | 1.18969 | 1.25605 | 1.62448 | 1.79198 |
| 0.780 | 0.55436 | 0.78001 | 1.21311 | 1.26825 | 1.64803 | 1.80299 |
| 0.800 | 0.57421 | 0.80001 | 1.23568 | 1.28065 | 1.67166 | 1.81518 |
| 0.820 | 0.59410 | 0.82001 | 1.25736 | 1.29323 | 1.69530 | 1.82858 |
| 0.840 | 0.61404 | 0.84001 | 1.27815 | 1.30601 | 1.71896 | 1.84318 |
| 0.860 | 0.63401 | 0.86001 | 1.29803 | 1.31896 | 1.74261 | 1.85902 |
| 0.880 | 0.65402 | 0.88001 | 1.31703 | 1.33209 | 1.76626 | 1.87604 |
| 0.900 | 0.67406 | 0.90001 | 1.33517 | 1.34538 | 1.78989 | 1.89424 |
| 0.920 | 0.69412 | 0.92001 | 1.35246 | 1.35885 | 1.81349 | 1.91357 |
| 0.940 | 0.71419 | 0.94001 | 1.36896 | 1.37246 | 1.83703 | 1.93399 |
| 0.960 | 0.73429 | 0.96001 | 1.38473 | 1.38624 | 1.86053 | 1.95545 |
| 0.980 | 0.75440 | 0.98001 | 1.39979 | 1.40016 | 1.88394 | 1.97789 |
| 1.000 | 0.77452 | 1.00001 | 1.41422 | 1.41424 | 1.90727 | 2.00125 |

$H/R = 0.020$ $n = 1.$ $H/R = 0.030$

| $H/L$ | $\Omega_1$ | $\Omega_2$ | $\Omega_3$ | $\Omega_1$ | $\Omega_2$ | $\Omega_3$ | $H/L$ |
|---|---|---|---|---|---|---|---|
| 0. | 0. | 0.00637 | 0.01522 | 0. | 0.00955 | 0.02282 | 0. |
| 0.001 | 0.000173 | 0.00656 | 0.01526 | 0.000118 | 0.00968 | 0.02285 | 0.001 |
| 0.005 | 0.00276 | 0.00956 | 0.01642 | 0.00227 | 0.01216 | 0.02357 | 0.005 |
| 0.010 | 0.00616 | 0.01343 | 0.02096 | 0.00602 | 0.01646 | 0.02628 | 0.010 |
| 0.020 | 0.00911 | 0.02137 | 0.03571 | 0.01150 | 0.02371 | 0.03822 | 0.020 |
| 0.030 | 0.00984 | 0.03079 | 0.05196 | 0.01371 | 0.03206 | 0.05357 | 0.030 |
| 0.040 | 0.01028 | 0.04055 | 0.06854 | 0.01465 | 0.04138 | 0.06973 | 0.040 |
| 0.050 | 0.01079 | 0.05043 | 0.08525 | 0.01528 | 0.05104 | 0.08619 | 0.050 |
| 0.060 | 0.01153 | 0.06035 | 0.10201 | 0.01596 | 0.06083 | 0.10280 | 0.060 |
| 0.070 | 0.01261 | 0.07030 | 0.11881 | 0.01683 | 0.07070 | 0.11948 | 0.070 |
| 0.080 | 0.01408 | 0.08026 | 0.13563 | 0.01801 | 0.08060 | 0.13621 | 0.080 |
| 0.090 | 0.01595 | 0.09023 | 0.15245 | 0.01954 | 0.09053 | 0.15297 | 0.090 |
| 0.100 | 0.01821 | 0.10021 | 0.16928 | 0.02145 | 0.10047 | 0.16974 | 0.100 |
| 0.120 | 0.02385 | 0.12017 | 0.20293 | 0.02643 | 0.12039 | 0.20332 | 0.120 |
| 0.140 | 0.03083 | 0.14015 | 0.23656 | 0.03288 | 0.14033 | 0.23689 | 0.140 |
| 0.160 | 0.03899 | 0.16013 | 0.27015 | 0.04064 | 0.16029 | 0.27044 | 0.160 |
| 0.180 | 0.04822 | 0.18011 | 0.30370 | 0.04956 | 0.18026 | 0.30396 | 0.180 |
| 0.200 | 0.05842 | 0.20010 | 0.33719 | 0.05953 | 0.20023 | 0.33743 | 0.200 |
| 0.220 | 0.06951 | 0.22009 | 0.37062 | 0.07045 | 0.22021 | 0.37083 | 0.220 |
| 0.240 | 0.08142 | 0.24009 | 0.40398 | 0.08222 | 0.24019 | 0.40417 | 0.240 |
| 0.260 | 0.09407 | 0.26008 | 0.43725 | 0.09477 | 0.26018 | 0.43743 | 0.260 |
| 0.280 | 0.10742 | 0.28007 | 0.47042 | 0.10803 | 0.28017 | 0.47059 | 0.280 |
| 0.300 | 0.12140 | 0.30007 | 0.50349 | 0.12195 | 0.30015 | 0.50365 | 0.300 |
| 0.320 | 0.13597 | 0.32006 | 0.53645 | 0.13645 | 0.32014 | 0.53659 | 0.320 |
| 0.340 | 0.15106 | 0.34006 | 0.56927 | 0.15150 | 0.34014 | 0.56941 | 0.340 |
| 0.360 | 0.16664 | 0.36006 | 0.60195 | 0.16704 | 0.36013 | 0.60208 | 0.360 |
| 0.380 | 0.18267 | 0.38005 | 0.63447 | 0.18303 | 0.38012 | 0.63460 | 0.380 |
| 0.400 | 0.19910 | 0.40005 | 0.66683 | 0.19944 | 0.40011 | 0.66694 | 0.400 |
| 0.420 | 0.21591 | 0.42005 | 0.69898 | 0.21621 | 0.42011 | 0.69909 | 0.420 |
| 0.440 | 0.23305 | 0.44005 | 0.73093 | 0.23333 | 0.44010 | 0.73104 | 0.440 |
| 0.460 | 0.25049 | 0.46004 | 0.76265 | 0.25076 | 0.46010 | 0.76275 | 0.460 |
| 0.480 | 0.26822 | 0.48004 | 0.79411 | 0.26846 | 0.48010 | 0.79420 | 0.480 |
| 0.500 | 0.28620 | 0.50004 | 0.82529 | 0.28643 | 0.50009 | 0.82539 | 0.500 |
| 0.520 | 0.30440 | 0.52004 | 0.85617 | 0.30462 | 0.52009 | 0.85626 | 0.520 |
| 0.540 | 0.32282 | 0.54004 | 0.88672 | 0.32302 | 0.54009 | 0.88680 | 0.540 |
| 0.560 | 0.34142 | 0.56004 | 0.91690 | 0.34162 | 0.56008 | 0.91698 | 0.560 |
| 0.580 | 0.36020 | 0.58004 | 0.94667 | 0.36038 | 0.58008 | 0.94675 | 0.580 |
| 0.600 | 0.37913 | 0.60003 | 0.97600 | 0.37931 | 0.60008 | 0.97608 | 0.600 |
| 0.620 | 0.39820 | 0.62003 | 1.00486 | 0.39837 | 0.62008 | 1.00494 | 0.620 |
| 0.640 | 0.41740 | 0.64004 | 1.03320 | 0.41756 | 0.64008 | 1.03327 | 0.640 |
| 0.660 | 0.43672 | 0.66004 | 1.06097 | 0.43686 | 0.66007 | 1.06104 | 0.660 |
| 0.680 | 0.45613 | 0.68003 | 1.08813 | 0.45628 | 0.68007 | 1.08820 | 0.680 |
| 0.700 | 0.47564 | 0.70003 | 1.11463 | 0.47578 | 0.70007 | 1.11470 | 0.700 |
| 0.720 | 0.49523 | 0.72003 | 1.14043 | 0.49536 | 0.72007 | 1.14049 | 0.720 |
| 0.740 | 0.51490 | 0.74003 | 1.16547 | 0.51503 | 0.74006 | 1.16553 | 0.740 |
| 0.760 | 0.53463 | 0.76003 | 1.18972 | 0.53476 | 0.76006 | 1.18978 | 0.760 |
| 0.780 | 0.55443 | 0.78003 | 1.21315 | 0.55454 | 0.78006 | 1.21321 | 0.780 |
| 0.800 | 0.57427 | 0.80003 | 1.23571 | 0.57439 | 0.80006 | 1.23577 | 0.800 |
| 0.820 | 0.59417 | 0.82003 | 1.25740 | 0.59428 | 0.82006 | 1.25745 | 0.820 |
| 0.840 | 0.61410 | 0.84003 | 1.27818 | 0.61421 | 0.84006 | 1.27823 | 0.840 |
| 0.860 | 0.63407 | 0.86003 | 1.29807 | 0.63418 | 0.86005 | 1.29812 | 0.860 |
| 0.880 | 0.65408 | 0.88002 | 1.31706 | 0.65418 | 0.88006 | 1.31712 | 0.880 |
| 0.900 | 0.67411 | 0.90003 | 1.33520 | 0.67421 | 0.90006 | 1.33524 | 0.900 |
| 0.920 | 0.69417 | 0.92003 | 1.35249 | 0.69426 | 0.92005 | 1.35254 | 0.920 |
| 0.940 | 0.71424 | 0.94002 | 1.36899 | 0.71434 | 0.94005 | 1.36904 | 0.940 |
| 0.960 | 0.73434 | 0.96002 | 1.38476 | 0.73443 | 0.96005 | 1.38479 | 0.960 |
| 0.980 | 0.75445 | 0.98002 | 1.39981 | 0.75453 | 0.98005 | 1.39985 | 0.980 |
| 1.000 | 0.77457 | 1.00003 | 1.41424 | 0.77465 | 1.00005 | 1.41425 | 1.000 |

$H/R = 0.040$ $\quad n = 1. \quad$ $H/R = 0.050$

| $H/L$ | $\Omega_1$ | $\Omega_2$ | $\Omega_3$ | $\Omega_1$ | $\Omega_2$ | $\Omega_3$ | $H/L$ |
|---|---|---|---|---|---|---|---|
| 0. | 0. | 0.01273 | 0.03043 | 0. | 0.01592 | 0.03803 | 0. |
| 0.001 | 0.0000888 | 0.01283 | 0.03045 | 0.0000713 | 0.01600 | 0.03805 | 0.001 |
| 0.005 | 0.00188 | 0.01488 | 0.03097 | 0.00159 | 0.01773 | 0.03846 | 0.005 |
| 0.010 | 0.00553 | 0.01912 | 0.03283 | 0.00502 | 0.02170 | 0.03987 | 0.010 |
| 0.020 | 0.01234 | 0.02686 | 0.04192 | 0.01236 | 0.03002 | 0.04679 | 0.020 |
| 0.030 | 0.01640 | 0.03420 | 0.05588 | 0.01790 | 0.03707 | 0.05898 | 0.030 |
| 0.040 | 0.01834 | 0.04275 | 0.07142 | 0.02118 | 0.04478 | 0.07364 | 0.040 |
| 0.050 | 0.01943 | 0.05201 | 0.08753 | 0.02304 | 0.05344 | 0.08927 | 0.050 |
| 0.060 | 0.02026 | 0.06158 | 0.10390 | 0.02427 | 0.06265 | 0.10533 | 0.060 |
| 0.070 | 0.02115 | 0.07130 | 0.12042 | 0.02531 | 0.07216 | 0.12164 | 0.070 |
| 0.080 | 0.02221 | 0.08111 | 0.13703 | 0.02641 | 0.08182 | 0.13810 | 0.080 |
| 0.090 | 0.02354 | 0.09097 | 0.15370 | 0.02766 | 0.09157 | 0.15464 | 0.090 |
| 0.100 | 0.02519 | 0.10086 | 0.17040 | 0.02917 | 0.10139 | 0.17125 | 0.100 |
| 0.120 | 0.02960 | 0.12070 | 0.20386 | 0.03315 | 0.12113 | 0.20456 | 0.120 |
| 0.140 | 0.03551 | 0.14060 | 0.23735 | 0.03857 | 0.14095 | 0.23796 | 0.140 |
| 0.160 | 0.04281 | 0.16052 | 0.27085 | 0.04541 | 0.16082 | 0.27138 | 0.160 |
| 0.180 | 0.05137 | 0.18046 | 0.30432 | 0.05357 | 0.18073 | 0.30479 | 0.180 |
| 0.200 | 0.06105 | 0.20041 | 0.33775 | 0.06293 | 0.20065 | 0.33817 | 0.200 |
| 0.220 | 0.07174 | 0.22037 | 0.37113 | 0.07335 | 0.22059 | 0.37151 | 0.220 |
| 0.240 | 0.08333 | 0.24034 | 0.40444 | 0.08473 | 0.24054 | 0.40479 | 0.240 |
| 0.260 | 0.09574 | 0.26031 | 0.43768 | 0.09696 | 0.26050 | 0.43800 | 0.260 |
| 0.280 | 0.10888 | 0.28029 | 0.47082 | 0.10996 | 0.28046 | 0.47112 | 0.280 |
| 0.300 | 0.12270 | 0.30027 | 0.50387 | 0.12366 | 0.30043 | 0.50415 | 0.300 |
| 0.320 | 0.13713 | 0.32026 | 0.53680 | 0.13799 | 0.32040 | 0.53706 | 0.320 |
| 0.340 | 0.15211 | 0.34024 | 0.56960 | 0.15288 | 0.34038 | 0.56985 | 0.340 |
| 0.360 | 0.16759 | 0.36023 | 0.60226 | 0.16830 | 0.36036 | 0.60250 | 0.360 |
| 0.380 | 0.18354 | 0.38022 | 0.63477 | 0.18418 | 0.38034 | 0.63499 | 0.380 |
| 0.400 | 0.19990 | 0.40020 | 0.66711 | 0.20049 | 0.40032 | 0.66731 | 0.400 |
| 0.420 | 0.21664 | 0.42019 | 0.69925 | 0.21719 | 0.42030 | 0.69945 | 0.420 |
| 0.440 | 0.23373 | 0.44019 | 0.73118 | 0.23423 | 0.44029 | 0.73137 | 0.440 |
| 0.460 | 0.25113 | 0.46018 | 0.76289 | 0.25160 | 0.46028 | 0.76307 | 0.460 |
| 0.480 | 0.26881 | 0.48017 | 0.79434 | 0.26925 | 0.48027 | 0.79452 | 0.480 |
| 0.500 | 0.28675 | 0.50016 | 0.82552 | 0.28716 | 0.50026 | 0.82568 | 0.500 |
| 0.520 | 0.30492 | 0.52016 | 0.85639 | 0.30531 | 0.52025 | 0.85655 | 0.520 |
| 0.540 | 0.32331 | 0.54015 | 0.88692 | 0.32367 | 0.54024 | 0.88708 | 0.540 |
| 0.560 | 0.34189 | 0.56015 | 0.91709 | 0.34223 | 0.56023 | 0.91724 | 0.560 |
| 0.580 | 0.36064 | 0.58014 | 0.94686 | 0.36096 | 0.58022 | 0.94700 | 0.580 |
| 0.600 | 0.37955 | 0.60014 | 0.97619 | 0.37986 | 0.60021 | 0.97633 | 0.600 |
| 0.620 | 0.39860 | 0.62013 | 1.00504 | 0.39889 | 0.62021 | 1.00518 | 0.620 |
| 0.640 | 0.41778 | 0.64013 | 1.03337 | 0.41806 | 0.64020 | 1.03351 | 0.640 |
| 0.660 | 0.43707 | 0.66012 | 1.06114 | 0.43734 | 0.66020 | 1.06127 | 0.660 |
| 0.680 | 0.45647 | 0.68012 | 1.08829 | 0.45673 | 0.68019 | 1.08842 | 0.680 |
| 0.700 | 0.47597 | 0.70012 | 1.11479 | 0.47621 | 0.70018 | 1.11491 | 0.700 |
| 0.720 | 0.49555 | 0.72012 | 1.14058 | 0.49579 | 0.72018 | 1.14069 | 0.720 |
| 0.740 | 0.51520 | 0.74011 | 1.16562 | 0.51543 | 0.74017 | 1.16574 | 0.740 |
| 0.760 | 0.53492 | 0.76011 | 1.18987 | 0.53514 | 0.76017 | 1.18998 | 0.760 |
| 0.780 | 0.55471 | 0.78011 | 1.21329 | 0.55492 | 0.78016 | 1.21339 | 0.780 |
| 0.800 | 0.57454 | 0.80010 | 1.23585 | 0.57475 | 0.80016 | 1.23595 | 0.800 |
| 0.820 | 0.59443 | 0.82010 | 1.25753 | 0.59462 | 0.82016 | 1.25763 | 0.820 |
| 0.840 | 0.61435 | 0.84010 | 1.27831 | 0.61454 | 0.84015 | 1.27840 | 0.840 |
| 0.860 | 0.63432 | 0.86010 | 1.29819 | 0.63450 | 0.86015 | 1.29829 | 0.860 |
| 0.880 | 0.65431 | 0.88010 | 1.31719 | 0.65449 | 0.88015 | 1.31728 | 0.880 |
| 0.900 | 0.67434 | 0.90009 | 1.33532 | 0.67451 | 0.90014 | 1.33541 | 0.900 |
| 0.920 | 0.69439 | 0.92009 | 1.35262 | 0.69455 | 0.92014 | 1.35270 | 0.920 |
| 0.940 | 0.71446 | 0.94009 | 1.36911 | 0.71462 | 0.94014 | 1.36919 | 0.940 |
| 0.960 | 0.73455 | 0.96009 | 1.38485 | 0.73470 | 0.96013 | 1.38492 | 0.960 |
| 0.980 | 0.75465 | 0.98009 | 1.39989 | 0.75480 | 0.98013 | 1.39993 | 0.980 |
| 1.000 | 0.77477 | 1.00009 | 1.41426 | 0.77491 | 1.00013 | 1.41429 | 1.000 |

$H/R = 0.060$ $n = 1.$ $H/R = 0.070$

| $H/L$ | $\Omega_1$ | $\Omega_2$ | $\Omega_3$ | $\Omega_1$ | $\Omega_2$ | $\Omega_3$ | $H/L$ |
|---|---|---|---|---|---|---|---|
| 0. | 0. | 0.01910 | 0.04563 | 0. | 0.02229 | 0.05323 | 0. |
| 0.001 | 0.0000595 | 0.01917 | 0.04565 | 0.0000511 | 0.02234 | 0.05324 | 0.001 |
| 0.005 | 0.00137 | 0.02066 | 0.04599 | 0.00120 | 0.02365 | 0.05353 | 0.005 |
| 0.010 | 0.00454 | 0.02432 | 0.04712 | 0.00413 | 0.02701 | 0.05449 | 0.010 |
| 0.020 | 0.01204 | 0.03292 | 0.05255 | 0.01158 | 0.03563 | 0.05890 | 0.020 |
| 0.030 | 0.01853 | 0.04029 | 0.06288 | 0.01864 | 0.04349 | 0.06756 | 0.030 |
| 0.040 | 0.02309 | 0.04742 | 0.07643 | 0.02421 | 0.05048 | 0.07982 | 0.040 |
| 0.050 | 0.02595 | 0.05540 | 0.09143 | 0.02812 | 0.05788 | 0.09405 | 0.050 |
| 0.060 | 0.02781 | 0.06413 | 0.10711 | 0.03076 | 0.06604 | 0.10924 | 0.060 |
| 0.070 | 0.02920 | 0.07331 | 0.12315 | 0.03266 | 0.07482 | 0.12495 | 0.070 |
| 0.080 | 0.03050 | 0.08276 | 0.13941 | 0.03420 | 0.08398 | 0.14096 | 0.080 |
| 0.090 | 0.03171 | 0.09237 | 0.15580 | 0.03562 | 0.09338 | 0.15717 | 0.090 |
| 0.100 | 0.03320 | 0.10207 | 0.17228 | 0.03715 | 0.10294 | 0.17352 | 0.100 |
| 0.120 | 0.03691 | 0.12167 | 0.20543 | 0.04076 | 0.12234 | 0.20645 | 0.120 |
| 0.140 | 0.04193 | 0.14140 | 0.23869 | 0.04548 | 0.14194 | 0.23957 | 0.140 |
| 0.160 | 0.04835 | 0.16120 | 0.27202 | 0.05153 | 0.16167 | 0.27278 | 0.160 |
| 0.180 | 0.05611 | 0.18106 | 0.30536 | 0.05892 | 0.18146 | 0.30604 | 0.180 |
| 0.200 | 0.06512 | 0.20094 | 0.33869 | 0.06759 | 0.20130 | 0.33930 | 0.200 |
| 0.220 | 0.07525 | 0.22085 | 0.37198 | 0.07742 | 0.22117 | 0.37253 | 0.220 |
| 0.240 | 0.08639 | 0.24078 | 0.40522 | 0.08829 | 0.24107 | 0.40573 | 0.240 |
| 0.260 | 0.09842 | 0.26072 | 0.43840 | 0.10010 | 0.26098 | 0.43886 | 0.260 |
| 0.280 | 0.11126 | 0.28067 | 0.47149 | 0.11276 | 0.28091 | 0.47192 | 0.280 |
| 0.300 | 0.12481 | 0.30062 | 0.50449 | 0.12616 | 0.30085 | 0.50489 | 0.300 |
| 0.320 | 0.13902 | 0.32058 | 0.53738 | 0.14024 | 0.32079 | 0.53776 | 0.320 |
| 0.340 | 0.15382 | 0.34055 | 0.57015 | 0.15492 | 0.34074 | 0.57051 | 0.340 |
| 0.360 | 0.16915 | 0.36051 | 0.60278 | 0.17015 | 0.36070 | 0.60312 | 0.360 |
| 0.380 | 0.18496 | 0.38049 | 0.63526 | 0.18588 | 0.38066 | 0.63558 | 0.380 |
| 0.400 | 0.20121 | 0.40046 | 0.66757 | 0.20205 | 0.40063 | 0.66787 | 0.400 |
| 0.420 | 0.21785 | 0.42044 | 0.69969 | 0.21863 | 0.42060 | 0.69998 | 0.420 |
| 0.440 | 0.23485 | 0.44042 | 0.73160 | 0.23557 | 0.44057 | 0.73188 | 0.440 |
| 0.460 | 0.25217 | 0.46040 | 0.76329 | 0.25284 | 0.46055 | 0.76355 | 0.460 |
| 0.480 | 0.26978 | 0.48038 | 0.79473 | 0.27041 | 0.48052 | 0.79498 | 0.480 |
| 0.500 | 0.28766 | 0.50037 | 0.82589 | 0.28825 | 0.50050 | 0.82613 | 0.500 |
| 0.520 | 0.30578 | 0.52035 | 0.85674 | 0.30634 | 0.52049 | 0.85697 | 0.520 |
| 0.540 | 0.32412 | 0.54034 | 0.88726 | 0.32464 | 0.54047 | 0.88749 | 0.540 |
| 0.560 | 0.34265 | 0.56033 | 0.91742 | 0.34315 | 0.56045 | 0.91763 | 0.560 |
| 0.580 | 0.36136 | 0.58032 | 0.94718 | 0.36183 | 0.58043 | 0.94738 | 0.580 |
| 0.600 | 0.38024 | 0.60031 | 0.97650 | 0.38068 | 0.60042 | 0.97670 | 0.600 |
| 0.620 | 0.39925 | 0.62030 | 1.00534 | 0.39968 | 0.62041 | 1.00553 | 0.620 |
| 0.640 | 0.41840 | 0.64029 | 1.03367 | 0.41881 | 0.64039 | 1.03385 | 0.640 |
| 0.660 | 0.43767 | 0.66028 | 1.06142 | 0.43805 | 0.66038 | 1.06160 | 0.660 |
| 0.680 | 0.45704 | 0.68027 | 1.08856 | 0.45741 | 0.68037 | 1.08874 | 0.680 |
| 0.700 | 0.47651 | 0.70027 | 1.11505 | 0.47687 | 0.70036 | 1.11522 | 0.700 |
| 0.720 | 0.49607 | 0.72026 | 1.14084 | 0.49641 | 0.72035 | 1.14100 | 0.720 |
| 0.740 | 0.51570 | 0.74025 | 1.16587 | 0.51603 | 0.74034 | 1.16603 | 0.740 |
| 0.760 | 0.53541 | 0.76024 | 1.19011 | 0.53572 | 0.76033 | 1.19027 | 0.760 |
| 0.780 | 0.55517 | 0.78024 | 1.21353 | 0.55547 | 0.78032 | 1.21368 | 0.780 |
| 0.800 | 0.57499 | 0.80023 | 1.23608 | 0.57528 | 0.80031 | 1.23623 | 0.800 |
| 0.820 | 0.59486 | 0.82023 | 1.25775 | 0.59514 | 0.82030 | 1.25789 | 0.820 |
| 0.840 | 0.61477 | 0.84022 | 1.27853 | 0.61504 | 0.84030 | 1.27867 | 0.840 |
| 0.860 | 0.63472 | 0.86022 | 1.29841 | 0.63498 | 0.86029 | 1.29854 | 0.860 |
| 0.880 | 0.65471 | 0.88021 | 1.31740 | 0.65496 | 0.88029 | 1.31753 | 0.880 |
| 0.900 | 0.67472 | 0.90021 | 1.33552 | 0.67496 | 0.90028 | 1.33564 | 0.900 |
| 0.920 | 0.69475 | 0.92020 | 1.35280 | 0.69499 | 0.92028 | 1.35291 | 0.920 |
| 0.940 | 0.71481 | 0.94020 | 1.36929 | 0.71505 | 0.94027 | 1.36938 | 0.940 |
| 0.960 | 0.73489 | 0.96019 | 1.38500 | 0.73512 | 0.96026 | 1.38507 | 0.960 |
| 0.980 | 0.75498 | 0.98019 | 1.39997 | 0.75520 | 0.98026 | 1.40001 | 0.980 |
| 1.000 | 0.77509 | 1.00019 | 1.41432 | 0.77530 | 1.00025 | 1.41435 | 1.000 |

$H/R = 0.080$ $\quad n = 1. \quad$ $H/R = 0.090$

| $H/L$ | $\Omega_1$ | $\Omega_2$ | $\Omega_3$ | $\Omega_1$ | $\Omega_2$ | $\Omega_3$ | $H/L$ |
|---|---|---|---|---|---|---|---|
| 0. | 0. | 0.02547 | 0.06082 | 0. | 0.02866 | 0.06840 | 0. |
| 0.001 | 0.0000447 | 0.02552 | 0.06083 | 0.0000397 | 0.02870 | 0.06841 | 0.001 |
| 0.005 | 0.00107 | 0.02668 | 0.06108 | 0.000956 | 0.02974 | 0.06864 | 0.005 |
| 0.010 | 0.00377 | 0.02977 | 0.06191 | 0.00345 | 0.03260 | 0.06937 | 0.010 |
| 0.020 | 0.01107 | 0.03824 | 0.06562 | 0.01055 | 0.04082 | 0.07257 | 0.020 |
| 0.030 | 0.01844 | 0.04653 | 0.07291 | 0.01809 | 0.04939 | 0.07879 | 0.030 |
| 0.040 | 0.02475 | 0.05372 | 0.08382 | 0.02490 | 0.05694 | 0.08841 | 0.040 |
| 0.050 | 0.02960 | 0.06077 | 0.09713 | 0.03051 | 0.06391 | 0.10070 | 0.050 |
| 0.060 | 0.03310 | 0.06840 | 0.11173 | 0.03484 | 0.07114 | 0.11462 | 0.060 |
| 0.070 | 0.03563 | 0.07670 | 0.12705 | 0.03809 | 0.07897 | 0.12946 | 0.070 |
| 0.080 | 0.03760 | 0.08551 | 0.14278 | 0.04059 | 0.08737 | 0.14486 | 0.080 |
| 0.090 | 0.03931 | 0.09465 | 0.15877 | 0.04266 | 0.09620 | 0.16060 | 0.090 |
| 0.100 | 0.04094 | 0.10402 | 0.17495 | 0.04453 | 0.10533 | 0.17658 | 0.100 |
| 0.120 | 0.04457 | 0.12316 | 0.20763 | 0.04835 | 0.12415 | 0.20898 | 0.120 |
| 0.140 | 0.04913 | 0.14261 | 0.24058 | 0.05279 | 0.14340 | 0.24172 | 0.140 |
| 0.160 | 0.05489 | 0.16222 | 0.27366 | 0.05834 | 0.16288 | 0.27466 | 0.160 |
| 0.180 | 0.06195 | 0.18194 | 0.30682 | 0.06513 | 0.18250 | 0.30770 | 0.180 |
| 0.200 | 0.07028 | 0.20172 | 0.34000 | 0.07316 | 0.20222 | 0.34079 | 0.200 |
| 0.220 | 0.07981 | 0.22155 | 0.37317 | 0.08239 | 0.22199 | 0.37389 | 0.220 |
| 0.240 | 0.09042 | 0.24141 | 0.40631 | 0.09273 | 0.24181 | 0.40697 | 0.240 |
| 0.260 | 0.10199 | 0.26130 | 0.43940 | 0.10407 | 0.26166 | 0.44001 | 0.260 |
| 0.280 | 0.11445 | 0.28120 | 0.47242 | 0.11631 | 0.28153 | 0.47299 | 0.280 |
| 0.300 | 0.12768 | 0.30111 | 0.50536 | 0.12936 | 0.30142 | 0.50589 | 0.300 |
| 0.320 | 0.14161 | 0.32104 | 0.53820 | 0.14314 | 0.32133 | 0.53869 | 0.320 |
| 0.340 | 0.15617 | 0.34098 | 0.57092 | 0.15756 | 0.34124 | 0.57138 | 0.340 |
| 0.360 | 0.17129 | 0.36092 | 0.60351 | 0.17257 | 0.36117 | 0.60394 | 0.360 |
| 0.380 | 0.18693 | 0.38087 | 0.63595 | 0.18810 | 0.38111 | 0.63636 | 0.380 |
| 0.400 | 0.20302 | 0.40083 | 0.66822 | 0.20410 | 0.40105 | 0.66862 | 0.400 |
| 0.420 | 0.21952 | 0.42079 | 0.70031 | 0.22052 | 0.42100 | 0.70068 | 0.420 |
| 0.440 | 0.23640 | 0.44075 | 0.73220 | 0.23733 | 0.44095 | 0.73255 | 0.440 |
| 0.460 | 0.25361 | 0.46071 | 0.76386 | 0.25448 | 0.46091 | 0.76420 | 0.460 |
| 0.480 | 0.27113 | 0.48069 | 0.79527 | 0.27195 | 0.48087 | 0.79560 | 0.480 |
| 0.500 | 0.28893 | 0.50066 | 0.82641 | 0.28969 | 0.50084 | 0.82672 | 0.500 |
| 0.520 | 0.30697 | 0.52063 | 0.85724 | 0.30769 | 0.52081 | 0.85754 | 0.520 |
| 0.540 | 0.32524 | 0.54061 | 0.88774 | 0.32592 | 0.54077 | 0.88803 | 0.540 |
| 0.560 | 0.34371 | 0.56059 | 0.91788 | 0.34435 | 0.56074 | 0.91816 | 0.560 |
| 0.580 | 0.36237 | 0.58057 | 0.94762 | 0.36298 | 0.58072 | 0.94790 | 0.580 |
| 0.600 | 0.38119 | 0.60055 | 0.97693 | 0.38177 | 0.60069 | 0.97719 | 0.600 |
| 0.620 | 0.40016 | 0.62053 | 1.00575 | 0.40071 | 0.62067 | 1.00601 | 0.620 |
| 0.640 | 0.41927 | 0.64051 | 1.03406 | 0.41979 | 0.64065 | 1.03431 | 0.640 |
| 0.660 | 0.43850 | 0.66050 | 1.06181 | 0.43900 | 0.66063 | 1.06205 | 0.660 |
| 0.680 | 0.45784 | 0.68048 | 1.08895 | 0.45831 | 0.68061 | 1.08917 | 0.680 |
| 0.700 | 0.47727 | 0.70047 | 1.11542 | 0.47773 | 0.70060 | 1.11564 | 0.700 |
| 0.720 | 0.49680 | 0.72046 | 1.14120 | 0.49724 | 0.72058 | 1.14141 | 0.720 |
| 0.740 | 0.51641 | 0.74044 | 1.16622 | 0.51683 | 0.74056 | 1.16642 | 0.740 |
| 0.760 | 0.53608 | 0.76043 | 1.19045 | 0.53649 | 0.76055 | 1.19065 | 0.760 |
| 0.780 | 0.55582 | 0.78042 | 1.21385 | 0.55621 | 0.78053 | 1.21405 | 0.780 |
| 0.800 | 0.57561 | 0.80041 | 1.23640 | 0.57600 | 0.80052 | 1.23659 | 0.800 |
| 0.820 | 0.59546 | 0.82040 | 1.25806 | 0.59583 | 0.82050 | 1.25825 | 0.820 |
| 0.840 | 0.61536 | 0.84039 | 1.27883 | 0.61571 | 0.84050 | 1.27900 | 0.840 |
| 0.860 | 0.63529 | 0.86038 | 1.29869 | 0.63562 | 0.86048 | 1.29887 | 0.860 |
| 0.880 | 0.65525 | 0.88037 | 1.31768 | 0.65558 | 0.88047 | 1.31783 | 0.880 |
| 0.900 | 0.67525 | 0.90036 | 1.33578 | 0.67557 | 0.90046 | 1.33594 | 0.900 |
| 0.920 | 0.69527 | 0.92036 | 1.35305 | 0.69558 | 0.92045 | 1.35318 | 0.920 |
| 0.940 | 0.71531 | 0.94035 | 1.36949 | 0.71561 | 0.94044 | 1.36960 | 0.940 |
| 0.960 | 0.73537 | 0.96034 | 1.38515 | 0.73567 | 0.96043 | 1.38523 | 0.960 |
| 0.980 | 0.75545 | 0.98033 | 1.40005 | 0.75574 | 0.98043 | 1.40010 | 0.980 |
| 1.000 | 0.77555 | 1.00033 | 1.41440 | 0.77582 | 1.00042 | 1.41444 | 1.000 |

$H/R = 0.100 \qquad n = 1.$

| $H/L$ | $\Omega_1$ | $\Omega_2$ | $\Omega_3$ | $\Omega_4$ | $\Omega_5$ | $\Omega_6$ |
|---|---|---|---|---|---|---|
| 0. | 0. | 0.03184 | 0.07598 | 1.00089 | 1.00366 | 1.86911 |
| 0.001 | 0.0000358 | 0.03188 | 0.07599 | 1.00089 | 1.00366 | 1.86910 |
| 0.005 | 0.000867 | 0.03283 | 0.07620 | 1.00092 | 1.00369 | 1.86904 |
| 0.010 | 0.00318 | 0.03547 | 0.07685 | 1.00101 | 1.00377 | 1.86886 |
| 0.020 | 0.01004 | 0.04340 | 0.07966 | 1.00134 | 1.00410 | 1.86813 |
| 0.030 | 0.01764 | 0.05212 | 0.08505 | 1.00184 | 1.00471 | 1.86692 |
| 0.040 | 0.02480 | 0.06004 | 0.09351 | 1.00244 | 1.00565 | 1.86528 |
| 0.050 | 0.03100 | 0.06715 | 0.10475 | 1.00309 | 1.00698 | 1.86326 |
| 0.060 | 0.03606 | 0.07415 | 0.11790 | 1.00379 | 1.00871 | 1.86088 |
| 0.070 | 0.04002 | 0.08158 | 0.13220 | 1.00454 | 1.01080 | 1.85820 |
| 0.080 | 0.04312 | 0.08957 | 0.14721 | 1.00536 | 1.01327 | 1.85528 |
| 0.090 | 0.04563 | 0.09804 | 0.16266 | 1.00626 | 1.01607 | 1.85212 |
| 0.100 | 0.04782 | 0.10689 | 0.17842 | 1.00723 | 1.01920 | 1.84880 |
| 0.120 | 0.05194 | 0.12532 | 0.21049 | 1.00947 | 1.02643 | 1.84176 |
| 0.140 | 0.05648 | 0.14432 | 0.24301 | 1.01207 | 1.03489 | 1.83439 |
| 0.160 | 0.06185 | 0.16365 | 0.27578 | 1.01505 | 1.04450 | 1.82684 |
| 0.180 | 0.06841 | 0.18315 | 0.30870 | 1.01841 | 1.05522 | 1.81927 |
| 0.200 | 0.07618 | 0.20278 | 0.34168 | 1.02213 | 1.06698 | 1.81178 |
| 0.220 | 0.08513 | 0.22249 | 0.37470 | 1.02624 | 1.07972 | 1.80443 |
| 0.240 | 0.09521 | 0.24226 | 0.40771 | 1.03072 | 1.09339 | 1.79729 |
| 0.260 | 0.10631 | 0.26207 | 0.44069 | 1.03556 | 1.10791 | 1.79041 |
| 0.280 | 0.11834 | 0.28191 | 0.47362 | 1.04077 | 1.12324 | 1.78383 |
| 0.300 | 0.13120 | 0.30177 | 0.50648 | 1.04633 | 1.13932 | 1.77758 |
| 0.320 | 0.14481 | 0.32165 | 0.53924 | 1.05224 | 1.15610 | 1.77169 |
| 0.340 | 0.15909 | 0.34155 | 0.57190 | 1.05849 | 1.17352 | 1.76621 |
| 0.360 | 0.17397 | 0.36145 | 0.60443 | 1.06509 | 1.19155 | 1.76113 |
| 0.380 | 0.18939 | 0.38137 | 0.63683 | 1.07202 | 1.21012 | 1.75652 |
| 0.400 | 0.20529 | 0.40130 | 0.66905 | 1.07927 | 1.22922 | 1.75237 |
| 0.420 | 0.22163 | 0.42124 | 0.70110 | 1.08684 | 1.24879 | 1.74872 |
| 0.440 | 0.23836 | 0.44118 | 0.73295 | 1.09473 | 1.26880 | 1.74559 |
| 0.460 | 0.25544 | 0.46113 | 0.76458 | 1.10292 | 1.28921 | 1.74301 |
| 0.480 | 0.27285 | 0.48108 | 0.79596 | 1.11141 | 1.30999 | 1.74100 |
| 0.500 | 0.29054 | 0.50103 | 0.82707 | 1.12019 | 1.33110 | 1.73959 |
| 0.520 | 0.30849 | 0.52099 | 0.85788 | 1.12927 | 1.35254 | 1.73883 |
| 0.540 | 0.32667 | 0.54095 | 0.88836 | 1.13861 | 1.37425 | 1.73873 |
| 0.560 | 0.34507 | 0.56092 | 0.91848 | 1.14823 | 1.39624 | 1.73932 |
| 0.580 | 0.36365 | 0.58089 | 0.94819 | 1.15812 | 1.41845 | 1.74065 |
| 0.600 | 0.38241 | 0.60086 | 0.97748 | 1.16827 | 1.44088 | 1.74276 |
| 0.620 | 0.40132 | 0.62083 | 1.00629 | 1.17866 | 1.46351 | 1.74567 |
| 0.640 | 0.42038 | 0.64081 | 1.03457 | 1.18931 | 1.48632 | 1.74945 |
| 0.660 | 0.43956 | 0.66078 | 1.06231 | 1.20019 | 1.50928 | 1.75414 |
| 0.680 | 0.45885 | 0.68076 | 1.08942 | 1.21131 | 1.53237 | 1.75976 |
| 0.700 | 0.47824 | 0.70074 | 1.11589 | 1.22264 | 1.55560 | 1.76638 |
| 0.720 | 0.49773 | 0.72072 | 1.14165 | 1.23421 | 1.57892 | 1.77404 |
| 0.740 | 0.51730 | 0.74069 | 1.16665 | 1.24599 | 1.60234 | 1.78277 |
| 0.760 | 0.53694 | 0.76068 | 1.19087 | 1.25797 | 1.62582 | 1.79264 |
| 0.780 | 0.55665 | 0.78066 | 1.21427 | 1.27016 | 1.64936 | 1.80365 |
| 0.800 | 0.57642 | 0.80064 | 1.23679 | 1.28254 | 1.67295 | 1.81586 |
| 0.820 | 0.59623 | 0.82062 | 1.25845 | 1.29513 | 1.69657 | 1.82927 |
| 0.840 | 0.61610 | 0.84061 | 1.27919 | 1.30789 | 1.72021 | 1.84389 |
| 0.860 | 0.63600 | 0.86060 | 1.29906 | 1.32083 | 1.74385 | 1.85972 |
| 0.880 | 0.65595 | 0.88058 | 1.31801 | 1.33396 | 1.76747 | 1.87675 |
| 0.900 | 0.67592 | 0.90057 | 1.33609 | 1.34727 | 1.79108 | 1.89497 |
| 0.920 | 0.69592 | 0.92056 | 1.35332 | 1.36076 | 1.81464 | 1.91431 |
| 0.940 | 0.71595 | 0.94055 | 1.36972 | 1.37446 | 1.83816 | 1.93475 |
| 0.960 | 0.73599 | 0.96053 | 1.38530 | 1.38838 | 1.86163 | 1.95622 |
| 0.980 | 0.75605 | 0.98052 | 1.40015 | 1.40250 | 1.88501 | 1.97867 |
| 1.000 | 0.77613 | 1.00051 | 1.41448 | 1.41664 | 1.90832 | 2.00203 |

$H/R = 0.120$ $n = 1.$ $H/R = 0.140$

| $H/L$ | $\Omega_1$ | $\Omega_2$ | $\Omega_3$ | $\Omega_1$ | $\Omega_2$ | $\Omega_3$ | $H/L$ |
|---|---|---|---|---|---|---|---|
| 0. | 0. | 0.03822 | 0.09112 | 0. | 0.04460 | 0.10623 | 0. |
| 0.001 | 0.0000299 | 0.03825 | 0.09113 | 0.0000256 | 0.04463 | 0.10624 | 0.001 |
| 0.005 | 0.000730 | 0.03905 | 0.09130 | 0.000630 | 0.04531 | 0.10638 | 0.005 |
| 0.010 | 0.00274 | 0.04133 | 0.09184 | 0.00240 | 0.04732 | 0.10684 | 0.010 |
| 0.020 | 0.00910 | 0.04866 | 0.09411 | 0.00827 | 0.05406 | 0.10875 | 0.020 |
| 0.030 | 0.01663 | 0.05737 | 0.09833 | 0.01560 | 0.06254 | 0.11222 | 0.030 |
| 0.040 | 0.02416 | 0.06586 | 0.10498 | 0.02324 | 0.07128 | 0.11760 | 0.040 |
| 0.050 | 0.03114 | 0.07352 | 0.11421 | 0.03064 | 0.07950 | 0.12519 | 0.050 |
| 0.060 | 0.03730 | 0.08058 | 0.12566 | 0.03750 | 0.08699 | 0.13496 | 0.060 |
| 0.070 | 0.04252 | 0.08757 | 0.13869 | 0.04364 | 0.09402 | 0.14655 | 0.070 |
| 0.080 | 0.04683 | 0.09486 | 0.15277 | 0.04901 | 0.10099 | 0.15950 | 0.080 |
| 0.090 | 0.05039 | 0.10262 | 0.16752 | 0.05361 | 0.10820 | 0.17338 | 0.090 |
| 0.100 | 0.05340 | 0.11082 | 0.18273 | 0.05756 | 0.11579 | 0.18792 | 0.100 |
| 0.120 | 0.05853 | 0.12828 | 0.21402 | 0.06409 | 0.13213 | 0.21825 | 0.120 |
| 0.140 | 0.06343 | 0.14665 | 0.24600 | 0.06978 | 0.14967 | 0.24958 | 0.140 |
| 0.160 | 0.06883 | 0.16554 | 0.27839 | 0.07549 | 0.16799 | 0.28148 | 0.160 |
| 0.180 | 0.07510 | 0.18475 | 0.31100 | 0.08175 | 0.18680 | 0.31373 | 0.180 |
| 0.200 | 0.08248 | 0.20416 | 0.34375 | 0.08896 | 0.20591 | 0.34620 | 0.200 |
| 0.220 | 0.09097 | 0.22371 | 0.37657 | 0.09707 | 0.22524 | 0.37879 | 0.220 |
| 0.240 | 0.10057 | 0.24334 | 0.40943 | 0.10630 | 0.24470 | 0.41146 | 0.240 |
| 0.260 | 0.11121 | 0.26305 | 0.44227 | 0.11653 | 0.26427 | 0.44414 | 0.260 |
| 0.280 | 0.12281 | 0.28280 | 0.47509 | 0.12773 | 0.28391 | 0.47682 | 0.280 |
| 0.300 | 0.13529 | 0.30259 | 0.50785 | 0.13983 | 0.30361 | 0.50946 | 0.300 |
| 0.320 | 0.14855 | 0.32242 | 0.54052 | 0.15275 | 0.32335 | 0.54203 | 0.320 |
| 0.340 | 0.16252 | 0.34226 | 0.57310 | 0.16640 | 0.34313 | 0.57453 | 0.340 |
| 0.360 | 0.17713 | 0.36212 | 0.60557 | 0.18072 | 0.36294 | 0.60691 | 0.360 |
| 0.380 | 0.19231 | 0.38200 | 0.63790 | 0.19564 | 0.38277 | 0.63917 | 0.380 |
| 0.400 | 0.20800 | 0.40190 | 0.67007 | 0.21110 | 0.40262 | 0.67128 | 0.400 |
| 0.420 | 0.22415 | 0.42180 | 0.70208 | 0.22704 | 0.42248 | 0.70322 | 0.420 |
| 0.440 | 0.24071 | 0.44171 | 0.73388 | 0.24341 | 0.44236 | 0.73497 | 0.440 |
| 0.460 | 0.25764 | 0.46164 | 0.76547 | 0.26018 | 0.46225 | 0.76651 | 0.460 |
| 0.480 | 0.27491 | 0.48156 | 0.79681 | 0.27729 | 0.48215 | 0.79780 | 0.480 |
| 0.500 | 0.29247 | 0.50150 | 0.82788 | 0.29472 | 0.50206 | 0.82883 | 0.500 |
| 0.520 | 0.31031 | 0.52144 | 0.85865 | 0.31243 | 0.52198 | 0.85957 | 0.520 |
| 0.540 | 0.32840 | 0.54138 | 0.88911 | 0.33040 | 0.54190 | 0.88999 | 0.540 |
| 0.560 | 0.34670 | 0.56133 | 0.91920 | 0.34860 | 0.56183 | 0.92004 | 0.560 |
| 0.580 | 0.36520 | 0.58129 | 0.94889 | 0.36701 | 0.58176 | 0.94970 | 0.580 |
| 0.600 | 0.38389 | 0.60124 | 0.97815 | 0.38560 | 0.60170 | 0.97894 | 0.600 |
| 0.620 | 0.40273 | 0.62120 | 1.00693 | 0.40437 | 0.62165 | 1.00769 | 0.620 |
| 0.640 | 0.42172 | 0.64116 | 1.03520 | 0.42328 | 0.64159 | 1.03593 | 0.640 |
| 0.660 | 0.44084 | 0.66113 | 1.06291 | 0.44233 | 0.66154 | 1.06362 | 0.660 |
| 0.680 | 0.46007 | 0.68109 | 1.09001 | 0.46151 | 0.68150 | 1.09069 | 0.680 |
| 0.700 | 0.47942 | 0.70106 | 1.11645 | 0.48079 | 0.70145 | 1.11711 | 0.700 |
| 0.720 | 0.49886 | 0.72103 | 1.14219 | 0.50018 | 0.72141 | 1.14283 | 0.720 |
| 0.740 | 0.51838 | 0.74100 | 1.16718 | 0.51965 | 0.74137 | 1.16780 | 0.740 |
| 0.760 | 0.53798 | 0.76097 | 1.19139 | 0.53921 | 0.76133 | 1.19199 | 0.760 |
| 0.780 | 0.55766 | 0.78095 | 1.21477 | 0.55883 | 0.78130 | 1.21533 | 0.780 |
| 0.800 | 0.57739 | 0.80092 | 1.23728 | 0.57852 | 0.80127 | 1.23783 | 0.800 |
| 0.820 | 0.59717 | 0.82091 | 1.25891 | 0.59827 | 0.82123 | 1.25943 | 0.820 |
| 0.840 | 0.61701 | 0.84088 | 1.27964 | 0.61807 | 0.84120 | 1.28014 | 0.840 |
| 0.860 | 0.63688 | 0.86086 | 1.29947 | 0.63791 | 0.86117 | 1.29994 | 0.860 |
| 0.880 | 0.65680 | 0.88084 | 1.31840 | 0.65779 | 0.88115 | 1.31883 | 0.880 |
| 0.900 | 0.67674 | 0.90082 | 1.33645 | 0.67771 | 0.90112 | 1.33682 | 0.900 |
| 0.920 | 0.69672 | 0.92080 | 1.35362 | 0.69766 | 0.92110 | 1.35394 | 0.920 |
| 0.940 | 0.71672 | 0.94079 | 1.36995 | 0.71763 | 0.94107 | 1.37018 | 0.940 |
| 0.960 | 0.73675 | 0.96077 | 1.38546 | 0.73763 | 0.96105 | 1.38562 | 0.960 |
| 0.980 | 0.75679 | 0.98076 | 1.40028 | 0.75765 | 0.98103 | 1.40041 | 0.980 |
| 1.000 | 0.77684 | 1.00074 | 1.41461 | 0.77768 | 1.00101 | 1.41474 | 1.000 |

$H/R = 0.160$ $n = 1.$ $H/R = 0.180$

| $H/L$ | $\Omega_1$ | $\Omega_2$ | $\Omega_3$ | $\Omega_1$ | $\Omega_2$ | $\Omega_3$ | $H/L$ |
|---|---|---|---|---|---|---|---|
| 0. | 0. | 0.05098 | 0.12130 | 0. | 0.05737 | 0.13633 | 0. |
| 0.001 | 0.0000224 | 0.05101 | 0.12131 | 0.0000200 | 0.05740 | 0.13634 | 0.001 |
| 0.005 | 0.000554 | 0.05161 | 0.12144 | 0.000494 | 0.05793 | 0.13645 | 0.005 |
| 0.010 | 0.00214 | 0.05340 | 0.12184 | 0.00192 | 0.05954 | 0.13681 | 0.010 |
| 0.020 | 0.00755 | 0.05959 | 0.12349 | 0.00693 | 0.06525 | 0.13826 | 0.020 |
| 0.030 | 0.01461 | 0.06774 | 0.12643 | 0.01369 | 0.07304 | 0.14082 | 0.030 |
| 0.040 | 0.02223 | 0.07652 | 0.13093 | 0.02119 | 0.08170 | 0.14468 | 0.040 |
| 0.050 | 0.02983 | 0.08509 | 0.13726 | 0.02888 | 0.09045 | 0.15007 | 0.050 |
| 0.060 | 0.03712 | 0.09307 | 0.14556 | 0.03641 | 0.09882 | 0.15717 | 0.060 |
| 0.070 | 0.04389 | 0.10044 | 0.15572 | 0.04360 | 0.10661 | 0.16602 | 0.070 |
| 0.080 | 0.05005 | 0.10746 | 0.16742 | 0.05033 | 0.11390 | 0.17647 | 0.080 |
| 0.090 | 0.05554 | 0.11442 | 0.18030 | 0.05652 | 0.12089 | 0.18825 | 0.090 |
| 0.100 | 0.06039 | 0.12157 | 0.19403 | 0.06214 | 0.12785 | 0.20108 | 0.100 |
| 0.120 | 0.06850 | 0.13685 | 0.22320 | 0.07178 | 0.14234 | 0.22889 | 0.120 |
| 0.140 | 0.07528 | 0.15346 | 0.25374 | 0.07982 | 0.15801 | 0.25851 | 0.140 |
| 0.160 | 0.08160 | 0.17107 | 0.28508 | 0.08700 | 0.17482 | 0.28919 | 0.160 |
| 0.180 | 0.08813 | 0.18935 | 0.31690 | 0.09402 | 0.19248 | 0.32052 | 0.180 |
| 0.200 | 0.09527 | 0.20809 | 0.34904 | 0.10137 | 0.21073 | 0.35226 | 0.200 |
| 0.220 | 0.10328 | 0.22712 | 0.38136 | 0.10935 | 0.22940 | 0.38428 | 0.220 |
| 0.240 | 0.11222 | 0.24636 | 0.41380 | 0.11814 | 0.24836 | 0.41646 | 0.240 |
| 0.260 | 0.12210 | 0.26575 | 0.44630 | 0.12785 | 0.26753 | 0.44875 | 0.260 |
| 0.280 | 0.13297 | 0.28525 | 0.47882 | 0.13841 | 0.28685 | 0.48108 | 0.280 |
| 0.300 | 0.14473 | 0.30483 | 0.51132 | 0.14985 | 0.30628 | 0.51343 | 0.300 |
| 0.320 | 0.15730 | 0.32448 | 0.54378 | 0.16214 | 0.32581 | 0.54575 | 0.320 |
| 0.340 | 0.17065 | 0.34417 | 0.57616 | 0.17519 | 0.34540 | 0.57801 | 0.340 |
| 0.360 | 0.18468 | 0.36391 | 0.60845 | 0.18894 | 0.36505 | 0.61020 | 0.360 |
| 0.380 | 0.19933 | 0.38368 | 0.64063 | 0.20333 | 0.38474 | 0.64228 | 0.380 |
| 0.400 | 0.21455 | 0.40347 | 0.67266 | 0.21831 | 0.40447 | 0.67422 | 0.400 |
| 0.420 | 0.23027 | 0.42329 | 0.70453 | 0.23380 | 0.42423 | 0.70602 | 0.420 |
| 0.440 | 0.24644 | 0.44312 | 0.73622 | 0.24977 | 0.44401 | 0.73764 | 0.440 |
| 0.460 | 0.26303 | 0.46298 | 0.76770 | 0.26616 | 0.46382 | 0.76906 | 0.460 |
| 0.480 | 0.27997 | 0.48284 | 0.79895 | 0.28294 | 0.48364 | 0.80024 | 0.480 |
| 0.500 | 0.29725 | 0.50272 | 0.82993 | 0.30006 | 0.50349 | 0.83117 | 0.500 |
| 0.520 | 0.31483 | 0.52261 | 0.86063 | 0.31749 | 0.52334 | 0.86181 | 0.520 |
| 0.540 | 0.33267 | 0.54251 | 0.89100 | 0.33520 | 0.54320 | 0.89214 | 0.540 |
| 0.560 | 0.35076 | 0.56241 | 0.92102 | 0.35316 | 0.56308 | 0.92211 | 0.560 |
| 0.580 | 0.36906 | 0.58232 | 0.95064 | 0.37135 | 0.58297 | 0.95169 | 0.580 |
| 0.600 | 0.38756 | 0.60224 | 0.97984 | 0.38974 | 0.60286 | 0.98085 | 0.600 |
| 0.620 | 0.40623 | 0.62216 | 1.00856 | 0.40831 | 0.62277 | 1.00954 | 0.620 |
| 0.640 | 0.42507 | 0.64209 | 1.03678 | 0.42706 | 0.64267 | 1.03772 | 0.640 |
| 0.660 | 0.44404 | 0.66203 | 1.06443 | 0.44595 | 0.66259 | 1.06533 | 0.660 |
| 0.680 | 0.46315 | 0.68197 | 1.09148 | 0.46497 | 0.68251 | 1.09235 | 0.680 |
| 0.700 | 0.48236 | 0.70191 | 1.11786 | 0.48412 | 0.70243 | 1.11871 | 0.700 |
| 0.720 | 0.50169 | 0.72185 | 1.14356 | 0.50338 | 0.72236 | 1.14437 | 0.720 |
| 0.740 | 0.52111 | 0.74180 | 1.16851 | 0.52273 | 0.74229 | 1.16928 | 0.740 |
| 0.760 | 0.54061 | 0.76175 | 1.19266 | 0.54218 | 0.76223 | 1.19340 | 0.760 |
| 0.780 | 0.56018 | 0.78171 | 1.21598 | 0.56170 | 0.78217 | 1.21669 | 0.780 |
| 0.800 | 0.57983 | 0.80166 | 1.23845 | 0.58129 | 0.80212 | 1.23912 | 0.800 |
| 0.820 | 0.59953 | 0.82162 | 1.26001 | 0.60094 | 0.82206 | 1.26065 | 0.820 |
| 0.840 | 0.61929 | 0.84158 | 1.28069 | 0.62065 | 0.84201 | 1.28127 | 0.840 |
| 0.860 | 0.63909 | 0.86154 | 1.30044 | 0.64041 | 0.86196 | 1.30098 | 0.860 |
| 0.880 | 0.65893 | 0.88151 | 1.31929 | 0.66022 | 0.88192 | 1.31976 | 0.880 |
| 0.900 | 0.67882 | 0.90147 | 1.33721 | 0.68006 | 0.90187 | 1.33761 | 0.900 |
| 0.920 | 0.69873 | 0.92144 | 1.35425 | 0.69994 | 0.92183 | 1.35456 | 0.920 |
| 0.940 | 0.71868 | 0.94141 | 1.37041 | 0.71985 | 0.94179 | 1.37065 | 0.940 |
| 0.960 | 0.73864 | 0.96138 | 1.38580 | 0.73978 | 0.96175 | 1.38599 | 0.960 |
| 0.980 | 0.75863 | 0.98135 | 1.40057 | 0.75974 | 0.98172 | 1.40075 | 0.980 |
| 1.000 | 0.77864 | 1.00133 | 1.41491 | 0.77972 | 1.00168 | 1.41509 | 1.000 |

$H/R = 0.200 \qquad n = 1.$

| $H/L$ | $\Omega_1$ | $\Omega_2$ | $\Omega_3$ | $\Omega_4$ | $\Omega_5$ | $\Omega_6$ |
|---|---|---|---|---|---|---|
| 0. | 0. | 0.06377 | 0.15131 | 1.00359 | 1.01482 | 1.86454 |
| 0.001 | 0.0000180 | 0.06379 | 0.15132 | 1.00359 | 1.01483 | 1.86454 |
| 0.005 | 0.000446 | 0.06427 | 0.15142 | 1.00362 | 1.01485 | 1.86449 |
| 0.010 | 0.00174 | 0.06573 | 0.15174 | 1.00371 | 1.01492 | 1.86432 |
| 0.020 | 0.00639 | 0.07102 | 0.15304 | 1.00405 | 1.01525 | 1.86367 |
| 0.030 | 0.01285 | 0.07842 | 0.15531 | 1.00461 | 1.01579 | 1.86260 |
| 0.040 | 0.02019 | 0.08689 | 0.15869 | 1.00537 | 1.01657 | 1.86114 |
| 0.050 | 0.02786 | 0.09568 | 0.16337 | 1.00629 | 1.01761 | 1.85933 |
| 0.060 | 0.03553 | 0.10430 | 0.16951 | 1.00735 | 1.01895 | 1.85718 |
| 0.070 | 0.04299 | 0.11247 | 0.17724 | 1.00851 | 1.02062 | 1.85475 |
| 0.080 | 0.05012 | 0.12012 | 0.18652 | 1.00975 | 1.02265 | 1.85206 |
| 0.090 | 0.05682 | 0.12735 | 0.19720 | 1.01103 | 1.02503 | 1.84916 |
| 0.100 | 0.06306 | 0.13434 | 0.20906 | 1.01235 | 1.02780 | 1.84609 |
| 0.120 | 0.07407 | 0.14838 | 0.23535 | 1.01513 | 1.03442 | 1.83953 |
| 0.140 | 0.08339 | 0.16326 | 0.26391 | 1.01810 | 1.04243 | 1.83258 |
| 0.160 | 0.09159 | 0.17924 | 0.29383 | 1.02133 | 1.05171 | 1.82543 |
| 0.180 | 0.09930 | 0.19620 | 0.32458 | 1.02486 | 1.06218 | 1.81821 |
| 0.200 | 0.10704 | 0.21389 | 0.35588 | 1.02871 | 1.07372 | 1.81101 |
| 0.220 | 0.11518 | 0.23212 | 0.38754 | 1.03290 | 1.08628 | 1.80394 |
| 0.240 | 0.12397 | 0.25073 | 0.41944 | 1.03743 | 1.09979 | 1.79704 |
| 0.260 | 0.13352 | 0.26963 | 0.45148 | 1.04231 | 1.11417 | 1.79038 |
| 0.280 | 0.14392 | 0.28873 | 0.48361 | 1.04753 | 1.12937 | 1.78399 |
| 0.300 | 0.15516 | 0.30799 | 0.51578 | 1.05309 | 1.14534 | 1.77791 |
| 0.320 | 0.16717 | 0.32736 | 0.54795 | 1.05900 | 1.16200 | 1.77219 |
| 0.340 | 0.17994 | 0.34683 | 0.58008 | 1.06524 | 1.17931 | 1.76684 |
| 0.360 | 0.19344 | 0.36637 | 0.61214 | 1.07183 | 1.19723 | 1.76191 |
| 0.380 | 0.20759 | 0.38597 | 0.64411 | 1.07873 | 1.21572 | 1.75741 |
| 0.400 | 0.22232 | 0.40562 | 0.67597 | 1.08595 | 1.23471 | 1.75336 |
| 0.420 | 0.23759 | 0.42531 | 0.70767 | 1.09350 | 1.25419 | 1.74982 |
| 0.440 | 0.25335 | 0.44503 | 0.73922 | 1.10135 | 1.27410 | 1.74678 |
| 0.460 | 0.26955 | 0.46478 | 0.77056 | 1.10951 | 1.29443 | 1.74428 |
| 0.480 | 0.28615 | 0.48456 | 0.80168 | 1.11796 | 1.31513 | 1.74235 |
| 0.500 | 0.30310 | 0.50436 | 0.83254 | 1.12671 | 1.33616 | 1.74102 |
| 0.520 | 0.32038 | 0.52417 | 0.86313 | 1.13574 | 1.35752 | 1.74032 |
| 0.540 | 0.33795 | 0.54400 | 0.89340 | 1.14505 | 1.37917 | 1.74028 |
| 0.560 | 0.35578 | 0.56385 | 0.92332 | 1.15463 | 1.40107 | 1.74093 |
| 0.580 | 0.37384 | 0.58370 | 0.95286 | 1.16448 | 1.42322 | 1.74232 |
| 0.600 | 0.39213 | 0.60357 | 0.98197 | 1.17459 | 1.44558 | 1.74447 |
| 0.620 | 0.41060 | 0.62344 | 1.01062 | 1.18494 | 1.46815 | 1.74744 |
| 0.640 | 0.42924 | 0.64333 | 1.03876 | 1.19555 | 1.49088 | 1.75126 |
| 0.660 | 0.44805 | 0.66322 | 1.06634 | 1.20640 | 1.51377 | 1.75598 |
| 0.680 | 0.46699 | 0.68312 | 1.09330 | 1.21748 | 1.53681 | 1.76165 |
| 0.700 | 0.48606 | 0.70302 | 1.11963 | 1.22879 | 1.55996 | 1.76830 |
| 0.720 | 0.50524 | 0.72294 | 1.14525 | 1.24032 | 1.58322 | 1.77598 |
| 0.740 | 0.52453 | 0.74285 | 1.17013 | 1.25209 | 1.60658 | 1.78476 |
| 0.760 | 0.54391 | 0.76277 | 1.19421 | 1.26406 | 1.63000 | 1.79466 |
| 0.780 | 0.56337 | 0.78270 | 1.21746 | 1.27625 | 1.65347 | 1.80571 |
| 0.800 | 0.58290 | 0.80263 | 1.23984 | 1.28864 | 1.67700 | 1.81794 |
| 0.820 | 0.60250 | 0.82256 | 1.26133 | 1.30126 | 1.70054 | 1.83138 |
| 0.840 | 0.62217 | 0.84250 | 1.28190 | 1.31408 | 1.72410 | 1.84603 |
| 0.860 | 0.64188 | 0.86243 | 1.30153 | 1.32713 | 1.74765 | 1.86191 |
| 0.880 | 0.66164 | 0.88238 | 1.32024 | 1.34040 | 1.77120 | 1.87898 |
| 0.900 | 0.68144 | 0.90232 | 1.33800 | 1.35392 | 1.79472 | 1.89723 |
| 0.920 | 0.70128 | 0.92227 | 1.35487 | 1.36768 | 1.81819 | 1.91661 |
| 0.940 | 0.72115 | 0.94222 | 1.37089 | 1.38165 | 1.84162 | 1.93709 |
| 0.960 | 0.74105 | 0.96217 | 1.38620 | 1.39575 | 1.86499 | 1.95859 |
| 0.980 | 0.76097 | 0.98213 | 1.40096 | 1.40988 | 1.88828 | 1.98108 |
| 1.000 | 0.78092 | 1.00209 | 1.41530 | 1.42393 | 1.91150 | 2.00448 |

$H/R = 0.250$ $n = 1.$

| $H/L$ | $\Omega_1$ | $\Omega_2$ | $\Omega_3$ |
|---|---|---|---|
| 0. | 0. | 0.07978 | 0.18850 |
| 0.001 | 0.0000144 | 0.07980 | 0.18851 |
| 0.005 | 0.000359 | 0.08018 | 0.18859 |
| 0.010 | 0.00141 | 0.08137 | 0.18884 |
| 0.020 | 0.00533 | 0.08580 | 0.18987 |
| 0.030 | 0.01106 | 0.09232 | 0.19164 |
| 0.040 | 0.01789 | 0.10016 | 0.19423 |
| 0.050 | 0.02533 | 0.10869 | 0.19774 |
| 0.060 | 0.03303 | 0.11748 | 0.20230 |
| 0.070 | 0.04079 | 0.12619 | 0.20801 |
| 0.080 | 0.04845 | 0.13459 | 0.21496 |
| 0.090 | 0.05592 | 0.14260 | 0.22320 |
| 0.100 | 0.06313 | 0.15019 | 0.23266 |
| 0.120 | 0.07659 | 0.16447 | 0.25480 |
| 0.140 | 0.08869 | 0.17842 | 0.28019 |
| 0.160 | 0.09953 | 0.19282 | 0.30776 |
| 0.180 | 0.10942 | 0.20804 | 0.33676 |
| 0.200 | 0.11876 | 0.22415 | 0.36670 |
| 0.220 | 0.12795 | 0.24102 | 0.39727 |
| 0.240 | 0.13731 | 0.25852 | 0.42828 |
| 0.260 | 0.14708 | 0.27650 | 0.45959 |
| 0.280 | 0.15739 | 0.29486 | 0.49110 |
| 0.300 | 0.16835 | 0.31350 | 0.52274 |
| 0.320 | 0.17998 | 0.33237 | 0.55444 |
| 0.340 | 0.19231 | 0.35141 | 0.58617 |
| 0.360 | 0.20529 | 0.37059 | 0.61787 |
| 0.380 | 0.21894 | 0.38988 | 0.64953 |
| 0.400 | 0.23315 | 0.40926 | 0.68109 |
| 0.420 | 0.24791 | 0.42872 | 0.71254 |
| 0.440 | 0.26317 | 0.44824 | 0.74384 |
| 0.460 | 0.27889 | 0.46781 | 0.77497 |
| 0.480 | 0.29505 | 0.48742 | 0.80588 |
| 0.500 | 0.31160 | 0.50708 | 0.83657 |
| 0.520 | 0.32849 | 0.52676 | 0.86698 |
| 0.540 | 0.34569 | 0.54647 | 0.89709 |
| 0.560 | 0.36318 | 0.56621 | 0.92686 |
| 0.580 | 0.38093 | 0.58596 | 0.95625 |
| 0.600 | 0.39891 | 0.60574 | 0.98523 |
| 0.620 | 0.41711 | 0.62553 | 1.01374 |
| 0.640 | 0.43549 | 0.64534 | 1.04175 |
| 0.660 | 0.45405 | 0.66516 | 1.06921 |
| 0.680 | 0.47277 | 0.68499 | 1.09607 |
| 0.700 | 0.49162 | 0.70484 | 1.12227 |
| 0.720 | 0.51060 | 0.72469 | 1.14776 |
| 0.740 | 0.52970 | 0.74455 | 1.17252 |
| 0.760 | 0.54890 | 0.76442 | 1.19647 |
| 0.780 | 0.56820 | 0.78430 | 1.21959 |
| 0.800 | 0.58757 | 0.80418 | 1.24182 |
| 0.820 | 0.60703 | 0.82407 | 1.26314 |
| 0.840 | 0.62655 | 0.84397 | 1.28353 |
| 0.860 | 0.64613 | 0.86387 | 1.30297 |
| 0.880 | 0.66576 | 0.88378 | 1.32144 |
| 0.900 | 0.68544 | 0.90369 | 1.33899 |
| 0.920 | 0.70517 | 0.92360 | 1.35566 |
| 0.940 | 0.72494 | 0.94352 | 1.37156 |
| 0.960 | 0.74473 | 0.96344 | 1.38681 |
| 0.980 | 0.76455 | 0.98337 | 1.40155 |
| 1.000 | 0.78441 | 1.00330 | 1.41591 |

$H/R = 0.300 \qquad n = 1.$

| $H/L$ | $\Omega_1$ | $\Omega_2$ | $\Omega_3$ | $\Omega_4$ | $\Omega_5$ | $\Omega_6$ |
|---|---|---|---|---|---|---|
| 0. | 0. | 0.09584 | 0.22524 | 1.00819 | 1.03403 | 1.85845 |
| 0.001 | 0.0000122 | 0.09586 | 0.22524 | 1.00819 | 1.03404 | 1.85844 |
| 0.005 | 0.000301 | 0.09618 | 0.22531 | 1.00822 | 1.03406 | 1.85840 |
| 0.010 | 0.00119 | 0.09718 | 0.22552 | 1.00830 | 1.03413 | 1.85825 |
| 0.020 | 0.00456 | 0.10097 | 0.22638 | 1.00865 | 1.03444 | 1.85770 |
| 0.030 | 0.00965 | 0.10673 | 0.22783 | 1.00922 | 1.03496 | 1.85679 |
| 0.040 | 0.01595 | 0.11390 | 0.22994 | 1.01000 | 1.03570 | 1.85553 |
| 0.050 | 0.02301 | 0.12197 | 0.23276 | 1.01098 | 1.03667 | 1.85397 |
| 0.060 | 0.03053 | 0.13054 | 0.23636 | 1.01215 | 1.03787 | 1.85210 |
| 0.070 | 0.03828 | 0.13932 | 0.24084 | 1.01348 | 1.03934 | 1.84996 |
| 0.080 | 0.04610 | 0.14807 | 0.24628 | 1.01496 | 1.04108 | 1.84761 |
| 0.090 | 0.05389 | 0.15661 | 0.25274 | 1.01656 | 1.04311 | 1.84503 |
| 0.100 | 0.06158 | 0.16484 | 0.26027 | 1.01826 | 1.04544 | 1.84229 |
| 0.120 | 0.07643 | 0.18027 | 0.27845 | 1.02190 | 1.05111 | 1.83636 |
| 0.140 | 0.09037 | 0.19462 | 0.30034 | 1.02574 | 1.05812 | 1.83002 |
| 0.160 | 0.10332 | 0.20856 | 0.32507 | 1.02975 | 1.06650 | 1.82343 |
| 0.180 | 0.11535 | 0.22271 | 0.35185 | 1.03391 | 1.07617 | 1.81672 |
| 0.200 | 0.12663 | 0.23746 | 0.38005 | 1.03827 | 1.08706 | 1.80998 |
| 0.220 | 0.13741 | 0.25293 | 0.40924 | 1.04285 | 1.09905 | 1.80331 |
| 0.240 | 0.14798 | 0.26912 | 0.43913 | 1.04768 | 1.11209 | 1.79678 |
| 0.260 | 0.15857 | 0.28594 | 0.46951 | 1.05278 | 1.12605 | 1.79046 |
| 0.280 | 0.16940 | 0.30330 | 0.50025 | 1.05817 | 1.14090 | 1.78437 |
| 0.300 | 0.18060 | 0.32109 | 0.53121 | 1.06387 | 1.15653 | 1.77858 |
| 0.320 | 0.19228 | 0.33924 | 0.56234 | 1.06987 | 1.17290 | 1.77310 |
| 0.340 | 0.20449 | 0.35767 | 0.59356 | 1.07618 | 1.18995 | 1.76799 |
| 0.360 | 0.21725 | 0.37633 | 0.62482 | 1.08281 | 1.20763 | 1.76326 |
| 0.380 | 0.23058 | 0.39517 | 0.65607 | 1.08974 | 1.22588 | 1.75895 |
| 0.400 | 0.24444 | 0.41417 | 0.68728 | 1.09699 | 1.24467 | 1.75509 |
| 0.420 | 0.25884 | 0.43329 | 0.71840 | 1.10452 | 1.26394 | 1.75170 |
| 0.440 | 0.27372 | 0.45251 | 0.74941 | 1.11238 | 1.28366 | 1.74881 |
| 0.460 | 0.28907 | 0.47182 | 0.78026 | 1.12051 | 1.30381 | 1.74646 |
| 0.480 | 0.30485 | 0.49121 | 0.81093 | 1.12895 | 1.32433 | 1.74465 |
| 0.500 | 0.32101 | 0.51065 | 0.84138 | 1.13767 | 1.34521 | 1.74344 |
| 0.520 | 0.33753 | 0.53015 | 0.87158 | 1.14668 | 1.36640 | 1.74285 |
| 0.540 | 0.35439 | 0.54970 | 0.90148 | 1.15596 | 1.38790 | 1.74290 |
| 0.560 | 0.37155 | 0.56929 | 0.93107 | 1.16552 | 1.40966 | 1.74366 |
| 0.580 | 0.38897 | 0.58891 | 0.96028 | 1.17533 | 1.43166 | 1.74513 |
| 0.600 | 0.40665 | 0.60856 | 0.98908 | 1.18542 | 1.45388 | 1.74737 |
| 0.620 | 0.42456 | 0.62823 | 1.01743 | 1.19575 | 1.47631 | 1.75041 |
| 0.640 | 0.44267 | 0.64794 | 1.04527 | 1.20634 | 1.49893 | 1.75431 |
| 0.660 | 0.46098 | 0.66766 | 1.07257 | 1.21718 | 1.52169 | 1.75910 |
| 0.680 | 0.47945 | 0.68740 | 1.09926 | 1.22826 | 1.54461 | 1.76482 |
| 0.700 | 0.49808 | 0.70716 | 1.12530 | 1.23958 | 1.56763 | 1.77154 |
| 0.720 | 0.51684 | 0.72694 | 1.15063 | 1.25114 | 1.59077 | 1.77929 |
| 0.740 | 0.53574 | 0.74673 | 1.17522 | 1.26295 | 1.61399 | 1.78812 |
| 0.760 | 0.55475 | 0.76653 | 1.19900 | 1.27499 | 1.63729 | 1.79807 |
| 0.780 | 0.57386 | 0.78634 | 1.22192 | 1.28728 | 1.66063 | 1.80918 |
| 0.800 | 0.59306 | 0.80616 | 1.24395 | 1.29981 | 1.68401 | 1.82148 |
| 0.820 | 0.61234 | 0.82600 | 1.26506 | 1.31260 | 1.70742 | 1.83498 |
| 0.840 | 0.63171 | 0.84584 | 1.28520 | 1.32564 | 1.73083 | 1.84971 |
| 0.860 | 0.65114 | 0.86569 | 1.30440 | 1.33895 | 1.75422 | 1.86565 |
| 0.880 | 0.67063 | 0.88555 | 1.32265 | 1.35251 | 1.77759 | 1.88280 |
| 0.900 | 0.69018 | 0.90542 | 1.33999 | 1.36628 | 1.80094 | 1.90112 |
| 0.920 | 0.70978 | 0.92529 | 1.35650 | 1.38023 | 1.82424 | 1.92058 |
| 0.940 | 0.72942 | 0.94516 | 1.37231 | 1.39426 | 1.84748 | 1.94114 |
| 0.960 | 0.74910 | 0.96505 | 1.38753 | 1.40831 | 1.87067 | 1.96273 |
| 0.980 | 0.76882 | 0.98494 | 1.40228 | 1.42229 | 1.89379 | 1.98528 |
| 1.000 | 0.78856 | 1.00483 | 1.41665 | 1.43618 | 1.91684 | 2.00872 |

$H/R = 0.350$ $n = 1.$

| $H/L$ | $\Omega_1$ | $\Omega_2$ | $\Omega_3$ |
|---|---|---|---|
| 0. | 0. | 0.11196 | 0.26140 |
| 0.001 | 0.0000106 | 0.11197 | 0.26140 |
| 0.005 | 0.000259 | 0.11225 | 0.26146 |
| 0.010 | 0.00103 | 0.11311 | 0.26164 |
| 0.020 | 0.00398 | 0.11642 | 0.26238 |
| 0.030 | 0.00855 | 0.12155 | 0.26362 |
| 0.040 | 0.01433 | 0.12810 | 0.26540 |
| 0.050 | 0.02098 | 0.13564 | 0.26776 |
| 0.060 | 0.02821 | 0.14385 | 0.27075 |
| 0.070 | 0.03580 | 0.15245 | 0.27443 |
| 0.080 | 0.04360 | 0.16122 | 0.27887 |
| 0.090 | 0.05148 | 0.16998 | 0.28412 |
| 0.100 | 0.05938 | 0.17860 | 0.29025 |
| 0.120 | 0.07497 | 0.19507 | 0.30524 |
| 0.140 | 0.09005 | 0.21037 | 0.32381 |
| 0.160 | 0.10445 | 0.22476 | 0.34554 |
| 0.180 | 0.11813 | 0.23874 | 0.36979 |
| 0.200 | 0.13113 | 0.25278 | 0.39592 |
| 0.220 | 0.14355 | 0.26723 | 0.42345 |
| 0.240 | 0.15557 | 0.28225 | 0.45197 |
| 0.260 | 0.16737 | 0.29790 | 0.48122 |
| 0.280 | 0.17913 | 0.31415 | 0.51100 |
| 0.300 | 0.19101 | 0.33094 | 0.54116 |
| 0.320 | 0.20313 | 0.34819 | 0.57159 |
| 0.340 | 0.21560 | 0.36584 | 0.60220 |
| 0.360 | 0.22847 | 0.38381 | 0.63292 |
| 0.380 | 0.24178 | 0.40206 | 0.66370 |
| 0.400 | 0.25554 | 0.42054 | 0.69448 |
| 0.420 | 0.26975 | 0.43921 | 0.72521 |
| 0.440 | 0.28441 | 0.45803 | 0.75586 |
| 0.460 | 0.29949 | 0.47699 | 0.78639 |
| 0.480 | 0.31498 | 0.49607 | 0.81676 |
| 0.500 | 0.33086 | 0.51523 | 0.84693 |
| 0.520 | 0.34709 | 0.53449 | 0.87687 |
| 0.540 | 0.36364 | 0.55381 | 0.90653 |
| 0.560 | 0.38052 | 0.57319 | 0.93588 |
| 0.580 | 0.39767 | 0.59263 | 0.96486 |
| 0.600 | 0.41507 | 0.61211 | 0.99345 |
| 0.620 | 0.43271 | 0.63164 | 1.02158 |
| 0.640 | 0.45056 | 0.65120 | 1.04923 |
| 0.660 | 0.46862 | 0.67079 | 1.07632 |
| 0.680 | 0.48685 | 0.69042 | 1.10281 |
| 0.700 | 0.50525 | 0.71007 | 1.12865 |
| 0.720 | 0.52380 | 0.72974 | 1.15377 |
| 0.740 | 0.54248 | 0.74943 | 1.17813 |
| 0.760 | 0.56128 | 0.76914 | 1.20168 |
| 0.780 | 0.58020 | 0.78887 | 1.22436 |
| 0.800 | 0.59923 | 0.80862 | 1.24615 |
| 0.820 | 0.61834 | 0.82838 | 1.26698 |
| 0.840 | 0.63754 | 0.84815 | 1.28688 |
| 0.860 | 0.65681 | 0.86794 | 1.30582 |
| 0.880 | 0.67616 | 0.88774 | 1.32385 |
| 0.900 | 0.69556 | 0.90754 | 1.34102 |
| 0.920 | 0.71502 | 0.92736 | 1.35743 |
| 0.940 | 0.73453 | 0.94718 | 1.37318 |
| 0.960 | 0.75409 | 0.96702 | 1.38838 |
| 0.980 | 0.77368 | 0.98686 | 1.40315 |
| 1.000 | 0.79331 | 1.00671 | 1.41753 |

$H/R = 0.400 \qquad n = 1.$

| $H/L$ | $\Omega_1$ | $\Omega_2$ | $\Omega_3$ | $\Omega_4$ | $\Omega_5$ | $\Omega_6$ |
|---|---|---|---|---|---|---|
| 0. | 0. | 0.12813 | 0.29686 | 1.01485 | 1.06217 | 1.85200 |
| 0.001 | 0.00000913 | 0.12814 | 0.29686 | 1.01486 | 1.06216 | 1.85200 |
| 0.005 | 0.000228 | 0.12839 | 0.29692 | 1.01488 | 1.06219 | 1.85197 |
| 0.010 | 0.000906 | 0.12914 | 0.29707 | 1.01497 | 1.06226 | 1.85184 |
| 0.020 | 0.00354 | 0.13208 | 0.29772 | 1.01531 | 1.06256 | 1.85138 |
| 0.030 | 0.00766 | 0.13670 | 0.29880 | 1.01587 | 1.06307 | 1.85061 |
| 0.040 | 0.01299 | 0.14268 | 0.30035 | 1.01665 | 1.06378 | 1.84955 |
| 0.050 | 0.01923 | 0.14972 | 0.30238 | 1.01764 | 1.06470 | 1.84821 |
| 0.060 | 0.02613 | 0.15751 | 0.30495 | 1.01884 | 1.06584 | 1.84662 |
| 0.070 | 0.03349 | 0.16582 | 0.30808 | 1.02022 | 1.06721 | 1.84479 |
| 0.080 | 0.04115 | 0.17443 | 0.31183 | 1.02178 | 1.06881 | 1.84274 |
| 0.090 | 0.04900 | 0.18319 | 0.31625 | 1.02351 | 1.07066 | 1.84050 |
| 0.100 | 0.05695 | 0.19195 | 0.32138 | 1.02538 | 1.07276 | 1.83810 |
| 0.120 | 0.07291 | 0.20909 | 0.33396 | 1.02951 | 1.07777 | 1.83286 |
| 0.140 | 0.08867 | 0.22530 | 0.34975 | 1.03405 | 1.08392 | 1.82719 |
| 0.160 | 0.10402 | 0.24050 | 0.36864 | 1.03888 | 1.09127 | 1.82124 |
| 0.180 | 0.11887 | 0.25492 | 0.39028 | 1.04394 | 1.09984 | 1.81512 |
| 0.200 | 0.13319 | 0.26893 | 0.41416 | 1.04915 | 1.10964 | 1.80894 |
| 0.220 | 0.14700 | 0.28293 | 0.43978 | 1.05451 | 1.12061 | 1.80278 |
| 0.240 | 0.16039 | 0.29720 | 0.46672 | 1.06003 | 1.13271 | 1.79672 |
| 0.260 | 0.17346 | 0.31192 | 0.49465 | 1.06573 | 1.14582 | 1.79082 |
| 0.280 | 0.18634 | 0.32717 | 0.52332 | 1.07161 | 1.15990 | 1.78512 |
| 0.300 | 0.19915 | 0.34295 | 0.55253 | 1.07772 | 1.17485 | 1.77968 |
| 0.320 | 0.21202 | 0.35925 | 0.58214 | 1.08406 | 1.19061 | 1.77453 |
| 0.340 | 0.22505 | 0.37602 | 0.61204 | 1.09065 | 1.20710 | 1.76973 |
| 0.360 | 0.23832 | 0.39319 | 0.64213 | 1.09750 | 1.22427 | 1.76527 |
| 0.380 | 0.25188 | 0.41072 | 0.67235 | 1.10462 | 1.24206 | 1.76123 |
| 0.400 | 0.26578 | 0.42856 | 0.70262 | 1.11201 | 1.26041 | 1.75760 |
| 0.420 | 0.28004 | 0.44665 | 0.73290 | 1.11967 | 1.27930 | 1.75443 |
| 0.440 | 0.29466 | 0.46498 | 0.76314 | 1.12761 | 1.29866 | 1.75175 |
| 0.460 | 0.30966 | 0.48348 | 0.79329 | 1.13584 | 1.31845 | 1.74958 |
| 0.480 | 0.32501 | 0.50215 | 0.82331 | 1.14433 | 1.33866 | 1.74796 |
| 0.500 | 0.34071 | 0.52096 | 0.85315 | 1.15311 | 1.35922 | 1.74691 |
| 0.520 | 0.35675 | 0.53989 | 0.88278 | 1.16216 | 1.38013 | 1.74647 |
| 0.540 | 0.37310 | 0.55892 | 0.91215 | 1.17148 | 1.40135 | 1.74667 |
| 0.560 | 0.38974 | 0.57804 | 0.94121 | 1.18107 | 1.42285 | 1.74755 |
| 0.580 | 0.40666 | 0.59724 | 0.96993 | 1.19093 | 1.44461 | 1.74915 |
| 0.600 | 0.42383 | 0.61651 | 0.99826 | 1.20105 | 1.46659 | 1.75151 |
| 0.620 | 0.44124 | 0.63583 | 1.02614 | 1.21143 | 1.48878 | 1.75466 |
| 0.640 | 0.45887 | 0.65521 | 1.05353 | 1.22209 | 1.51117 | 1.75865 |
| 0.660 | 0.47671 | 0.67464 | 1.08038 | 1.23300 | 1.53371 | 1.76354 |
| 0.680 | 0.49473 | 0.69411 | 1.10662 | 1.24416 | 1.55640 | 1.76936 |
| 0.700 | 0.51292 | 0.71362 | 1.13221 | 1.25560 | 1.57921 | 1.77618 |
| 0.720 | 0.53126 | 0.73316 | 1.15707 | 1.26730 | 1.60213 | 1.78402 |
| 0.740 | 0.54975 | 0.75273 | 1.18117 | 1.27927 | 1.62514 | 1.79295 |
| 0.760 | 0.56836 | 0.77233 | 1.20444 | 1.29151 | 1.64820 | 1.80299 |
| 0.780 | 0.58709 | 0.79195 | 1.22685 | 1.30403 | 1.67131 | 1.81420 |
| 0.800 | 0.60594 | 0.81160 | 1.24834 | 1.31681 | 1.69444 | 1.82660 |
| 0.820 | 0.62488 | 0.83127 | 1.26891 | 1.32988 | 1.71759 | 1.84022 |
| 0.840 | 0.64391 | 0.85095 | 1.28853 | 1.34319 | 1.74073 | 1.85507 |
| 0.860 | 0.66303 | 0.87066 | 1.30725 | 1.35674 | 1.76384 | 1.87115 |
| 0.880 | 0.68222 | 0.89037 | 1.32508 | 1.37047 | 1.78692 | 1.88843 |
| 0.900 | 0.70148 | 0.91010 | 1.34212 | 1.38433 | 1.80996 | 1.90690 |
| 0.920 | 0.72079 | 0.92986 | 1.35845 | 1.39825 | 1.83294 | 1.92652 |
| 0.940 | 0.74017 | 0.94961 | 1.37417 | 1.41218 | 1.85588 | 1.94721 |
| 0.960 | 0.75959 | 0.96938 | 1.38937 | 1.42605 | 1.87876 | 1.96892 |
| 0.980 | 0.77907 | 0.98917 | 1.40415 | 1.43982 | 1.90159 | 1.99160 |
| 1.000 | 0.79858 | 1.00896 | 1.41855 | 1.45351 | 1.92435 | 2.01516 |

$H/R = 0.450$ $n = 1.$

| $H/L$ | $\Omega_1$ | $\Omega_2$ | $\Omega_3$ |
|---|---|---|---|
| 0. | 0. | 0.14438 | 0.33149 |
| 0.001 | 0.00000816 | 0.14439 | 0.33149 |
| 0.005 | 0.000204 | 0.14460 | 0.33153 |
| 0.010 | 0.000810 | 0.14528 | 0.33168 |
| 0.020 | 0.00318 | 0.14791 | 0.33225 |
| 0.030 | 0.00694 | 0.15210 | 0.33321 |
| 0.040 | 0.01187 | 0.15760 | 0.33458 |
| 0.050 | 0.01772 | 0.16416 | 0.33638 |
| 0.060 | 0.02429 | 0.17153 | 0.33863 |
| 0.070 | 0.03138 | 0.17949 | 0.34137 |
| 0.080 | 0.03885 | 0.18786 | 0.34462 |
| 0.090 | 0.04659 | 0.19648 | 0.34843 |
| 0.100 | 0.05450 | 0.20523 | 0.35285 |
| 0.120 | 0.07059 | 0.22267 | 0.36363 |
| 0.140 | 0.08673 | 0.23955 | 0.37720 |
| 0.160 | 0.10270 | 0.25555 | 0.39365 |
| 0.180 | 0.11835 | 0.27066 | 0.41283 |
| 0.200 | 0.13364 | 0.28509 | 0.43443 |
| 0.220 | 0.14854 | 0.29914 | 0.45803 |
| 0.240 | 0.16307 | 0.31313 | 0.48324 |
| 0.260 | 0.17728 | 0.32730 | 0.50969 |
| 0.280 | 0.19125 | 0.34181 | 0.53709 |
| 0.300 | 0.20506 | 0.35676 | 0.56523 |
| 0.320 | 0.21881 | 0.37219 | 0.59390 |
| 0.340 | 0.23258 | 0.38808 | 0.62298 |
| 0.360 | 0.24646 | 0.40443 | 0.65236 |
| 0.380 | 0.26049 | 0.42117 | 0.68193 |
| 0.400 | 0.27475 | 0.43829 | 0.71163 |
| 0.420 | 0.28925 | 0.45573 | 0.74139 |
| 0.440 | 0.30403 | 0.47345 | 0.77116 |
| 0.460 | 0.31911 | 0.49142 | 0.80087 |
| 0.480 | 0.33448 | 0.50960 | 0.83048 |
| 0.500 | 0.35015 | 0.52796 | 0.85994 |
| 0.520 | 0.36611 | 0.54649 | 0.88922 |
| 0.540 | 0.38236 | 0.56516 | 0.91824 |
| 0.560 | 0.39888 | 0.58395 | 0.94699 |
| 0.580 | 0.41565 | 0.60285 | 0.97540 |
| 0.600 | 0.43266 | 0.62185 | 1.00342 |
| 0.620 | 0.44991 | 0.64092 | 1.03101 |
| 0.640 | 0.46736 | 0.66008 | 1.05811 |
| 0.660 | 0.48501 | 0.67930 | 1.08466 |
| 0.680 | 0.50285 | 0.69857 | 1.11062 |
| 0.700 | 0.52086 | 0.71790 | 1.13590 |
| 0.720 | 0.53902 | 0.73728 | 1.16047 |
| 0.740 | 0.55733 | 0.75670 | 1.18426 |
| 0.760 | 0.57578 | 0.77615 | 1.20723 |
| 0.780 | 0.59434 | 0.79564 | 1.22933 |
| 0.800 | 0.61303 | 0.81516 | 1.25053 |
| 0.820 | 0.63181 | 0.83471 | 1.27082 |
| 0.840 | 0.65068 | 0.85429 | 1.29020 |
| 0.860 | 0.66965 | 0.87389 | 1.30870 |
| 0.880 | 0.68869 | 0.89351 | 1.32638 |
| 0.900 | 0.70781 | 0.91316 | 1.34332 |
| 0.920 | 0.72699 | 0.93281 | 1.35959 |
| 0.940 | 0.74623 | 0.95249 | 1.37529 |
| 0.960 | 0.76553 | 0.97219 | 1.39050 |
| 0.980 | 0.78487 | 0.99190 | 1.40530 |
| 1.000 | 0.80426 | 1.01162 | 1.41973 |

$H/R = 0.500 \qquad n = 1.$

| $H/L$ | $\Omega_1$ | $\Omega_2$ | $\Omega_3$ | $\Omega_4$ | $\Omega_5$ | $\Omega_6$ |
|---|---|---|---|---|---|---|
| 0. | 0. | 0.16069 | 0.36513 | 1.02382 | 1.10034 | 1.84602 |
| 0.010 | 0.000734 | 0.16150 | 0.36530 | 1.02394 | 1.10044 | 1.84589 |
| 0.050 | 0.01642 | 0.17892 | 0.36954 | 1.02658 | 1.10279 | 1.84283 |
| 0.100 | 0.05213 | 0.21862 | 0.38412 | 1.03443 | 1.11037 | 1.83414 |
| 0.120 | 0.06819 | 0.23610 | 0.39354 | 1.03877 | 1.11497 | 1.82958 |
| 0.140 | 0.08453 | 0.25337 | 0.40540 | 1.04366 | 1.12053 | 1.82458 |
| 0.160 | 0.10087 | 0.26999 | 0.41983 | 1.04900 | 1.12711 | 1.81928 |
| 0.180 | 0.11708 | 0.28581 | 0.43684 | 1.05472 | 1.13475 | 1.81379 |
| 0.200 | 0.13306 | 0.30084 | 0.45629 | 1.06070 | 1.14347 | 1.80821 |
| 0.220 | 0.14877 | 0.31527 | 0.47788 | 1.06691 | 1.15328 | 1.80261 |
| 0.240 | 0.16419 | 0.32936 | 0.50128 | 1.07329 | 1.16417 | 1.79708 |
| 0.260 | 0.17935 | 0.34335 | 0.52615 | 1.07983 | 1.17610 | 1.79166 |
| 0.280 | 0.19428 | 0.35747 | 0.55217 | 1.08653 | 1.18902 | 1.78642 |
| 0.300 | 0.20903 | 0.37185 | 0.57911 | 1.09336 | 1.20288 | 1.78139 |
| 0.320 | 0.22365 | 0.38659 | 0.60675 | 1.10036 | 1.21761 | 1.77665 |
| 0.340 | 0.23822 | 0.40173 | 0.63492 | 1.10754 | 1.23316 | 1.77220 |
| 0.360 | 0.25280 | 0.41730 | 0.66349 | 1.11490 | 1.24945 | 1.76809 |
| 0.380 | 0.26744 | 0.43329 | 0.69235 | 1.12247 | 1.26643 | 1.76435 |
| 0.400 | 0.28220 | 0.44967 | 0.72141 | 1.13025 | 1.28403 | 1.76102 |
| 0.420 | 0.29711 | 0.46641 | 0.75059 | 1.13827 | 1.30221 | 1.75812 |
| 0.440 | 0.31221 | 0.48348 | 0.77981 | 1.14651 | 1.32092 | 1.75569 |
| 0.460 | 0.32752 | 0.50084 | 0.80903 | 1.15500 | 1.34012 | 1.75376 |
| 0.480 | 0.34306 | 0.51846 | 0.83819 | 1.16374 | 1.35976 | 1.75237 |
| 0.500 | 0.35884 | 0.53632 | 0.86723 | 1.17273 | 1.37980 | 1.75153 |
| 0.520 | 0.37485 | 0.55438 | 0.89609 | 1.18197 | 1.40021 | 1.75128 |
| 0.540 | 0.39111 | 0.57262 | 0.92474 | 1.19148 | 1.42096 | 1.75166 |
| 0.560 | 0.40760 | 0.59102 | 0.95312 | 1.20125 | 1.44201 | 1.75271 |
| 0.580 | 0.42432 | 0.60956 | 0.98118 | 1.21128 | 1.46335 | 1.75448 |
| 0.600 | 0.44126 | 0.62823 | 1.00886 | 1.22158 | 1.48492 | 1.75698 |
| 0.620 | 0.45840 | 0.64700 | 1.03611 | 1.23215 | 1.50672 | 1.76027 |
| 0.640 | 0.47575 | 0.66588 | 1.06287 | 1.24300 | 1.52872 | 1.76442 |
| 0.660 | 0.49328 | 0.68484 | 1.08909 | 1.25412 | 1.55090 | 1.76943 |
| 0.680 | 0.51099 | 0.70388 | 1.11471 | 1.26551 | 1.57321 | 1.77540 |
| 0.700 | 0.52886 | 0.72299 | 1.13966 | 1.27719 | 1.59566 | 1.78233 |
| 0.720 | 0.54689 | 0.74217 | 1.16390 | 1.28916 | 1.61820 | 1.79031 |
| 0.740 | 0.56506 | 0.76140 | 1.18737 | 1.30141 | 1.64083 | 1.79939 |
| 0.760 | 0.58336 | 0.78068 | 1.21001 | 1.31393 | 1.66351 | 1.80958 |
| 0.780 | 0.60178 | 0.80001 | 1.23180 | 1.32675 | 1.68621 | 1.82096 |
| 0.800 | 0.62032 | 0.81938 | 1.25271 | 1.33981 | 1.70891 | 1.83355 |
| 0.820 | 0.63896 | 0.83878 | 1.27273 | 1.35312 | 1.73161 | 1.84736 |
| 0.840 | 0.65770 | 0.85823 | 1.29188 | 1.36662 | 1.75429 | 1.86243 |
| 0.860 | 0.67653 | 0.87771 | 1.31021 | 1.38028 | 1.77690 | 1.87875 |
| 0.880 | 0.69544 | 0.89721 | 1.32777 | 1.39404 | 1.79948 | 1.89629 |
| 0.900 | 0.71443 | 0.91674 | 1.34462 | 1.40784 | 1.82199 | 1.91503 |
| 0.920 | 0.73348 | 0.93629 | 1.36086 | 1.42164 | 1.84446 | 1.93489 |
| 0.940 | 0.75260 | 0.95587 | 1.37654 | 1.43537 | 1.86688 | 1.95582 |
| 0.960 | 0.77177 | 0.97547 | 1.39177 | 1.44901 | 1.88927 | 1.97776 |
| 0.980 | 0.79100 | 0.99509 | 1.40658 | 1.46256 | 1.91162 | 2.00062 |
| 1.000 | 0.81027 | 1.01473 | 1.42103 | 1.47600 | 1.93396 | 2.02431 |

$H/R = 0.600 \qquad n = 1.$

| $H/L$ | $\Omega_1$ | $\Omega_2$ | $\Omega_3$ | $\Omega_4$ | $\Omega_5$ | $\Omega_6$ |
|---|---|---|---|---|---|---|
| 0. | 0. | 0.19354 | 0.42894 | 1.03543 | 1.14972 | 1.84102 |
| 0.010 | 0.000621 | 0.19422 | 0.42909 | 1.03554 | 1.14982 | 1.84091 |
| 0.050 | 0.01432 | 0.20927 | 0.43265 | 1.03813 | 1.15207 | 1.83833 |
| 0.100 | 0.04776 | 0.24604 | 0.44462 | 1.04594 | 1.15924 | 1.83090 |
| 0.120 | 0.06351 | 0.26315 | 0.45221 | 1.05032 | 1.16353 | 1.82694 |
| 0.140 | 0.07987 | 0.28058 | 0.46167 | 1.05535 | 1.16867 | 1.82259 |
| 0.160 | 0.09655 | 0.29787 | 0.47317 | 1.06093 | 1.17469 | 1.81793 |
| 0.180 | 0.11336 | 0.31470 | 0.48682 | 1.06699 | 1.18161 | 1.81307 |
| 0.200 | 0.13019 | 0.33088 | 0.50267 | 1.07346 | 1.18947 | 1.80810 |
| 0.220 | 0.14694 | 0.34637 | 0.52065 | 1.08027 | 1.19828 | 1.80309 |
| 0.240 | 0.16358 | 0.36125 | 0.54057 | 1.08735 | 1.20805 | 1.79811 |
| 0.260 | 0.18008 | 0.37566 | 0.56223 | 1.09467 | 1.21876 | 1.79322 |
| 0.280 | 0.19644 | 0.38979 | 0.58536 | 1.10218 | 1.23042 | 1.78847 |
| 0.300 | 0.21267 | 0.40382 | 0.60971 | 1.10986 | 1.24299 | 1.78392 |
| 0.320 | 0.22878 | 0.41789 | 0.63506 | 1.11770 | 1.25643 | 1.77960 |
| 0.340 | 0.24479 | 0.43212 | 0.66119 | 1.12569 | 1.27070 | 1.77557 |
| 0.360 | 0.26074 | 0.44660 | 0.68795 | 1.13383 | 1.28577 | 1.77185 |
| 0.380 | 0.27666 | 0.46139 | 0.71519 | 1.14212 | 1.30157 | 1.76848 |
| 0.400 | 0.29257 | 0.47651 | 0.74277 | 1.15058 | 1.31806 | 1.76549 |
| 0.420 | 0.30852 | 0.49197 | 0.77062 | 1.15921 | 1.33520 | 1.76291 |
| 0.440 | 0.32452 | 0.50777 | 0.79863 | 1.16804 | 1.35291 | 1.76079 |
| 0.460 | 0.34060 | 0.52390 | 0.82672 | 1.17706 | 1.37116 | 1.75913 |
| 0.480 | 0.35679 | 0.54034 | 0.85483 | 1.18629 | 1.38991 | 1.75800 |
| 0.500 | 0.37310 | 0.55708 | 0.88288 | 1.19574 | 1.40911 | 1.75741 |
| 0.520 | 0.38954 | 0.57408 | 0.91082 | 1.20542 | 1.42873 | 1.75740 |
| 0.540 | 0.40612 | 0.59133 | 0.93858 | 1.21533 | 1.44873 | 1.75800 |
| 0.560 | 0.42286 | 0.60881 | 0.96611 | 1.22550 | 1.46905 | 1.75927 |
| 0.580 | 0.43975 | 0.62648 | 0.99334 | 1.23591 | 1.48970 | 1.76122 |
| 0.600 | 0.45679 | 0.64435 | 1.02022 | 1.24658 | 1.51062 | 1.76392 |
| 0.620 | 0.47398 | 0.66238 | 1.04669 | 1.25752 | 1.53179 | 1.76741 |
| 0.640 | 0.49132 | 0.68056 | 1.07269 | 1.26874 | 1.55316 | 1.77172 |
| 0.660 | 0.50882 | 0.69888 | 1.09815 | 1.28024 | 1.57471 | 1.77693 |
| 0.680 | 0.52645 | 0.71732 | 1.12303 | 1.29200 | 1.59643 | 1.78307 |
| 0.700 | 0.54422 | 0.73587 | 1.14725 | 1.30406 | 1.61826 | 1.79021 |
| 0.720 | 0.56212 | 0.75453 | 1.17078 | 1.31640 | 1.64019 | 1.79840 |
| 0.740 | 0.58014 | 0.77327 | 1.19356 | 1.32901 | 1.66217 | 1.80770 |
| 0.760 | 0.59828 | 0.79210 | 1.21555 | 1.34189 | 1.68418 | 1.81815 |
| 0.780 | 0.61653 | 0.81100 | 1.23674 | 1.35500 | 1.70619 | 1.82982 |
| 0.800 | 0.63488 | 0.82998 | 1.25711 | 1.36832 | 1.72817 | 1.84274 |
| 0.820 | 0.65333 | 0.84902 | 1.27667 | 1.38182 | 1.75010 | 1.85692 |
| 0.840 | 0.67188 | 0.86811 | 1.29546 | 1.39544 | 1.77195 | 1.87240 |
| 0.860 | 0.69050 | 0.88726 | 1.31352 | 1.40913 | 1.79374 | 1.88913 |
| 0.880 | 0.70921 | 0.90645 | 1.33089 | 1.42286 | 1.81545 | 1.90711 |
| 0.900 | 0.72799 | 0.92569 | 1.34762 | 1.43656 | 1.83710 | 1.92628 |
| 0.920 | 0.74684 | 0.94497 | 1.36380 | 1.45019 | 1.85874 | 1.94657 |
| 0.940 | 0.76576 | 0.96429 | 1.37948 | 1.46372 | 1.88037 | 1.96787 |
| 0.960 | 0.78472 | 0.98365 | 1.39473 | 1.47715 | 1.90201 | 1.99012 |
| 0.980 | 0.80375 | 1.00303 | 1.40960 | 1.49046 | 1.92370 | 2.01323 |
| 1.000 | 0.82283 | 1.02245 | 1.42411 | 1.50367 | 1.94542 | 2.03709 |

$H/R = 0.700$ $n = 1.$

| $H/L$ | $\Omega_1$ | $\Omega_2$ | $\Omega_3$ | $\Omega_4$ | $\Omega_5$ | $\Omega_6$ |
|---|---|---|---|---|---|---|
| 0. | 0. | 0.22668 | 0.48735 | 1.05012 | 1.21115 | 1.83731 |
| 0.010 | 0.000541 | 0.22726 | 0.48747 | 1.05022 | 1.21124 | 1.83722 |
| 0.050 | 0.01272 | 0.24050 | 0.49059 | 1.05273 | 1.21338 | 1.83505 |
| 0.100 | 0.04396 | 0.27446 | 0.50090 | 1.06038 | 1.22016 | 1.82875 |
| 0.120 | 0.05922 | 0.29090 | 0.50733 | 1.06474 | 1.22418 | 1.82534 |
| 0.140 | 0.07534 | 0.30804 | 0.51527 | 1.06977 | 1.22897 | 1.82158 |
| 0.160 | 0.09202 | 0.32542 | 0.52487 | 1.07540 | 1.23453 | 1.81753 |
| 0.180 | 0.10903 | 0.34269 | 0.53625 | 1.08159 | 1.24089 | 1.81327 |
| 0.200 | 0.12625 | 0.35959 | 0.54950 | 1.08828 | 1.24807 | 1.80890 |
| 0.220 | 0.14356 | 0.37596 | 0.56464 | 1.09541 | 1.25607 | 1.80449 |
| 0.240 | 0.16090 | 0.39175 | 0.58165 | 1.10291 | 1.26491 | 1.80007 |
| 0.260 | 0.17823 | 0.40696 | 0.60042 | 1.11075 | 1.27458 | 1.79574 |
| 0.280 | 0.19551 | 0.42168 | 0.62079 | 1.11887 | 1.28508 | 1.79152 |
| 0.300 | 0.21273 | 0.43605 | 0.64255 | 1.12724 | 1.29641 | 1.78748 |
| 0.320 | 0.22988 | 0.45019 | 0.66552 | 1.13582 | 1.30856 | 1.78364 |
| 0.340 | 0.24698 | 0.46423 | 0.68949 | 1.14460 | 1.32150 | 1.78005 |
| 0.360 | 0.26402 | 0.47829 | 0.71429 | 1.15355 | 1.33520 | 1.77677 |
| 0.380 | 0.28101 | 0.49247 | 0.73973 | 1.16267 | 1.34964 | 1.77380 |
| 0.400 | 0.29797 | 0.50682 | 0.76570 | 1.17194 | 1.36477 | 1.77119 |
| 0.420 | 0.31492 | 0.52140 | 0.79205 | 1.18138 | 1.38056 | 1.76898 |
| 0.440 | 0.33187 | 0.53624 | 0.81870 | 1.19099 | 1.39698 | 1.76720 |
| 0.460 | 0.34884 | 0.55136 | 0.84551 | 1.20077 | 1.41397 | 1.76587 |
| 0.480 | 0.36583 | 0.56677 | 0.87243 | 1.21073 | 1.43150 | 1.76505 |
| 0.500 | 0.38287 | 0.58247 | 0.89937 | 1.22088 | 1.44953 | 1.76475 |
| 0.520 | 0.39997 | 0.59844 | 0.92625 | 1.23125 | 1.46801 | 1.76501 |
| 0.540 | 0.41713 | 0.61469 | 0.95301 | 1.24182 | 1.48692 | 1.76588 |
| 0.560 | 0.43438 | 0.63118 | 0.97957 | 1.25261 | 1.50621 | 1.76739 |
| 0.580 | 0.45170 | 0.64793 | 1.00588 | 1.26364 | 1.52582 | 1.76961 |
| 0.600 | 0.46911 | 0.66489 | 1.03186 | 1.27491 | 1.54576 | 1.77255 |
| 0.620 | 0.48662 | 0.68206 | 1.05746 | 1.28644 | 1.56595 | 1.77626 |
| 0.640 | 0.50421 | 0.69944 | 1.08262 | 1.29821 | 1.58637 | 1.78083 |
| 0.660 | 0.52191 | 0.71698 | 1.10727 | 1.31026 | 1.60696 | 1.78630 |
| 0.680 | 0.53970 | 0.73469 | 1.13136 | 1.32257 | 1.62772 | 1.79272 |
| 0.700 | 0.55759 | 0.75256 | 1.15483 | 1.33515 | 1.64857 | 1.80015 |
| 0.720 | 0.57556 | 0.77056 | 1.17765 | 1.34797 | 1.66948 | 1.80869 |
| 0.740 | 0.59364 | 0.78869 | 1.19977 | 1.36103 | 1.69041 | 1.81839 |
| 0.760 | 0.61179 | 0.80695 | 1.22117 | 1.37431 | 1.71129 | 1.82931 |
| 0.780 | 0.63004 | 0.82531 | 1.24182 | 1.38776 | 1.73212 | 1.84150 |
| 0.800 | 0.64837 | 0.84377 | 1.26174 | 1.40136 | 1.75284 | 1.85502 |
| 0.820 | 0.66677 | 0.86232 | 1.28094 | 1.41507 | 1.77344 | 1.86987 |
| 0.840 | 0.68526 | 0.88095 | 1.29944 | 1.42882 | 1.79392 | 1.88606 |
| 0.860 | 0.70381 | 0.89967 | 1.31729 | 1.44258 | 1.81430 | 1.90354 |
| 0.880 | 0.72243 | 0.91845 | 1.33451 | 1.45629 | 1.83463 | 1.92226 |
| 0.900 | 0.74112 | 0.93731 | 1.35118 | 1.46993 | 1.85496 | 1.94209 |
| 0.920 | 0.75987 | 0.95622 | 1.36734 | 1.48346 | 1.87535 | 1.96295 |
| 0.940 | 0.77866 | 0.97520 | 1.38303 | 1.49687 | 1.89585 | 1.98473 |
| 0.960 | 0.79752 | 0.99422 | 1.39831 | 1.51014 | 1.91650 | 2.00733 |

$H/R = 0.800 \qquad n = 1.$

| $H/L$ | $\Omega_1$ | $\Omega_2$ | $\Omega_3$ | $\Omega_4$ | $\Omega_5$ | $\Omega_6$ |
|---|---|---|---|---|---|---|
| 0. | 0. | 0.26005 | 0.53979 | 1.06843 | 1.28472 | 1.83516 |
| 0.010 | 0.000481 | 0.26056 | 0.53991 | 1.06854 | 1.28482 | 1.83508 |
| 0.050 | 0.01147 | 0.27239 | 0.54272 | 1.07096 | 1.28684 | 1.83326 |
| 0.100 | 0.04072 | 0.30382 | 0.55191 | 1.07838 | 1.29321 | 1.82794 |
| 0.120 | 0.05542 | 0.31949 | 0.55758 | 1.08264 | 1.29696 | 1.82505 |
| 0.140 | 0.07116 | 0.33612 | 0.56452 | 1.08757 | 1.30142 | 1.82184 |
| 0.160 | 0.08764 | 0.35327 | 0.57285 | 1.09315 | 1.30659 | 1.81836 |
| 0.180 | 0.10464 | 0.37060 | 0.58268 | 1.09932 | 1.31247 | 1.81470 |
| 0.200 | 0.12199 | 0.38781 | 0.59410 | 1.10604 | 1.31907 | 1.81093 |
| 0.220 | 0.13957 | 0.40472 | 0.60717 | 1.11326 | 1.32641 | 1.80709 |
| 0.240 | 0.15730 | 0.42118 | 0.62193 | 1.12093 | 1.33448 | 1.80326 |
| 0.260 | 0.17512 | 0.43712 | 0.63833 | 1.12902 | 1.34327 | 1.79950 |
| 0.280 | 0.19298 | 0.45254 | 0.65632 | 1.13748 | 1.35282 | 1.79582 |
| 0.300 | 0.21085 | 0.46749 | 0.67576 | 1.14626 | 1.36309 | 1.79231 |
| 0.320 | 0.22872 | 0.48206 | 0.69649 | 1.15534 | 1.37408 | 1.78898 |
| 0.340 | 0.24657 | 0.49635 | 0.71837 | 1.16468 | 1.38579 | 1.78589 |
| 0.360 | 0.26440 | 0.51046 | 0.74122 | 1.17426 | 1.39821 | 1.78306 |
| 0.380 | 0.28220 | 0.52451 | 0.76486 | 1.18407 | 1.41130 | 1.78055 |
| 0.400 | 0.29998 | 0.53856 | 0.78916 | 1.19407 | 1.42505 | 1.77838 |
| 0.420 | 0.31775 | 0.55270 | 0.81397 | 1.20428 | 1.43943 | 1.77657 |
| 0.440 | 0.33550 | 0.56698 | 0.83917 | 1.21468 | 1.45443 | 1.77517 |
| 0.460 | 0.35325 | 0.58144 | 0.86466 | 1.22526 | 1.47000 | 1.77422 |
| 0.480 | 0.37100 | 0.59611 | 0.89032 | 1.23603 | 1.48612 | 1.77375 |
| 0.500 | 0.38875 | 0.61101 | 0.91609 | 1.24700 | 1.50273 | 1.77379 |
| 0.520 | 0.40653 | 0.62616 | 0.94185 | 1.25817 | 1.51983 | 1.77438 |
| 0.540 | 0.42433 | 0.64155 | 0.96755 | 1.26955 | 1.53735 | 1.77557 |
| 0.560 | 0.44216 | 0.65718 | 0.99311 | 1.28114 | 1.55528 | 1.77739 |
| 0.580 | 0.46002 | 0.67306 | 1.01845 | 1.29294 | 1.57354 | 1.77991 |
| 0.600 | 0.47793 | 0.68917 | 1.04350 | 1.30498 | 1.59212 | 1.78316 |
| 0.620 | 0.49588 | 0.70551 | 1.06822 | 1.31726 | 1.61096 | 1.78722 |
| 0.640 | 0.51388 | 0.72206 | 1.09253 | 1.32978 | 1.63001 | 1.79215 |
| 0.660 | 0.53194 | 0.73880 | 1.11637 | 1.34254 | 1.64920 | 1.79800 |
| 0.680 | 0.55005 | 0.75575 | 1.13970 | 1.35555 | 1.66848 | 1.80488 |
| 0.700 | 0.56821 | 0.77287 | 1.16246 | 1.36877 | 1.68781 | 1.81285 |
| 0.720 | 0.58643 | 0.79016 | 1.18462 | 1.38221 | 1.70709 | 1.82204 |
| 0.740 | 0.60471 | 0.80761 | 1.20615 | 1.39584 | 1.72626 | 1.83248 |
| 0.760 | 0.62305 | 0.82521 | 1.22702 | 1.40963 | 1.74526 | 1.84429 |
| 0.780 | 0.64145 | 0.84294 | 1.24723 | 1.42353 | 1.76406 | 1.85751 |
| 0.800 | 0.65989 | 0.86080 | 1.26678 | 1.43753 | 1.78265 | 1.87216 |
| 0.820 | 0.67840 | 0.87878 | 1.28567 | 1.45155 | 1.80106 | 1.88820 |
| 0.840 | 0.69696 | 0.89687 | 1.30395 | 1.46556 | 1.81936 | 1.90557 |
| 0.860 | 0.71557 | 0.91507 | 1.32164 | 1.47952 | 1.83763 | 1.92416 |
| 0.880 | 0.73422 | 0.93336 | 1.33876 | 1.49337 | 1.85600 | 1.94381 |
| 0.900 | 0.75293 | 0.95174 | 1.35537 | 1.50711 | 1.87456 | 1.96443 |
| 0.920 | 0.77168 | 0.97021 | 1.37151 | 1.52070 | 1.89337 | 1.98585 |
| 0.940 | 0.79047 | 0.98875 | 1.38721 | 1.53413 | 1.91250 | 2.00801 |
| 0.960 | 0.80931 | 1.00737 | 1.40252 | 1.54740 | 1.93197 | 2.03078 |
| 0.980 | 0.82818 | 1.02605 | 1.41747 | 1.56051 | 1.95176 | 2.05409 |
| 1.000 | 0.84710 | 1.04480 | 1.43212 | 1.57349 | 1.97186 | 2.07791 |

$H/R = 0.900 \qquad n = 1.$

| $H/L$ | $\Omega_1$ | $\Omega_2$ | $\Omega_3$ | $\Omega_4$ | $\Omega_5$ | $\Omega_6$ |
|---|---|---|---|---|---|---|
| 0. | 0. | 0.29354 | 0.58628 | 1.09112 | 1.36944 | 1.83473 |
| 0.010 | 0.000436 | 0.29401 | 0.58639 | 1.09122 | 1.36952 | 1.83468 |
| 0.050 | 0.01048 | 0.30471 | 0.58899 | 1.09353 | 1.37142 | 1.83317 |
| 0.100 | 0.03797 | 0.33393 | 0.59740 | 1.10065 | 1.37738 | 1.82874 |
| 0.120 | 0.05209 | 0.34885 | 0.60254 | 1.10475 | 1.38088 | 1.82634 |
| 0.140 | 0.06739 | 0.36488 | 0.60879 | 1.10953 | 1.38503 | 1.82364 |
| 0.160 | 0.08358 | 0.38163 | 0.61625 | 1.11496 | 1.38983 | 1.82072 |
| 0.180 | 0.10042 | 0.39878 | 0.62500 | 1.12099 | 1.39526 | 1.81763 |
| 0.200 | 0.11775 | 0.41604 | 0.63512 | 1.12760 | 1.40136 | 1.81444 |
| 0.220 | 0.13542 | 0.43319 | 0.64669 | 1.13474 | 1.40812 | 1.81121 |
| 0.240 | 0.15334 | 0.45006 | 0.65976 | 1.14239 | 1.41552 | 1.80796 |
| 0.260 | 0.17143 | 0.46652 | 0.67433 | 1.15050 | 1.42358 | 1.80478 |
| 0.280 | 0.18965 | 0.48253 | 0.69038 | 1.15904 | 1.43229 | 1.80168 |
| 0.300 | 0.20794 | 0.49805 | 0.70786 | 1.16797 | 1.44166 | 1.79872 |
| 0.320 | 0.22629 | 0.51314 | 0.72666 | 1.17727 | 1.45167 | 1.79594 |
| 0.340 | 0.24467 | 0.52784 | 0.74664 | 1.18690 | 1.46231 | 1.79338 |
| 0.360 | 0.26305 | 0.54225 | 0.76768 | 1.19684 | 1.47358 | 1.79106 |
| 0.380 | 0.28145 | 0.55644 | 0.78962 | 1.20707 | 1.48547 | 1.78903 |
| 0.400 | 0.29984 | 0.57050 | 0.81232 | 1.21757 | 1.49796 | 1.78734 |
| 0.420 | 0.31823 | 0.58451 | 0.83563 | 1.22832 | 1.51103 | 1.78599 |
| 0.440 | 0.33661 | 0.59854 | 0.85943 | 1.23931 | 1.52465 | 1.78503 |
| 0.460 | 0.35499 | 0.61265 | 0.88361 | 1.25053 | 1.53881 | 1.78450 |
| 0.480 | 0.37336 | 0.62687 | 0.90804 | 1.26199 | 1.55348 | 1.78444 |
| 0.500 | 0.39172 | 0.64125 | 0.93264 | 1.27367 | 1.56863 | 1.78488 |
| 0.520 | 0.41009 | 0.65581 | 0.95730 | 1.28558 | 1.58421 | 1.78587 |
| 0.540 | 0.42847 | 0.67056 | 0.98196 | 1.29773 | 1.60020 | 1.78745 |
| 0.560 | 0.44685 | 0.68553 | 1.00652 | 1.31010 | 1.61655 | 1.78968 |
| 0.580 | 0.46524 | 0.70070 | 1.03092 | 1.32270 | 1.63319 | 1.79261 |
| 0.600 | 0.48365 | 0.71609 | 1.05509 | 1.33554 | 1.65010 | 1.79633 |
| 0.620 | 0.50207 | 0.73169 | 1.07896 | 1.34862 | 1.66718 | 1.80091 |
| 0.640 | 0.52052 | 0.74751 | 1.10246 | 1.36193 | 1.68436 | 1.80644 |
| 0.660 | 0.53899 | 0.76353 | 1.12556 | 1.37548 | 1.70156 | 1.81303 |
| 0.680 | 0.55748 | 0.77975 | 1.14819 | 1.38926 | 1.71866 | 1.82081 |
| 0.700 | 0.57600 | 0.79616 | 1.17031 | 1.40323 | 1.73557 | 1.82989 |
| 0.720 | 0.59455 | 0.81275 | 1.19189 | 1.41740 | 1.75217 | 1.84041 |
| 0.740 | 0.61312 | 0.82951 | 1.21290 | 1.43171 | 1.76842 | 1.85246 |
| 0.760 | 0.63173 | 0.84644 | 1.23332 | 1.44616 | 1.78428 | 1.86604 |
| 0.780 | 0.65037 | 0.86353 | 1.25315 | 1.46067 | 1.79982 | 1.88112 |
| 0.800 | 0.66904 | 0.88076 | 1.27238 | 1.47522 | 1.81518 | 1.89760 |
| 0.820 | 0.68774 | 0.89814 | 1.29103 | 1.48976 | 1.83051 | 1.91530 |
| 0.840 | 0.70647 | 0.91564 | 1.30910 | 1.50424 | 1.84598 | 1.93406 |
| 0.860 | 0.72523 | 0.93327 | 1.32665 | 1.51862 | 1.86176 | 1.95370 |
| 0.880 | 0.74403 | 0.95102 | 1.34369 | 1.53286 | 1.87794 | 1.97410 |
| 0.900 | 0.76285 | 0.96887 | 1.36025 | 1.54692 | 1.89461 | 1.99512 |
| 0.920 | 0.78171 | 0.98683 | 1.37637 | 1.56080 | 1.91183 | 2.01669 |
| 0.940 | 0.80058 | 1.00489 | 1.39209 | 1.57448 | 1.92958 | 2.03875 |
| 0.960 | 0.81949 | 1.02303 | 1.40743 | 1.58796 | 1.94787 | 2.06123 |
| 0.980 | 0.83842 | 1.04127 | 1.42244 | 1.60127 | 1.96666 | 2.08408 |
| 1.000 | 0.85738 | 1.05958 | 1.43714 | 1.61439 | 1.98590 | 2.10729 |

$H/R = 1.000 \qquad n = 1.$

| $H/L$ | $\Omega_1$ | $\Omega_2$ | $\Omega_3$ | $\Omega_4$ | $\Omega_5$ | $\Omega_6$ |
|---|---|---|---|---|---|---|
| 0. | 0. | 0.32698 | 0.62732 | 1.11903 | 1.46294 | 1.83627 |
| 0.010 | 0.000400 | 0.32741 | 0.62741 | 1.11913 | 1.46302 | 1.83622 |
| 0.050 | 0.00969 | 0.33721 | 0.62986 | 1.12131 | 1.46479 | 1.83501 |
| 0.100 | 0.03563 | 0.36452 | 0.63771 | 1.12806 | 1.47034 | 1.83143 |
| 0.120 | 0.04919 | 0.37872 | 0.64247 | 1.13196 | 1.47358 | 1.82947 |
| 0.140 | 0.06404 | 0.39415 | 0.64823 | 1.13652 | 1.47744 | 1.82727 |
| 0.160 | 0.07988 | 0.41045 | 0.65506 | 1.14171 | 1.48187 | 1.82490 |
| 0.180 | 0.09648 | 0.42730 | 0.66303 | 1.14752 | 1.48691 | 1.82239 |
| 0.200 | 0.11368 | 0.44444 | 0.67222 | 1.15390 | 1.49253 | 1.81978 |
| 0.220 | 0.13132 | 0.46163 | 0.68268 | 1.16084 | 1.49876 | 1.81714 |
| 0.240 | 0.14930 | 0.47870 | 0.69448 | 1.16830 | 1.50557 | 1.81450 |
| 0.260 | 0.16753 | 0.49549 | 0.70764 | 1.17625 | 1.51297 | 1.81192 |
| 0.280 | 0.18595 | 0.51191 | 0.72217 | 1.18467 | 1.52095 | 1.80943 |
| 0.300 | 0.20451 | 0.52790 | 0.73806 | 1.19352 | 1.52953 | 1.80706 |
| 0.320 | 0.22318 | 0.54346 | 0.75522 | 1.20279 | 1.53867 | 1.80487 |
| 0.340 | 0.24191 | 0.55859 | 0.77359 | 1.21244 | 1.54839 | 1.80288 |
| 0.360 | 0.26070 | 0.57335 | 0.79305 | 1.22246 | 1.55866 | 1.80112 |
| 0.380 | 0.27952 | 0.58781 | 0.81346 | 1.23283 | 1.56948 | 1.79965 |
| 0.400 | 0.29836 | 0.60204 | 0.83470 | 1.24352 | 1.58083 | 1.79847 |
| 0.420 | 0.31721 | 0.61611 | 0.85663 | 1.25451 | 1.59270 | 1.79764 |
| 0.440 | 0.33607 | 0.63010 | 0.87913 | 1.26581 | 1.60506 | 1.79720 |
| 0.460 | 0.35493 | 0.64406 | 0.90207 | 1.27741 | 1.61790 | 1.79717 |
| 0.480 | 0.37379 | 0.65805 | 0.92535 | 1.28928 | 1.63118 | 1.79760 |
| 0.500 | 0.39265 | 0.67212 | 0.94885 | 1.30142 | 1.64485 | 1.79854 |
| 0.520 | 0.41151 | 0.68630 | 0.97249 | 1.31384 | 1.65889 | 1.80005 |
| 0.540 | 0.43037 | 0.70061 | 0.99617 | 1.32654 | 1.67325 | 1.80217 |
| 0.560 | 0.44922 | 0.71508 | 1.01981 | 1.33950 | 1.68784 | 1.80501 |
| 0.580 | 0.46808 | 0.72972 | 1.04334 | 1.35272 | 1.70259 | 1.80865 |
| 0.600 | 0.48693 | 0.74455 | 1.06669 | 1.36622 | 1.71739 | 1.81323 |
| 0.620 | 0.50579 | 0.75957 | 1.08978 | 1.37999 | 1.73209 | 1.81888 |
| 0.640 | 0.52465 | 0.77477 | 1.11257 | 1.39401 | 1.74653 | 1.82580 |
| 0.660 | 0.54351 | 0.79017 | 1.13498 | 1.40827 | 1.76055 | 1.83415 |
| 0.680 | 0.56239 | 0.80576 | 1.15699 | 1.42277 | 1.77398 | 1.84414 |
| 0.700 | 0.58127 | 0.82153 | 1.17855 | 1.43748 | 1.78677 | 1.85581 |
| 0.720 | 0.60016 | 0.83749 | 1.19962 | 1.45239 | 1.79896 | 1.86916 |
| 0.740 | 0.61906 | 0.85363 | 1.22018 | 1.46744 | 1.81076 | 1.88401 |
| 0.760 | 0.63797 | 0.86993 | 1.24021 | 1.48262 | 1.82237 | 1.90015 |
| 0.780 | 0.65689 | 0.88640 | 1.25970 | 1.49786 | 1.83407 | 1.91734 |
| 0.800 | 0.67583 | 0.90303 | 1.27867 | 1.51313 | 1.84606 | 1.93539 |
| 0.820 | 0.69477 | 0.91981 | 1.29710 | 1.52836 | 1.85849 | 1.95415 |
| 0.840 | 0.71373 | 0.93673 | 1.31501 | 1.54352 | 1.87150 | 1.97349 |
| 0.860 | 0.73270 | 0.95379 | 1.33243 | 1.55855 | 1.88518 | 1.99332 |
| 0.880 | 0.75169 | 0.97098 | 1.34938 | 1.57342 | 1.89955 | 2.01357 |
| 0.900 | 0.77069 | 0.98830 | 1.36590 | 1.58809 | 1.91466 | 2.03420 |
| 0.920 | 0.78971 | 1.00573 | 1.38201 | 1.60254 | 1.93049 | 2.05516 |
| 0.940 | 0.80874 | 1.02327 | 1.39773 | 1.61675 | 1.94703 | 2.07642 |
| 0.960 | 0.82777 | 1.04092 | 1.41310 | 1.63074 | 1.96426 | 2.09795 |
| 0.980 | 0.84683 | 1.05867 | 1.42815 | 1.64450 | 1.98208 | 2.11974 |
| 1.000 | 0.86589 | 1.07651 | 1.44292 | 1.65803 | 2.00047 | 2.14177 |

$H/R = 1.100 \qquad n = 1.$

| $H/L$ | $\Omega_1$ | $\Omega_2$ | $\Omega_3$ | $\Omega_4$ | $\Omega_5$ | $\Omega_6$ |
|---|---|---|---|---|---|---|
| 0. | 0. | 0.36011 | 0.66369 | 1.15320 | 1.56164 | 1.83998 |
| 0.010 | 0.000371 | 0.36050 | 0.66378 | 1.15329 | 1.56170 | 1.83993 |
| 0.050 | 0.00904 | 0.36958 | 0.66611 | 1.15532 | 1.56335 | 1.83900 |
| 0.100 | 0.03363 | 0.39526 | 0.67354 | 1.16163 | 1.56847 | 1.83625 |
| 0.120 | 0.04667 | 0.40881 | 0.67801 | 1.16528 | 1.57147 | 1.83473 |
| 0.140 | 0.06106 | 0.42366 | 0.68340 | 1.16957 | 1.57502 | 1.83304 |
| 0.160 | 0.07653 | 0.43948 | 0.68977 | 1.17446 | 1.57911 | 1.83121 |
| 0.180 | 0.09286 | 0.45599 | 0.69716 | 1.17995 | 1.58376 | 1.82927 |
| 0.200 | 0.10986 | 0.47291 | 0.70564 | 1.18601 | 1.58893 | 1.82728 |
| 0.220 | 0.12740 | 0.49003 | 0.71527 | 1.19261 | 1.59465 | 1.82526 |
| 0.240 | 0.14534 | 0.50715 | 0.72610 | 1.19974 | 1.60090 | 1.82326 |
| 0.260 | 0.16361 | 0.52412 | 0.73818 | 1.20737 | 1.60769 | 1.82133 |
| 0.280 | 0.18213 | 0.54081 | 0.75152 | 1.21549 | 1.61501 | 1.81948 |
| 0.300 | 0.20085 | 0.55715 | 0.76612 | 1.22407 | 1.62284 | 1.81777 |
| 0.320 | 0.21971 | 0.57308 | 0.78196 | 1.23308 | 1.63119 | 1.81621 |
| 0.340 | 0.23868 | 0.58859 | 0.79897 | 1.24252 | 1.64005 | 1.81486 |
| 0.360 | 0.25774 | 0.60371 | 0.81707 | 1.25236 | 1.64941 | 1.81374 |
| 0.380 | 0.27687 | 0.61846 | 0.83616 | 1.26258 | 1.65923 | 1.81289 |
| 0.400 | 0.29604 | 0.63292 | 0.85611 | 1.27317 | 1.66953 | 1.81234 |
| 0.420 | 0.31524 | 0.64714 | 0.87682 | 1.28414 | 1.68027 | 1.81213 |
| 0.440 | 0.33446 | 0.66119 | 0.89815 | 1.29544 | 1.69143 | 1.81230 |
| 0.460 | 0.35369 | 0.67514 | 0.91999 | 1.30707 | 1.70295 | 1.81292 |
| 0.480 | 0.37294 | 0.68903 | 0.94222 | 1.31905 | 1.71481 | 1.81404 |
| 0.500 | 0.39219 | 0.70293 | 0.96474 | 1.33135 | 1.72694 | 1.81572 |
| 0.520 | 0.41144 | 0.71687 | 0.98744 | 1.34397 | 1.73922 | 1.81811 |
| 0.540 | 0.43068 | 0.73089 | 1.01025 | 1.35691 | 1.75153 | 1.82132 |
| 0.560 | 0.44993 | 0.74501 | 1.03306 | 1.37017 | 1.76368 | 1.82555 |
| 0.580 | 0.46916 | 0.75927 | 1.05581 | 1.38375 | 1.77541 | 1.83111 |
| 0.600 | 0.48840 | 0.77367 | 1.07843 | 1.39763 | 1.78643 | 1.83826 |
| 0.620 | 0.50763 | 0.78822 | 1.10084 | 1.41182 | 1.79650 | 1.84727 |
| 0.640 | 0.52685 | 0.80295 | 1.12299 | 1.42631 | 1.80561 | 1.85816 |
| 0.660 | 0.54607 | 0.81784 | 1.14481 | 1.44110 | 1.81394 | 1.87076 |
| 0.680 | 0.56529 | 0.83290 | 1.16628 | 1.45615 | 1.82184 | 1.88474 |
| 0.700 | 0.58450 | 0.84814 | 1.18735 | 1.47145 | 1.82965 | 1.89979 |
| 0.720 | 0.60371 | 0.86355 | 1.20798 | 1.48698 | 1.83759 | 1.91568 |
| 0.740 | 0.62291 | 0.87913 | 1.22815 | 1.50269 | 1.84587 | 1.93223 |
| 0.760 | 0.64212 | 0.89488 | 1.24784 | 1.51855 | 1.85464 | 1.94934 |
| 0.780 | 0.66132 | 0.91079 | 1.26705 | 1.53453 | 1.86399 | 1.96690 |
| 0.800 | 0.68052 | 0.92686 | 1.28576 | 1.55055 | 1.87400 | 1.98485 |
| 0.820 | 0.69973 | 0.94309 | 1.30399 | 1.56656 | 1.88473 | 2.00315 |
| 0.840 | 0.71893 | 0.95947 | 1.32175 | 1.58253 | 1.89622 | 2.02177 |
| 0.860 | 0.73813 | 0.97599 | 1.33907 | 1.59839 | 1.90850 | 2.04066 |
| 0.880 | 0.75733 | 0.99264 | 1.35594 | 1.61409 | 1.92159 | 2.05983 |
| 0.900 | 0.77654 | 1.00943 | 1.37240 | 1.62959 | 1.93550 | 2.07924 |
| 0.920 | 0.79574 | 1.02635 | 1.38847 | 1.64487 | 1.95020 | 2.09888 |
| 0.940 | 0.81495 | 1.04338 | 1.40420 | 1.65988 | 1.96565 | 2.11871 |
| 0.960 | 0.83416 | 1.06053 | 1.41960 | 1.67465 | 1.98186 | 2.13877 |
| 0.980 | 0.85337 | 1.07780 | 1.43469 | 1.68913 | 1.99874 | 2.15900 |
| 1.000 | 0.87257 | 1.09517 | 1.44951 | 1.70336 | 2.01625 | 2.17941 |

$H/R = 1.200 \qquad n = 1.$

| $H/L$ | $\Omega_1$ | $\Omega_2$ | $\Omega_3$ | $\Omega_4$ | $\Omega_5$ | $\Omega_6$ |
|---|---|---|---|---|---|---|
| 0. | 0. | 0.39260 | 0.69631 | 1.19473 | 1.66112 | 1.84615 |
| 0.010 | 0.000348 | 0.39296 | 0.69640 | 1.19481 | 1.66119 | 1.84613 |
| 0.050 | 0.00850 | 0.40144 | 0.69863 | 1.19667 | 1.66269 | 1.84548 |
| 0.100 | 0.03191 | 0.42573 | 0.70573 | 1.20246 | 1.66739 | 1.84354 |
| 0.120 | 0.04448 | 0.43869 | 0.70997 | 1.20583 | 1.67013 | 1.84249 |
| 0.140 | 0.05843 | 0.45301 | 0.71507 | 1.20978 | 1.67338 | 1.84132 |
| 0.160 | 0.07353 | 0.46838 | 0.72107 | 1.21430 | 1.67712 | 1.84006 |
| 0.180 | 0.08956 | 0.48452 | 0.72802 | 1.21939 | 1.68137 | 1.83872 |
| 0.200 | 0.10633 | 0.50119 | 0.73596 | 1.22502 | 1.68610 | 1.83737 |
| 0.220 | 0.12370 | 0.51816 | 0.74494 | 1.23117 | 1.69131 | 1.83604 |
| 0.240 | 0.14155 | 0.53525 | 0.75502 | 1.23784 | 1.69701 | 1.83473 |
| 0.260 | 0.15979 | 0.55230 | 0.76624 | 1.24500 | 1.70319 | 1.83352 |
| 0.280 | 0.17833 | 0.56916 | 0.77863 | 1.25265 | 1.70984 | 1.83239 |
| 0.300 | 0.19712 | 0.58574 | 0.79221 | 1.26077 | 1.71695 | 1.83141 |
| 0.320 | 0.21609 | 0.60196 | 0.80695 | 1.26933 | 1.72452 | 1.83061 |
| 0.340 | 0.23522 | 0.61778 | 0.82283 | 1.27832 | 1.73252 | 1.83000 |
| 0.360 | 0.25446 | 0.63321 | 0.83979 | 1.28775 | 1.74094 | 1.82965 |
| 0.380 | 0.27379 | 0.64825 | 0.85775 | 1.29758 | 1.74977 | 1.82958 |
| 0.400 | 0.29320 | 0.66294 | 0.87660 | 1.30782 | 1.75893 | 1.82986 |
| 0.420 | 0.31265 | 0.67734 | 0.89624 | 1.31844 | 1.76842 | 1.83056 |
| 0.440 | 0.33215 | 0.69152 | 0.91655 | 1.32945 | 1.77810 | 1.83177 |
| 0.460 | 0.35167 | 0.70551 | 0.93742 | 1.34083 | 1.78786 | 1.83366 |
| 0.480 | 0.37121 | 0.71939 | 0.95873 | 1.35258 | 1.79743 | 1.83646 |
| 0.500 | 0.39076 | 0.73321 | 0.98038 | 1.36471 | 1.80644 | 1.84061 |
| 0.520 | 0.41032 | 0.74701 | 1.00227 | 1.37721 | 1.81441 | 1.84657 |
| 0.540 | 0.42988 | 0.76085 | 1.02431 | 1.39006 | 1.82105 | 1.85461 |
| 0.560 | 0.44943 | 0.77473 | 1.04640 | 1.40329 | 1.82656 | 1.86458 |
| 0.580 | 0.46899 | 0.78870 | 1.06847 | 1.41688 | 1.83142 | 1.87601 |
| 0.600 | 0.48854 | 0.80278 | 1.09044 | 1.43083 | 1.83605 | 1.88847 |
| 0.620 | 0.50808 | 0.81698 | 1.11226 | 1.44513 | 1.84072 | 1.90171 |
| 0.640 | 0.52761 | 0.83131 | 1.13386 | 1.45979 | 1.84561 | 1.91554 |
| 0.660 | 0.54713 | 0.84579 | 1.15518 | 1.47479 | 1.85084 | 1.92987 |
| 0.680 | 0.56664 | 0.86043 | 1.17618 | 1.49013 | 1.85648 | 1.94462 |
| 0.700 | 0.58614 | 0.87522 | 1.19683 | 1.50576 | 1.86260 | 1.95978 |
| 0.720 | 0.60563 | 0.89017 | 1.21708 | 1.52169 | 1.86924 | 1.97528 |
| 0.740 | 0.62512 | 0.90527 | 1.23692 | 1.53787 | 1.87646 | 1.99109 |
| 0.760 | 0.64459 | 0.92054 | 1.25632 | 1.55426 | 1.88428 | 2.00720 |
| 0.780 | 0.66405 | 0.93597 | 1.27527 | 1.57083 | 1.89275 | 2.02360 |
| 0.800 | 0.68350 | 0.95155 | 1.29377 | 1.58751 | 1.90189 | 2.04025 |
| 0.820 | 0.70294 | 0.96728 | 1.31183 | 1.60427 | 1.91173 | 2.05715 |
| 0.840 | 0.72238 | 0.98316 | 1.32944 | 1.62103 | 1.92232 | 2.07429 |
| 0.860 | 0.74180 | 0.99918 | 1.34665 | 1.63774 | 1.93366 | 2.09164 |
| 0.880 | 0.76122 | 1.01535 | 1.36344 | 1.65436 | 1.94577 | 2.10920 |
| 0.900 | 0.78062 | 1.03164 | 1.37984 | 1.67081 | 1.95866 | 2.12696 |
| 0.920 | 0.80002 | 1.04807 | 1.39589 | 1.68706 | 1.97231 | 2.14492 |
| 0.940 | 0.81942 | 1.06463 | 1.41162 | 1.70308 | 1.98674 | 2.16307 |
| 0.960 | 0.83880 | 1.08130 | 1.42702 | 1.71884 | 2.00189 | 2.18139 |
| 0.980 | 0.85818 | 1.09810 | 1.44215 | 1.73431 | 2.01775 | 2.19990 |
| 1.000 | 0.87755 | 1.11501 | 1.45702 | 1.74949 | 2.03427 | 2.21857 |

$H/R = 1.300 \qquad n = 1.$

| $H/L$ | $\Omega_1$ | $\Omega_2$ | $\Omega_3$ | $\Omega_4$ | $\Omega_5$ | $\Omega_6$ |
|---|---|---|---|---|---|---|
| 0. | 0. | 0.42403 | 0.72602 | 1.24465 | 1.75711 | 1.85525 |
| 0.010 | 0.000328 | 0.42437 | 0.72611 | 1.24472 | 1.75717 | 1.85523 |
| 0.050 | 0.00805 | 0.43237 | 0.72826 | 1.24638 | 1.75852 | 1.85488 |
| 0.100 | 0.03044 | 0.45547 | 0.73507 | 1.25157 | 1.76278 | 1.85383 |
| 0.120 | 0.04257 | 0.46792 | 0.73913 | 1.25460 | 1.76526 | 1.85327 |
| 0.140 | 0.05611 | 0.48175 | 0.74400 | 1.25814 | 1.76820 | 1.85265 |
| 0.160 | 0.07085 | 0.49670 | 0.74971 | 1.26223 | 1.77158 | 1.85202 |
| 0.180 | 0.08656 | 0.51248 | 0.75629 | 1.26681 | 1.77540 | 1.85138 |
| 0.200 | 0.10308 | 0.52888 | 0.76380 | 1.27191 | 1.77966 | 1.85077 |
| 0.220 | 0.12026 | 0.54568 | 0.77227 | 1.27751 | 1.78434 | 1.85022 |
| 0.240 | 0.13797 | 0.56268 | 0.78175 | 1.28359 | 1.78943 | 1.84975 |
| 0.260 | 0.15612 | 0.57974 | 0.79228 | 1.29013 | 1.79493 | 1.84941 |
| 0.280 | 0.17463 | 0.59669 | 0.80390 | 1.29714 | 1.80081 | 1.84922 |
| 0.300 | 0.19342 | 0.61344 | 0.81663 | 1.30461 | 1.80704 | 1.84925 |
| 0.320 | 0.21244 | 0.62988 | 0.83047 | 1.31252 | 1.81355 | 1.84958 |
| 0.340 | 0.23165 | 0.64596 | 0.84541 | 1.32087 | 1.82024 | 1.85034 |
| 0.360 | 0.25100 | 0.66165 | 0.86141 | 1.32964 | 1.82691 | 1.85172 |
| 0.380 | 0.27048 | 0.67695 | 0.87840 | 1.33883 | 1.83317 | 1.85414 |
| 0.400 | 0.29004 | 0.69188 | 0.89630 | 1.34844 | 1.83847 | 1.85816 |
| 0.420 | 0.30968 | 0.70647 | 0.91503 | 1.35847 | 1.84247 | 1.86413 |
| 0.440 | 0.32938 | 0.72078 | 0.93445 | 1.36888 | 1.84548 | 1.87175 |
| 0.460 | 0.34911 | 0.73487 | 0.95448 | 1.37972 | 1.84802 | 1.88052 |
| 0.480 | 0.36888 | 0.74879 | 0.97500 | 1.39095 | 1.85043 | 1.89009 |
| 0.500 | 0.38866 | 0.76258 | 0.99589 | 1.40258 | 1.85291 | 1.90027 |
| 0.520 | 0.40846 | 0.77632 | 1.01708 | 1.41462 | 1.85559 | 1.91095 |
| 0.540 | 0.42827 | 0.79003 | 1.03845 | 1.42706 | 1.85852 | 1.92207 |
| 0.560 | 0.44808 | 0.80374 | 1.05992 | 1.43991 | 1.86174 | 1.93361 |
| 0.580 | 0.46789 | 0.81751 | 1.08141 | 1.45317 | 1.86530 | 1.94552 |
| 0.600 | 0.48769 | 0.83135 | 1.10285 | 1.46683 | 1.86922 | 1.95779 |
| 0.620 | 0.50749 | 0.84527 | 1.12417 | 1.48092 | 1.87351 | 1.97040 |
| 0.640 | 0.52727 | 0.85930 | 1.14529 | 1.49541 | 1.87821 | 1.98334 |
| 0.660 | 0.54705 | 0.87345 | 1.16619 | 1.51030 | 1.88333 | 1.99657 |
| 0.680 | 0.56681 | 0.88774 | 1.18681 | 1.52560 | 1.88890 | 2.01011 |
| 0.700 | 0.58656 | 0.90216 | 1.20709 | 1.54127 | 1.89494 | 2.02394 |
| 0.720 | 0.60630 | 0.91672 | 1.22703 | 1.55729 | 1.90146 | 2.03803 |
| 0.740 | 0.62602 | 0.93143 | 1.24658 | 1.57366 | 1.90849 | 2.05239 |
| 0.760 | 0.64572 | 0.94629 | 1.26573 | 1.59031 | 1.91607 | 2.06698 |
| 0.780 | 0.66541 | 0.96129 | 1.28446 | 1.60724 | 1.92418 | 2.08183 |
| 0.800 | 0.68509 | 0.97645 | 1.30279 | 1.62439 | 1.93289 | 2.09692 |
| 0.820 | 0.70475 | 0.99174 | 1.32068 | 1.64169 | 1.94220 | 2.11221 |
| 0.840 | 0.72439 | 1.00718 | 1.33818 | 1.65911 | 1.95215 | 2.12773 |
| 0.860 | 0.74402 | 1.02276 | 1.35528 | 1.67657 | 1.96274 | 2.14346 |
| 0.880 | 0.76363 | 1.03849 | 1.37199 | 1.69404 | 1.97401 | 2.15939 |
| 0.900 | 0.78323 | 1.05434 | 1.38835 | 1.71143 | 1.98595 | 2.17552 |
| 0.920 | 0.80282 | 1.07032 | 1.40436 | 1.72870 | 1.99858 | 2.19186 |
| 0.940 | 0.82239 | 1.08644 | 1.42007 | 1.74579 | 2.01191 | 2.20839 |
| 0.960 | 0.84194 | 1.10267 | 1.43549 | 1.76265 | 2.02593 | 2.22512 |
| 0.980 | 0.86148 | 1.11903 | 1.45063 | 1.77927 | 2.04062 | 2.24204 |
| 1.000 | 0.88101 | 1.13550 | 1.46553 | 1.79559 | 2.05595 | 2.25917 |

$H/R = 1.400 \qquad n = 1.$

| $H/L$ | $\Omega_1$ | $\Omega_2$ | $\Omega_3$ | $\Omega_4$ | $\Omega_5$ | $\Omega_6$ |
|---|---|---|---|---|---|---|
| 0. | 0. | 0.45399 | 0.75354 | 1.30353 | 1.84505 | 1.86931 |
| 0.010 | 0.000312 | 0.45431 | 0.75362 | 1.30359 | 1.84509 | 1.86931 |
| 0.050 | 0.00767 | 0.46189 | 0.75570 | 1.30504 | 1.84618 | 1.86941 |
| 0.100 | 0.02916 | 0.48399 | 0.76227 | 1.30952 | 1.84948 | 1.86986 |
| 0.120 | 0.04089 | 0.49599 | 0.76617 | 1.31214 | 1.85131 | 1.87025 |
| 0.140 | 0.05406 | 0.50939 | 0.77085 | 1.31523 | 1.85334 | 1.87087 |
| 0.160 | 0.06844 | 0.52394 | 0.77631 | 1.31877 | 1.85549 | 1.87183 |
| 0.180 | 0.08385 | 0.53939 | 0.78259 | 1.32278 | 1.85760 | 1.87329 |
| 0.200 | 0.10010 | 0.55552 | 0.78974 | 1.32723 | 1.85952 | 1.87538 |
| 0.220 | 0.11707 | 0.57212 | 0.79778 | 1.33214 | 1.86117 | 1.87826 |
| 0.240 | 0.13462 | 0.58902 | 0.80677 | 1.33749 | 1.86256 | 1.88189 |
| 0.260 | 0.15265 | 0.60604 | 0.81673 | 1.34326 | 1.86374 | 1.88622 |
| 0.280 | 0.17107 | 0.62305 | 0.82772 | 1.34948 | 1.86482 | 1.89120 |
| 0.300 | 0.18982 | 0.63990 | 0.83975 | 1.35611 | 1.86589 | 1.89671 |
| 0.320 | 0.20884 | 0.65651 | 0.85284 | 1.36317 | 1.86701 | 1.90273 |
| 0.340 | 0.22808 | 0.67280 | 0.86698 | 1.37065 | 1.86823 | 1.90920 |
| 0.360 | 0.24749 | 0.68872 | 0.88217 | 1.37856 | 1.86959 | 1.91612 |
| 0.380 | 0.26704 | 0.70425 | 0.89833 | 1.38687 | 1.87111 | 1.92344 |
| 0.400 | 0.28671 | 0.71940 | 0.91542 | 1.39559 | 1.87282 | 1.93117 |
| 0.420 | 0.30648 | 0.73419 | 0.93334 | 1.40473 | 1.87474 | 1.93928 |
| 0.440 | 0.32631 | 0.74866 | 0.95200 | 1.41429 | 1.87688 | 1.94776 |
| 0.460 | 0.34620 | 0.76286 | 0.97130 | 1.42426 | 1.87926 | 1.95662 |
| 0.480 | 0.36613 | 0.77685 | 0.99113 | 1.43467 | 1.88192 | 1.96582 |
| 0.500 | 0.38610 | 0.79067 | 1.01138 | 1.44549 | 1.88482 | 1.97537 |
| 0.520 | 0.40608 | 0.80438 | 1.03196 | 1.45675 | 1.88802 | 1.98526 |
| 0.540 | 0.42608 | 0.81802 | 1.05277 | 1.46844 | 1.89152 | 1.99546 |
| 0.560 | 0.44609 | 0.83163 | 1.07372 | 1.48058 | 1.89531 | 2.00599 |
| 0.580 | 0.46610 | 0.84525 | 1.09472 | 1.49318 | 1.89944 | 2.01682 |
| 0.600 | 0.48611 | 0.85891 | 1.11570 | 1.50624 | 1.90389 | 2.02797 |
| 0.620 | 0.50611 | 0.87263 | 1.13660 | 1.51975 | 1.90868 | 2.03938 |
| 0.640 | 0.52611 | 0.88643 | 1.15736 | 1.53373 | 1.91383 | 2.05110 |
| 0.660 | 0.54609 | 0.90032 | 1.17790 | 1.54817 | 1.91934 | 2.06309 |
| 0.680 | 0.56606 | 0.91433 | 1.19819 | 1.56308 | 1.92522 | 2.07533 |
| 0.700 | 0.58602 | 0.92844 | 1.21820 | 1.57845 | 1.93150 | 2.08783 |
| 0.720 | 0.60596 | 0.94269 | 1.23788 | 1.59424 | 1.93818 | 2.10060 |
| 0.740 | 0.62589 | 0.95707 | 1.25720 | 1.61047 | 1.94528 | 2.11360 |
| 0.760 | 0.64580 | 0.97158 | 1.27614 | 1.62707 | 1.95281 | 2.12684 |
| 0.780 | 0.66568 | 0.98623 | 1.29470 | 1.64405 | 1.96080 | 2.14031 |
| 0.800 | 0.68555 | 1.00102 | 1.31288 | 1.66136 | 1.96924 | 2.15400 |
| 0.820 | 0.70540 | 1.01593 | 1.33065 | 1.67895 | 1.97817 | 2.16793 |
| 0.840 | 0.72523 | 1.03100 | 1.34803 | 1.69676 | 1.98760 | 2.18206 |
| 0.860 | 0.74504 | 1.04619 | 1.36505 | 1.71475 | 1.99756 | 2.19642 |
| 0.880 | 0.76483 | 1.06151 | 1.38170 | 1.73287 | 2.00805 | 2.21099 |
| 0.900 | 0.78461 | 1.07697 | 1.39801 | 1.75103 | 2.01910 | 2.22577 |
| 0.920 | 0.80436 | 1.09257 | 1.41400 | 1.76919 | 2.03071 | 2.24077 |
| 0.940 | 0.82410 | 1.10828 | 1.42968 | 1.78728 | 2.04292 | 2.25599 |
| 0.960 | 0.84381 | 1.12411 | 1.44510 | 1.80525 | 2.05570 | 2.27143 |
| 0.980 | 0.86351 | 1.14007 | 1.46026 | 1.82304 | 2.06908 | 2.28710 |
| 1.000 | 0.88318 | 1.15614 | 1.47519 | 1.84061 | 2.08305 | 2.30299 |

$H/R = 1.500 \qquad n = 1.$

| $H/L$ | $\Omega_1$ | $\Omega_2$ | $\Omega_3$ | $\Omega_4$ | $\Omega_5$ | $\Omega_6$ |
|---|---|---|---|---|---|---|
| 0. | 0. | 0.48205 | 0.77943 | 1.37090 | 1.88106 | 1.93161 |
| 0.010 | 0.000298 | 0.48236 | 0.77952 | 1.37096 | 1.88107 | 1.93166 |
| 0.050 | 0.00734 | 0.48960 | 0.78153 | 1.37214 | 1.88133 | 1.93283 |
| 0.100 | 0.02805 | 0.51083 | 0.78788 | 1.37580 | 1.88216 | 1.93650 |
| 0.120 | 0.03943 | 0.52243 | 0.79164 | 1.37795 | 1.88266 | 1.93864 |
| 0.140 | 0.05224 | 0.53545 | 0.79614 | 1.38049 | 1.88325 | 1.94117 |
| 0.160 | 0.06629 | 0.54964 | 0.80139 | 1.38342 | 1.88397 | 1.94409 |
| 0.180 | 0.08140 | 0.56477 | 0.80742 | 1.38674 | 1.88480 | 1.94738 |
| 0.200 | 0.09739 | 0.58065 | 0.81426 | 1.39043 | 1.88575 | 1.95107 |
| 0.220 | 0.11413 | 0.59706 | 0.82195 | 1.39451 | 1.88684 | 1.95513 |
| 0.240 | 0.13149 | 0.61383 | 0.83052 | 1.39899 | 1.88807 | 1.95955 |
| 0.260 | 0.14937 | 0.63079 | 0.84001 | 1.40385 | 1.88944 | 1.96436 |
| 0.280 | 0.16769 | 0.64780 | 0.85046 | 1.40909 | 1.89098 | 1.96953 |
| 0.300 | 0.18637 | 0.66473 | 0.86190 | 1.41471 | 1.89269 | 1.97507 |
| 0.320 | 0.20534 | 0.68147 | 0.87435 | 1.42073 | 1.89456 | 1.98095 |
| 0.340 | 0.22456 | 0.69794 | 0.88782 | 1.42715 | 1.89663 | 1.98720 |
| 0.360 | 0.24398 | 0.71406 | 0.90230 | 1.43394 | 1.89888 | 1.99380 |
| 0.380 | 0.26357 | 0.72980 | 0.91775 | 1.44115 | 1.90133 | 2.00073 |
| 0.400 | 0.28330 | 0.74516 | 0.93412 | 1.44875 | 1.90400 | 2.00800 |
| 0.420 | 0.30313 | 0.76015 | 0.95135 | 1.45676 | 1.90688 | 2.01560 |
| 0.440 | 0.32306 | 0.77479 | 0.96934 | 1.46519 | 1.90999 | 2.02353 |
| 0.460 | 0.34305 | 0.78912 | 0.98800 | 1.47404 | 1.91332 | 2.03177 |
| 0.480 | 0.36310 | 0.80321 | 1.00723 | 1.48333 | 1.91690 | 2.04033 |
| 0.500 | 0.38320 | 0.81709 | 1.02693 | 1.49307 | 1.92073 | 2.04919 |
| 0.520 | 0.40332 | 0.83081 | 1.04698 | 1.50326 | 1.92479 | 2.05835 |
| 0.540 | 0.42347 | 0.84443 | 1.06731 | 1.51391 | 1.92913 | 2.06780 |
| 0.560 | 0.44363 | 0.85799 | 1.08782 | 1.52505 | 1.93372 | 2.07753 |
| 0.580 | 0.46380 | 0.87152 | 1.10842 | 1.53666 | 1.93859 | 2.08754 |
| 0.600 | 0.48397 | 0.88506 | 1.12903 | 1.54879 | 1.94373 | 2.09782 |
| 0.620 | 0.50414 | 0.89862 | 1.14960 | 1.56142 | 1.94915 | 2.10836 |
| 0.640 | 0.52430 | 0.91225 | 1.17004 | 1.57457 | 1.95487 | 2.11917 |
| 0.660 | 0.54446 | 0.92594 | 1.19032 | 1.58824 | 1.96088 | 2.13021 |
| 0.680 | 0.56460 | 0.93973 | 1.21037 | 1.60244 | 1.96719 | 2.14152 |
| 0.700 | 0.58473 | 0.95361 | 1.23016 | 1.61716 | 1.97381 | 2.15305 |
| 0.720 | 0.60484 | 0.96760 | 1.24964 | 1.63240 | 1.98076 | 2.16482 |
| 0.740 | 0.62493 | 0.98171 | 1.26879 | 1.64814 | 1.98802 | 2.17682 |
| 0.760 | 0.64501 | 0.99594 | 1.28760 | 1.66437 | 1.99562 | 2.18905 |
| 0.780 | 0.66506 | 1.01029 | 1.30603 | 1.68107 | 2.00356 | 2.20151 |
| 0.800 | 0.68510 | 1.02477 | 1.32409 | 1.69820 | 2.01186 | 2.21418 |
| 0.820 | 0.70512 | 1.03938 | 1.34178 | 1.71574 | 2.02052 | 2.22709 |
| 0.840 | 0.72511 | 1.05412 | 1.35909 | 1.73362 | 2.02955 | 2.24019 |
| 0.860 | 0.74508 | 1.06898 | 1.37605 | 1.75182 | 2.03898 | 2.25354 |
| 0.880 | 0.76503 | 1.08397 | 1.39266 | 1.77026 | 2.04881 | 2.26709 |
| 0.900 | 0.78495 | 1.09909 | 1.40895 | 1.78891 | 2.05907 | 2.28087 |
| 0.920 | 0.80485 | 1.11433 | 1.42491 | 1.80768 | 2.06975 | 2.29487 |
| 0.940 | 0.82473 | 1.12969 | 1.44059 | 1.82654 | 2.08088 | 2.30911 |
| 0.960 | 0.84459 | 1.14517 | 1.45601 | 1.84542 | 2.09247 | 2.32358 |
| 0.980 | 0.86442 | 1.16078 | 1.47118 | 1.86425 | 2.10452 | 2.33830 |
| 1.000 | 0.88424 | 1.17649 | 1.48615 | 1.88297 | 2.11707 | 2.35326 |

$H/R = 1.600 \qquad n = 1.$

| $H/L$ | $\Omega_1$ | $\Omega_2$ | $\Omega_3$ | $\Omega_4$ | $\Omega_5$ | $\Omega_6$ |
|---|---|---|---|---|---|---|
| 0. | 0. | 0.50787 | 0.80415 | 1.44434 | 1.90161 | 2.00670 |
| 0.010 | 0.000287 | 0.50817 | 0.80423 | 1.44439 | 1.90163 | 2.00675 |
| 0.050 | 0.00706 | 0.51512 | 0.80618 | 1.44526 | 1.90220 | 2.00794 |
| 0.100 | 0.02708 | 0.53560 | 0.81233 | 1.44801 | 1.90398 | 2.01164 |
| 0.120 | 0.03813 | 0.54685 | 0.81598 | 1.44963 | 1.90502 | 2.01380 |
| 0.140 | 0.05062 | 0.55953 | 0.82032 | 1.45155 | 1.90625 | 2.01635 |
| 0.160 | 0.06437 | 0.57339 | 0.82539 | 1.45377 | 1.90766 | 2.01929 |
| 0.180 | 0.07918 | 0.58824 | 0.83119 | 1.45630 | 1.90927 | 2.02261 |
| 0.200 | 0.09491 | 0.60387 | 0.83777 | 1.45914 | 1.91107 | 2.02631 |
| 0.220 | 0.11142 | 0.62008 | 0.84515 | 1.46229 | 1.91306 | 2.03039 |
| 0.240 | 0.12859 | 0.63671 | 0.85336 | 1.46577 | 1.91523 | 2.03483 |
| 0.260 | 0.14631 | 0.65360 | 0.86245 | 1.46957 | 1.91760 | 2.03963 |
| 0.280 | 0.16449 | 0.67059 | 0.87244 | 1.47370 | 1.92016 | 2.04480 |
| 0.300 | 0.18306 | 0.68757 | 0.88338 | 1.47817 | 1.92292 | 2.05033 |
| 0.320 | 0.20196 | 0.70441 | 0.89527 | 1.48300 | 1.92586 | 2.05618 |
| 0.340 | 0.22113 | 0.72101 | 0.90815 | 1.48818 | 1.92900 | 2.06240 |
| 0.360 | 0.24053 | 0.73731 | 0.92201 | 1.49371 | 1.93234 | 2.06893 |
| 0.380 | 0.26012 | 0.75326 | 0.93683 | 1.49963 | 1.93587 | 2.07580 |
| 0.400 | 0.27986 | 0.76882 | 0.95258 | 1.50594 | 1.93961 | 2.08298 |
| 0.420 | 0.29973 | 0.78400 | 0.96918 | 1.51264 | 1.94355 | 2.09047 |
| 0.440 | 0.31970 | 0.79883 | 0.98658 | 1.51975 | 1.94769 | 2.09826 |
| 0.460 | 0.33976 | 0.81332 | 1.00467 | 1.52730 | 1.95204 | 2.10634 |
| 0.480 | 0.35988 | 0.82752 | 1.02337 | 1.53528 | 1.95659 | 2.11470 |
| 0.500 | 0.38006 | 0.84149 | 1.04257 | 1.54372 | 1.96135 | 2.12334 |
| 0.520 | 0.40028 | 0.85528 | 1.06218 | 1.55263 | 1.96632 | 2.13226 |
| 0.540 | 0.42053 | 0.86891 | 1.08210 | 1.56204 | 1.97150 | 2.14142 |
| 0.560 | 0.44081 | 0.88246 | 1.10224 | 1.57194 | 1.97691 | 2.15084 |
| 0.580 | 0.46110 | 0.89595 | 1.12250 | 1.58237 | 1.98254 | 2.16051 |
| 0.600 | 0.48139 | 0.90941 | 1.14282 | 1.59334 | 1.98838 | 2.17043 |
| 0.620 | 0.50168 | 0.92288 | 1.16312 | 1.60485 | 1.99444 | 2.18056 |
| 0.640 | 0.52198 | 0.93638 | 1.18333 | 1.61693 | 2.00073 | 2.19092 |
| 0.660 | 0.54227 | 0.94993 | 1.20341 | 1.62959 | 2.00724 | 2.20151 |
| 0.680 | 0.56254 | 0.96355 | 1.22329 | 1.64282 | 2.01400 | 2.21231 |
| 0.700 | 0.58281 | 0.97726 | 1.24293 | 1.65664 | 2.02098 | 2.22333 |
| 0.720 | 0.60306 | 0.99106 | 1.26228 | 1.67105 | 2.02821 | 2.23455 |
| 0.740 | 0.62330 | 1.00495 | 1.28134 | 1.68604 | 2.03568 | 2.24598 |
| 0.760 | 0.64351 | 1.01896 | 1.30006 | 1.70160 | 2.04341 | 2.25762 |
| 0.780 | 0.66371 | 1.03308 | 1.31843 | 1.71772 | 2.05138 | 2.26946 |
| 0.800 | 0.68388 | 1.04730 | 1.33645 | 1.73437 | 2.05960 | 2.28149 |
| 0.820 | 0.70403 | 1.06166 | 1.35410 | 1.75150 | 2.06810 | 2.29373 |
| 0.840 | 0.72416 | 1.07612 | 1.37139 | 1.76912 | 2.07688 | 2.30617 |
| 0.860 | 0.74427 | 1.09072 | 1.38834 | 1.78715 | 2.08593 | 2.31881 |
| 0.880 | 0.76435 | 1.10543 | 1.40495 | 1.80557 | 2.09527 | 2.33166 |
| 0.900 | 0.78441 | 1.12026 | 1.42123 | 1.82432 | 2.10490 | 2.34473 |
| 0.920 | 0.80444 | 1.13521 | 1.43720 | 1.84336 | 2.11484 | 2.35801 |
| 0.940 | 0.82445 | 1.15027 | 1.45290 | 1.86261 | 2.12510 | 2.37151 |
| 0.960 | 0.84443 | 1.16545 | 1.46834 | 1.88203 | 2.13568 | 2.38524 |
| 0.980 | 0.86439 | 1.18074 | 1.48353 | 1.90155 | 2.14661 | 2.39921 |
| 1.000 | 0.88433 | 1.19615 | 1.49852 | 1.92110 | 2.15789 | 2.41341 |

$H/R = 1.700 \qquad n = 1.$

| $H/L$ | $\Omega_1$ | $\Omega_2$ | $\Omega_3$ | $\Omega_4$ | $\Omega_5$ | $\Omega_6$ |
|---|---|---|---|---|---|---|
| 0. | 0. | 0.53122 | 0.82802 | 1.51906 | 1.92682 | 2.07630 |
| 0.010 | 0.000276 | 0.53151 | 0.82810 | 1.51909 | 1.92686 | 2.07636 |
| 0.050 | 0.00682 | 0.53821 | 0.83000 | 1.51965 | 1.92774 | 2.07760 |
| 0.100 | 0.02623 | 0.55805 | 0.83597 | 1.52142 | 1.93045 | 2.08146 |
| 0.120 | 0.03699 | 0.56900 | 0.83949 | 1.52249 | 1.93201 | 2.08373 |
| 0.140 | 0.04919 | 0.58136 | 0.84370 | 1.52375 | 1.93387 | 2.08640 |
| 0.160 | 0.06264 | 0.59493 | 0.84860 | 1.52523 | 1.93597 | 2.08948 |
| 0.180 | 0.07718 | 0.60951 | 0.85421 | 1.52694 | 1.93833 | 2.09297 |
| 0.200 | 0.09266 | 0.62490 | 0.86056 | 1.52887 | 1.94094 | 2.09685 |
| 0.220 | 0.10894 | 0.64092 | 0.86766 | 1.53106 | 1.94380 | 2.10113 |
| 0.240 | 0.12590 | 0.65741 | 0.87557 | 1.53350 | 1.94688 | 2.10580 |
| 0.260 | 0.14344 | 0.67421 | 0.88430 | 1.53620 | 1.95019 | 2.11086 |
| 0.280 | 0.16148 | 0.69117 | 0.89390 | 1.53919 | 1.95371 | 2.11629 |
| 0.300 | 0.17993 | 0.70816 | 0.90439 | 1.54246 | 1.95746 | 2.12210 |
| 0.320 | 0.19873 | 0.72507 | 0.91581 | 1.54605 | 1.96139 | 2.12828 |
| 0.340 | 0.21782 | 0.74179 | 0.92817 | 1.54997 | 1.96552 | 2.13481 |
| 0.360 | 0.23716 | 0.75825 | 0.94148 | 1.55422 | 1.96985 | 2.14167 |
| 0.380 | 0.25671 | 0.77437 | 0.95574 | 1.55882 | 1.97437 | 2.14888 |
| 0.400 | 0.27644 | 0.79013 | 0.97092 | 1.56379 | 1.97906 | 2.15641 |
| 0.420 | 0.29630 | 0.80552 | 0.98696 | 1.56917 | 1.98394 | 2.16424 |
| 0.440 | 0.31629 | 0.82053 | 1.00381 | 1.57493 | 1.98899 | 2.17238 |
| 0.460 | 0.33638 | 0.83519 | 1.02139 | 1.58114 | 1.99422 | 2.18080 |
| 0.480 | 0.35654 | 0.84955 | 1.03960 | 1.58779 | 1.99961 | 2.18949 |
| 0.500 | 0.37677 | 0.86364 | 1.05836 | 1.59489 | 2.00517 | 2.19845 |
| 0.520 | 0.39705 | 0.87751 | 1.07757 | 1.60249 | 2.01091 | 2.20766 |
| 0.540 | 0.41737 | 0.89120 | 1.09712 | 1.61060 | 2.01682 | 2.21709 |
| 0.560 | 0.43771 | 0.90477 | 1.11694 | 1.61923 | 2.02289 | 2.22676 |
| 0.580 | 0.45808 | 0.91826 | 1.13693 | 1.62841 | 2.02914 | 2.23664 |
| 0.600 | 0.47846 | 0.93169 | 1.15701 | 1.63815 | 2.03554 | 2.24672 |
| 0.620 | 0.49885 | 0.94511 | 1.17711 | 1.64847 | 2.04213 | 2.25701 |
| 0.640 | 0.51924 | 0.95854 | 1.19716 | 1.65939 | 2.04890 | 2.26747 |
| 0.660 | 0.53963 | 0.97199 | 1.21710 | 1.67093 | 2.05582 | 2.27811 |
| 0.680 | 0.56001 | 0.98550 | 1.23688 | 1.68309 | 2.06294 | 2.28892 |
| 0.700 | 0.58038 | 0.99907 | 1.25644 | 1.69589 | 2.07022 | 2.29991 |
| 0.720 | 0.60074 | 1.01272 | 1.27574 | 1.70932 | 2.07769 | 2.31104 |
| 0.740 | 0.62109 | 1.02646 | 1.29478 | 1.72340 | 2.08534 | 2.32235 |
| 0.760 | 0.64141 | 1.04030 | 1.31348 | 1.73810 | 2.09317 | 2.33381 |
| 0.780 | 0.66172 | 1.05423 | 1.33187 | 1.75343 | 2.10120 | 2.34542 |
| 0.800 | 0.68201 | 1.06827 | 1.34992 | 1.76936 | 2.10940 | 2.35719 |
| 0.820 | 0.70227 | 1.08242 | 1.36760 | 1.78586 | 2.11781 | 2.36913 |
| 0.840 | 0.72251 | 1.09667 | 1.38494 | 1.80294 | 2.12642 | 2.38121 |
| 0.860 | 0.74273 | 1.11104 | 1.40193 | 1.82053 | 2.13523 | 2.39346 |
| 0.880 | 0.76292 | 1.12553 | 1.41859 | 1.83860 | 2.14425 | 2.40586 |
| 0.900 | 0.78309 | 1.14012 | 1.43493 | 1.85711 | 2.15348 | 2.41846 |
| 0.920 | 0.80323 | 1.15483 | 1.45097 | 1.87601 | 2.16293 | 2.43122 |
| 0.940 | 0.82335 | 1.16965 | 1.46671 | 1.89526 | 2.17260 | 2.44418 |
| 0.960 | 0.84344 | 1.18458 | 1.48220 | 1.91480 | 2.18251 | 2.45733 |
| 0.980 | 0.86351 | 1.19962 | 1.49746 | 1.93458 | 2.19265 | 2.47068 |
| 1.000 | 0.88354 | 1.21476 | 1.51250 | 1.95454 | 2.20304 | 2.48424 |

$H/R = 1.800 \qquad n = 1.$

| $H/L$ | $\Omega_1$ | $\Omega_2$ | $\Omega_3$ | $\Omega_4$ | $\Omega_5$ | $\Omega_6$ |
|---|---|---|---|---|---|---|
| 0. | 0. | 0.55201 | 0.85130 | 1.58887 | 1.95806 | 2.14123 |
| 0.010 | 0.000268 | 0.55228 | 0.85138 | 1.58888 | 1.95810 | 2.14128 |
| 0.050 | 0.00661 | 0.55878 | 0.85322 | 1.58916 | 1.95922 | 2.14258 |
| 0.100 | 0.02549 | 0.57805 | 0.85902 | 1.59005 | 1.96270 | 2.14663 |
| 0.120 | 0.03598 | 0.58873 | 0.86245 | 1.59060 | 1.96469 | 2.14900 |
| 0.140 | 0.04791 | 0.60081 | 0.86653 | 1.59128 | 1.96703 | 2.15180 |
| 0.160 | 0.06109 | 0.61412 | 0.87128 | 1.59210 | 1.96969 | 2.15505 |
| 0.180 | 0.07537 | 0.62844 | 0.87671 | 1.59307 | 1.97265 | 2.15871 |
| 0.200 | 0.09060 | 0.64361 | 0.88285 | 1.59421 | 1.97591 | 2.16280 |
| 0.220 | 0.10666 | 0.65944 | 0.88972 | 1.59554 | 1.97944 | 2.16733 |
| 0.240 | 0.12342 | 0.67579 | 0.89735 | 1.59707 | 1.98323 | 2.17226 |
| 0.260 | 0.14078 | 0.69248 | 0.90578 | 1.59881 | 1.98727 | 2.17761 |
| 0.280 | 0.15865 | 0.70939 | 0.91502 | 1.60079 | 1.99154 | 2.18338 |
| 0.300 | 0.17696 | 0.72638 | 0.92513 | 1.60303 | 1.99603 | 2.18955 |
| 0.320 | 0.19564 | 0.74334 | 0.93612 | 1.60555 | 2.00071 | 2.19611 |
| 0.340 | 0.21464 | 0.76016 | 0.94802 | 1.60836 | 2.00559 | 2.20306 |
| 0.360 | 0.23390 | 0.77675 | 0.96084 | 1.61151 | 2.01065 | 2.21037 |
| 0.380 | 0.25339 | 0.79304 | 0.97459 | 1.61497 | 2.01587 | 2.21805 |
| 0.400 | 0.27307 | 0.80898 | 0.98924 | 1.61880 | 2.02126 | 2.22607 |
| 0.420 | 0.29291 | 0.82457 | 1.00477 | 1.62302 | 2.02682 | 2.23443 |
| 0.440 | 0.31287 | 0.83977 | 1.02111 | 1.62764 | 2.03250 | 2.24310 |
| 0.460 | 0.33295 | 0.85462 | 1.03821 | 1.63267 | 2.03834 | 2.25207 |
| 0.480 | 0.35312 | 0.86914 | 1.05598 | 1.63816 | 2.04432 | 2.26133 |
| 0.500 | 0.37337 | 0.88337 | 1.07432 | 1.64412 | 2.05042 | 2.27085 |
| 0.520 | 0.39367 | 0.89736 | 1.09316 | 1.65057 | 2.05668 | 2.28064 |
| 0.540 | 0.41402 | 0.91115 | 1.11239 | 1.65753 | 2.06307 | 2.29064 |
| 0.560 | 0.43441 | 0.92478 | 1.13193 | 1.66503 | 2.06958 | 2.30086 |
| 0.580 | 0.45483 | 0.93830 | 1.15168 | 1.67307 | 2.07622 | 2.31127 |
| 0.600 | 0.47526 | 0.95174 | 1.17157 | 1.68171 | 2.08300 | 2.32189 |
| 0.620 | 0.49571 | 0.96514 | 1.19152 | 1.69094 | 2.08992 | 2.33265 |
| 0.640 | 0.51616 | 0.97854 | 1.21145 | 1.70079 | 2.09697 | 2.34358 |
| 0.660 | 0.53662 | 0.99194 | 1.23131 | 1.71128 | 2.10415 | 2.35465 |
| 0.680 | 0.55707 | 1.00537 | 1.25106 | 1.72240 | 2.11148 | 2.36585 |
| 0.700 | 0.57752 | 1.01886 | 1.27061 | 1.73419 | 2.11895 | 2.37717 |
| 0.720 | 0.59795 | 1.03240 | 1.28994 | 1.74665 | 2.12655 | 2.38860 |
| 0.740 | 0.61838 | 1.04602 | 1.30900 | 1.75978 | 2.13431 | 2.40014 |
| 0.760 | 0.63879 | 1.05973 | 1.32779 | 1.77359 | 2.14220 | 2.41178 |
| 0.780 | 0.65918 | 1.07352 | 1.34626 | 1.78805 | 2.15024 | 2.42353 |
| 0.800 | 0.67955 | 1.08741 | 1.36441 | 1.80316 | 2.15844 | 2.43537 |
| 0.820 | 0.69990 | 1.10140 | 1.38221 | 1.81892 | 2.16679 | 2.44731 |
| 0.840 | 0.72023 | 1.11549 | 1.39968 | 1.83529 | 2.17529 | 2.45934 |
| 0.860 | 0.74053 | 1.12969 | 1.41679 | 1.85226 | 2.18396 | 2.47148 |
| 0.880 | 0.76081 | 1.14399 | 1.43358 | 1.86977 | 2.19278 | 2.48372 |
| 0.900 | 0.78107 | 1.15840 | 1.45004 | 1.88780 | 2.20176 | 2.49607 |
| 0.920 | 0.80131 | 1.17292 | 1.46620 | 1.90631 | 2.21092 | 2.50854 |
| 0.940 | 0.82151 | 1.18754 | 1.48207 | 1.92525 | 2.22024 | 2.52114 |
| 0.960 | 0.84169 | 1.20227 | 1.49768 | 1.94458 | 2.22975 | 2.53387 |
| 0.980 | 0.86184 | 1.21710 | 1.51303 | 1.96425 | 2.23943 | 2.54677 |
| 1.000 | 0.88197 | 1.23205 | 1.52816 | 1.98419 | 2.24930 | 2.55981 |

$H/R = 1.900$ $n = 1.$

| $H/L$ | $\Omega_1$ | $\Omega_2$ | $\Omega_3$ | $\Omega_4$ | $\Omega_5$ | $\Omega_6$ |
|---|---|---|---|---|---|---|
| 0. | 0. | 0.57025 | 0.87415 | 1.64883 | 1.99648 | 2.20236 |
| 0.010 | 0.000260 | 0.57052 | 0.87423 | 1.64883 | 1.99654 | 2.20241 |
| 0.050 | 0.00642 | 0.57682 | 0.87602 | 1.64890 | 1.99781 | 2.20374 |
| 0.100 | 0.02482 | 0.59561 | 0.88167 | 1.64920 | 2.00173 | 2.20786 |
| 0.120 | 0.03508 | 0.60604 | 0.88500 | 1.64941 | 2.00399 | 2.21029 |
| 0.140 | 0.04676 | 0.61788 | 0.88897 | 1.64969 | 2.00661 | 2.21316 |
| 0.160 | 0.05969 | 0.63093 | 0.89358 | 1.65006 | 2.00959 | 2.21647 |
| 0.180 | 0.07373 | 0.64503 | 0.89886 | 1.65054 | 2.01290 | 2.22022 |
| 0.200 | 0.08873 | 0.65999 | 0.90481 | 1.65113 | 2.01652 | 2.22440 |
| 0.220 | 0.10457 | 0.67564 | 0.91147 | 1.65188 | 2.02045 | 2.22904 |
| 0.240 | 0.12112 | 0.69184 | 0.91886 | 1.65279 | 2.02465 | 2.23410 |
| 0.260 | 0.13830 | 0.70843 | 0.92700 | 1.65389 | 2.02909 | 2.23961 |
| 0.280 | 0.15601 | 0.72527 | 0.93594 | 1.65520 | 2.03378 | 2.24553 |
| 0.300 | 0.17417 | 0.74224 | 0.94570 | 1.65675 | 2.03869 | 2.25189 |
| 0.320 | 0.19272 | 0.75923 | 0.95631 | 1.65854 | 2.04380 | 2.25865 |
| 0.340 | 0.21160 | 0.77612 | 0.96780 | 1.66062 | 2.04909 | 2.26582 |
| 0.360 | 0.23077 | 0.79282 | 0.98018 | 1.66298 | 2.05454 | 2.27339 |
| 0.380 | 0.25017 | 0.80925 | 0.99346 | 1.66568 | 2.06016 | 2.28132 |
| 0.400 | 0.26978 | 0.82538 | 1.00764 | 1.66873 | 2.06593 | 2.28964 |
| 0.420 | 0.28956 | 0.84115 | 1.02268 | 1.67215 | 2.07183 | 2.29831 |
| 0.440 | 0.30948 | 0.85656 | 1.03855 | 1.67595 | 2.07786 | 2.30731 |
| 0.460 | 0.32953 | 0.87160 | 1.05519 | 1.68017 | 2.08402 | 2.31664 |
| 0.480 | 0.34968 | 0.88631 | 1.07253 | 1.68484 | 2.09029 | 2.32626 |
| 0.500 | 0.36991 | 0.90070 | 1.09049 | 1.68995 | 2.09668 | 2.33616 |
| 0.520 | 0.39021 | 0.91483 | 1.10897 | 1.69556 | 2.10318 | 2.34633 |
| 0.540 | 0.41056 | 0.92873 | 1.12790 | 1.70167 | 2.10979 | 2.35674 |
| 0.560 | 0.43096 | 0.94246 | 1.14718 | 1.70833 | 2.11652 | 2.36739 |
| 0.580 | 0.45139 | 0.95604 | 1.16673 | 1.71552 | 2.12336 | 2.37822 |
| 0.600 | 0.47184 | 0.96952 | 1.18645 | 1.72329 | 2.13030 | 2.38925 |
| 0.620 | 0.49232 | 0.98294 | 1.20628 | 1.73165 | 2.13736 | 2.40044 |
| 0.640 | 0.51281 | 0.99633 | 1.22614 | 1.74064 | 2.14454 | 2.41177 |
| 0.660 | 0.53330 | 1.00971 | 1.24598 | 1.75024 | 2.15183 | 2.42324 |
| 0.680 | 0.55379 | 1.02310 | 1.26572 | 1.76051 | 2.15923 | 2.43483 |
| 0.700 | 0.57428 | 1.03652 | 1.28533 | 1.77143 | 2.16675 | 2.44651 |
| 0.720 | 0.59477 | 1.05000 | 1.30474 | 1.78302 | 2.17440 | 2.45830 |
| 0.740 | 0.61524 | 1.06354 | 1.32393 | 1.79531 | 2.18216 | 2.47014 |
| 0.760 | 0.63571 | 1.07714 | 1.34286 | 1.80827 | 2.19005 | 2.48206 |
| 0.780 | 0.65615 | 1.09083 | 1.36149 | 1.82191 | 2.19807 | 2.49403 |
| 0.800 | 0.67657 | 1.10461 | 1.37982 | 1.83622 | 2.20622 | 2.50608 |
| 0.820 | 0.69698 | 1.11848 | 1.39783 | 1.85121 | 2.21449 | 2.51817 |
| 0.840 | 0.71738 | 1.13244 | 1.41550 | 1.86684 | 2.22291 | 2.53030 |
| 0.860 | 0.73774 | 1.14650 | 1.43283 | 1.88310 | 2.23145 | 2.54250 |
| 0.880 | 0.75809 | 1.16066 | 1.44983 | 1.89996 | 2.24013 | 2.55475 |
| 0.900 | 0.77841 | 1.17492 | 1.46651 | 1.91738 | 2.24895 | 2.56706 |
| 0.920 | 0.79871 | 1.18928 | 1.48288 | 1.93533 | 2.25793 | 2.57946 |
| 0.940 | 0.81898 | 1.20375 | 1.49895 | 1.95379 | 2.26704 | 2.59191 |
| 0.960 | 0.83922 | 1.21832 | 1.51474 | 1.97269 | 2.27630 | 2.60444 |
| 0.980 | 0.85944 | 1.23299 | 1.53027 | 1.99201 | 2.28571 | 2.61707 |
| 1.000 | 0.87963 | 1.24776 | 1.54558 | 2.01168 | 2.29528 | 2.62982 |

$H/R = 2.000 \qquad n = 1.$

| $H/L$ | $\Omega_1$ | $\Omega_2$ | $\Omega_3$ | $\Omega_4$ | $\Omega_5$ | $\Omega_6$ |
|---|---|---|---|---|---|---|
| 0. | 0. | 0.58607 | 0.89670 | 1.69706 | 2.04320 | 2.26031 |
| 0.010 | 0.000253 | 0.58633 | 0.89677 | 1.69705 | 2.04326 | 2.26038 |
| 0.050 | 0.00626 | 0.59247 | 0.89852 | 1.69706 | 2.04456 | 2.26168 |
| 0.100 | 0.02423 | 0.61082 | 0.90402 | 1.69715 | 2.04855 | 2.26576 |
| 0.120 | 0.03428 | 0.62104 | 0.90727 | 1.69725 | 2.05085 | 2.26815 |
| 0.140 | 0.04573 | 0.63265 | 0.91113 | 1.69739 | 2.05353 | 2.27099 |
| 0.160 | 0.05843 | 0.64548 | 0.91562 | 1.69759 | 2.05657 | 2.27427 |
| 0.180 | 0.07224 | 0.65937 | 0.92075 | 1.69789 | 2.05994 | 2.27798 |
| 0.200 | 0.08703 | 0.67413 | 0.92654 | 1.69829 | 2.06364 | 2.28214 |
| 0.220 | 0.10265 | 0.68961 | 0.93301 | 1.69882 | 2.06764 | 2.28671 |
| 0.240 | 0.11901 | 0.70566 | 0.94018 | 1.69950 | 2.07191 | 2.29174 |
| 0.260 | 0.13600 | 0.72214 | 0.94808 | 1.70034 | 2.07645 | 2.29719 |
| 0.280 | 0.15354 | 0.73891 | 0.95674 | 1.70140 | 2.08122 | 2.30307 |
| 0.300 | 0.17154 | 0.75585 | 0.96620 | 1.70265 | 2.08621 | 2.30937 |
| 0.320 | 0.18995 | 0.77284 | 0.97646 | 1.70414 | 2.09138 | 2.31608 |
| 0.340 | 0.20871 | 0.78978 | 0.98758 | 1.70589 | 2.09676 | 2.32320 |
| 0.360 | 0.22776 | 0.80657 | 0.99955 | 1.70793 | 2.10229 | 2.33073 |
| 0.380 | 0.24706 | 0.82314 | 1.01241 | 1.71029 | 2.10799 | 2.33863 |
| 0.400 | 0.26658 | 0.83942 | 1.02614 | 1.71296 | 2.11383 | 2.34692 |
| 0.420 | 0.28629 | 0.85537 | 1.04073 | 1.71598 | 2.11979 | 2.35556 |
| 0.440 | 0.30614 | 0.87098 | 1.05616 | 1.71938 | 2.12589 | 2.36454 |
| 0.460 | 0.32613 | 0.88622 | 1.07237 | 1.72319 | 2.13210 | 2.37386 |
| 0.480 | 0.34623 | 0.90112 | 1.08930 | 1.72740 | 2.13842 | 2.38348 |
| 0.500 | 0.36642 | 0.91570 | 1.10688 | 1.73206 | 2.14486 | 2.39339 |
| 0.520 | 0.38669 | 0.92999 | 1.12502 | 1.73719 | 2.15140 | 2.40358 |
| 0.540 | 0.40702 | 0.94403 | 1.14366 | 1.74280 | 2.15804 | 2.41402 |
| 0.560 | 0.42740 | 0.95787 | 1.16269 | 1.74892 | 2.16478 | 2.42469 |
| 0.580 | 0.44782 | 0.97154 | 1.18203 | 1.75558 | 2.17162 | 2.43558 |
| 0.600 | 0.46827 | 0.98510 | 1.20161 | 1.76278 | 2.17858 | 2.44666 |
| 0.620 | 0.48874 | 0.99856 | 1.22134 | 1.77055 | 2.18563 | 2.45791 |
| 0.640 | 0.50923 | 1.01196 | 1.24116 | 1.77893 | 2.19278 | 2.46933 |
| 0.660 | 0.52973 | 1.02534 | 1.26100 | 1.78791 | 2.20004 | 2.48089 |
| 0.680 | 0.55024 | 1.03872 | 1.28080 | 1.79752 | 2.20741 | 2.49256 |
| 0.700 | 0.57074 | 1.05211 | 1.30049 | 1.80778 | 2.21489 | 2.50433 |
| 0.720 | 0.59124 | 1.06554 | 1.32003 | 1.81868 | 2.22248 | 2.51619 |
| 0.740 | 0.61174 | 1.07901 | 1.33939 | 1.83026 | 2.23018 | 2.52812 |
| 0.760 | 0.63223 | 1.09255 | 1.35852 | 1.84250 | 2.23799 | 2.54013 |
| 0.780 | 0.65270 | 1.10616 | 1.37739 | 1.85543 | 2.24592 | 2.55219 |
| 0.800 | 0.67316 | 1.11984 | 1.39598 | 1.86903 | 2.25398 | 2.56430 |
| 0.820 | 0.69360 | 1.13361 | 1.41426 | 1.88330 | 2.26216 | 2.57643 |
| 0.840 | 0.71402 | 1.14747 | 1.43222 | 1.89823 | 2.27045 | 2.58863 |
| 0.860 | 0.73442 | 1.16142 | 1.44986 | 1.91379 | 2.27887 | 2.60086 |
| 0.880 | 0.75481 | 1.17546 | 1.46718 | 1.92997 | 2.28743 | 2.61311 |
| 0.900 | 0.77516 | 1.18961 | 1.48418 | 1.94675 | 2.29611 | 2.62542 |
| 0.920 | 0.79550 | 1.20385 | 1.50084 | 1.96410 | 2.30492 | 2.63775 |
| 0.940 | 0.81581 | 1.21819 | 1.51722 | 1.98198 | 2.31388 | 2.65014 |
| 0.960 | 0.83610 | 1.23264 | 1.53329 | 2.00035 | 2.32295 | 2.66261 |
| 0.980 | 0.85636 | 1.24716 | 1.54909 | 2.01918 | 2.33218 | 2.67511 |
| 1.000 | 0.87660 | 1.26181 | 1.56464 | 2.03841 | 2.34155 | 2.68771 |

$H/R = 0.010 \qquad n = 2.$

| $H/L$ | $\Omega_1$ | $\Omega_2$ | $\Omega_3$ | $\Omega_4$ | $\Omega_5$ | $\Omega_6$ |
|---|---|---|---|---|---|---|
| 0. | 0.0000417 | 0.00637 | 0.01203 | 1.00003 | 1.00009 | 1.87074 |
| 0.001 | 0.000118 | 0.00650 | 0.01212 | 1.00003 | 1.00010 | 1.87074 |
| 0.005 | 0.00176 | 0.00862 | 0.01429 | 1.00004 | 1.00013 | 1.87068 |
| 0.010 | 0.00351 | 0.01222 | 0.02023 | 1.00009 | 1.00026 | 1.87047 |
| 0.020 | 0.00468 | 0.02107 | 0.03553 | 1.00024 | 1.00077 | 1.86971 |
| 0.030 | 0.00511 | 0.03070 | 0.05187 | 1.00049 | 1.00162 | 1.86846 |
| 0.040 | 0.00559 | 0.04052 | 0.06848 | 1.00084 | 1.00282 | 1.86676 |
| 0.050 | 0.00636 | 0.05041 | 0.08520 | 1.00129 | 1.00436 | 1.86465 |
| 0.060 | 0.00750 | 0.06034 | 0.10197 | 1.00184 | 1.00624 | 1.86218 |
| 0.070 | 0.00906 | 0.07029 | 0.11878 | 1.00249 | 1.00844 | 1.85941 |
| 0.080 | 0.01100 | 0.08025 | 0.13560 | 1.00323 | 1.01098 | 1.85638 |
| 0.090 | 0.01331 | 0.09023 | 0.15242 | 1.00409 | 1.01384 | 1.85314 |
| 0.100 | 0.01596 | 0.10020 | 0.16925 | 1.00503 | 1.01702 | 1.84972 |
| 0.120 | 0.02219 | 0.12017 | 0.20291 | 1.00722 | 1.02432 | 1.84252 |
| 0.140 | 0.02957 | 0.14015 | 0.23654 | 1.00979 | 1.03281 | 1.83499 |
| 0.160 | 0.03801 | 0.16013 | 0.27013 | 1.01276 | 1.04247 | 1.82731 |
| 0.180 | 0.04744 | 0.18011 | 0.30369 | 1.01611 | 1.05322 | 1.81963 |
| 0.200 | 0.05778 | 0.20010 | 0.33718 | 1.01984 | 1.06501 | 1.81202 |
| 0.220 | 0.06898 | 0.22009 | 0.37061 | 1.02396 | 1.07779 | 1.80458 |
| 0.240 | 0.08097 | 0.24009 | 0.40397 | 1.02844 | 1.09148 | 1.79736 |
| 0.260 | 0.09369 | 0.26008 | 0.43724 | 1.03329 | 1.10603 | 1.79041 |
| 0.280 | 0.10709 | 0.28007 | 0.47042 | 1.03851 | 1.12138 | 1.78376 |
| 0.300 | 0.12111 | 0.30007 | 0.50348 | 1.04407 | 1.13749 | 1.77745 |
| 0.320 | 0.13571 | 0.32006 | 0.53644 | 1.04999 | 1.15429 | 1.77152 |
| 0.340 | 0.15083 | 0.34006 | 0.56927 | 1.05626 | 1.17175 | 1.76598 |
| 0.360 | 0.16644 | 0.36006 | 0.60194 | 1.06287 | 1.18980 | 1.76088 |
| 0.380 | 0.18249 | 0.38006 | 0.63447 | 1.06980 | 1.20840 | 1.75621 |
| 0.400 | 0.19894 | 0.40005 | 0.66681 | 1.07707 | 1.22752 | 1.75204 |
| 0.420 | 0.21576 | 0.42005 | 0.69897 | 1.08466 | 1.24711 | 1.74835 |
| 0.440 | 0.23291 | 0.44005 | 0.73092 | 1.09256 | 1.26714 | 1.74519 |
| 0.460 | 0.25036 | 0.46005 | 0.76264 | 1.10077 | 1.28758 | 1.74259 |
| 0.480 | 0.26810 | 0.48004 | 0.79410 | 1.10928 | 1.30838 | 1.74055 |
| 0.500 | 0.28608 | 0.50004 | 0.82528 | 1.11808 | 1.32952 | 1.73912 |
| 0.520 | 0.30430 | 0.52004 | 0.85616 | 1.12716 | 1.35098 | 1.73834 |
| 0.540 | 0.32272 | 0.54004 | 0.88671 | 1.13652 | 1.37272 | 1.73822 |
| 0.560 | 0.34133 | 0.56004 | 0.91688 | 1.14616 | 1.39472 | 1.73879 |
| 0.580 | 0.36012 | 0.58004 | 0.94666 | 1.15606 | 1.41696 | 1.74011 |
| 0.600 | 0.37905 | 0.60003 | 0.97600 | 1.16623 | 1.43941 | 1.74220 |
| 0.620 | 0.39813 | 0.62003 | 1.00486 | 1.17665 | 1.46205 | 1.74510 |
| 0.640 | 0.41733 | 0.64003 | 1.03319 | 1.18730 | 1.48488 | 1.74886 |
| 0.660 | 0.43665 | 0.66003 | 1.06096 | 1.19820 | 1.50786 | 1.75354 |
| 0.680 | 0.45607 | 0.68003 | 1.08812 | 1.20934 | 1.53097 | 1.75914 |
| 0.700 | 0.47558 | 0.70003 | 1.11462 | 1.22069 | 1.55421 | 1.76576 |
| 0.720 | 0.49518 | 0.72003 | 1.14041 | 1.23228 | 1.57756 | 1.77340 |
| 0.740 | 0.51485 | 0.74003 | 1.16546 | 1.24406 | 1.60099 | 1.78214 |
| 0.760 | 0.53458 | 0.76003 | 1.18971 | 1.25606 | 1.62450 | 1.79199 |
| 0.780 | 0.55438 | 0.78003 | 1.21314 | 1.26826 | 1.64806 | 1.80299 |
| 0.800 | 0.57423 | 0.80003 | 1.23569 | 1.28066 | 1.67167 | 1.81519 |
| 0.820 | 0.59412 | 0.82002 | 1.25738 | 1.29325 | 1.69531 | 1.82859 |
| 0.840 | 0.61406 | 0.84003 | 1.27816 | 1.30602 | 1.71897 | 1.84320 |
| 0.860 | 0.63403 | 0.86003 | 1.29805 | 1.31898 | 1.74263 | 1.85902 |
| 0.880 | 0.65404 | 0.88003 | 1.31705 | 1.33210 | 1.76628 | 1.87605 |
| 0.900 | 0.67407 | 0.90002 | 1.33518 | 1.34539 | 1.78992 | 1.89425 |
| 0.920 | 0.69413 | 0.92003 | 1.35248 | 1.35886 | 1.81350 | 1.91359 |
| 0.940 | 0.71421 | 0.94003 | 1.36898 | 1.37248 | 1.83705 | 1.93401 |
| 0.960 | 0.73431 | 0.96003 | 1.38473 | 1.38625 | 1.86055 | 1.95547 |
| 0.980 | 0.75441 | 0.98002 | 1.39980 | 1.40018 | 1.88396 | 1.97791 |
| 1.000 | 0.77454 | 1.00002 | 1.41422 | 1.41426 | 1.90729 | 2.00125 |

$H/R = 0.020$ $n = 2$: $H/R = 0.030$

| $H/L$ | $\Omega_1$ | $\Omega_2$ | $\Omega_3$ | $\Omega_1$ | $\Omega_2$ | $\Omega_3$ | $H/L$ |
|---|---|---|---|---|---|---|---|
| 0. | 0.000167 | 0.01273 | 0.02406 | 0.000375 | 0.01910 | 0.03608 | 0. |
| 0.001 | 0.000177 | 0.01280 | 0.02410 | 0.000378 | 0.01914 | 0.03611 | 0.001 |
| 0.005 | 0.00123 | 0.01421 | 0.02518 | 0.000963 | 0.02015 | 0.03683 | 0.005 |
| 0.010 | 0.00353 | 0.01724 | 0.02858 | 0.00297 | 0.02271 | 0.03909 | 0.010 |
| 0.020 | 0.00706 | 0.02444 | 0.04046 | 0.00748 | 0.02925 | 0.04795 | 0.020 |
| 0.030 | 0.00871 | 0.03293 | 0.05526 | 0.01066 | 0.03667 | 0.06069 | 0.030 |
| 0.040 | 0.00963 | 0.04215 | 0.07106 | 0.01263 | 0.04499 | 0.07525 | 0.040 |
| 0.050 | 0.01041 | 0.05169 | 0.08727 | 0.01395 | 0.05392 | 0.09068 | 0.050 |
| 0.060 | 0.01133 | 0.06139 | 0.10371 | 0.01509 | 0.06322 | 0.10657 | 0.060 |
| 0.070 | 0.01252 | 0.07119 | 0.12027 | 0.01629 | 0.07273 | 0.12273 | 0.070 |
| 0.080 | 0.01407 | 0.08103 | 0.13690 | 0.01770 | 0.08237 | 0.13906 | 0.080 |
| 0.090 | 0.01599 | 0.09091 | 0.15358 | 0.01940 | 0.09209 | 0.15551 | 0.090 |
| 0.100 | 0.01829 | 0.10082 | 0.17030 | 0.02144 | 0.10187 | 0.17203 | 0.100 |
| 0.120 | 0.02398 | 0.12068 | 0.20378 | 0.02659 | 0.12155 | 0.20523 | 0.120 |
| 0.140 | 0.03098 | 0.14058 | 0.23728 | 0.03312 | 0.14132 | 0.23853 | 0.140 |
| 0.160 | 0.03915 | 0.16051 | 0.27079 | 0.04093 | 0.16115 | 0.27187 | 0.160 |
| 0.180 | 0.04838 | 0.18045 | 0.30426 | 0.04988 | 0.18102 | 0.30523 | 0.180 |
| 0.200 | 0.05858 | 0.20041 | 0.33770 | 0.05986 | 0.20092 | 0.33857 | 0.200 |
| 0.220 | 0.06966 | 0.22037 | 0.37108 | 0.07078 | 0.22084 | 0.37187 | 0.220 |
| 0.240 | 0.08157 | 0.24034 | 0.40440 | 0.08255 | 0.24076 | 0.40512 | 0.240 |
| 0.260 | 0.09422 | 0.26031 | 0.43764 | 0.09510 | 0.26070 | 0.43830 | 0.260 |
| 0.280 | 0.10757 | 0.28029 | 0.47078 | 0.10836 | 0.28065 | 0.47140 | 0.280 |
| 0.300 | 0.12155 | 0.30027 | 0.50383 | 0.12226 | 0.30061 | 0.50440 | 0.300 |
| 0.320 | 0.13611 | 0.32025 | 0.53676 | 0.13676 | 0.32057 | 0.53730 | 0.320 |
| 0.340 | 0.15120 | 0.34024 | 0.56957 | 0.15180 | 0.34054 | 0.57007 | 0.340 |
| 0.360 | 0.16678 | 0.36023 | 0.60223 | 0.16734 | 0.36051 | 0.60270 | 0.360 |
| 0.380 | 0.18280 | 0.38022 | 0.63473 | 0.18332 | 0.38048 | 0.63518 | 0.380 |
| 0.400 | 0.19923 | 0.40021 | 0.66707 | 0.19972 | 0.40046 | 0.66749 | 0.400 |
| 0.420 | 0.21603 | 0.42019 | 0.69922 | 0.21649 | 0.42044 | 0.69961 | 0.420 |
| 0.440 | 0.23317 | 0.44019 | 0.73115 | 0.23360 | 0.44042 | 0.73153 | 0.440 |
| 0.460 | 0.25061 | 0.46018 | 0.76285 | 0.25102 | 0.46040 | 0.76322 | 0.460 |
| 0.480 | 0.26833 | 0.48017 | 0.79431 | 0.26872 | 0.48038 | 0.79465 | 0.480 |
| 0.500 | 0.28631 | 0.50016 | 0.82548 | 0.28667 | 0.50037 | 0.82581 | 0.500 |
| 0.520 | 0.30451 | 0.52016 | 0.85635 | 0.30486 | 0.52035 | 0.85667 | 0.520 |
| 0.540 | 0.32292 | 0.54015 | 0.88688 | 0.32326 | 0.54034 | 0.88719 | 0.540 |
| 0.560 | 0.34152 | 0.56015 | 0.91706 | 0.34184 | 0.56033 | 0.91734 | 0.560 |
| 0.580 | 0.36030 | 0.58014 | 0.94682 | 0.36060 | 0.58032 | 0.94709 | 0.580 |
| 0.600 | 0.37923 | 0.60014 | 0.97615 | 0.37952 | 0.60031 | 0.97641 | 0.600 |
| 0.620 | 0.39830 | 0.62013 | 1.00501 | 0.39858 | 0.62030 | 1.00525 | 0.620 |
| 0.640 | 0.41750 | 0.64013 | 1.03333 | 0.41777 | 0.64029 | 1.03357 | 0.640 |
| 0.660 | 0.43681 | 0.66013 | 1.06110 | 0.43707 | 0.66028 | 1.06133 | 0.660 |
| 0.680 | 0.45622 | 0.68012 | 1.08825 | 0.45647 | 0.68027 | 1.08847 | 0.680 |
| 0.700 | 0.47573 | 0.70012 | 1.11474 | 0.47597 | 0.70026 | 1.11495 | 0.700 |
| 0.720 | 0.49532 | 0.72012 | 1.14053 | 0.49555 | 0.72025 | 1.14073 | 0.720 |
| 0.740 | 0.51498 | 0.74011 | 1.16557 | 0.51521 | 0.74025 | 1.16576 | 0.740 |
| 0.760 | 0.53471 | 0.76011 | 1.18982 | 0.53493 | 0.76024 | 1.19000 | 0.760 |
| 0.780 | 0.55451 | 0.78011 | 1.21323 | 0.55472 | 0.78024 | 1.21341 | 0.780 |
| 0.800 | 0.57435 | 0.80011 | 1.23580 | 0.57456 | 0.80023 | 1.23596 | 0.800 |
| 0.820 | 0.59424 | 0.82011 | 1.25747 | 0.59444 | 0.82023 | 1.25762 | 0.820 |
| 0.840 | 0.61418 | 0.84010 | 1.27825 | 0.61437 | 0.84022 | 1.27839 | 0.840 |
| 0.860 | 0.63415 | 0.86010 | 1.29813 | 0.63434 | 0.86022 | 1.29827 | 0.860 |
| 0.880 | 0.65415 | 0.88009 | 1.31712 | 0.65433 | 0.88021 | 1.31726 | 0.880 |
| 0.900 | 0.67418 | 0.90010 | 1.33525 | 0.67436 | 0.90020 | 1.33537 | 0.900 |
| 0.920 | 0.69423 | 0.92009 | 1.35255 | 0.69441 | 0.92020 | 1.35266 | 0.920 |
| 0.940 | 0.71431 | 0.94009 | 1.36905 | 0.71448 | 0.94020 | 1.36916 | 0.940 |
| 0.960 | 0.73440 | 0.96009 | 1.38480 | 0.73457 | 0.96020 | 1.38489 | 0.960 |
| 0.980 | 0.75452 | 0.98009 | 1.39985 | 0.75468 | 0.98019 | 1.39990 | 0.980 |
| 1.000 | 0.77464 | 1.00008 | 1.41424 | 0.77479 | 1.00019 | 1.41425 | 1.000 |

$H/R = 0.040$ $\quad n = 2. \quad$ $H/R = 0.050$

| $H/L$ | $\Omega_1$ | $\Omega_2$ | $\Omega_3$ | $\Omega_1$ | $\Omega_2$ | $\Omega_3$ | $H/L$ |
|---|---|---|---|---|---|---|---|
| 0. | 0.000666 | 0.02547 | 0.04810 | 0.00104 | 0.03183 | 0.06011 | 0. |
| 0.001 | 0.000668 | 0.02550 | 0.04813 | 0.00104 | 0.03186 | 0.06013 | 0.001 |
| 0.005 | 0.000967 | 0.02628 | 0.04866 | 0.00120 | 0.03249 | 0.06056 | 0.005 |
| 0.010 | 0.00254 | 0.02841 | 0.05035 | 0.00234 | 0.03430 | 0.06191 | 0.010 |
| 0.020 | 0.00714 | 0.03449 | 0.05715 | 0.00663 | 0.03989 | 0.06736 | 0.020 |
| 0.030 | 0.01130 | 0.04135 | 0.06782 | 0.01124 | 0.04646 | 0.07628 | 0.030 |
| 0.040 | 0.01436 | 0.04889 | 0.08091 | 0.01513 | 0.05351 | 0.08785 | 0.040 |
| 0.050 | 0.01661 | 0.05712 | 0.09532 | 0.01817 | 0.06112 | 0.10112 | 0.050 |
| 0.060 | 0.01821 | 0.06586 | 0.11050 | 0.02060 | 0.06928 | 0.11544 | 0.060 |
| 0.070 | 0.01973 | 0.07496 | 0.12613 | 0.02255 | 0.07789 | 0.13043 | 0.070 |
| 0.080 | 0.02125 | 0.08429 | 0.14206 | 0.02438 | 0.08684 | 0.14585 | 0.080 |
| 0.090 | 0.02293 | 0.09378 | 0.15818 | 0.02625 | 0.09602 | 0.16158 | 0.090 |
| 0.100 | 0.02486 | 0.10338 | 0.17444 | 0.02823 | 0.10537 | 0.17752 | 0.100 |
| 0.120 | 0.02966 | 0.12279 | 0.20724 | 0.03290 | 0.12441 | 0.20982 | 0.120 |
| 0.140 | 0.03578 | 0.14237 | 0.24026 | 0.03875 | 0.14375 | 0.24247 | 0.140 |
| 0.160 | 0.04322 | 0.16206 | 0.27339 | 0.04587 | 0.16326 | 0.27533 | 0.160 |
| 0.180 | 0.05185 | 0.18183 | 0.30658 | 0.05420 | 0.18288 | 0.30830 | 0.180 |
| 0.200 | 0.06158 | 0.20164 | 0.33978 | 0.06366 | 0.20258 | 0.34133 | 0.200 |
| 0.220 | 0.07229 | 0.22149 | 0.37297 | 0.07414 | 0.22234 | 0.37438 | 0.220 |
| 0.240 | 0.08389 | 0.24136 | 0.40613 | 0.08555 | 0.24214 | 0.40742 | 0.240 |
| 0.260 | 0.09630 | 0.26126 | 0.43923 | 0.09779 | 0.26197 | 0.44042 | 0.260 |
| 0.280 | 0.10944 | 0.28117 | 0.47226 | 0.11080 | 0.28183 | 0.47336 | 0.280 |
| 0.300 | 0.12325 | 0.30109 | 0.50520 | 0.12449 | 0.30170 | 0.50623 | 0.300 |
| 0.320 | 0.13767 | 0.32102 | 0.53805 | 0.13881 | 0.32160 | 0.53901 | 0.320 |
| 0.340 | 0.15264 | 0.34096 | 0.57077 | 0.15370 | 0.34150 | 0.57167 | 0.340 |
| 0.360 | 0.16811 | 0.36091 | 0.60336 | 0.16910 | 0.36142 | 0.60421 | 0.360 |
| 0.380 | 0.18405 | 0.38086 | 0.63581 | 0.18496 | 0.38134 | 0.63661 | 0.380 |
| 0.400 | 0.20039 | 0.40082 | 0.66808 | 0.20125 | 0.40127 | 0.66884 | 0.400 |
| 0.420 | 0.21712 | 0.42078 | 0.70017 | 0.21793 | 0.42121 | 0.70090 | 0.420 |
| 0.440 | 0.23420 | 0.44074 | 0.73206 | 0.23496 | 0.44116 | 0.73274 | 0.440 |
| 0.460 | 0.25158 | 0.46071 | 0.76372 | 0.25230 | 0.46111 | 0.76437 | 0.460 |
| 0.480 | 0.26925 | 0.48068 | 0.79513 | 0.26994 | 0.48106 | 0.79575 | 0.480 |
| 0.500 | 0.28718 | 0.50065 | 0.82627 | 0.28783 | 0.50102 | 0.82686 | 0.500 |
| 0.520 | 0.30535 | 0.52063 | 0.85710 | 0.30597 | 0.52098 | 0.85767 | 0.520 |
| 0.540 | 0.32372 | 0.54060 | 0.88760 | 0.32432 | 0.54094 | 0.88814 | 0.540 |
| 0.560 | 0.34229 | 0.56058 | 0.91774 | 0.34285 | 0.56091 | 0.91826 | 0.560 |
| 0.580 | 0.36103 | 0.58056 | 0.94748 | 0.36157 | 0.58088 | 0.94797 | 0.580 |
| 0.600 | 0.37993 | 0.60054 | 0.97678 | 0.38045 | 0.60085 | 0.97725 | 0.600 |
| 0.620 | 0.39897 | 0.62052 | 1.00560 | 0.39947 | 0.62083 | 1.00605 | 0.620 |
| 0.640 | 0.41814 | 0.64051 | 1.03391 | 0.41862 | 0.64080 | 1.03433 | 0.640 |
| 0.660 | 0.43743 | 0.66049 | 1.06164 | 0.43789 | 0.66077 | 1.06205 | 0.660 |
| 0.680 | 0.45682 | 0.68048 | 1.08877 | 0.45727 | 0.68075 | 1.08916 | 0.680 |
| 0.700 | 0.47631 | 0.70046 | 1.11524 | 0.47674 | 0.70073 | 1.11561 | 0.700 |
| 0.720 | 0.49588 | 0.72045 | 1.14101 | 0.49630 | 0.72071 | 1.14136 | 0.720 |
| 0.740 | 0.51552 | 0.74044 | 1.16603 | 0.51593 | 0.74069 | 1.16636 | 0.740 |
| 0.760 | 0.53524 | 0.76043 | 1.19025 | 0.53563 | 0.76067 | 1.19057 | 0.760 |
| 0.780 | 0.55501 | 0.78042 | 1.21365 | 0.55539 | 0.78065 | 1.21395 | 0.780 |
| 0.800 | 0.57485 | 0.80041 | 1.23618 | 0.57521 | 0.80063 | 1.23647 | 0.800 |
| 0.820 | 0.59472 | 0.82040 | 1.25784 | 0.59508 | 0.82062 | 1.25812 | 0.820 |
| 0.840 | 0.61464 | 0.84039 | 1.27860 | 0.61499 | 0.84061 | 1.27886 | 0.840 |
| 0.860 | 0.63460 | 0.86038 | 1.29846 | 0.63494 | 0.86060 | 1.29871 | 0.860 |
| 0.880 | 0.65459 | 0.88037 | 1.31744 | 0.65492 | 0.88058 | 1.31767 | 0.880 |
| 0.900 | 0.67461 | 0.90036 | 1.33555 | 0.67493 | 0.90057 | 1.33576 | 0.900 |
| 0.920 | 0.69465 | 0.92036 | 1.35282 | 0.69497 | 0.92056 | 1.35302 | 0.920 |
| 0.940 | 0.71472 | 0.94035 | 1.36930 | 0.71503 | 0.94055 | 1.36945 | 0.940 |
| 0.960 | 0.73481 | 0.96034 | 1.38500 | 0.73510 | 0.96053 | 1.38512 | 0.960 |
| 0.980 | 0.75490 | 0.98034 | 1.39995 | 0.75519 | 0.98052 | 1.40000 | 0.980 |
| 1.000 | 0.77501 | 1.00033 | 1.41427 | 0.77529 | 1.00051 | 1.41431 | 1.000 |

$H/R = 0.060$ $n = 2.$ $H/R = 0.070$

| $H/L$ | $\Omega_1$ | $\Omega_2$ | $\Omega_3$ | $\Omega_1$ | $\Omega_2$ | $\Omega_3$ | $H/L$ |
|---|---|---|---|---|---|---|---|
| 0. | 0.00150 | 0.03820 | 0.07212 | 0.00204 | 0.04457 | 0.08411 | 0. |
| 0.001 | 0.00150 | 0.03823 | 0.07213 | 0.00204 | 0.04459 | 0.08412 | 0.001 |
| 0.005 | 0.00159 | 0.03876 | 0.07249 | 0.00210 | 0.04505 | 0.08442 | 0.005 |
| 0.010 | 0.00237 | 0.04031 | 0.07361 | 0.00262 | 0.04641 | 0.08538 | 0.010 |
| 0.020 | 0.00618 | 0.04542 | 0.07814 | 0.00587 | 0.05108 | 0.08926 | 0.020 |
| 0.030 | 0.01089 | 0.05174 | 0.08570 | 0.01047 | 0.05712 | 0.09576 | 0.030 |
| 0.040 | 0.01530 | 0.05850 | 0.09586 | 0.01514 | 0.06369 | 0.10471 | 0.040 |
| 0.050 | 0.01904 | 0.06568 | 0.10794 | 0.01939 | 0.07060 | 0.11565 | 0.050 |
| 0.060 | 0.02213 | 0.07334 | 0.12132 | 0.02307 | 0.07787 | 0.12806 | 0.060 |
| 0.070 | 0.02467 | 0.08147 | 0.13558 | 0.02624 | 0.08557 | 0.14152 | 0.070 |
| 0.080 | 0.02703 | 0.08998 | 0.15042 | 0.02902 | 0.09366 | 0.15572 | 0.080 |
| 0.090 | 0.02914 | 0.09880 | 0.16568 | 0.03150 | 0.10210 | 0.17045 | 0.090 |
| 0.100 | 0.03132 | 0.10785 | 0.18123 | 0.03406 | 0.11082 | 0.18556 | 0.100 |
| 0.120 | 0.03611 | 0.12645 | 0.21294 | 0.03914 | 0.12889 | 0.21659 | 0.120 |
| 0.140 | 0.04184 | 0.14546 | 0.24516 | 0.04492 | 0.14752 | 0.24832 | 0.140 |
| 0.160 | 0.04874 | 0.16474 | 0.27769 | 0.05170 | 0.16652 | 0.28046 | 0.160 |
| 0.180 | 0.05681 | 0.18418 | 0.31040 | 0.05959 | 0.18575 | 0.31288 | 0.180 |
| 0.200 | 0.06602 | 0.20374 | 0.34323 | 0.06858 | 0.20514 | 0.34545 | 0.200 |
| 0.220 | 0.07627 | 0.22339 | 0.37610 | 0.07863 | 0.22465 | 0.37813 | 0.220 |
| 0.240 | 0.08748 | 0.24310 | 0.40900 | 0.08964 | 0.24424 | 0.41085 | 0.240 |
| 0.260 | 0.09955 | 0.26285 | 0.44187 | 0.10154 | 0.26390 | 0.44358 | 0.260 |
| 0.280 | 0.11241 | 0.28264 | 0.47471 | 0.11423 | 0.28362 | 0.47630 | 0.280 |
| 0.300 | 0.12597 | 0.30246 | 0.50749 | 0.12766 | 0.30337 | 0.50897 | 0.300 |
| 0.320 | 0.14017 | 0.32231 | 0.54018 | 0.14174 | 0.32315 | 0.54156 | 0.320 |
| 0.340 | 0.15496 | 0.34217 | 0.57277 | 0.15642 | 0.34296 | 0.57407 | 0.340 |
| 0.360 | 0.17027 | 0.36205 | 0.60525 | 0.17164 | 0.36279 | 0.60647 | 0.360 |
| 0.380 | 0.18607 | 0.38194 | 0.63759 | 0.18735 | 0.38265 | 0.63874 | 0.380 |
| 0.400 | 0.20229 | 0.40184 | 0.66977 | 0.20350 | 0.40251 | 0.67086 | 0.400 |
| 0.420 | 0.21891 | 0.42175 | 0.70177 | 0.22005 | 0.42239 | 0.70281 | 0.420 |
| 0.440 | 0.23588 | 0.44167 | 0.73358 | 0.23696 | 0.44228 | 0.73456 | 0.440 |
| 0.460 | 0.25318 | 0.46160 | 0.76516 | 0.25420 | 0.46218 | 0.76610 | 0.460 |
| 0.480 | 0.27077 | 0.48153 | 0.79651 | 0.27174 | 0.48209 | 0.79740 | 0.480 |
| 0.500 | 0.28862 | 0.50147 | 0.82758 | 0.28955 | 0.50200 | 0.82843 | 0.500 |
| 0.520 | 0.30672 | 0.52141 | 0.85835 | 0.30760 | 0.52192 | 0.85916 | 0.520 |
| 0.540 | 0.32503 | 0.54136 | 0.88880 | 0.32588 | 0.54185 | 0.88957 | 0.540 |
| 0.560 | 0.34354 | 0.56131 | 0.91888 | 0.34435 | 0.56179 | 0.91962 | 0.560 |
| 0.580 | 0.36223 | 0.58127 | 0.94856 | 0.36301 | 0.58172 | 0.94927 | 0.580 |
| 0.600 | 0.38108 | 0.60122 | 0.97782 | 0.38183 | 0.60167 | 0.97849 | 0.600 |
| 0.620 | 0.40008 | 0.62118 | 1.00660 | 0.40080 | 0.62161 | 1.00724 | 0.620 |
| 0.640 | 0.41921 | 0.64115 | 1.03485 | 0.41990 | 0.64157 | 1.03547 | 0.640 |
| 0.660 | 0.43846 | 0.66111 | 1.06255 | 0.43913 | 0.66151 | 1.06313 | 0.660 |
| 0.680 | 0.45781 | 0.68108 | 1.08963 | 0.45846 | 0.68147 | 1.09019 | 0.680 |
| 0.700 | 0.47727 | 0.70105 | 1.11606 | 0.47789 | 0.70143 | 1.11660 | 0.700 |
| 0.720 | 0.49681 | 0.72102 | 1.14179 | 0.49741 | 0.72139 | 1.14230 | 0.720 |
| 0.740 | 0.51642 | 0.74099 | 1.16677 | 0.51700 | 0.74135 | 1.16725 | 0.740 |
| 0.760 | 0.53611 | 0.76097 | 1.19096 | 0.53667 | 0.76131 | 1.19142 | 0.760 |
| 0.780 | 0.55586 | 0.78094 | 1.21432 | 0.55640 | 0.78128 | 1.21475 | 0.780 |
| 0.800 | 0.57566 | 0.80092 | 1.23682 | 0.57619 | 0.80125 | 1.23723 | 0.800 |
| 0.820 | 0.59552 | 0.82090 | 1.25844 | 0.59603 | 0.82122 | 1.25883 | 0.820 |
| 0.840 | 0.61542 | 0.84088 | 1.27916 | 0.61592 | 0.84119 | 1.27954 | 0.840 |
| 0.860 | 0.63535 | 0.86085 | 1.29899 | 0.63584 | 0.86116 | 1.29933 | 0.860 |
| 0.880 | 0.65532 | 0.88084 | 1.31794 | 0.65580 | 0.88113 | 1.31825 | 0.880 |
| 0.900 | 0.67532 | 0.90082 | 1.33601 | 0.67579 | 0.90111 | 1.33629 | 0.900 |
| 0.920 | 0.69535 | 0.92079 | 1.35323 | 0.69580 | 0.92109 | 1.35347 | 0.920 |
| 0.940 | 0.71540 | 0.94078 | 1.36963 | 0.71583 | 0.94106 | 1.36982 | 0.940 |
| 0.960 | 0.73546 | 0.96077 | 1.38522 | 0.73589 | 0.96104 | 1.38533 | 0.960 |
| 0.980 | 0.75555 | 0.98075 | 1.40005 | 0.75596 | 0.98102 | 1.40010 | 0.980 |
| 1.000 | 0.77564 | 1.00073 | 1.41434 | 0.77605 | 1.00100 | 1.41439 | 1.000 |

$H/R = 0.080$ $n = 2.$ $H/R = 0.090$

| $H/L$ | $\Omega_1$ | $\Omega_2$ | $\Omega_3$ | $\Omega_1$ | $\Omega_2$ | $\Omega_3$ | $H/L$ |
|---|---|---|---|---|---|---|---|
| 0. | 0.00266 | 0.05094 | 0.09608 | 0.00336 | 0.05731 | 0.10804 | 0. |
| 0.001 | 0.00266 | 0.05096 | 0.09609 | 0.00336 | 0.05733 | 0.10805 | 0.001 |
| 0.005 | 0.00270 | 0.05136 | 0.09636 | 0.00340 | 0.05769 | 0.10829 | 0.005 |
| 0.010 | 0.00307 | 0.05257 | 0.09720 | 0.00366 | 0.05877 | 0.10903 | 0.010 |
| 0.020 | 0.00575 | 0.05684 | 0.10058 | 0.00582 | 0.06270 | 0.11204 | 0.020 |
| 0.030 | 0.01009 | 0.06259 | 0.10628 | 0.00982 | 0.06815 | 0.11709 | 0.030 |
| 0.040 | 0.01485 | 0.06899 | 0.11420 | 0.01453 | 0.07436 | 0.12418 | 0.040 |
| 0.050 | 0.01942 | 0.07571 | 0.12409 | 0.01928 | 0.08094 | 0.13314 | 0.050 |
| 0.060 | 0.02357 | 0.08272 | 0.13555 | 0.02376 | 0.08777 | 0.14368 | 0.060 |
| 0.070 | 0.02725 | 0.09008 | 0.14818 | 0.02787 | 0.09487 | 0.15551 | 0.070 |
| 0.080 | 0.03053 | 0.09780 | 0.16170 | 0.03161 | 0.10228 | 0.16831 | 0.080 |
| 0.090 | 0.03353 | 0.10587 | 0.17585 | 0.03503 | 0.11003 | 0.18186 | 0.090 |
| 0.100 | 0.03633 | 0.11425 | 0.19049 | 0.03823 | 0.11808 | 0.19598 | 0.100 |
| 0.120 | 0.04192 | 0.13174 | 0.22077 | 0.04427 | 0.13499 | 0.22544 | 0.120 |
| 0.140 | 0.04788 | 0.14994 | 0.25193 | 0.05063 | 0.15271 | 0.25598 | 0.140 |
| 0.160 | 0.05466 | 0.16860 | 0.28364 | 0.05754 | 0.17100 | 0.28722 | 0.160 |
| 0.180 | 0.06245 | 0.18758 | 0.31571 | 0.06531 | 0.18968 | 0.31890 | 0.180 |
| 0.200 | 0.07128 | 0.20677 | 0.34801 | 0.07406 | 0.20864 | 0.35089 | 0.200 |
| 0.220 | 0.08116 | 0.22612 | 0.38046 | 0.08379 | 0.22780 | 0.38308 | 0.220 |
| 0.240 | 0.09199 | 0.24558 | 0.41299 | 0.09447 | 0.24711 | 0.41539 | 0.240 |
| 0.260 | 0.10371 | 0.26513 | 0.44555 | 0.10605 | 0.26654 | 0.44778 | 0.260 |
| 0.280 | 0.11626 | 0.28475 | 0.47812 | 0.11844 | 0.28605 | 0.48018 | 0.280 |
| 0.300 | 0.12954 | 0.30442 | 0.51067 | 0.13159 | 0.30563 | 0.51259 | 0.300 |
| 0.320 | 0.14350 | 0.32414 | 0.54315 | 0.14542 | 0.32526 | 0.54495 | 0.320 |
| 0.340 | 0.15806 | 0.34389 | 0.57557 | 0.15987 | 0.34494 | 0.57726 | 0.340 |
| 0.360 | 0.17318 | 0.36366 | 0.60788 | 0.17488 | 0.36466 | 0.60947 | 0.360 |
| 0.380 | 0.18879 | 0.38347 | 0.64007 | 0.19039 | 0.38440 | 0.64157 | 0.380 |
| 0.400 | 0.20486 | 0.40329 | 0.67211 | 0.20637 | 0.40418 | 0.67354 | 0.400 |
| 0.420 | 0.22134 | 0.42313 | 0.70400 | 0.22277 | 0.42397 | 0.70535 | 0.420 |
| 0.440 | 0.23818 | 0.44298 | 0.73569 | 0.23954 | 0.44379 | 0.73697 | 0.440 |
| 0.460 | 0.25536 | 0.46285 | 0.76718 | 0.25665 | 0.46362 | 0.76839 | 0.460 |
| 0.480 | 0.27284 | 0.48273 | 0.79843 | 0.27408 | 0.48347 | 0.79958 | 0.480 |
| 0.500 | 0.29060 | 0.50262 | 0.82940 | 0.29178 | 0.50333 | 0.83051 | 0.500 |
| 0.520 | 0.30861 | 0.52252 | 0.86009 | 0.30973 | 0.52319 | 0.86115 | 0.520 |
| 0.540 | 0.32684 | 0.54242 | 0.89046 | 0.32792 | 0.54308 | 0.89147 | 0.540 |
| 0.560 | 0.34527 | 0.56234 | 0.92047 | 0.34631 | 0.56296 | 0.92143 | 0.560 |
| 0.580 | 0.36390 | 0.58225 | 0.95009 | 0.36489 | 0.58286 | 0.95100 | 0.580 |
| 0.600 | 0.38268 | 0.60218 | 0.97927 | 0.38364 | 0.60276 | 0.98014 | 0.600 |
| 0.620 | 0.40162 | 0.62211 | 1.00798 | 0.40254 | 0.62267 | 1.00881 | 0.620 |
| 0.640 | 0.42069 | 0.64204 | 1.03618 | 0.42158 | 0.64259 | 1.03697 | 0.640 |
| 0.660 | 0.43989 | 0.66198 | 1.06381 | 0.44074 | 0.66251 | 1.06457 | 0.660 |
| 0.680 | 0.45919 | 0.68192 | 1.09083 | 0.46002 | 0.68243 | 1.09156 | 0.680 |
| 0.700 | 0.47860 | 0.70187 | 1.11721 | 0.47940 | 0.70237 | 1.11790 | 0.700 |
| 0.720 | 0.49809 | 0.72181 | 1.14288 | 0.49887 | 0.72230 | 1.14354 | 0.720 |
| 0.740 | 0.51767 | 0.74176 | 1.16780 | 0.51842 | 0.74224 | 1.16842 | 0.740 |
| 0.760 | 0.53732 | 0.76172 | 1.19194 | 0.53805 | 0.76218 | 1.19253 | 0.760 |
| 0.780 | 0.55703 | 0.78167 | 1.21525 | 0.55774 | 0.78212 | 1.21580 | 0.780 |
| 0.800 | 0.57680 | 0.80163 | 1.23770 | 0.57749 | 0.80207 | 1.23822 | 0.800 |
| 0.820 | 0.59662 | 0.82159 | 1.25927 | 0.59729 | 0.82202 | 1.25975 | 0.820 |
| 0.840 | 0.61649 | 0.84155 | 1.27994 | 0.61714 | 0.84197 | 1.28039 | 0.840 |
| 0.860 | 0.63640 | 0.86152 | 1.29971 | 0.63703 | 0.86192 | 1.30012 | 0.860 |
| 0.880 | 0.65634 | 0.88148 | 1.31859 | 0.65696 | 0.88188 | 1.31896 | 0.880 |
| 0.900 | 0.67631 | 0.90145 | 1.33659 | 0.67691 | 0.90184 | 1.33691 | 0.900 |
| 0.920 | 0.69631 | 0.92142 | 1.35372 | 0.69690 | 0.92180 | 1.35398 | 0.920 |
| 0.940 | 0.71634 | 0.94139 | 1.37000 | 0.71691 | 0.94176 | 1.37018 | 0.940 |
| 0.960 | 0.73638 | 0.96136 | 1.38543 | 0.73694 | 0.96172 | 1.38553 | 0.960 |
| 0.980 | 0.75644 | 0.98133 | 1.40016 | 0.75698 | 0.98168 | 1.40023 | 0.980 |
| 1.000 | 0.77651 | 1.00131 | 1.41446 | 0.77704 | 1.00165 | 1.41452 | 1.000 |

$H/R = 0.100 \qquad n = 2.$

| $H/L$ | $\Omega_1$ | $\Omega_2$ | $\Omega_3$ | $\Omega_4$ | $\Omega_2$ | $\Omega_6$ |
|---|---|---|---|---|---|---|
| 0. | 0.00415 | 0.06369 | 0.11998 | 1.00242 | 1.00891 | 1.86209 |
| 0.001 | 0.00415 | 0.06370 | 0.11999 | 1.00242 | 1.00891 | 1.86208 |
| 0.005 | 0.00418 | 0.06402 | 0.12021 | 1.00244 | 1.00895 | 1.86202 |
| 0.010 | 0.00432 | 0.06501 | 0.12088 | 1.00250 | 1.00905 | 1.86186 |
| 0.020 | 0.00610 | 0.06863 | 0.12357 | 1.00272 | 1.00948 | 1.86123 |
| 0.030 | 0.00971 | 0.07379 | 0.12812 | 1.00310 | 1.01020 | 1.86017 |
| 0.040 | 0.01426 | 0.07980 | 0.13451 | 1.00358 | 1.01125 | 1.85873 |
| 0.050 | 0.01907 | 0.08625 | 0.14266 | 1.00418 | 1.01263 | 1.85693 |
| 0.060 | 0.02377 | 0.09294 | 0.15237 | 1.00487 | 1.01434 | 1.85482 |
| 0.070 | 0.02821 | 0.09986 | 0.16341 | 1.00564 | 1.01640 | 1.85242 |
| 0.080 | 0.03233 | 0.10704 | 0.17550 | 1.00649 | 1.01879 | 1.84976 |
| 0.090 | 0.03615 | 0.11450 | 0.18843 | 1.00743 | 1.02152 | 1.84689 |
| 0.100 | 0.03972 | 0.12226 | 0.20201 | 1.00845 | 1.02459 | 1.84384 |
| 0.120 | 0.04649 | 0.13859 | 0.23059 | 1.01074 | 1.03169 | 1.83733 |
| 0.140 | 0.05319 | 0.15582 | 0.26047 | 1.01340 | 1.04002 | 1.83045 |
| 0.160 | 0.06027 | 0.17370 | 0.29118 | 1.01641 | 1.04952 | 1.82335 |
| 0.180 | 0.06813 | 0.19206 | 0.32245 | 1.01979 | 1.06012 | 1.81616 |
| 0.200 | 0.07684 | 0.21076 | 0.35410 | 1.02354 | 1.07177 | 1.80901 |
| 0.220 | 0.08649 | 0.22971 | 0.38600 | 1.02766 | 1.08439 | 1.80197 |
| 0.240 | 0.09705 | 0.24885 | 0.41807 | 1.03215 | 1.09794 | 1.79510 |
| 0.260 | 0.10849 | 0.26812 | 0.45025 | 1.03699 | 1.11236 | 1.78846 |
| 0.280 | 0.12076 | 0.28751 | 0.48248 | 1.04220 | 1.12758 | 1.78210 |
| 0.300 | 0.13377 | 0.30698 | 0.51473 | 1.04776 | 1.14355 | 1.77605 |
| 0.320 | 0.14748 | 0.32653 | 0.54696 | 1.05367 | 1.16022 | 1.77035 |
| 0.340 | 0.16181 | 0.34612 | 0.57914 | 1.05992 | 1.17754 | 1.76503 |
| 0.360 | 0.17671 | 0.36577 | 0.61124 | 1.06651 | 1.19546 | 1.76013 |
| 0.380 | 0.19214 | 0.38546 | 0.64324 | 1.07343 | 1.21394 | 1.75565 |
| 0.400 | 0.20802 | 0.40518 | 0.67512 | 1.08068 | 1.23294 | 1.75164 |
| 0.420 | 0.22434 | 0.42492 | 0.70685 | 1.08824 | 1.25242 | 1.74811 |
| 0.440 | 0.24103 | 0.44469 | 0.73840 | 1.09612 | 1.27233 | 1.74510 |
| 0.460 | 0.25808 | 0.46448 | 0.76975 | 1.10431 | 1.29265 | 1.74263 |
| 0.480 | 0.27543 | 0.48429 | 0.80087 | 1.11279 | 1.31334 | 1.74073 |
| 0.500 | 0.29308 | 0.50411 | 0.83174 | 1.12156 | 1.33437 | 1.73944 |
| 0.520 | 0.31098 | 0.52395 | 0.86232 | 1.13062 | 1.35572 | 1.73877 |
| 0.540 | 0.32911 | 0.54381 | 0.89259 | 1.13997 | 1.37736 | 1.73876 |
| 0.560 | 0.34745 | 0.56367 | 0.92249 | 1.14958 | 1.39926 | 1.73945 |
| 0.580 | 0.36599 | 0.58354 | 0.95202 | 1.15946 | 1.42141 | 1.74088 |
| 0.600 | 0.38470 | 0.60342 | 0.98112 | 1.16959 | 1.44376 | 1.74308 |
| 0.620 | 0.40356 | 0.62331 | 1.00974 | 1.17998 | 1.46633 | 1.74609 |
| 0.640 | 0.42256 | 0.64320 | 1.03786 | 1.19062 | 1.48906 | 1.74996 |
| 0.660 | 0.44169 | 0.66310 | 1.06541 | 1.20149 | 1.51196 | 1.75473 |
| 0.680 | 0.46094 | 0.68301 | 1.09236 | 1.21260 | 1.53499 | 1.76044 |
| 0.700 | 0.48029 | 0.70293 | 1.11866 | 1.22394 | 1.55815 | 1.76716 |
| 0.720 | 0.49973 | 0.72284 | 1.14426 | 1.23549 | 1.58140 | 1.77491 |
| 0.740 | 0.51925 | 0.74276 | 1.16911 | 1.24727 | 1.60476 | 1.78374 |
| 0.760 | 0.53885 | 0.76269 | 1.19318 | 1.25925 | 1.62819 | 1.79368 |
| 0.780 | 0.55852 | 0.78262 | 1.21641 | 1.27144 | 1.65168 | 1.80479 |
| 0.800 | 0.57825 | 0.80256 | 1.23880 | 1.28383 | 1.67522 | 1.81708 |
| 0.820 | 0.59803 | 0.82249 | 1.26028 | 1.29641 | 1.69878 | 1.83058 |
| 0.840 | 0.61786 | 0.84243 | 1.28088 | 1.30920 | 1.72236 | 1.84528 |
| 0.860 | 0.63773 | 0.86238 | 1.30058 | 1.32218 | 1.74594 | 1.86119 |
| 0.880 | 0.65764 | 0.88232 | 1.31937 | 1.33535 | 1.76952 | 1.87829 |
| 0.900 | 0.67758 | 0.90227 | 1.33725 | 1.34874 | 1.79307 | 1.89656 |
| 0.920 | 0.69755 | 0.92222 | 1.35424 | 1.36236 | 1.81659 | 1.91596 |
| 0.940 | 0.71754 | 0.94217 | 1.37034 | 1.37624 | 1.84007 | 1.93645 |
| 0.960 | 0.73756 | 0.96213 | 1.38564 | 1.39035 | 1.86348 | 1.95798 |
| 0.980 | 0.75759 | 0.98208 | 1.40030 | 1.40456 | 1.88683 | 1.98046 |
| 1.000 | 0.77763 | 1.00204 | 1.41458 | 1.41869 | 1.91008 | 2.00384 |

$H/R = 0.120$ $n = 2.$ $H/R = 0.140$

| $H/L$ | $\Omega_1$ | $\Omega_2$ | $\Omega_3$ | $\Omega_1$ | $\Omega_2$ | $\Omega_3$ | $H/L$ |
|---|---|---|---|---|---|---|---|
| 0. | 0.00596 | 0.07644 | 0.14381 | 0.00809 | 0.08920 | 0.16753 | 0. |
| 0.001 | 0.00596 | 0.07645 | 0.14381 | 0.00809 | 0.08921 | 0.16754 | 0.001 |
| 0.005 | 0.00598 | 0.07672 | 0.14399 | 0.00811 | 0.08944 | 0.16769 | 0.005 |
| 0.010 | 0.00612 | 0.07755 | 0.14455 | 0.00821 | 0.09015 | 0.16817 | 0.010 |
| 0.020 | 0.00723 | 0.08067 | 0.14680 | 0.00899 | 0.09288 | 0.17010 | 0.020 |
| 0.030 | 0.01002 | 0.08529 | 0.15057 | 0.01106 | 0.09704 | 0.17333 | 0.030 |
| 0.040 | 0.01405 | 0.09089 | 0.15590 | 0.01446 | 0.10222 | 0.17788 | 0.040 |
| 0.050 | 0.01872 | 0.09704 | 0.16275 | 0.01875 | 0.10806 | 0.18377 | 0.050 |
| 0.060 | 0.02359 | 0.10351 | 0.17105 | 0.02351 | 0.11430 | 0.19095 | 0.060 |
| 0.070 | 0.02842 | 0.11018 | 0.18065 | 0.02846 | 0.12078 | 0.19936 | 0.070 |
| 0.080 | 0.03309 | 0.11705 | 0.19139 | 0.03341 | 0.12744 | 0.20888 | 0.080 |
| 0.090 | 0.03755 | 0.12411 | 0.20307 | 0.03828 | 0.13426 | 0.21939 | 0.090 |
| 0.100 | 0.04180 | 0.13141 | 0.21554 | 0.04303 | 0.14125 | 0.23076 | 0.100 |
| 0.120 | 0.04978 | 0.14673 | 0.24225 | 0.05209 | 0.15581 | 0.25556 | 0.120 |
| 0.140 | 0.05744 | 0.16298 | 0.27067 | 0.06075 | 0.17121 | 0.28240 | 0.140 |
| 0.160 | 0.06515 | 0.18001 | 0.30022 | 0.06931 | 0.18740 | 0.31068 | 0.160 |
| 0.180 | 0.07345 | 0.19765 | 0.33056 | 0.07809 | 0.20428 | 0.33996 | 0.180 |
| 0.200 | 0.08226 | 0.21575 | 0.36144 | 0.08727 | 0.22172 | 0.36997 | 0.200 |
| 0.220 | 0.09190 | 0.23421 | 0.39270 | 0.09711 | 0.23961 | 0.40050 | 0.220 |
| 0.240 | 0.10234 | 0.25293 | 0.42423 | 0.10758 | 0.25785 | 0.43141 | 0.240 |
| 0.260 | 0.11361 | 0.27186 | 0.45594 | 0.11881 | 0.27637 | 0.46258 | 0.260 |
| 0.280 | 0.12566 | 0.29096 | 0.48776 | 0.13076 | 0.29511 | 0.49394 | 0.280 |
| 0.300 | 0.13846 | 0.31018 | 0.51965 | 0.14341 | 0.31403 | 0.52542 | 0.300 |
| 0.320 | 0.15194 | 0.32950 | 0.55157 | 0.15672 | 0.33309 | 0.55697 | 0.320 |
| 0.340 | 0.16606 | 0.34891 | 0.58347 | 0.17066 | 0.35227 | 0.58854 | 0.340 |
| 0.360 | 0.18076 | 0.36839 | 0.61532 | 0.18518 | 0.37154 | 0.62011 | 0.360 |
| 0.380 | 0.19599 | 0.38793 | 0.64710 | 0.20023 | 0.39090 | 0.65162 | 0.380 |
| 0.400 | 0.21169 | 0.40751 | 0.67877 | 0.21577 | 0.41032 | 0.68304 | 0.400 |
| 0.420 | 0.22783 | 0.42714 | 0.71030 | 0.23174 | 0.42981 | 0.71436 | 0.420 |
| 0.440 | 0.24437 | 0.44681 | 0.74168 | 0.24812 | 0.44934 | 0.74553 | 0.440 |
| 0.460 | 0.26127 | 0.46650 | 0.77287 | 0.26487 | 0.46891 | 0.77653 | 0.460 |
| 0.480 | 0.27849 | 0.48622 | 0.80385 | 0.28195 | 0.48853 | 0.80733 | 0.480 |
| 0.500 | 0.29600 | 0.50596 | 0.83457 | 0.29933 | 0.50817 | 0.83789 | 0.500 |
| 0.520 | 0.31378 | 0.52573 | 0.86502 | 0.31698 | 0.52785 | 0.86819 | 0.520 |
| 0.540 | 0.33181 | 0.54551 | 0.89516 | 0.33489 | 0.54755 | 0.89818 | 0.540 |
| 0.560 | 0.35005 | 0.56531 | 0.92496 | 0.35302 | 0.56727 | 0.92784 | 0.560 |
| 0.580 | 0.36849 | 0.58512 | 0.95437 | 0.37135 | 0.58701 | 0.95711 | 0.580 |
| 0.600 | 0.38710 | 0.60494 | 0.98336 | 0.38987 | 0.60677 | 0.98597 | 0.600 |
| 0.620 | 0.40588 | 0.62478 | 1.01188 | 0.40855 | 0.62654 | 1.01437 | 0.620 |
| 0.640 | 0.42480 | 0.64463 | 1.03989 | 0.42738 | 0.64633 | 1.04227 | 0.640 |
| 0.660 | 0.44386 | 0.66449 | 1.06735 | 0.44636 | 0.66614 | 1.06961 | 0.660 |
| 0.680 | 0.46303 | 0.68435 | 1.09421 | 0.46545 | 0.68596 | 1.09635 | 0.680 |
| 0.700 | 0.48231 | 0.70423 | 1.12041 | 0.48466 | 0.70578 | 1.12244 | 0.700 |
| 0.720 | 0.50169 | 0.72411 | 1.14591 | 0.50397 | 0.72561 | 1.14783 | 0.720 |
| 0.740 | 0.52116 | 0.74399 | 1.17068 | 0.52337 | 0.74546 | 1.17248 | 0.740 |
| 0.760 | 0.54070 | 0.76389 | 1.19465 | 0.54285 | 0.76531 | 1.19634 | 0.760 |
| 0.780 | 0.56032 | 0.78379 | 1.21779 | 0.56240 | 0.78517 | 1.21936 | 0.780 |
| 0.800 | 0.57999 | 0.80369 | 1.24007 | 0.58202 | 0.80504 | 1.24154 | 0.800 |
| 0.820 | 0.59973 | 0.82360 | 1.26147 | 0.60170 | 0.82492 | 1.26280 | 0.820 |
| 0.840 | 0.61951 | 0.84351 | 1.28197 | 0.62143 | 0.84480 | 1.28316 | 0.840 |
| 0.860 | 0.63934 | 0.86343 | 1.30155 | 0.64121 | 0.86468 | 1.30258 | 0.860 |
| 0.880 | 0.65921 | 0.88335 | 1.32020 | 0.66103 | 0.88458 | 1.32107 | 0.880 |
| 0.900 | 0.67910 | 0.90327 | 1.33793 | 0.68088 | 0.90447 | 1.33861 | 0.900 |
| 0.920 | 0.69904 | 0.92320 | 1.35474 | 0.70077 | 0.92437 | 1.35523 | 0.920 |
| 0.940 | 0.71899 | 0.94314 | 1.37068 | 0.72069 | 0.94428 | 1.37100 | 0.940 |
| 0.960 | 0.73897 | 0.96307 | 1.38585 | 0.74063 | 0.96419 | 1.38608 | 0.960 |
| 0.980 | 0.75897 | 0.98300 | 1.40048 | 0.76059 | 0.98410 | 1.40068 | 0.980 |
| 1.000 | 0.77899 | 1.00294 | 1.41474 | 0.78057 | 1.00402 | 1.41494 | 1.000 |

$H/R = 0.160$ $n = 2.$ $H/R = 0.180$

| $H/L$ | $\Omega_1$ | $\Omega_2$ | $\Omega_3$ | $\Omega_1$ | $\Omega_2$ | $\Omega_3$ | $H/L$ |
|---|---|---|---|---|---|---|---|
| 0. | 0.01054 | 0.10196 | 0.19114 | 0.01329 | 0.11474 | 0.21462 | 0. |
| 0.001 | 0.01054 | 0.10197 | 0.19115 | 0.01329 | 0.11475 | 0.21462 | 0.001 |
| 0.005 | 0.01056 | 0.10218 | 0.19128 | 0.01331 | 0.11493 | 0.21474 | 0.005 |
| 0.010 | 0.01064 | 0.10280 | 0.19170 | 0.01339 | 0.11549 | 0.21512 | 0.010 |
| 0.020 | 0.01123 | 0.10523 | 0.19339 | 0.01385 | 0.11767 | 0.21662 | 0.020 |
| 0.030 | 0.01280 | 0.10899 | 0.19621 | 0.01505 | 0.12109 | 0.21913 | 0.030 |
| 0.040 | 0.01555 | 0.11377 | 0.20020 | 0.01730 | 0.12552 | 0.22267 | 0.040 |
| 0.050 | 0.01933 | 0.11928 | 0.20534 | 0.02056 | 0.13070 | 0.22724 | 0.050 |
| 0.060 | 0.02380 | 0.12527 | 0.21165 | 0.02461 | 0.13643 | 0.23286 | 0.060 |
| 0.070 | 0.02865 | 0.13157 | 0.21908 | 0.02922 | 0.14251 | 0.23949 | 0.070 |
| 0.080 | 0.03368 | 0.13806 | 0.22757 | 0.03416 | 0.14885 | 0.24711 | 0.080 |
| 0.090 | 0.03877 | 0.14471 | 0.23704 | 0.03929 | 0.15535 | 0.25567 | 0.090 |
| 0.100 | 0.04382 | 0.15150 | 0.24738 | 0.04450 | 0.16200 | 0.26509 | 0.100 |
| 0.120 | 0.05370 | 0.16554 | 0.27029 | 0.05492 | 0.17566 | 0.28622 | 0.120 |
| 0.140 | 0.06326 | 0.18025 | 0.29551 | 0.06520 | 0.18987 | 0.30985 | 0.140 |
| 0.160 | 0.07265 | 0.19570 | 0.32243 | 0.07533 | 0.20471 | 0.33538 | 0.160 |
| 0.180 | 0.08207 | 0.21185 | 0.35059 | 0.08545 | 0.22021 | 0.36234 | 0.180 |
| 0.200 | 0.09181 | 0.22861 | 0.37964 | 0.09571 | 0.23632 | 0.39038 | 0.200 |
| 0.220 | 0.10187 | 0.24589 | 0.40936 | 0.10629 | 0.25299 | 0.41922 | 0.220 |
| 0.240 | 0.11264 | 0.26360 | 0.43957 | 0.11727 | 0.27014 | 0.44868 | 0.240 |
| 0.260 | 0.12391 | 0.28165 | 0.47015 | 0.12874 | 0.28769 | 0.47859 | 0.260 |
| 0.280 | 0.13584 | 0.29999 | 0.50098 | 0.14084 | 0.30558 | 0.50885 | 0.280 |
| 0.300 | 0.14844 | 0.31855 | 0.53200 | 0.15344 | 0.32374 | 0.53935 | 0.300 |
| 0.320 | 0.16167 | 0.33730 | 0.56314 | 0.16662 | 0.34215 | 0.57004 | 0.320 |
| 0.340 | 0.17548 | 0.35621 | 0.59434 | 0.18038 | 0.36074 | 0.60083 | 0.340 |
| 0.360 | 0.18986 | 0.37524 | 0.62557 | 0.19467 | 0.37950 | 0.63169 | 0.360 |
| 0.380 | 0.20476 | 0.39438 | 0.65678 | 0.20947 | 0.39840 | 0.66256 | 0.380 |
| 0.400 | 0.22014 | 0.41362 | 0.68793 | 0.22473 | 0.41741 | 0.69341 | 0.400 |
| 0.420 | 0.23597 | 0.43293 | 0.71899 | 0.24043 | 0.43652 | 0.72419 | 0.420 |
| 0.440 | 0.25220 | 0.45231 | 0.74994 | 0.25653 | 0.45572 | 0.75487 | 0.440 |
| 0.460 | 0.26880 | 0.47174 | 0.78072 | 0.27300 | 0.47500 | 0.78541 | 0.460 |
| 0.480 | 0.28574 | 0.49123 | 0.81131 | 0.28981 | 0.49433 | 0.81578 | 0.480 |
| 0.500 | 0.30299 | 0.51076 | 0.84169 | 0.30694 | 0.51373 | 0.84593 | 0.500 |
| 0.520 | 0.32052 | 0.53033 | 0.87180 | 0.32434 | 0.53317 | 0.87585 | 0.520 |
| 0.540 | 0.33830 | 0.54993 | 0.90163 | 0.34201 | 0.55266 | 0.90548 | 0.540 |
| 0.560 | 0.35632 | 0.56955 | 0.93112 | 0.35991 | 0.57218 | 0.93479 | 0.560 |
| 0.580 | 0.37454 | 0.58921 | 0.96024 | 0.37802 | 0.59174 | 0.96374 | 0.580 |
| 0.600 | 0.39295 | 0.60889 | 0.98896 | 0.39633 | 0.61133 | 0.99229 | 0.600 |
| 0.620 | 0.41154 | 0.62860 | 1.01722 | 0.41481 | 0.63095 | 1.02037 | 0.620 |
| 0.640 | 0.43028 | 0.64832 | 1.04497 | 0.43346 | 0.65059 | 1.04796 | 0.640 |
| 0.660 | 0.44917 | 0.66806 | 1.07216 | 0.45225 | 0.67026 | 1.07501 | 0.660 |
| 0.680 | 0.46818 | 0.68782 | 1.09877 | 0.47117 | 0.68994 | 1.10145 | 0.680 |
| 0.700 | 0.48730 | 0.70758 | 1.12473 | 0.49022 | 0.70965 | 1.12726 | 0.700 |
| 0.720 | 0.50654 | 0.72737 | 1.14999 | 0.50937 | 0.72938 | 1.15236 | 0.720 |
| 0.740 | 0.52586 | 0.74716 | 1.17450 | 0.52862 | 0.74911 | 1.17671 | 0.740 |
| 0.760 | 0.54527 | 0.76697 | 1.19822 | 0.54795 | 0.76886 | 1.20027 | 0.760 |
| 0.780 | 0.56476 | 0.78679 | 1.22111 | 0.56737 | 0.78863 | 1.22298 | 0.780 |
| 0.800 | 0.58432 | 0.80661 | 1.24312 | 0.58687 | 0.80841 | 1.24482 | 0.800 |
| 0.820 | 0.60394 | 0.82645 | 1.26424 | 0.60642 | 0.82820 | 1.26573 | 0.820 |
| 0.840 | 0.62361 | 0.84629 | 1.28442 | 0.62603 | 0.84800 | 1.28572 | 0.840 |
| 0.860 | 0.64333 | 0.86614 | 1.30365 | 0.64569 | 0.86780 | 1.30473 | 0.860 |
| 0.880 | 0.66310 | 0.88600 | 1.32194 | 0.66540 | 0.88762 | 1.32278 | 0.880 |
| 0.900 | 0.68291 | 0.90586 | 1.33926 | 0.68516 | 0.90745 | 1.33990 | 0.900 |
| 0.920 | 0.70274 | 0.92573 | 1.35570 | 0.70494 | 0.92728 | 1.35616 | 0.920 |
| 0.940 | 0.72261 | 0.94561 | 1.37134 | 0.72476 | 0.94712 | 1.37169 | 0.940 |
| 0.960 | 0.74251 | 0.96549 | 1.38635 | 0.74461 | 0.96697 | 1.38665 | 0.960 |
| 0.980 | 0.76243 | 0.98538 | 1.40092 | 0.76448 | 0.98683 | 1.40118 | 0.980 |
| 1.000 | 0.78237 | 1.00526 | 1.41516 | 0.78438 | 1.00669 | 1.41541 | 1.000 |

$H/R = 0.200 \qquad n = 2.$

| $H/L$ | $\Omega_1$ | $\Omega_2$ | $\Omega_3$ | $\Omega_4$ | $\Omega_5$ | $\Omega_6$ |
|---|---|---|---|---|---|---|
| 0. | 0.01635 | 0.12752 | 0.23794 | 1.00975 | 1.03594 | 1.84276 |
| 0.001 | 0.01635 | 0.12753 | 0.23794 | 1.00975 | 1.03595 | 1.84276 |
| 0.005 | 0.01637 | 0.12769 | 0.23805 | 1.00977 | 1.03598 | 1.84272 |
| 0.010 | 0.01644 | 0.12820 | 0.23839 | 1.00983 | 1.03608 | 1.84262 |
| 0.020 | 0.01686 | 0.13017 | 0.23974 | 1.01007 | 1.03649 | 1.84216 |
| 0.030 | 0.01781 | 0.13331 | 0.24201 | 1.01045 | 1.03717 | 1.84144 |
| 0.040 | 0.01964 | 0.13742 | 0.24519 | 1.01098 | 1.03813 | 1.84042 |
| 0.050 | 0.02239 | 0.14230 | 0.24931 | 1.01165 | 1.03937 | 1.83914 |
| 0.060 | 0.02601 | 0.14775 | 0.25437 | 1.01245 | 1.04090 | 1.83761 |
| 0.070 | 0.03028 | 0.15362 | 0.26035 | 1.01337 | 1.04273 | 1.83585 |
| 0.080 | 0.03501 | 0.15978 | 0.26725 | 1.01440 | 1.04486 | 1.83387 |
| 0.090 | 0.04005 | 0.16614 | 0.27503 | 1.01554 | 1.04730 | 1.83171 |
| 0.100 | 0.04528 | 0.17266 | 0.28365 | 1.01677 | 1.05003 | 1.82938 |
| 0.120 | 0.05600 | 0.18604 | 0.30314 | 1.01951 | 1.05643 | 1.82432 |
| 0.140 | 0.06679 | 0.19988 | 0.32522 | 1.02257 | 1.06406 | 1.81879 |
| 0.160 | 0.07754 | 0.21425 | 0.34936 | 1.02594 | 1.07287 | 1.81299 |
| 0.180 | 0.08828 | 0.22920 | 0.37511 | 1.02964 | 1.08281 | 1.80700 |
| 0.200 | 0.09912 | 0.24473 | 0.40208 | 1.03364 | 1.09384 | 1.80096 |
| 0.220 | 0.11017 | 0.26081 | 0.43001 | 1.03796 | 1.10587 | 1.79491 |
| 0.240 | 0.12153 | 0.27741 | 0.45866 | 1.04260 | 1.11885 | 1.78896 |
| 0.260 | 0.13330 | 0.29444 | 0.48787 | 1.04758 | 1.13273 | 1.78316 |
| 0.280 | 0.14550 | 0.31186 | 0.51750 | 1.05288 | 1.14743 | 1.77756 |
| 0.300 | 0.15826 | 0.32960 | 0.54745 | 1.05851 | 1.16291 | 1.77222 |
| 0.320 | 0.17155 | 0.34762 | 0.57764 | 1.06447 | 1.17910 | 1.76717 |
| 0.340 | 0.18528 | 0.36587 | 0.60799 | 1.07075 | 1.19595 | 1.76245 |
| 0.360 | 0.19952 | 0.38433 | 0.63844 | 1.07736 | 1.21343 | 1.75809 |
| 0.380 | 0.21425 | 0.40295 | 0.66895 | 1.08429 | 1.23148 | 1.75413 |
| 0.400 | 0.22943 | 0.42171 | 0.69945 | 1.09153 | 1.25005 | 1.75060 |
| 0.420 | 0.24504 | 0.44060 | 0.72992 | 1.09908 | 1.26913 | 1.74754 |
| 0.440 | 0.26103 | 0.45959 | 0.76032 | 1.10694 | 1.28865 | 1.74496 |
| 0.460 | 0.27740 | 0.47868 | 0.79058 | 1.11510 | 1.30859 | 1.74290 |
| 0.480 | 0.29409 | 0.49785 | 0.82070 | 1.12356 | 1.32892 | 1.74140 |
| 0.500 | 0.31110 | 0.51709 | 0.85062 | 1.13231 | 1.34960 | 1.74047 |
| 0.520 | 0.32840 | 0.53639 | 0.88031 | 1.14133 | 1.37061 | 1.74016 |
| 0.540 | 0.34595 | 0.55575 | 0.90973 | 1.15064 | 1.39192 | 1.74051 |
| 0.560 | 0.36374 | 0.57515 | 0.93883 | 1.16022 | 1.41350 | 1.74155 |
| 0.580 | 0.38174 | 0.59460 | 0.96758 | 1.17007 | 1.43533 | 1.74330 |
| 0.600 | 0.39995 | 0.61409 | 0.99593 | 1.18017 | 1.45739 | 1.74582 |
| 0.620 | 0.41833 | 0.63361 | 1.02384 | 1.19054 | 1.47966 | 1.74916 |
| 0.640 | 0.43688 | 0.65317 | 1.05125 | 1.20116 | 1.50210 | 1.75334 |
| 0.660 | 0.45558 | 0.67275 | 1.07811 | 1.21202 | 1.52473 | 1.75844 |
| 0.680 | 0.47442 | 0.69236 | 1.10438 | 1.22314 | 1.54748 | 1.76447 |
| 0.700 | 0.49337 | 0.71199 | 1.13000 | 1.23449 | 1.57037 | 1.77149 |
| 0.720 | 0.51244 | 0.73164 | 1.15491 | 1.24608 | 1.59338 | 1.77954 |
| 0.740 | 0.53161 | 0.75132 | 1.17908 | 1.25792 | 1.61647 | 1.78868 |
| 0.760 | 0.55087 | 0.77100 | 1.20244 | 1.26998 | 1.63964 | 1.79894 |
| 0.780 | 0.57022 | 0.79071 | 1.22496 | 1.28231 | 1.66288 | 1.81035 |
| 0.800 | 0.58964 | 0.81044 | 1.24659 | 1.29488 | 1.68616 | 1.82294 |
| 0.820 | 0.60912 | 0.83017 | 1.26729 | 1.30770 | 1.70947 | 1.83672 |
| 0.840 | 0.62867 | 0.84992 | 1.28703 | 1.32081 | 1.73279 | 1.85171 |
| 0.860 | 0.64827 | 0.86968 | 1.30580 | 1.33420 | 1.75612 | 1.86790 |
| 0.880 | 0.66793 | 0.88945 | 1.32362 | 1.34787 | 1.77944 | 1.88525 |
| 0.900 | 0.68762 | 0.90924 | 1.34053 | 1.36178 | 1.80273 | 1.90376 |
| 0.920 | 0.70735 | 0.92903 | 1.35663 | 1.37588 | 1.82600 | 1.92339 |
| 0.940 | 0.72712 | 0.94884 | 1.37206 | 1.39004 | 1.84921 | 1.94409 |
| 0.960 | 0.74692 | 0.96865 | 1.38697 | 1.40417 | 1.87238 | 1.96578 |
| 0.980 | 0.76674 | 0.98847 | 1.40149 | 1.41820 | 1.89548 | 1.98842 |
| 1.000 | 0.78659 | 1.00829 | 1.41570 | 1.43207 | 1.91850 | 2.01193 |

$H/R = 0.250$    $n = 2.$

| $H/L$ | $\Omega_1$ | $\Omega_2$ | $\Omega_3$ |
|---|---|---|---|
| 0. | 0.02528 | 0.15953 | 0.29543 |
| 0.001 | 0.02528 | 0.15954 | 0.29543 |
| 0.005 | 0.02530 | 0.15954 | 0.29552 |
| 0.010 | 0.02539 | 0.16007 | 0.29579 |
| 0.020 | 0.02566 | 0.16168 | 0.29688 |
| 0.030 | 0.02634 | 0.16426 | 0.29871 |
| 0.040 | 0.02751 | 0.16773 | 0.30127 |
| 0.050 | 0.02938 | 0.17193 | 0.30457 |
| 0.060 | 0.03198 | 0.17675 | 0.30863 |
| 0.070 | 0.03530 | 0.18205 | 0.31344 |
| 0.080 | 0.03925 | 0.18773 | 0.31900 |
| 0.090 | 0.04371 | 0.19370 | 0.32530 |
| 0.100 | 0.04858 | 0.19988 | 0.33232 |
| 0.120 | 0.05917 | 0.21269 | 0.34846 |
| 0.140 | 0.07045 | 0.22593 | 0.36713 |
| 0.160 | 0.08210 | 0.23953 | 0.38802 |
| 0.180 | 0.09396 | 0.25353 | 0.41077 |
| 0.200 | 0.10597 | 0.26797 | 0.43504 |
| 0.220 | 0.11814 | 0.28286 | 0.46055 |
| 0.240 | 0.13051 | 0.29823 | 0.48705 |
| 0.260 | 0.14313 | 0.31405 | 0.51434 |
| 0.280 | 0.15604 | 0.33030 | 0.54226 |
| 0.300 | 0.16930 | 0.34695 | 0.57067 |
| 0.320 | 0.18292 | 0.36394 | 0.59947 |
| 0.340 | 0.19691 | 0.38125 | 0.62857 |
| 0.360 | 0.21129 | 0.39884 | 0.65788 |
| 0.380 | 0.22605 | 0.41667 | 0.68733 |
| 0.400 | 0.24121 | 0.43471 | 0.71687 |
| 0.420 | 0.25673 | 0.45294 | 0.74644 |
| 0.440 | 0.27262 | 0.47133 | 0.77600 |
| 0.460 | 0.28880 | 0.48986 | 0.80549 |
| 0.480 | 0.30530 | 0.50852 | 0.83487 |
| 0.500 | 0.32211 | 0.52729 | 0.86409 |
| 0.520 | 0.33917 | 0.54617 | 0.89312 |
| 0.540 | 0.35650 | 0.56512 | 0.92191 |
| 0.560 | 0.37406 | 0.58416 | 0.95041 |
| 0.580 | 0.39184 | 0.60327 | 0.97856 |
| 0.600 | 0.40981 | 0.62244 | 1.00634 |
| 0.620 | 0.42797 | 0.64167 | 1.03368 |
| 0.640 | 0.44630 | 0.66095 | 1.06053 |
| 0.660 | 0.46477 | 0.68028 | 1.08684 |
| 0.680 | 0.48339 | 0.69965 | 1.11255 |
| 0.700 | 0.50214 | 0.71905 | 1.13761 |
| 0.720 | 0.52100 | 0.73849 | 1.16195 |
| 0.740 | 0.53997 | 0.75796 | 1.18553 |
| 0.760 | 0.55904 | 0.77746 | 1.20828 |
| 0.780 | 0.57820 | 0.79699 | 1.23018 |
| 0.800 | 0.59744 | 0.81655 | 1.25115 |
| 0.820 | 0.61676 | 0.83612 | 1.27117 |
| 0.840 | 0.63613 | 0.85572 | 1.29024 |
| 0.860 | 0.65557 | 0.87533 | 1.30840 |
| 0.880 | 0.67507 | 0.89497 | 1.32565 |
| 0.900 | 0.69461 | 0.91462 | 1.34212 |
| 0.920 | 0.71420 | 0.93429 | 1.35790 |
| 0.940 | 0.73382 | 0.95397 | 1.37314 |
| 0.960 | 0.75348 | 0.97367 | 1.38795 |
| 0.980 | 0.77318 | 0.99338 | 1.40241 |
| 1.000 | 0.79290 | 1.01310 | 1.41659 |

$H/R = 0.300$ $n = 2.$

| $H/L$ | $\Omega_1$ | $\Omega_2$ | $\Omega_3$ | $\Omega_4$ | $\Omega_5$ | $\Omega_6$ |
|---|---|---|---|---|---|---|
| 0. | 0.03596 | 0.19161 | 0.35149 | 1.02230 | 1.08213 | 1.82136 |
| 0.001 | 0.03596 | 0.19161 | 0.35150 | 1.02230 | 1.08214 | 1.82135 |
| 0.005 | 0.03598 | 0.19172 | 0.35157 | 1.02232 | 1.08217 | 1.82133 |
| 0.010 | 0.03604 | 0.19206 | 0.35180 | 1.02238 | 1.08225 | 1.82124 |
| 0.020 | 0.03629 | 0.19341 | 0.35272 | 1.02261 | 1.08263 | 1.82095 |
| 0.030 | 0.03681 | 0.19560 | 0.35425 | 1.02300 | 1.08327 | 1.82043 |
| 0.040 | 0.03772 | 0.19858 | 0.35640 | 1.02354 | 1.08415 | 1.81974 |
| 0.050 | 0.03905 | 0.20225 | 0.35917 | 1.02421 | 1.08529 | 1.81885 |
| 0.060 | 0.04097 | 0.20652 | 0.36257 | 1.02504 | 1.08670 | 1.81779 |
| 0.070 | 0.04350 | 0.21130 | 0.36660 | 1.02601 | 1.08835 | 1.81654 |
| 0.080 | 0.04664 | 0.21649 | 0.37126 | 1.02710 | 1.09027 | 1.81515 |
| 0.090 | 0.05034 | 0.22203 | 0.37655 | 1.02832 | 1.09247 | 1.81360 |
| 0.100 | 0.05456 | 0.22784 | 0.38247 | 1.02966 | 1.09492 | 1.81191 |
| 0.120 | 0.06424 | 0.24005 | 0.39613 | 1.03269 | 1.10064 | 1.80820 |
| 0.140 | 0.07514 | 0.25281 | 0.41213 | 1.03612 | 1.10744 | 1.80406 |
| 0.160 | 0.08686 | 0.26595 | 0.43027 | 1.03993 | 1.11532 | 1.79963 |
| 0.180 | 0.09911 | 0.27940 | 0.45031 | 1.04408 | 1.12425 | 1.79500 |
| 0.200 | 0.11174 | 0.29317 | 0.47202 | 1.04855 | 1.13423 | 1.79024 |
| 0.220 | 0.12466 | 0.30728 | 0.49513 | 1.05332 | 1.14520 | 1.78542 |
| 0.240 | 0.13781 | 0.32176 | 0.51943 | 1.05837 | 1.15713 | 1.78064 |
| 0.260 | 0.15120 | 0.33663 | 0.54471 | 1.06373 | 1.16997 | 1.77594 |
| 0.280 | 0.16483 | 0.35189 | 0.57079 | 1.06937 | 1.18366 | 1.77138 |
| 0.300 | 0.17872 | 0.36754 | 0.59752 | 1.07531 | 1.19815 | 1.76699 |
| 0.320 | 0.19287 | 0.38355 | 0.62479 | 1.08153 | 1.21340 | 1.76286 |
| 0.340 | 0.20731 | 0.39991 | 0.65247 | 1.08804 | 1.22934 | 1.75900 |
| 0.360 | 0.22204 | 0.41659 | 0.68048 | 1.09485 | 1.24594 | 1.75546 |
| 0.380 | 0.23706 | 0.43356 | 0.70874 | 1.10195 | 1.26314 | 1.75228 |
| 0.400 | 0.25240 | 0.45079 | 0.73716 | 1.10934 | 1.28091 | 1.74948 |
| 0.420 | 0.26802 | 0.46827 | 0.76569 | 1.11702 | 1.29920 | 1.74710 |
| 0.440 | 0.28393 | 0.48595 | 0.79428 | 1.12499 | 1.31797 | 1.74520 |
| 0.460 | 0.30011 | 0.50383 | 0.82285 | 1.13325 | 1.33719 | 1.74378 |
| 0.480 | 0.31659 | 0.52189 | 0.85136 | 1.14180 | 1.35682 | 1.74289 |
| 0.500 | 0.33332 | 0.54009 | 0.87975 | 1.15063 | 1.37683 | 1.74257 |
| 0.520 | 0.35030 | 0.55844 | 0.90799 | 1.15973 | 1.39719 | 1.74284 |
| 0.540 | 0.36750 | 0.57691 | 0.93601 | 1.16913 | 1.41786 | 1.74375 |
| 0.560 | 0.38491 | 0.59549 | 0.96376 | 1.17879 | 1.43883 | 1.74533 |
| 0.580 | 0.40254 | 0.61418 | 0.99119 | 1.18874 | 1.46008 | 1.74765 |
| 0.600 | 0.42036 | 0.63295 | 1.01825 | 1.19895 | 1.48156 | 1.75071 |
| 0.620 | 0.43834 | 0.65181 | 1.04488 | 1.20944 | 1.50326 | 1.75459 |
| 0.640 | 0.45648 | 0.67074 | 1.07103 | 1.22021 | 1.52516 | 1.75931 |
| 0.660 | 0.47477 | 0.68974 | 1.09663 | 1.23126 | 1.54725 | 1.76493 |
| 0.680 | 0.49321 | 0.70880 | 1.12163 | 1.24258 | 1.56947 | 1.77151 |
| 0.700 | 0.51178 | 0.72792 | 1.14596 | 1.25418 | 1.59184 | 1.77906 |
| 0.720 | 0.53046 | 0.74709 | 1.16957 | 1.26608 | 1.61431 | 1.78767 |
| 0.740 | 0.54926 | 0.76631 | 1.19239 | 1.27827 | 1.63689 | 1.79736 |
| 0.760 | 0.56815 | 0.78557 | 1.21438 | 1.29078 | 1.65954 | 1.80816 |
| 0.780 | 0.58713 | 0.80487 | 1.23548 | 1.30358 | 1.68224 | 1.82012 |
| 0.800 | 0.60620 | 0.82421 | 1.25569 | 1.31667 | 1.70499 | 1.83326 |
| 0.820 | 0.62535 | 0.84358 | 1.27498 | 1.33006 | 1.72777 | 1.84758 |
| 0.840 | 0.64457 | 0.86298 | 1.29337 | 1.34369 | 1.75055 | 1.86310 |
| 0.860 | 0.66384 | 0.88242 | 1.31093 | 1.35750 | 1.77333 | 1.87980 |
| 0.880 | 0.68318 | 0.90187 | 1.32771 | 1.37146 | 1.79611 | 1.89764 |
| 0.900 | 0.70257 | 0.92136 | 1.34383 | 1.38546 | 1.81887 | 1.91662 |
| 0.920 | 0.72201 | 0.94087 | 1.35938 | 1.39944 | 1.84161 | 1.93666 |
| 0.940 | 0.74149 | 0.96040 | 1.37448 | 1.41333 | 1.86432 | 1.95773 |
| 0.960 | 0.76101 | 0.97995 | 1.38919 | 1.42710 | 1.88699 | 1.97976 |
| 0.980 | 0.78057 | 0.99952 | 1.40358 | 1.44072 | 1.90962 | 2.00266 |
| 1.000 | 0.80015 | 1.01911 | 1.41769 | 1.45421 | 1.93221 | 2.02639 |

$H/R = 0.350$ $n = 2.$

| $H/L$ | $\Omega_1$ | $\Omega_2$ | $\Omega_3$ |
|---|---|---|---|
| 0. | 0.04828 | 0.22374 | 0.40578 |
| 0.001 | 0.04828 | 0.22375 | 0.40578 |
| 0.005 | 0.04829 | 0.22384 | 0.40584 |
| 0.010 | 0.04835 | 0.22414 | 0.40604 |
| 0.020 | 0.04858 | 0.22530 | 0.40684 |
| 0.030 | 0.04902 | 0.22721 | 0.40817 |
| 0.040 | 0.04973 | 0.22981 | 0.41003 |
| 0.050 | 0.05064 | 0.23306 | 0.41243 |
| 0.060 | 0.05229 | 0.23687 | 0.41537 |
| 0.070 | 0.05427 | 0.24119 | 0.41885 |
| 0.080 | 0.05677 | 0.24594 | 0.42288 |
| 0.090 | 0.05979 | 0.25106 | 0.42745 |
| 0.100 | 0.06333 | 0.25649 | 0.43257 |
| 0.120 | 0.07180 | 0.26806 | 0.44441 |
| 0.140 | 0.08182 | 0.28031 | 0.45835 |
| 0.160 | 0.09302 | 0.29303 | 0.47428 |
| 0.180 | 0.10508 | 0.30608 | 0.49204 |
| 0.200 | 0.11779 | 0.31942 | 0.51148 |
| 0.220 | 0.13097 | 0.33302 | 0.53239 |
| 0.240 | 0.14453 | 0.34691 | 0.55459 |
| 0.260 | 0.15841 | 0.36111 | 0.57788 |
| 0.280 | 0.17255 | 0.37563 | 0.60211 |
| 0.300 | 0.18695 | 0.39048 | 0.62712 |
| 0.320 | 0.20160 | 0.40567 | 0.65279 |
| 0.340 | 0.21649 | 0.42119 | 0.67897 |
| 0.360 | 0.23162 | 0.43704 | 0.70559 |
| 0.380 | 0.24699 | 0.45319 | 0.73254 |
| 0.400 | 0.26261 | 0.46962 | 0.75974 |
| 0.420 | 0.27846 | 0.48633 | 0.78713 |
| 0.440 | 0.29456 | 0.50328 | 0.81462 |
| 0.460 | 0.31088 | 0.52046 | 0.84215 |
| 0.480 | 0.32744 | 0.53785 | 0.86968 |
| 0.500 | 0.34420 | 0.55543 | 0.89712 |
| 0.520 | 0.36119 | 0.57319 | 0.92443 |
| 0.540 | 0.37836 | 0.59110 | 0.95156 |
| 0.560 | 0.39573 | 0.60916 | 0.97844 |
| 0.580 | 0.41329 | 0.62735 | 1.00501 |
| 0.600 | 0.43102 | 0.64566 | 1.03122 |
| 0.620 | 0.44890 | 0.66408 | 1.05701 |
| 0.640 | 0.46694 | 0.68260 | 1.08231 |
| 0.660 | 0.48511 | 0.70121 | 1.10706 |
| 0.680 | 0.50342 | 0.71991 | 1.13120 |
| 0.700 | 0.52185 | 0.73868 | 1.15466 |
| 0.720 | 0.54040 | 0.75753 | 1.17739 |
| 0.740 | 0.55905 | 0.77643 | 1.19934 |
| 0.760 | 0.57780 | 0.79540 | 1.22046 |
| 0.780 | 0.59664 | 0.81442 | 1.24073 |
| 0.800 | 0.61557 | 0.83350 | 1.26015 |
| 0.820 | 0.63458 | 0.85263 | 1.27872 |
| 0.840 | 0.65364 | 0.87179 | 1.29649 |
| 0.860 | 0.67278 | 0.89100 | 1.31355 |
| 0.880 | 0.69197 | 0.91024 | 1.32995 |
| 0.900 | 0.71122 | 0.92952 | 1.34579 |
| 0.920 | 0.73052 | 0.94884 | 1.36114 |
| 0.940 | 0.74986 | 0.96818 | 1.37608 |
| 0.960 | 0.76925 | 0.98755 | 1.39070 |
| 0.980 | 0.78867 | 1.00695 | 1.40501 |
| 1.000 | 0.80813 | 1.02638 | 1.41906 |

$H/R = 0.400 \qquad n = 2.$

| $H/L$ | $\Omega_1$ | $\Omega_2$ | $\Omega_3$ | $\Omega_4$ | $\Omega_5$ | $\Omega_6$ |
|---|---|---|---|---|---|---|
| 0. | 0.06211 | 0.25593 | 0.45793 | 1.04057 | 1.14894 | 1.80159 |
| 0.001 | 0.06211 | 0.25593 | 0.45793 | 1.04058 | 1.14894 | 1.80159 |
| 0.005 | 0.06212 | 0.25602 | 0.45798 | 1.04060 | 1.14897 | 1.80157 |
| 0.010 | 0.06217 | 0.25628 | 0.45816 | 1.04065 | 1.14905 | 1.80151 |
| 0.020 | 0.06239 | 0.25730 | 0.45887 | 1.04088 | 1.14940 | 1.80131 |
| 0.030 | 0.06277 | 0.25899 | 0.46004 | 1.04126 | 1.14997 | 1.80096 |
| 0.040 | 0.06338 | 0.26131 | 0.46169 | 1.04180 | 1.15078 | 1.80048 |
| 0.050 | 0.06426 | 0.26421 | 0.46382 | 1.04247 | 1.15180 | 1.79987 |
| 0.060 | 0.06547 | 0.26765 | 0.46642 | 1.04330 | 1.15307 | 1.79913 |
| 0.070 | 0.06710 | 0.27158 | 0.46950 | 1.04427 | 1.15457 | 1.79826 |
| 0.080 | 0.06912 | 0.27594 | 0.47306 | 1.04538 | 1.15630 | 1.79729 |
| 0.090 | 0.07161 | 0.28067 | 0.47710 | 1.04663 | 1.15826 | 1.79620 |
| 0.100 | 0.07456 | 0.28573 | 0.48162 | 1.04801 | 1.16046 | 1.79501 |
| 0.120 | 0.08184 | 0.29665 | 0.49210 | 1.05115 | 1.16555 | 1.79236 |
| 0.140 | 0.09078 | 0.30837 | 0.50446 | 1.05478 | 1.17160 | 1.78938 |
| 0.160 | 0.10113 | 0.32064 | 0.51863 | 1.05884 | 1.17857 | 1.78616 |
| 0.180 | 0.11259 | 0.33332 | 0.53454 | 1.06332 | 1.18649 | 1.78273 |
| 0.200 | 0.12494 | 0.34630 | 0.55205 | 1.06817 | 1.19533 | 1.77921 |
| 0.220 | 0.13798 | 0.35954 | 0.57102 | 1.07338 | 1.20508 | 1.77560 |
| 0.240 | 0.15157 | 0.37302 | 0.59130 | 1.07892 | 1.21571 | 1.77201 |
| 0.260 | 0.16559 | 0.38675 | 0.61275 | 1.08476 | 1.22721 | 1.76845 |
| 0.280 | 0.17998 | 0.40073 | 0.63521 | 1.09090 | 1.23954 | 1.76502 |
| 0.300 | 0.19468 | 0.41500 | 0.65853 | 1.09733 | 1.25265 | 1.76173 |
| 0.320 | 0.20965 | 0.42955 | 0.68258 | 1.10404 | 1.26652 | 1.75863 |
| 0.340 | 0.22488 | 0.44440 | 0.70726 | 1.11102 | 1.28111 | 1.75578 |
| 0.360 | 0.24034 | 0.45954 | 0.73244 | 1.11828 | 1.29637 | 1.75321 |
| 0.380 | 0.25603 | 0.47497 | 0.75803 | 1.12581 | 1.31227 | 1.75095 |
| 0.400 | 0.27193 | 0.49069 | 0.78395 | 1.13361 | 1.32876 | 1.74907 |
| 0.420 | 0.28804 | 0.50668 | 0.81010 | 1.14168 | 1.34581 | 1.74756 |
| 0.440 | 0.30435 | 0.52293 | 0.83642 | 1.15002 | 1.36338 | 1.74651 |
| 0.460 | 0.32086 | 0.53943 | 0.86283 | 1.15864 | 1.38142 | 1.74592 |
| 0.480 | 0.33757 | 0.55615 | 0.88927 | 1.16754 | 1.39992 | 1.74584 |
| 0.500 | 0.35445 | 0.57309 | 0.91566 | 1.17671 | 1.41882 | 1.74632 |
| 0.520 | 0.37152 | 0.59023 | 0.94195 | 1.18617 | 1.43811 | 1.74738 |
| 0.540 | 0.38876 | 0.60755 | 0.96807 | 1.19591 | 1.45774 | 1.74907 |
| 0.560 | 0.40617 | 0.62505 | 0.99397 | 1.20595 | 1.47769 | 1.75144 |
| 0.580 | 0.42373 | 0.64270 | 1.01957 | 1.21628 | 1.49793 | 1.75451 |
| 0.600 | 0.44144 | 0.66051 | 1.04481 | 1.22690 | 1.51843 | 1.75836 |
| 0.620 | 0.45930 | 0.67844 | 1.06963 | 1.23783 | 1.53915 | 1.76302 |
| 0.640 | 0.47729 | 0.69650 | 1.09396 | 1.24906 | 1.56009 | 1.76855 |
| 0.660 | 0.49540 | 0.71468 | 1.11775 | 1.26060 | 1.58121 | 1.77498 |
| 0.680 | 0.51364 | 0.73296 | 1.14092 | 1.27246 | 1.60248 | 1.78238 |
| 0.700 | 0.53199 | 0.75133 | 1.16341 | 1.28463 | 1.62388 | 1.79079 |
| 0.720 | 0.55045 | 0.76981 | 1.18520 | 1.29712 | 1.64538 | 1.80027 |
| 0.740 | 0.56901 | 0.78836 | 1.20622 | 1.30990 | 1.66696 | 1.81084 |
| 0.760 | 0.58766 | 0.80699 | 1.22646 | 1.32295 | 1.68861 | 1.82255 |
| 0.780 | 0.60639 | 0.82569 | 1.24590 | 1.33626 | 1.71030 | 1.83544 |
| 0.800 | 0.62520 | 0.84446 | 1.26458 | 1.34977 | 1.73201 | 1.84949 |
| 0.820 | 0.64409 | 0.86329 | 1.28252 | 1.36344 | 1.75374 | 1.86475 |
| 0.840 | 0.66304 | 0.88218 | 1.29976 | 1.37719 | 1.77547 | 1.88117 |
| 0.860 | 0.68206 | 0.90112 | 1.31640 | 1.39096 | 1.79722 | 1.89873 |
| 0.880 | 0.70114 | 0.92011 | 1.33246 | 1.40471 | 1.81899 | 1.91739 |
| 0.900 | 0.72027 | 0.93915 | 1.34804 | 1.41838 | 1.84077 | 1.93712 |
| 0.920 | 0.73945 | 0.95823 | 1.36322 | 1.43192 | 1.86258 | 1.95782 |
| 0.940 | 0.75868 | 0.97735 | 1.37802 | 1.44533 | 1.88443 | 1.97943 |
| 0.960 | 0.77795 | 0.99651 | 1.39253 | 1.45861 | 1.90630 | 2.00191 |
| 0.980 | 0.79725 | 1.01571 | 1.40675 | 1.47176 | 1.92820 | 2.02515 |
| 1.000 | 0.81660 | 1.03494 | 1.42072 | 1.48478 | 1.95012 | 2.04913 |

$H/R = 0.450$ $n = 2.$

| $H/L$ | $\Omega_1$ | $\Omega_2$ | $\Omega_3$ |
|---|---|---|---|
| 0. | 0.07733 | 0.28814 | 0.50762 |
| 0.001 | 0.07733 | 0.28815 | 0.50763 |
| 0.005 | 0.07734 | 0.28822 | 0.50768 |
| 0.010 | 0.07739 | 0.28845 | 0.50784 |
| 0.020 | 0.07759 | 0.28937 | 0.50847 |
| 0.030 | 0.07793 | 0.29089 | 0.50954 |
| 0.040 | 0.07847 | 0.29297 | 0.51102 |
| 0.050 | 0.07923 | 0.29560 | 0.51294 |
| 0.060 | 0.08025 | 0.29874 | 0.51528 |
| 0.070 | 0.08160 | 0.30233 | 0.51805 |
| 0.080 | 0.08326 | 0.30635 | 0.52126 |
| 0.090 | 0.08539 | 0.31074 | 0.52489 |
| 0.100 | 0.08787 | 0.31546 | 0.52896 |
| 0.120 | 0.09410 | 0.32575 | 0.53838 |
| 0.140 | 0.10197 | 0.33692 | 0.54950 |
| 0.160 | 0.11132 | 0.34873 | 0.56229 |
| 0.180 | 0.12196 | 0.36102 | 0.57667 |
| 0.200 | 0.13366 | 0.37366 | 0.59257 |
| 0.220 | 0.14624 | 0.38658 | 0.60989 |
| 0.240 | 0.15953 | 0.39973 | 0.62849 |
| 0.260 | 0.17340 | 0.41311 | 0.64827 |
| 0.280 | 0.18775 | 0.42670 | 0.66908 |
| 0.300 | 0.20250 | 0.44053 | 0.69081 |
| 0.320 | 0.21760 | 0.45460 | 0.71333 |
| 0.340 | 0.23299 | 0.46892 | 0.73652 |
| 0.360 | 0.24864 | 0.48350 | 0.76027 |
| 0.380 | 0.26453 | 0.49835 | 0.78450 |
| 0.400 | 0.28064 | 0.51345 | 0.80910 |
| 0.420 | 0.29696 | 0.52882 | 0.83399 |
| 0.440 | 0.31346 | 0.54444 | 0.85909 |
| 0.460 | 0.33015 | 0.56031 | 0.88431 |
| 0.480 | 0.34701 | 0.57641 | 0.90960 |
| 0.500 | 0.36404 | 0.59274 | 0.93487 |
| 0.520 | 0.38122 | 0.60928 | 0.96006 |
| 0.540 | 0.39856 | 0.62602 | 0.98510 |
| 0.560 | 0.41604 | 0.64295 | 1.00992 |
| 0.580 | 0.43366 | 0.66006 | 1.03446 |
| 0.600 | 0.45141 | 0.67733 | 1.05864 |
| 0.620 | 0.46929 | 0.69476 | 1.08241 |
| 0.640 | 0.48728 | 0.71233 | 1.10570 |
| 0.660 | 0.50539 | 0.73005 | 1.12845 |
| 0.680 | 0.52361 | 0.74788 | 1.15059 |
| 0.700 | 0.54193 | 0.76583 | 1.17210 |
| 0.720 | 0.56034 | 0.78389 | 1.19292 |
| 0.740 | 0.57884 | 0.80205 | 1.21303 |
| 0.760 | 0.59743 | 0.82030 | 1.23244 |
| 0.780 | 0.61608 | 0.83865 | 1.25113 |
| 0.800 | 0.63483 | 0.85708 | 1.26914 |
| 0.820 | 0.65363 | 0.87558 | 1.28652 |
| 0.840 | 0.67251 | 0.89415 | 1.30331 |
| 0.860 | 0.69144 | 0.91279 | 1.31956 |
| 0.880 | 0.71043 | 0.93150 | 1.33532 |
| 0.900 | 0.72948 | 0.95026 | 1.35068 |
| 0.920 | 0.74856 | 0.96908 | 1.36565 |
| 0.940 | 0.76769 | 0.98794 | 1.38032 |
| 0.960 | 0.78687 | 1.00686 | 1.39469 |
| 0.980 | 0.80607 | 1.02582 | 1.40881 |
| 1.000 | 0.82532 | 1.04483 | 1.42270 |

$H/R = 0.500 \qquad n = 2.$

| $H/L$ | $\Omega_1$ | $\Omega_2$ | $\Omega_3$ | $\Omega_4$ | $\Omega_5$ | $\Omega_6$ |
|---|---|---|---|---|---|---|
| 0. | 0.09383 | 0.32034 | 0.55464 | 1.06533 | 1.23759 | 1.78511 |
| 0.010 | 0.09389 | 0.32062 | 0.55483 | 1.06541 | 1.23770 | 1.78507 |
| 0.050 | 0.09553 | 0.32714 | 0.55949 | 1.06720 | 1.24015 | 1.78395 |
| 0.100 | 0.10289 | 0.34556 | 0.57411 | 1.07271 | 1.24783 | 1.78065 |
| 0.120 | 0.10828 | 0.35526 | 0.58269 | 1.07588 | 1.25233 | 1.77884 |
| 0.140 | 0.11516 | 0.36589 | 0.59283 | 1.07956 | 1.25766 | 1.77678 |
| 0.160 | 0.12352 | 0.37724 | 0.60450 | 1.08374 | 1.26380 | 1.77456 |
| 0.180 | 0.13323 | 0.38912 | 0.61764 | 1.08837 | 1.27076 | 1.77220 |
| 0.200 | 0.14412 | 0.40142 | 0.63221 | 1.09345 | 1.27851 | 1.76975 |
| 0.220 | 0.15603 | 0.41402 | 0.64811 | 1.09893 | 1.28706 | 1.76726 |
| 0.240 | 0.16878 | 0.42687 | 0.66527 | 1.10481 | 1.29641 | 1.76478 |
| 0.260 | 0.18226 | 0.43995 | 0.68356 | 1.11106 | 1.30651 | 1.76236 |
| 0.280 | 0.19633 | 0.45323 | 0.70290 | 1.11765 | 1.31736 | 1.76004 |
| 0.300 | 0.21091 | 0.46672 | 0.72316 | 1.12457 | 1.32894 | 1.75785 |
| 0.320 | 0.22591 | 0.48041 | 0.74423 | 1.13180 | 1.34123 | 1.75586 |
| 0.340 | 0.24127 | 0.49432 | 0.76601 | 1.13934 | 1.35419 | 1.75408 |
| 0.360 | 0.25695 | 0.50846 | 0.78839 | 1.14718 | 1.36781 | 1.75258 |
| 0.380 | 0.27291 | 0.52283 | 0.81128 | 1.15530 | 1.38205 | 1.75138 |
| 0.400 | 0.28911 | 0.53743 | 0.83457 | 1.16372 | 1.39687 | 1.75052 |
| 0.420 | 0.30553 | 0.55228 | 0.85820 | 1.17241 | 1.41225 | 1.75005 |
| 0.440 | 0.32216 | 0.56736 | 0.88206 | 1.18138 | 1.42815 | 1.75001 |
| 0.460 | 0.33897 | 0.58268 | 0.90608 | 1.19064 | 1.44455 | 1.75044 |
| 0.480 | 0.35595 | 0.59822 | 0.93018 | 1.20019 | 1.46140 | 1.75138 |
| 0.500 | 0.37308 | 0.61400 | 0.95428 | 1.21002 | 1.47867 | 1.75288 |
| 0.520 | 0.39037 | 0.62998 | 0.97833 | 1.22015 | 1.49633 | 1.75496 |
| 0.540 | 0.40780 | 0.64617 | 1.00224 | 1.23058 | 1.51434 | 1.75769 |
| 0.560 | 0.42537 | 0.66256 | 1.02594 | 1.24131 | 1.53267 | 1.76113 |
| 0.580 | 0.44306 | 0.67914 | 1.04937 | 1.25235 | 1.55129 | 1.76529 |
| 0.600 | 0.46087 | 0.69590 | 1.07245 | 1.26369 | 1.57016 | 1.77028 |
| 0.620 | 0.47879 | 0.71282 | 1.09513 | 1.27536 | 1.58924 | 1.77610 |
| 0.640 | 0.49682 | 0.72990 | 1.11735 | 1.28731 | 1.60851 | 1.78285 |
| 0.660 | 0.51495 | 0.74714 | 1.13905 | 1.29959 | 1.62791 | 1.79056 |
| 0.680 | 0.53317 | 0.76452 | 1.16019 | 1.31214 | 1.64742 | 1.79929 |
| 0.700 | 0.55149 | 0.78202 | 1.18072 | 1.32497 | 1.66703 | 1.80911 |
| 0.720 | 0.56989 | 0.79966 | 1.20063 | 1.33803 | 1.68668 | 1.82004 |
| 0.740 | 0.58837 | 0.81740 | 1.21990 | 1.35131 | 1.70638 | 1.83214 |
| 0.760 | 0.60693 | 0.83526 | 1.23853 | 1.36473 | 1.72610 | 1.84540 |
| 0.780 | 0.62556 | 0.85322 | 1.25654 | 1.37829 | 1.74584 | 1.85985 |
| 0.800 | 0.64426 | 0.87127 | 1.27397 | 1.39188 | 1.76562 | 1.87545 |
| 0.820 | 0.66303 | 0.88943 | 1.29084 | 1.40550 | 1.78547 | 1.89218 |
| 0.840 | 0.68184 | 0.90766 | 1.30721 | 1.41905 | 1.80538 | 1.90999 |
| 0.860 | 0.70072 | 0.92597 | 1.32311 | 1.43253 | 1.82543 | 1.92880 |
| 0.880 | 0.71964 | 0.94436 | 1.33860 | 1.44590 | 1.84560 | 1.94855 |
| 0.900 | 0.73861 | 0.96281 | 1.35372 | 1.45912 | 1.86595 | 1.96916 |
| 0.920 | 0.75764 | 0.98134 | 1.36851 | 1.47220 | 1.88647 | 1.99058 |
| 0.940 | 0.77670 | 0.99993 | 1.38301 | 1.48514 | 1.90716 | 2.01271 |
| 0.960 | 0.79580 | 1.01858 | 1.39725 | 1.49794 | 1.92803 | 2.03551 |
| 0.980 | 0.81494 | 1.03728 | 1.41125 | 1.51062 | 1.94904 | 2.05892 |
| 1.000 | 0.83411 | 1.05603 | 1.42505 | 1.52322 | 1.97017 | 2.08290 |

$H/R = 0.600 \qquad n = 2.$

| $H/L$ | $\Omega_1$ | $\Omega_2$ | $\Omega_3$ | $\Omega_4$ | $\Omega_5$ | $\Omega_6$ |
|---|---|---|---|---|---|---|
| 0. | 0.13027 | 0.38446 | 0.64015 | 1.09765 | 1.34763 | 1.77315 |
| 0.010 | 0.13032 | 0.38469 | 0.64032 | 1.09772 | 1.34771 | 1.77313 |
| 0.050 | 0.13165 | 0.39025 | 0.64433 | 1.09947 | 1.34987 | 1.77243 |
| 0.100 | 0.13725 | 0.40635 | 0.65687 | 1.10486 | 1.35658 | 1.77037 |
| 0.120 | 0.14130 | 0.41503 | 0.66423 | 1.10798 | 1.36052 | 1.76923 |
| 0.140 | 0.14659 | 0.42469 | 0.67291 | 1.11163 | 1.36518 | 1.76797 |
| 0.160 | 0.15320 | 0.43514 | 0.68289 | 1.11580 | 1.37052 | 1.76660 |
| 0.180 | 0.16108 | 0.44622 | 0.69416 | 1.12046 | 1.37657 | 1.76516 |
| 0.200 | 0.17020 | 0.45779 | 0.70667 | 1.12560 | 1.38331 | 1.76369 |
| 0.220 | 0.18046 | 0.46975 | 0.72036 | 1.13120 | 1.39072 | 1.76224 |
| 0.240 | 0.19175 | 0.48202 | 0.73518 | 1.13724 | 1.39882 | 1.76081 |
| 0.260 | 0.20396 | 0.49455 | 0.75107 | 1.14369 | 1.40757 | 1.75948 |
| 0.280 | 0.21696 | 0.50729 | 0.76792 | 1.15055 | 1.41698 | 1.75827 |
| 0.300 | 0.23065 | 0.52023 | 0.78567 | 1.15779 | 1.42701 | 1.75723 |
| 0.320 | 0.24494 | 0.53335 | 0.80421 | 1.16542 | 1.43765 | 1.75637 |
| 0.340 | 0.25975 | 0.54665 | 0.82347 | 1.17340 | 1.44889 | 1.75577 |
| 0.360 | 0.27501 | 0.56012 | 0.84335 | 1.18173 | 1.46070 | 1.75545 |
| 0.380 | 0.29066 | 0.57379 | 0.86376 | 1.19040 | 1.47306 | 1.75545 |
| 0.400 | 0.30665 | 0.58764 | 0.88461 | 1.19941 | 1.48594 | 1.75582 |
| 0.420 | 0.32294 | 0.60169 | 0.90581 | 1.20875 | 1.49930 | 1.75661 |
| 0.440 | 0.33948 | 0.61593 | 0.92729 | 1.21840 | 1.51312 | 1.75785 |
| 0.460 | 0.35627 | 0.63038 | 0.94895 | 1.22839 | 1.52737 | 1.75960 |
| 0.480 | 0.37326 | 0.64503 | 0.97072 | 1.23869 | 1.54202 | 1.76190 |
| 0.500 | 0.39043 | 0.65988 | 0.99253 | 1.24932 | 1.55700 | 1.76483 |
| 0.520 | 0.40777 | 0.67493 | 1.01430 | 1.26026 | 1.57231 | 1.76841 |
| 0.540 | 0.42527 | 0.69018 | 1.03596 | 1.27152 | 1.58788 | 1.77271 |
| 0.560 | 0.44290 | 0.70562 | 1.05744 | 1.28310 | 1.60367 | 1.77782 |
| 0.580 | 0.46066 | 0.72125 | 1.07868 | 1.29499 | 1.61965 | 1.78377 |
| 0.600 | 0.47854 | 0.73707 | 1.09961 | 1.30718 | 1.63575 | 1.79065 |
| 0.620 | 0.49652 | 0.75306 | 1.12020 | 1.31966 | 1.65195 | 1.79851 |
| 0.640 | 0.51461 | 0.76922 | 1.14038 | 1.33241 | 1.66821 | 1.80742 |
| 0.660 | 0.53278 | 0.78555 | 1.16013 | 1.34542 | 1.68448 | 1.81742 |
| 0.680 | 0.55104 | 0.80204 | 1.17940 | 1.35863 | 1.70078 | 1.82855 |
| 0.700 | 0.56939 | 0.81867 | 1.19819 | 1.37203 | 1.71709 | 1.84084 |
| 0.720 | 0.58781 | 0.83545 | 1.21649 | 1.38557 | 1.73345 | 1.85425 |
| 0.740 | 0.60629 | 0.85237 | 1.23428 | 1.39919 | 1.74989 | 1.86878 |
| 0.760 | 0.62485 | 0.86942 | 1.25160 | 1.41286 | 1.76646 | 1.88435 |
| 0.780 | 0.64346 | 0.88659 | 1.26844 | 1.42650 | 1.78322 | 1.90092 |
| 0.800 | 0.66213 | 0.90389 | 1.28484 | 1.44011 | 1.80025 | 1.91842 |
| 0.820 | 0.68086 | 0.92129 | 1.30084 | 1.45362 | 1.81759 | 1.93676 |
| 0.840 | 0.69963 | 0.93880 | 1.31644 | 1.46701 | 1.83528 | 1.95585 |
| 0.860 | 0.71846 | 0.95641 | 1.33170 | 1.48025 | 1.85337 | 1.97566 |
| 0.880 | 0.73732 | 0.97412 | 1.34664 | 1.49332 | 1.87186 | 1.99611 |
| 0.900 | 0.75623 | 0.99193 | 1.36128 | 1.50624 | 1.89074 | 2.01712 |
| 0.920 | 0.77517 | 1.00981 | 1.37566 | 1.51898 | 1.91001 | 2.03868 |
| 0.940 | 0.79415 | 1.02779 | 1.38980 | 1.53160 | 1.92963 | 2.06072 |
| 0.960 | 0.81317 | 1.04583 | 1.40372 | 1.54408 | 1.94957 | 2.08322 |
| 0.980 | 0.83221 | 1.06396 | 1.41745 | 1.55645 | 1.96977 | 2.10614 |
| 1.000 | 0.85128 | 1.08215 | 1.43101 | 1.56875 | 1.99019 | 2.12946 |

$H/R = 0.700 \qquad n = 2.$

| $H/L$ | $\Omega_1$ | $\Omega_2$ | $\Omega_3$ | $\Omega_4$ | $\Omega_5$ | $\Omega_6$ |
|---|---|---|---|---|---|---|
| 0.000 | 0.17070 | 0.44764 | 0.71475 | 1.13888 | 1.47501 | 1.76792 |
| 0.010 | 0.17074 | 0.44784 | 0.71490 | 1.13894 | 1.47509 | 1.76790 |
| 0.050 | 0.17181 | 0.45273 | 0.71844 | 1.14061 | 1.47693 | 1.76757 |
| 0.100 | 0.17617 | 0.46709 | 0.72951 | 1.14580 | 1.48265 | 1.76663 |
| 0.120 | 0.17930 | 0.47495 | 0.73600 | 1.14882 | 1.48600 | 1.76613 |
| 0.140 | 0.18341 | 0.48379 | 0.74364 | 1.15236 | 1.48995 | 1.76560 |
| 0.160 | 0.18861 | 0.49344 | 0.75244 | 1.15642 | 1.49448 | 1.76506 |
| 0.180 | 0.19494 | 0.50377 | 0.76236 | 1.16099 | 1.49960 | 1.76453 |
| 0.200 | 0.20236 | 0.51465 | 0.77337 | 1.16605 | 1.50528 | 1.76405 |
| 0.220 | 0.21101 | 0.52598 | 0.78545 | 1.17158 | 1.51153 | 1.76365 |
| 0.240 | 0.22066 | 0.53767 | 0.79853 | 1.17759 | 1.51832 | 1.76338 |
| 0.260 | 0.23130 | 0.54966 | 0.81258 | 1.18406 | 1.52564 | 1.76325 |
| 0.280 | 0.24286 | 0.56189 | 0.82751 | 1.19096 | 1.53347 | 1.76332 |
| 0.300 | 0.25523 | 0.57434 | 0.84328 | 1.19831 | 1.54180 | 1.76362 |
| 0.320 | 0.26834 | 0.58698 | 0.85979 | 1.20607 | 1.55059 | 1.76419 |
| 0.340 | 0.28210 | 0.59978 | 0.87698 | 1.21425 | 1.55984 | 1.76510 |
| 0.360 | 0.29645 | 0.61276 | 0.89476 | 1.22284 | 1.56949 | 1.76638 |
| 0.380 | 0.31130 | 0.62589 | 0.91307 | 1.23182 | 1.57953 | 1.76807 |
| 0.400 | 0.32661 | 0.63919 | 0.93180 | 1.24119 | 1.58990 | 1.77026 |
| 0.420 | 0.34232 | 0.65265 | 0.95089 | 1.25094 | 1.60058 | 1.77297 |
| 0.440 | 0.35838 | 0.66628 | 0.97026 | 1.26108 | 1.61150 | 1.77629 |
| 0.460 | 0.37474 | 0.68008 | 0.98982 | 1.27158 | 1.62264 | 1.78029 |
| 0.480 | 0.39139 | 0.69406 | 1.00951 | 1.28247 | 1.63393 | 1.78502 |
| 0.500 | 0.40828 | 0.70821 | 1.02925 | 1.29370 | 1.64533 | 1.79056 |
| 0.520 | 0.42539 | 0.72254 | 1.04897 | 1.30530 | 1.65679 | 1.79697 |
| 0.540 | 0.44269 | 0.73706 | 1.06861 | 1.31725 | 1.66828 | 1.80431 |
| 0.560 | 0.46016 | 0.75174 | 1.08812 | 1.32952 | 1.67975 | 1.81264 |
| 0.580 | 0.47779 | 0.76661 | 1.10742 | 1.34213 | 1.69121 | 1.82196 |
| 0.600 | 0.49557 | 0.78165 | 1.12647 | 1.35502 | 1.70267 | 1.83233 |
| 0.620 | 0.51347 | 0.79686 | 1.14524 | 1.36819 | 1.71417 | 1.84370 |
| 0.640 | 0.53148 | 0.81224 | 1.16369 | 1.38160 | 1.72573 | 1.85606 |
| 0.660 | 0.54960 | 0.82777 | 1.18178 | 1.39521 | 1.73746 | 1.86935 |
| 0.680 | 0.56782 | 0.84348 | 1.19951 | 1.40898 | 1.74940 | 1.88352 |
| 0.700 | 0.58612 | 0.85933 | 1.21685 | 1.42287 | 1.76167 | 1.89852 |
| 0.720 | 0.60451 | 0.87534 | 1.23380 | 1.43682 | 1.77431 | 1.91428 |
| 0.740 | 0.62296 | 0.89150 | 1.25038 | 1.45079 | 1.78741 | 1.93071 |
| 0.760 | 0.64149 | 0.90779 | 1.26659 | 1.46471 | 1.80105 | 1.94779 |
| 0.780 | 0.66007 | 0.92421 | 1.28244 | 1.47857 | 1.81526 | 1.96543 |
| 0.800 | 0.67871 | 0.94077 | 1.29795 | 1.49231 | 1.83007 | 1.98359 |
| 0.820 | 0.69741 | 0.95746 | 1.31315 | 1.50590 | 1.84551 | 2.00224 |
| 0.840 | 0.71615 | 0.97426 | 1.32803 | 1.51932 | 1.86158 | 2.02133 |
| 0.860 | 0.73494 | 0.99117 | 1.34265 | 1.53256 | 1.87828 | 2.04083 |
| 0.880 | 0.75377 | 1.00820 | 1.35702 | 1.54561 | 1.89555 | 2.06070 |
| 0.900 | 0.77263 | 1.02534 | 1.37115 | 1.55847 | 1.91340 | 2.08092 |
| 0.920 | 0.79154 | 1.04258 | 1.38508 | 1.57117 | 1.93176 | 2.10145 |
| 0.940 | 0.81047 | 1.05991 | 1.39882 | 1.58371 | 1.95057 | 2.12228 |
| 0.960 | 0.82943 | 1.07734 | 1.41237 | 1.59611 | 1.96979 | 2.14340 |
| 0.980 | 0.84842 | 1.09485 | 1.42577 | 1.60841 | 1.98935 | 2.16474 |
| 1.000 | 0.86744 | 1.11245 | 1.43903 | 1.62063 | 2.00920 | 2.18631 |

$H/R = 0.800 \qquad n = 2.$

| $H/L$ | $\Omega_1$ | $\Omega_2$ | $\Omega_3$ | $\Omega_4$ | $\Omega_5$ | $\Omega_6$ |
|---|---|---|---|---|---|---|
| 0. | 0.21453 | 0.50900 | 0.78010 | 1.19055 | 1.60711 | 1.77762 |
| 0.010 | 0.21456 | 0.50919 | 0.78024 | 1.19062 | 1.60716 | 1.77763 |
| 0.050 | 0.21541 | 0.51358 | 0.78344 | 1.19219 | 1.60854 | 1.77776 |
| 0.100 | 0.21881 | 0.52663 | 0.79340 | 1.19707 | 1.61283 | 1.77829 |
| 0.120 | 0.22124 | 0.53385 | 0.79924 | 1.19992 | 1.61531 | 1.77865 |
| 0.140 | 0.22446 | 0.54202 | 0.80612 | 1.20327 | 1.61821 | 1.77914 |
| 0.160 | 0.22856 | 0.55101 | 0.81402 | 1.20713 | 1.62152 | 1.77979 |
| 0.180 | 0.23362 | 0.56069 | 0.82294 | 1.21148 | 1.62522 | 1.78061 |
| 0.200 | 0.23969 | 0.57095 | 0.83284 | 1.21632 | 1.62929 | 1.78165 |
| 0.220 | 0.24677 | 0.58169 | 0.84370 | 1.22165 | 1.63370 | 1.78296 |
| 0.240 | 0.25487 | 0.59284 | 0.85547 | 1.22744 | 1.63843 | 1.78456 |
| 0.260 | 0.26393 | 0.60432 | 0.86811 | 1.23372 | 1.64345 | 1.78652 |
| 0.280 | 0.27398 | 0.61607 | 0.88157 | 1.24045 | 1.64872 | 1.78887 |
| 0.300 | 0.28486 | 0.62806 | 0.89579 | 1.24763 | 1.65419 | 1.79167 |
| 0.320 | 0.29656 | 0.64025 | 0.91070 | 1.25528 | 1.65986 | 1.79499 |
| 0.340 | 0.30900 | 0.65262 | 0.92625 | 1.26336 | 1.66567 | 1.79886 |
| 0.360 | 0.32213 | 0.66515 | 0.94236 | 1.27190 | 1.67158 | 1.80334 |
| 0.380 | 0.33586 | 0.67783 | 0.95897 | 1.28086 | 1.67759 | 1.80847 |
| 0.400 | 0.35015 | 0.69068 | 0.97598 | 1.29026 | 1.68365 | 1.81429 |
| 0.420 | 0.36493 | 0.70367 | 0.99334 | 1.30008 | 1.68974 | 1.82083 |
| 0.440 | 0.38016 | 0.71681 | 1.01098 | 1.31034 | 1.69587 | 1.82812 |
| 0.460 | 0.39579 | 0.73010 | 1.02881 | 1.32101 | 1.70206 | 1.83617 |
| 0.480 | 0.41176 | 0.74356 | 1.04678 | 1.33210 | 1.70832 | 1.84495 |
| 0.500 | 0.42806 | 0.75716 | 1.06481 | 1.34360 | 1.71468 | 1.85448 |
| 0.520 | 0.44464 | 0.77093 | 1.08284 | 1.35549 | 1.72117 | 1.86472 |
| 0.540 | 0.46147 | 0.78485 | 1.10083 | 1.36777 | 1.72785 | 1.87565 |
| 0.560 | 0.47854 | 0.79894 | 1.11870 | 1.38040 | 1.73476 | 1.88723 |
| 0.580 | 0.49580 | 0.81318 | 1.13642 | 1.39339 | 1.74195 | 1.89944 |
| 0.600 | 0.51325 | 0.82760 | 1.15394 | 1.40668 | 1.74951 | 1.91223 |
| 0.620 | 0.53087 | 0.84217 | 1.17123 | 1.42026 | 1.75748 | 1.92557 |
| 0.640 | 0.54863 | 0.85689 | 1.18825 | 1.43409 | 1.76591 | 1.93942 |
| 0.660 | 0.56652 | 0.87178 | 1.20500 | 1.44812 | 1.77485 | 1.95375 |
| 0.680 | 0.58454 | 0.88682 | 1.22145 | 1.46231 | 1.78436 | 1.96854 |
| 0.700 | 0.60266 | 0.90202 | 1.23760 | 1.47659 | 1.79449 | 1.98373 |
| 0.720 | 0.62089 | 0.91736 | 1.25344 | 1.49093 | 1.80528 | 1.99933 |
| 0.740 | 0.63920 | 0.93285 | 1.26898 | 1.50527 | 1.81676 | 2.01530 |
| 0.760 | 0.65759 | 0.94848 | 1.28423 | 1.51954 | 1.82896 | 2.03159 |
| 0.780 | 0.67606 | 0.96424 | 1.29919 | 1.53373 | 1.84191 | 2.04821 |
| 0.800 | 0.69460 | 0.98014 | 1.31388 | 1.54779 | 1.85559 | 2.06512 |
| 0.820 | 0.71320 | 0.99617 | 1.32831 | 1.56168 | 1.87000 | 2.08229 |
| 0.840 | 0.73185 | 1.01233 | 1.34251 | 1.57540 | 1.88515 | 2.09972 |
| 0.860 | 0.75055 | 1.02860 | 1.35650 | 1.58890 | 1.90098 | 2.11738 |
| 0.880 | 0.76930 | 1.04500 | 1.37027 | 1.60220 | 1.91748 | 2.13524 |
| 0.900 | 0.78809 | 1.06151 | 1.38388 | 1.61532 | 1.93460 | 2.15330 |
| 0.920 | 0.80693 | 1.07813 | 1.39730 | 1.62825 | 1.95228 | 2.17153 |
| 0.940 | 0.82579 | 1.09486 | 1.41058 | 1.64100 | 1.97045 | 2.18990 |
| 0.960 | 0.84469 | 1.11168 | 1.42373 | 1.65361 | 1.98906 | 2.20841 |
| 0.980 | 0.86362 | 1.12861 | 1.43674 | 1.66609 | 2.00806 | 2.22701 |
| 1.000 | 0.88257 | 1.14562 | 1.44965 | 1.67847 | 2.02736 | 2.24569 |

$H/R = 0.900 \qquad n = 2.$

| $H/L$ | $\Omega_1$ | $\Omega_2$ | $\Omega_3$ | $\Omega_4$ | $\Omega_5$ | $\Omega_6$ |
|---|---|---|---|---|---|---|
| 0.000 | 0.26132 | 0.56753 | 0.83849 | 1.25418 | 1.69874 | 1.84088 |
| 0.010 | 0.26134 | 0.56771 | 0.83861 | 1.25424 | 1.69876 | 1.84093 |
| 0.050 | 0.26198 | 0.57175 | 0.84154 | 1.25566 | 1.69930 | 1.84186 |
| 0.100 | 0.26457 | 0.58382 | 0.85065 | 1.26013 | 1.70101 | 1.84486 |
| 0.120 | 0.26645 | 0.59054 | 0.85599 | 1.26274 | 1.70199 | 1.84665 |
| 0.140 | 0.26895 | 0.59818 | 0.86228 | 1.26583 | 1.70313 | 1.84881 |
| 0.160 | 0.27218 | 0.60663 | 0.86950 | 1.26939 | 1.70443 | 1.85133 |
| 0.180 | 0.27621 | 0.61578 | 0.87765 | 1.27341 | 1.70589 | 1.85424 |
| 0.200 | 0.28109 | 0.62552 | 0.88669 | 1.27790 | 1.70750 | 1.85756 |
| 0.220 | 0.28688 | 0.63576 | 0.89661 | 1.28286 | 1.70924 | 1.86129 |
| 0.240 | 0.29359 | 0.64642 | 0.90737 | 1.28827 | 1.71114 | 1.86545 |
| 0.260 | 0.30123 | 0.65744 | 0.91894 | 1.29415 | 1.71317 | 1.87006 |
| 0.280 | 0.30976 | 0.66876 | 0.93126 | 1.30048 | 1.71536 | 1.87510 |
| 0.300 | 0.31917 | 0.68033 | 0.94428 | 1.30727 | 1.71769 | 1.88062 |
| 0.320 | 0.32941 | 0.69211 | 0.95797 | 1.31453 | 1.72018 | 1.88659 |
| 0.340 | 0.34000 | 0.70409 | 0.97224 | 1.32223 | 1.72283 | 1.89302 |
| 0.360 | 0.35222 | 0.71624 | 0.98705 | 1.33039 | 1.72566 | 1.89992 |
| 0.380 | 0.36465 | 0.72853 | 1.00232 | 1.33902 | 1.72869 | 1.90728 |
| 0.400 | 0.37772 | 0.74098 | 1.01799 | 1.34810 | 1.73193 | 1.91511 |
| 0.420 | 0.39136 | 0.75357 | 1.03400 | 1.35763 | 1.73538 | 1.92337 |
| 0.440 | 0.40552 | 0.76630 | 1.05027 | 1.36764 | 1.73910 | 1.93209 |
| 0.460 | 0.42016 | 0.77917 | 1.06675 | 1.37808 | 1.74308 | 1.94124 |
| 0.480 | 0.43523 | 0.79219 | 1.08336 | 1.38898 | 1.74736 | 1.95080 |
| 0.500 | 0.45069 | 0.80535 | 1.10005 | 1.40032 | 1.75195 | 1.96078 |
| 0.520 | 0.46650 | 0.81865 | 1.11676 | 1.41210 | 1.75689 | 1.97115 |
| 0.540 | 0.48262 | 0.83210 | 1.13344 | 1.42430 | 1.76222 | 1.98191 |
| 0.560 | 0.49904 | 0.84569 | 1.15004 | 1.43692 | 1.76794 | 1.99302 |
| 0.580 | 0.51572 | 0.85944 | 1.16651 | 1.44991 | 1.77411 | 2.00449 |
| 0.600 | 0.53263 | 0.87333 | 1.18282 | 1.46326 | 1.78074 | 2.01631 |
| 0.620 | 0.54975 | 0.88738 | 1.19893 | 1.47695 | 1.78788 | 2.02843 |
| 0.640 | 0.56706 | 0.90158 | 1.21483 | 1.49092 | 1.79556 | 2.04086 |
| 0.660 | 0.58455 | 0.91592 | 1.23050 | 1.50514 | 1.80381 | 2.05358 |
| 0.680 | 0.60219 | 0.93041 | 1.24591 | 1.51954 | 1.81266 | 2.06658 |
| 0.700 | 0.61997 | 0.94505 | 1.26106 | 1.53410 | 1.82216 | 2.07983 |
| 0.720 | 0.63789 | 0.95984 | 1.27596 | 1.54874 | 1.83232 | 2.09333 |
| 0.740 | 0.65592 | 0.97476 | 1.29061 | 1.56344 | 1.84317 | 2.10706 |
| 0.760 | 0.67405 | 0.98982 | 1.30501 | 1.57811 | 1.85473 | 2.12101 |
| 0.780 | 0.69227 | 1.00502 | 1.31917 | 1.59271 | 1.86703 | 2.13516 |
| 0.800 | 0.71059 | 1.02035 | 1.33311 | 1.60721 | 1.88005 | 2.14949 |
| 0.820 | 0.72899 | 1.03582 | 1.34684 | 1.62157 | 1.89381 | 2.16401 |
| 0.840 | 0.74746 | 1.05141 | 1.36038 | 1.63574 | 1.90829 | 2.17870 |
| 0.860 | 0.76599 | 1.06712 | 1.37374 | 1.64973 | 1.92346 | 2.19354 |
| 0.880 | 0.78458 | 1.08295 | 1.38692 | 1.66352 | 1.93928 | 2.20855 |
| 0.900 | 0.80322 | 1.09890 | 1.39998 | 1.67710 | 1.95574 | 2.22368 |
| 0.920 | 0.82191 | 1.11496 | 1.41289 | 1.69049 | 1.97276 | 2.23895 |
| 0.940 | 0.84065 | 1.13114 | 1.42568 | 1.70369 | 1.99030 | 2.25435 |
| 0.960 | 0.85943 | 1.14742 | 1.43839 | 1.71673 | 2.00829 | 2.26987 |
| 0.980 | 0.87824 | 1.16380 | 1.45100 | 1.72960 | 2.02669 | 2.28551 |
| 1.000 | 0.89709 | 1.18029 | 1.46354 | 1.74237 | 2.04541 | 2.30126 |

$H/R = 1.000 \qquad n = 2.$

| $H/L$ | $\Omega_1$ | $\Omega_2$ | $\Omega_3$ | $\Omega_4$ | $\Omega_5$ | $\Omega_6$ |
|---|---|---|---|---|---|---|
| 0. | 0.31064 | 0.62230 | 0.89214 | 1.33060 | 1.72920 | 1.97090 |
| 0.010 | 0.31066 | 0.62245 | 0.89225 | 1.33066 | 1.72922 | 1.97095 |
| 0.050 | 0.31111 | 0.62624 | 0.89496 | 1.33192 | 1.72961 | 1.97193 |
| 0.100 | 0.31299 | 0.63760 | 0.90339 | 1.33586 | 1.73082 | 1.97500 |
| 0.120 | 0.31440 | 0.64395 | 0.90833 | 1.33817 | 1.73154 | 1.97682 |
| 0.140 | 0.31631 | 0.65120 | 0.91414 | 1.34090 | 1.73241 | 1.97896 |
| 0.160 | 0.31880 | 0.65923 | 0.92082 | 1.34406 | 1.73341 | 1.98144 |
| 0.180 | 0.32196 | 0.66796 | 0.92836 | 1.34763 | 1.73457 | 1.98426 |
| 0.200 | 0.32586 | 0.67729 | 0.93673 | 1.35164 | 1.73588 | 1.98743 |
| 0.220 | 0.33055 | 0.68713 | 0.94592 | 1.35608 | 1.73735 | 1.99093 |
| 0.240 | 0.33605 | 0.69740 | 0.95589 | 1.36094 | 1.73898 | 1.99479 |
| 0.260 | 0.34239 | 0.70804 | 0.96660 | 1.36623 | 1.74080 | 1.99897 |
| 0.280 | 0.34958 | 0.71900 | 0.97804 | 1.37196 | 1.74280 | 2.00350 |
| 0.300 | 0.35761 | 0.73022 | 0.99014 | 1.37814 | 1.74500 | 2.00837 |
| 0.320 | 0.36645 | 0.74166 | 1.00285 | 1.38475 | 1.74739 | 2.01358 |
| 0.340 | 0.37608 | 0.75329 | 1.01614 | 1.39181 | 1.75000 | 2.01914 |
| 0.360 | 0.38646 | 0.76511 | 1.02994 | 1.39932 | 1.75284 | 2.02501 |
| 0.380 | 0.39755 | 0.77708 | 1.04418 | 1.40730 | 1.75592 | 2.03123 |
| 0.400 | 0.40931 | 0.78919 | 1.05882 | 1.41574 | 1.75924 | 2.03777 |
| 0.420 | 0.42168 | 0.80144 | 1.07379 | 1.42464 | 1.76284 | 2.04463 |
| 0.440 | 0.43468 | 0.81383 | 1.08901 | 1.43402 | 1.76670 | 2.05180 |
| 0.460 | 0.44817 | 0.82635 | 1.10445 | 1.44388 | 1.77087 | 2.05929 |
| 0.480 | 0.46216 | 0.83901 | 1.12003 | 1.45423 | 1.77535 | 2.06707 |
| 0.500 | 0.47661 | 0.85179 | 1.13570 | 1.46504 | 1.78015 | 2.07516 |
| 0.520 | 0.49147 | 0.86471 | 1.15141 | 1.47633 | 1.78530 | 2.08352 |
| 0.540 | 0.50671 | 0.87777 | 1.16710 | 1.48808 | 1.79082 | 2.09218 |
| 0.560 | 0.52229 | 0.89096 | 1.18272 | 1.50029 | 1.79672 | 2.10110 |
| 0.580 | 0.53820 | 0.90429 | 1.19824 | 1.51295 | 1.80302 | 2.11029 |
| 0.600 | 0.55439 | 0.91776 | 1.21363 | 1.52603 | 1.80976 | 2.11974 |
| 0.620 | 0.57084 | 0.93137 | 1.22884 | 1.53950 | 1.81695 | 2.12943 |
| 0.640 | 0.58753 | 0.94512 | 1.24386 | 1.55333 | 1.82462 | 2.13937 |
| 0.660 | 0.60444 | 0.95902 | 1.25867 | 1.56749 | 1.83281 | 2.14955 |
| 0.680 | 0.62154 | 0.97305 | 1.27326 | 1.58192 | 1.84152 | 2.15995 |
| 0.700 | 0.63883 | 0.98723 | 1.28762 | 1.59658 | 1.85078 | 2.17059 |
| 0.720 | 0.65628 | 1.00154 | 1.30176 | 1.61142 | 1.86064 | 2.18143 |
| 0.740 | 0.67388 | 1.01599 | 1.31566 | 1.62637 | 1.87111 | 2.19250 |
| 0.760 | 0.69161 | 1.03057 | 1.32934 | 1.64139 | 1.88220 | 2.20378 |
| 0.780 | 0.70947 | 1.04529 | 1.34283 | 1.65643 | 1.89395 | 2.21526 |
| 0.800 | 0.72744 | 1.06014 | 1.35610 | 1.67142 | 1.90637 | 2.22695 |
| 0.820 | 0.74551 | 1.07511 | 1.36921 | 1.68632 | 1.91945 | 2.23884 |
| 0.840 | 0.76368 | 1.09022 | 1.38214 | 1.70109 | 1.93322 | 2.25093 |
| 0.860 | 0.78193 | 1.10544 | 1.39492 | 1.71570 | 1.94763 | 2.26322 |
| 0.880 | 0.80026 | 1.12078 | 1.40755 | 1.73013 | 1.96269 | 2.27572 |
| 0.900 | 0.81866 | 1.13625 | 1.42008 | 1.74435 | 1.97835 | 2.28843 |
| 0.920 | 0.83713 | 1.15183 | 1.43250 | 1.75837 | 1.99458 | 2.30134 |
| 0.940 | 0.85565 | 1.16751 | 1.44483 | 1.77219 | 2.01134 | 2.31446 |
| 0.960 | 0.87423 | 1.18331 | 1.45710 | 1.78581 | 2.02858 | 2.32778 |
| 0.980 | 0.89286 | 1.19921 | 1.46930 | 1.79926 | 2.04623 | 2.34133 |
| 1.000 | 0.91153 | 1.21522 | 1.48146 | 1.81254 | 2.06424 | 2.35509 |

$H/R = 1.100 \qquad n = 2.$

| $H/L$ | $\Omega_1$ | $\Omega_2$ | $\Omega_3$ | $\Omega_4$ | $\Omega_5$ | $\Omega_6$ |
|---|---|---|---|---|---|---|
| 0.000 | 0.36209 | 0.67265 | 0.94305 | 1.41936 | 1.74812 | 2.11177 |
| 0.010 | 0.36210 | 0.67280 | 0.94316 | 1.41941 | 1.74815 | 2.11180 |
| 0.050 | 0.36236 | 0.67642 | 0.94568 | 1.42045 | 1.74869 | 2.11225 |
| 0.100 | 0.36359 | 0.68728 | 0.95354 | 1.42372 | 1.75040 | 2.11374 |
| 0.120 | 0.36457 | 0.69338 | 0.95814 | 1.42565 | 1.75141 | 2.11465 |
| 0.140 | 0.36596 | 0.70035 | 0.96356 | 1.42793 | 1.75260 | 2.11576 |
| 0.160 | 0.36783 | 0.70810 | 0.96980 | 1.43058 | 1.75398 | 2.11708 |
| 0.180 | 0.37025 | 0.71653 | 0.97683 | 1.43359 | 1.75557 | 2.11862 |
| 0.200 | 0.37330 | 0.72557 | 0.98464 | 1.43697 | 1.75734 | 2.12039 |
| 0.220 | 0.37703 | 0.73511 | 0.99323 | 1.44072 | 1.75932 | 2.12241 |
| 0.240 | 0.38148 | 0.74510 | 1.00255 | 1.44485 | 1.76152 | 2.12468 |
| 0.260 | 0.38669 | 0.75546 | 1.01259 | 1.44938 | 1.76392 | 2.12720 |
| 0.280 | 0.39268 | 0.76614 | 1.02330 | 1.45431 | 1.76654 | 2.12999 |
| 0.300 | 0.39944 | 0.77709 | 1.03465 | 1.45962 | 1.76937 | 2.13307 |
| 0.320 | 0.40699 | 0.78828 | 1.04660 | 1.46536 | 1.77244 | 2.13641 |
| 0.340 | 0.41530 | 0.79966 | 1.05909 | 1.47152 | 1.77575 | 2.14005 |
| 0.360 | 0.42435 | 0.81121 | 1.07209 | 1.47812 | 1.77930 | 2.14396 |
| 0.380 | 0.43412 | 0.82293 | 1.08553 | 1.48516 | 1.78308 | 2.14816 |
| 0.400 | 0.44458 | 0.83478 | 1.09936 | 1.49266 | 1.78713 | 2.15265 |
| 0.420 | 0.45568 | 0.84677 | 1.11352 | 1.50062 | 1.79143 | 2.15742 |
| 0.440 | 0.46740 | 0.85888 | 1.12795 | 1.50907 | 1.79601 | 2.16249 |
| 0.460 | 0.47970 | 0.87112 | 1.14260 | 1.51800 | 1.80086 | 2.16784 |
| 0.480 | 0.49252 | 0.88349 | 1.15740 | 1.52744 | 1.80599 | 2.17346 |
| 0.500 | 0.50585 | 0.89597 | 1.17231 | 1.53737 | 1.81142 | 2.17937 |
| 0.520 | 0.52000 | 0.90858 | 1.18727 | 1.54783 | 1.81717 | 2.18557 |
| 0.540 | 0.53391 | 0.92131 | 1.20223 | 1.55879 | 1.82322 | 2.19203 |
| 0.560 | 0.54854 | 0.93417 | 1.21714 | 1.57026 | 1.82960 | 2.19876 |
| 0.580 | 0.56353 | 0.94716 | 1.23198 | 1.58224 | 1.83632 | 2.20578 |
| 0.600 | 0.57887 | 0.96028 | 1.24669 | 1.59471 | 1.84340 | 2.21307 |
| 0.620 | 0.59451 | 0.97353 | 1.26124 | 1.60766 | 1.85083 | 2.22061 |
| 0.640 | 0.61044 | 0.98692 | 1.27563 | 1.62105 | 1.85867 | 2.22842 |
| 0.660 | 0.62663 | 1.00043 | 1.28983 | 1.63488 | 1.86691 | 2.23649 |
| 0.680 | 0.64306 | 1.01408 | 1.30381 | 1.64909 | 1.87555 | 2.24482 |
| 0.700 | 0.65971 | 1.02787 | 1.31758 | 1.66365 | 1.88466 | 2.25340 |
| 0.720 | 0.67656 | 1.04178 | 1.33114 | 1.67849 | 1.89422 | 2.26225 |
| 0.740 | 0.69360 | 1.05583 | 1.34449 | 1.69360 | 1.90427 | 2.27135 |
| 0.760 | 0.71081 | 1.07000 | 1.35763 | 1.70889 | 1.91482 | 2.28071 |
| 0.780 | 0.72818 | 1.08431 | 1.37057 | 1.72431 | 1.92592 | 2.29031 |
| 0.800 | 0.74568 | 1.09874 | 1.38333 | 1.73980 | 1.93757 | 2.30018 |
| 0.820 | 0.76332 | 1.11330 | 1.39592 | 1.75529 | 1.94978 | 2.31031 |
| 0.840 | 0.78107 | 1.12798 | 1.40836 | 1.77075 | 1.96260 | 2.32068 |
| 0.860 | 0.79894 | 1.14278 | 1.42065 | 1.78611 | 1.97599 | 2.33132 |
| 0.880 | 0.81690 | 1.15771 | 1.43284 | 1.80134 | 1.98999 | 2.34222 |
| 0.900 | 0.83496 | 1.17275 | 1.44492 | 1.81639 | 2.00456 | 2.35339 |
| 0.920 | 0.85310 | 1.18790 | 1.45691 | 1.83124 | 2.01971 | 2.36483 |
| 0.940 | 0.87132 | 1.20317 | 1.46883 | 1.84587 | 2.03539 | 2.37654 |
| 0.960 | 0.88961 | 1.21854 | 1.48070 | 1.86028 | 2.05158 | 2.38852 |
| 0.980 | 0.90797 | 1.23402 | 1.49253 | 1.87448 | 2.06823 | 2.40078 |
| 1.000 | 0.92638 | 1.24960 | 1.50433 | 1.88846 | 2.08528 | 2.41333 |

$H/R = 1.200 \qquad n = 2.$

| $H/L$ | $\Omega_1$ | $\Omega_2$ | $\Omega_3$ | $\Omega_4$ | $\Omega_5$ | $\Omega_6$ |
|---|---|---|---|---|---|---|
| 0. | 0.41506 | 0.71842 | 0.99303 | 1.51769 | 1.77173 | 2.24477 |
| 0.010 | 0.41506 | 0.71857 | 0.99312 | 1.51772 | 1.77175 | 2.24474 |
| 0.050 | 0.41514 | 0.72208 | 0.99548 | 1.51848 | 1.77254 | 2.24400 |
| 0.100 | 0.41575 | 0.73267 | 1.00284 | 1.52091 | 1.77496 | 2.24233 |
| 0.120 | 0.41635 | 0.73862 | 1.00715 | 1.52235 | 1.77638 | 2.24169 |
| 0.140 | 0.41726 | 0.74543 | 1.01222 | 1.52405 | 1.77806 | 2.24116 |
| 0.160 | 0.41856 | 0.75301 | 1.01806 | 1.52605 | 1.77999 | 2.24080 |
| 0.180 | 0.42034 | 0.76126 | 1.02466 | 1.52832 | 1.78218 | 2.24066 |
| 0.200 | 0.42264 | 0.77012 | 1.03200 | 1.53090 | 1.78463 | 2.24073 |
| 0.220 | 0.42553 | 0.77949 | 1.04006 | 1.53379 | 1.78733 | 2.24107 |
| 0.240 | 0.42906 | 0.78930 | 1.04883 | 1.53699 | 1.79028 | 2.24169 |
| 0.260 | 0.43328 | 0.79950 | 1.05828 | 1.54053 | 1.79348 | 2.24258 |
| 0.280 | 0.43819 | 0.81002 | 1.06838 | 1.54441 | 1.79694 | 2.24376 |
| 0.300 | 0.44383 | 0.82081 | 1.07910 | 1.54865 | 1.80065 | 2.24525 |
| 0.320 | 0.45020 | 0.83183 | 1.09040 | 1.55326 | 1.80460 | 2.24704 |
| 0.340 | 0.45730 | 0.84305 | 1.10224 | 1.55826 | 1.80880 | 2.24912 |
| 0.360 | 0.46512 | 0.85444 | 1.11458 | 1.56367 | 1.81326 | 2.25153 |
| 0.380 | 0.47365 | 0.86599 | 1.12737 | 1.56951 | 1.81796 | 2.25423 |
| 0.400 | 0.48285 | 0.87766 | 1.14054 | 1.57578 | 1.82291 | 2.25723 |
| 0.420 | 0.49271 | 0.88947 | 1.15406 | 1.58251 | 1.82810 | 2.26054 |
| 0.440 | 0.50320 | 0.90139 | 1.16787 | 1.58972 | 1.83355 | 2.26414 |
| 0.460 | 0.51430 | 0.91343 | 1.18192 | 1.59742 | 1.83923 | 2.26806 |
| 0.480 | 0.52595 | 0.92558 | 1.19614 | 1.60565 | 1.84518 | 2.27227 |
| 0.500 | 0.53815 | 0.93783 | 1.21048 | 1.61439 | 1.85137 | 2.27677 |
| 0.520 | 0.55084 | 0.95020 | 1.22491 | 1.62368 | 1.85781 | 2.28157 |
| 0.540 | 0.56400 | 0.96269 | 1.23936 | 1.63353 | 1.86451 | 2.28666 |
| 0.560 | 0.57761 | 0.97528 | 1.25378 | 1.64393 | 1.87147 | 2.29203 |
| 0.580 | 0.59161 | 0.98800 | 1.26815 | 1.65490 | 1.87869 | 2.29770 |
| 0.600 | 0.60599 | 1.00083 | 1.28241 | 1.66644 | 1.88618 | 2.30365 |
| 0.620 | 0.62064 | 1.01378 | 1.29655 | 1.67852 | 1.89394 | 2.30988 |
| 0.640 | 0.63586 | 1.02686 | 1.31054 | 1.69117 | 1.90197 | 2.31639 |
| 0.660 | 0.65122 | 1.04006 | 1.32434 | 1.70435 | 1.91031 | 2.32319 |
| 0.680 | 0.66687 | 1.05338 | 1.33796 | 1.71802 | 1.91893 | 2.33026 |
| 0.700 | 0.68278 | 1.06684 | 1.35136 | 1.73218 | 1.92787 | 2.33761 |
| 0.720 | 0.69893 | 1.08041 | 1.36457 | 1.74678 | 1.93713 | 2.34524 |
| 0.740 | 0.71531 | 1.09411 | 1.37757 | 1.76179 | 1.94672 | 2.35314 |
| 0.760 | 0.73188 | 1.10794 | 1.39037 | 1.77714 | 1.95667 | 2.36133 |
| 0.780 | 0.74865 | 1.12188 | 1.40298 | 1.79279 | 1.96700 | 2.36979 |
| 0.800 | 0.76559 | 1.13595 | 1.41541 | 1.80867 | 1.97772 | 2.37854 |
| 0.820 | 0.78269 | 1.15015 | 1.42767 | 1.82474 | 1.98886 | 2.38756 |
| 0.840 | 0.79994 | 1.16446 | 1.43978 | 1.84092 | 2.00044 | 2.39688 |
| 0.860 | 0.81732 | 1.17889 | 1.45176 | 1.85714 | 2.01249 | 2.40648 |
| 0.880 | 0.83483 | 1.19344 | 1.46362 | 1.87334 | 2.02503 | 2.41635 |
| 0.900 | 0.85245 | 1.20810 | 1.47538 | 1.88947 | 2.03806 | 2.42653 |
| 0.920 | 0.87018 | 1.22288 | 1.48706 | 1.90545 | 2.05163 | 2.43701 |
| 0.940 | 0.88801 | 1.23777 | 1.49867 | 1.92125 | 2.06572 | 2.44779 |
| 0.960 | 0.90593 | 1.25277 | 1.51024 | 1.93682 | 2.08033 | 2.45886 |
| 0.980 | 0.92393 | 1.26787 | 1.52178 | 1.95213 | 2.09545 | 2.47024 |
| 1.000 | 0.94201 | 1.28307 | 1.53328 | 1.96717 | 2.11106 | 2.48195 |

$H/R = 1.300 \qquad n = 2.$

| $H/L$ | $\Omega_1$ | $\Omega_2$ | $\Omega_3$ | $\Omega_4$ | $\Omega_5$ | $\Omega_6$ |
|---|---|---|---|---|---|---|
| 0.000 | 0.46873 | 0.75985 | 1.04365 | 1.62011 | 1.80567 | 2.36654 |
| 0.010 | 0.46872 | 0.76000 | 1.04374 | 1.62013 | 1.80572 | 2.36642 |
| 0.050 | 0.46860 | 0.76348 | 1.04595 | 1.62053 | 1.80680 | 2.36414 |
| 0.100 | 0.46861 | 0.77397 | 1.05285 | 1.62185 | 1.81016 | 2.35974 |
| 0.120 | 0.46884 | 0.77986 | 1.05689 | 1.62265 | 1.81211 | 2.35809 |
| 0.140 | 0.46931 | 0.78661 | 1.06166 | 1.62362 | 1.81440 | 2.35665 |
| 0.160 | 0.47010 | 0.79414 | 1.06715 | 1.62478 | 1.81702 | 2.35544 |
| 0.180 | 0.47128 | 0.80233 | 1.07335 | 1.62614 | 1.81995 | 2.35450 |
| 0.200 | 0.47292 | 0.81113 | 1.08026 | 1.62772 | 1.82321 | 2.35385 |
| 0.220 | 0.47507 | 0.82044 | 1.08786 | 1.62953 | 1.82676 | 2.35348 |
| 0.240 | 0.47779 | 0.83020 | 1.09613 | 1.63160 | 1.83062 | 2.35341 |
| 0.260 | 0.48112 | 0.84034 | 1.10506 | 1.63393 | 1.83475 | 2.35364 |
| 0.280 | 0.48509 | 0.85080 | 1.11462 | 1.63657 | 1.83916 | 2.35418 |
| 0.300 | 0.48972 | 0.86154 | 1.12479 | 1.63951 | 1.84385 | 2.35502 |
| 0.320 | 0.49504 | 0.87251 | 1.13553 | 1.64279 | 1.84878 | 2.35615 |
| 0.340 | 0.50105 | 0.88367 | 1.14680 | 1.64644 | 1.85396 | 2.35759 |
| 0.360 | 0.50775 | 0.89501 | 1.15858 | 1.65046 | 1.85940 | 2.35933 |
| 0.380 | 0.51513 | 0.90648 | 1.17080 | 1.65489 | 1.86506 | 2.36136 |
| 0.400 | 0.52317 | 0.91808 | 1.18345 | 1.65975 | 1.87095 | 2.36370 |
| 0.420 | 0.53187 | 0.92979 | 1.19645 | 1.66506 | 1.87705 | 2.36633 |
| 0.440 | 0.54120 | 0.94161 | 1.20976 | 1.67086 | 1.88337 | 2.36924 |
| 0.460 | 0.55114 | 0.95353 | 1.22334 | 1.67714 | 1.88991 | 2.37245 |
| 0.480 | 0.56165 | 0.96554 | 1.23713 | 1.68396 | 1.89664 | 2.37594 |
| 0.500 | 0.57272 | 0.97765 | 1.25107 | 1.69130 | 1.90359 | 2.37972 |
| 0.520 | 0.58432 | 0.98986 | 1.26513 | 1.69923 | 1.91072 | 2.38379 |
| 0.540 | 0.59641 | 1.00216 | 1.27924 | 1.70773 | 1.91805 | 2.38813 |
| 0.560 | 0.60898 | 1.01456 | 1.29337 | 1.71683 | 1.92557 | 2.39276 |
| 0.580 | 0.62198 | 1.02707 | 1.30748 | 1.72653 | 1.93327 | 2.39766 |
| 0.600 | 0.63540 | 1.03968 | 1.32152 | 1.73685 | 1.94117 | 2.40283 |
| 0.620 | 0.64920 | 1.05239 | 1.33545 | 1.74780 | 1.94925 | 2.40829 |
| 0.640 | 0.66337 | 1.06522 | 1.34927 | 1.75938 | 1.95752 | 2.41402 |
| 0.660 | 0.67787 | 1.07816 | 1.36291 | 1.77156 | 1.96598 | 2.42003 |
| 0.680 | 0.69269 | 1.09121 | 1.37640 | 1.78435 | 1.97464 | 2.42631 |
| 0.700 | 0.70780 | 1.10437 | 1.38970 | 1.79772 | 1.98348 | 2.43286 |
| 0.720 | 0.72314 | 1.11765 | 1.40280 | 1.81168 | 1.99253 | 2.43970 |
| 0.740 | 0.73900 | 1.13105 | 1.41570 | 1.82616 | 2.00177 | 2.44681 |
| 0.760 | 0.75474 | 1.14456 | 1.42840 | 1.84115 | 2.01123 | 2.45419 |
| 0.780 | 0.77083 | 1.15819 | 1.44091 | 1.85660 | 2.02092 | 2.46185 |
| 0.800 | 0.78712 | 1.17194 | 1.45324 | 1.87248 | 2.03083 | 2.46979 |
| 0.820 | 0.80361 | 1.18581 | 1.46540 | 1.88873 | 2.04098 | 2.47803 |
| 0.840 | 0.82027 | 1.19978 | 1.47740 | 1.90529 | 2.05139 | 2.48654 |
| 0.860 | 0.83709 | 1.21388 | 1.48926 | 1.92212 | 2.06208 | 2.49532 |
| 0.880 | 0.85407 | 1.22808 | 1.50099 | 1.93914 | 2.07307 | 2.50441 |
| 0.900 | 0.87119 | 1.24240 | 1.51261 | 1.95629 | 2.08438 | 2.51379 |
| 0.920 | 0.88843 | 1.25683 | 1.52416 | 1.97351 | 2.09604 | 2.52347 |
| 0.940 | 0.90580 | 1.27136 | 1.53562 | 1.99072 | 2.10808 | 2.53346 |
| 0.960 | 0.92327 | 1.28601 | 1.54703 | 2.00783 | 2.12054 | 2.54377 |
| 0.980 | 0.94085 | 1.30075 | 1.55841 | 2.02479 | 2.13344 | 2.55438 |
| 1.000 | 0.95852 | 1.31560 | 1.56976 | 2.04154 | 2.14681 | 2.56531 |

$H/R = 1.400 \qquad n = 2.$

| $H/L$ | $\Omega_1$ | $\Omega_2$ | $\Omega_3$ | $\Omega_4$ | $\Omega_5$ | $\Omega_6$ |
|---|---|---|---|---|---|---|
| 0.000 | 0.52188 | 0.79747 | 1.09617 | 1.71948 | 1.85516 | 2.47788 |
| 0.010 | 0.52186 | 0.79762 | 1.09626 | 1.71948 | 1.85523 | 2.47783 |
| 0.050 | 0.52155 | 0.80111 | 1.09833 | 1.71945 | 1.85666 | 2.47659 |
| 0.100 | 0.52095 | 0.81165 | 1.10482 | 1.71947 | 1.86107 | 2.47383 |
| 0.120 | 0.52084 | 0.81758 | 1.10862 | 1.71955 | 1.86361 | 2.47273 |
| 0.140 | 0.52089 | 0.82436 | 1.11311 | 1.71969 | 1.86656 | 2.47175 |
| 0.160 | 0.52121 | 0.83191 | 1.11828 | 1.71995 | 1.86989 | 2.47093 |
| 0.180 | 0.52184 | 0.84014 | 1.12413 | 1.72032 | 1.87359 | 2.47028 |
| 0.200 | 0.52288 | 0.84898 | 1.13065 | 1.72085 | 1.87764 | 2.46987 |
| 0.220 | 0.52436 | 0.85833 | 1.13783 | 1.72156 | 1.88202 | 2.46968 |
| 0.240 | 0.52635 | 0.86813 | 1.14566 | 1.72248 | 1.88670 | 2.46975 |
| 0.260 | 0.52890 | 0.87832 | 1.15412 | 1.72362 | 1.89168 | 2.47007 |
| 0.280 | 0.53204 | 0.88883 | 1.16320 | 1.72504 | 1.89693 | 2.47063 |
| 0.300 | 0.53579 | 0.89962 | 1.17287 | 1.72676 | 1.90243 | 2.47146 |
| 0.320 | 0.54018 | 0.91063 | 1.18311 | 1.72877 | 1.90817 | 2.47256 |
| 0.340 | 0.54522 | 0.92183 | 1.19390 | 1.73115 | 1.91414 | 2.47390 |
| 0.360 | 0.55092 | 0.93319 | 1.20519 | 1.73389 | 1.92033 | 2.47550 |
| 0.380 | 0.55727 | 0.94468 | 1.21695 | 1.73703 | 1.92672 | 2.47737 |
| 0.400 | 0.56427 | 0.95630 | 1.22914 | 1.74058 | 1.93330 | 2.47949 |
| 0.420 | 0.57191 | 0.96801 | 1.24172 | 1.74459 | 1.94005 | 2.48189 |
| 0.440 | 0.58017 | 0.97981 | 1.25464 | 1.74907 | 1.94700 | 2.48451 |
| 0.460 | 0.58904 | 0.99170 | 1.26786 | 1.75403 | 1.95410 | 2.48743 |
| 0.480 | 0.59849 | 1.00367 | 1.28133 | 1.75953 | 1.96136 | 2.49058 |
| 0.500 | 0.60850 | 1.01570 | 1.29501 | 1.76556 | 1.96879 | 2.49397 |
| 0.520 | 0.61906 | 1.02782 | 1.30884 | 1.77216 | 1.97635 | 2.49763 |
| 0.540 | 0.63013 | 1.04002 | 1.32278 | 1.77934 | 1.98408 | 2.50154 |
| 0.560 | 0.64168 | 1.05229 | 1.33679 | 1.78712 | 1.99194 | 2.50572 |
| 0.580 | 0.65370 | 1.06465 | 1.35082 | 1.79553 | 1.99994 | 2.51013 |
| 0.600 | 0.66616 | 1.07710 | 1.36484 | 1.80456 | 2.00807 | 2.51479 |
| 0.620 | 0.67903 | 1.08963 | 1.37878 | 1.81425 | 2.01634 | 2.51972 |
| 0.640 | 0.69229 | 1.10227 | 1.39265 | 1.82458 | 2.02474 | 2.52487 |
| 0.660 | 0.70592 | 1.11499 | 1.40640 | 1.83555 | 2.03326 | 2.53029 |
| 0.680 | 0.71989 | 1.12781 | 1.41999 | 1.84718 | 2.04192 | 2.53597 |
| 0.700 | 0.73418 | 1.14073 | 1.43343 | 1.85944 | 2.05069 | 2.54189 |
| 0.720 | 0.74877 | 1.15376 | 1.44670 | 1.87235 | 2.05962 | 2.54808 |
| 0.740 | 0.76365 | 1.16689 | 1.45977 | 1.88587 | 2.06866 | 2.55454 |
| 0.760 | 0.77880 | 1.18013 | 1.47265 | 1.89997 | 2.07783 | 2.56123 |
| 0.780 | 0.79419 | 1.19348 | 1.48535 | 1.91466 | 2.08713 | 2.56819 |
| 0.800 | 0.80981 | 1.20692 | 1.49785 | 1.92988 | 2.09658 | 2.57541 |
| 0.820 | 0.82564 | 1.22049 | 1.51017 | 1.94560 | 2.10615 | 2.58290 |
| 0.840 | 0.84160 | 1.23416 | 1.52233 | 1.96180 | 2.11587 | 2.59067 |
| 0.860 | 0.85796 | 1.24793 | 1.53434 | 1.97841 | 2.12573 | 2.59871 |
| 0.880 | 0.87433 | 1.26182 | 1.54620 | 1.99542 | 2.13574 | 2.60701 |
| 0.900 | 0.89089 | 1.27581 | 1.55794 | 2.01275 | 2.14592 | 2.61561 |
| 0.920 | 0.90759 | 1.28991 | 1.56956 | 2.03037 | 2.15626 | 2.62451 |
| 0.940 | 0.92445 | 1.30411 | 1.58111 | 2.04823 | 2.16678 | 2.63367 |
| 0.960 | 0.94144 | 1.31841 | 1.59257 | 2.06625 | 2.17752 | 2.64316 |
| 0.980 | 0.95855 | 1.33282 | 1.60399 | 2.08440 | 2.18846 | 2.65295 |
| 1.000 | 0.97577 | 1.34732 | 1.61535 | 2.10260 | 2.19966 | 2.66304 |

$H/R = 1.500 \qquad n = 2.$

| $H/L$ | $\Omega_1$ | $\Omega_2$ | $\Omega_3$ | $\Omega_4$ | $\Omega_5$ | $\Omega_6$ |
|---|---|---|---|---|---|---|
| 0. | 0.57293 | 0.83187 | 1.15105 | 1.80987 | 1.92665 | 2.58064 |
| 0.010 | 0.57291 | 0.83202 | 1.15112 | 1.80985 | 1.92671 | 2.58063 |
| 0.050 | 0.57241 | 0.83557 | 1.15309 | 1.80954 | 1.92835 | 2.58066 |
| 0.100 | 0.57122 | 0.84626 | 1.15921 | 1.80874 | 1.93332 | 2.58084 |
| 0.120 | 0.57077 | 0.85227 | 1.16280 | 1.80836 | 1.93614 | 2.58102 |
| 0.140 | 0.57044 | 0.85914 | 1.16705 | 1.80801 | 1.93939 | 2.58128 |
| 0.160 | 0.57032 | 0.86678 | 1.17194 | 1.80771 | 1.94306 | 2.58163 |
| 0.180 | 0.57046 | 0.87511 | 1.17747 | 1.80749 | 1.94711 | 2.58208 |
| 0.200 | 0.57095 | 0.88405 | 1.18365 | 1.80740 | 1.95149 | 2.58269 |
| 0.220 | 0.57184 | 0.89351 | 1.19047 | 1.80745 | 1.95621 | 2.58342 |
| 0.240 | 0.57320 | 0.90342 | 1.19791 | 1.80769 | 1.96122 | 2.58431 |
| 0.260 | 0.57506 | 0.91372 | 1.20597 | 1.80814 | 1.96651 | 2.58535 |
| 0.280 | 0.57747 | 0.92434 | 1.21463 | 1.80883 | 1.97205 | 2.58657 |
| 0.300 | 0.58045 | 0.93523 | 1.22387 | 1.80980 | 1.97782 | 2.58796 |
| 0.320 | 0.58404 | 0.94636 | 1.23369 | 1.81106 | 1.98383 | 2.58955 |
| 0.340 | 0.58824 | 0.95766 | 1.24404 | 1.81264 | 1.99003 | 2.59131 |
| 0.360 | 0.59306 | 0.96912 | 1.25492 | 1.81459 | 1.99641 | 2.59326 |
| 0.380 | 0.59851 | 0.98071 | 1.26629 | 1.81689 | 2.00300 | 2.59539 |
| 0.400 | 0.60460 | 0.99240 | 1.27810 | 1.81959 | 2.00974 | 2.59773 |
| 0.420 | 0.61130 | 1.00418 | 1.29034 | 1.82272 | 2.01664 | 2.60025 |
| 0.440 | 0.61862 | 1.01604 | 1.30296 | 1.82628 | 2.02370 | 2.60297 |
| 0.460 | 0.62653 | 1.02796 | 1.31593 | 1.83031 | 2.03091 | 2.60590 |
| 0.480 | 0.63502 | 1.03994 | 1.32919 | 1.83482 | 2.03824 | 2.60901 |
| 0.500 | 0.64408 | 1.05198 | 1.34272 | 1.83985 | 2.04573 | 2.61233 |
| 0.520 | 0.65368 | 1.06408 | 1.35646 | 1.84541 | 2.05332 | 2.61586 |
| 0.540 | 0.66381 | 1.07624 | 1.37035 | 1.85151 | 2.06106 | 2.61957 |
| 0.560 | 0.67443 | 1.08845 | 1.38439 | 1.85818 | 2.06890 | 2.62350 |
| 0.580 | 0.68553 | 1.10072 | 1.39850 | 1.86545 | 2.07686 | 2.62764 |
| 0.600 | 0.69708 | 1.11307 | 1.41267 | 1.87331 | 2.08494 | 2.63198 |
| 0.620 | 0.70908 | 1.12548 | 1.42682 | 1.88178 | 2.09312 | 2.63652 |
| 0.640 | 0.72147 | 1.13796 | 1.44095 | 1.89089 | 2.10142 | 2.64130 |
| 0.660 | 0.73426 | 1.15052 | 1.45500 | 1.90064 | 2.10983 | 2.64627 |
| 0.680 | 0.74741 | 1.16316 | 1.46897 | 1.91102 | 2.11833 | 2.65148 |
| 0.700 | 0.76091 | 1.17588 | 1.48281 | 1.92203 | 2.12695 | 2.65688 |
| 0.720 | 0.77473 | 1.18870 | 1.49651 | 1.93370 | 2.13567 | 2.66251 |
| 0.740 | 0.78886 | 1.20160 | 1.51005 | 1.94600 | 2.14449 | 2.66838 |
| 0.760 | 0.80328 | 1.21459 | 1.52341 | 1.95891 | 2.15342 | 2.67445 |
| 0.780 | 0.81797 | 1.22769 | 1.53659 | 1.97243 | 2.16244 | 2.68076 |
| 0.800 | 0.83291 | 1.24087 | 1.54960 | 1.98654 | 2.17156 | 2.68733 |
| 0.820 | 0.84810 | 1.25415 | 1.56242 | 2.00122 | 2.18080 | 2.69412 |
| 0.840 | 0.86350 | 1.26753 | 1.57505 | 2.01641 | 2.19012 | 2.70115 |
| 0.860 | 0.87911 | 1.28101 | 1.58752 | 2.03214 | 2.19956 | 2.70843 |
| 0.880 | 0.89492 | 1.29459 | 1.59983 | 2.04832 | 2.20910 | 2.71597 |
| 0.900 | 0.91090 | 1.30828 | 1.61199 | 2.06494 | 2.21874 | 2.72379 |
| 0.920 | 0.92708 | 1.32205 | 1.62401 | 2.08197 | 2.22851 | 2.73184 |
| 0.940 | 0.94337 | 1.33593 | 1.63591 | 2.09935 | 2.23837 | 2.74019 |
| 0.960 | 0.96000 | 1.34991 | 1.64770 | 2.11707 | 2.24834 | 2.74880 |
| 0.980 | 0.97654 | 1.36398 | 1.65942 | 2.13504 | 2.25844 | 2.75770 |
| 1.000 | 0.99328 | 1.37815 | 1.67105 | 2.15327 | 2.26866 | 2.76689 |

$H/R = 1.600 \quad n = 2.$

| $H/L$ | $\Omega_1$ | $\Omega_2$ | $\Omega_3$ | $\Omega_4$ | $\Omega_5$ | $\Omega_6$ |
|---|---|---|---|---|---|---|
| 0. | 0.62012 | 0.86362 | 1.20734 | 1.88924 | 2.02695 | 2.67650 |
| 0.010 | 0.62009 | 0.86377 | 1.20742 | 1.88924 | 2.02700 | 2.67650 |
| 0.050 | 0.61942 | 0.86739 | 1.20928 | 1.88906 | 2.02842 | 2.67700 |
| 0.100 | 0.61770 | 0.87828 | 1.21510 | 1.88860 | 2.03279 | 2.67846 |
| 0.120 | 0.61694 | 0.88438 | 1.21851 | 1.88839 | 2.03529 | 2.67932 |
| 0.140 | 0.61626 | 0.89135 | 1.22255 | 1.88821 | 2.03819 | 2.68035 |
| 0.160 | 0.61575 | 0.89911 | 1.22720 | 1.88808 | 2.04147 | 2.68149 |
| 0.180 | 0.61546 | 0.90756 | 1.23248 | 1.88802 | 2.04510 | 2.68281 |
| 0.200 | 0.61548 | 0.91662 | 1.23837 | 1.88806 | 2.04908 | 2.68425 |
| 0.220 | 0.61587 | 0.92620 | 1.24488 | 1.88823 | 2.05337 | 2.68585 |
| 0.240 | 0.61668 | 0.93623 | 1.25198 | 1.88855 | 2.05796 | 2.68759 |
| 0.260 | 0.61796 | 0.94666 | 1.25970 | 1.88905 | 2.06280 | 2.68945 |
| 0.280 | 0.61975 | 0.95741 | 1.26801 | 1.88975 | 2.06792 | 2.69145 |
| 0.300 | 0.62208 | 0.96843 | 1.27688 | 1.89068 | 2.07327 | 2.69358 |
| 0.320 | 0.62498 | 0.97968 | 1.28633 | 1.89187 | 2.07884 | 2.69585 |
| 0.340 | 0.62847 | 0.99112 | 1.29633 | 1.89332 | 2.08463 | 2.69824 |
| 0.360 | 0.63256 | 1.00270 | 1.30687 | 1.89508 | 2.09062 | 2.70077 |
| 0.380 | 0.63724 | 1.01441 | 1.31790 | 1.89715 | 2.09679 | 2.70344 |
| 0.400 | 0.64254 | 1.02621 | 1.32942 | 1.89957 | 2.10313 | 2.70623 |
| 0.420 | 0.64844 | 1.03809 | 1.34139 | 1.90236 | 2.10964 | 2.70917 |
| 0.440 | 0.65495 | 1.05003 | 1.35377 | 1.90552 | 2.11631 | 2.71223 |
| 0.460 | 0.66203 | 1.06203 | 1.36655 | 1.90910 | 2.12312 | 2.71545 |
| 0.480 | 0.66969 | 1.07408 | 1.37968 | 1.91309 | 2.13009 | 2.71878 |
| 0.500 | 0.67791 | 1.08616 | 1.39312 | 1.91753 | 2.13718 | 2.72227 |
| 0.520 | 0.68668 | 1.09828 | 1.40684 | 1.92244 | 2.14441 | 2.72590 |
| 0.540 | 0.69597 | 1.11043 | 1.42079 | 1.92782 | 2.15176 | 2.72968 |
| 0.560 | 0.70577 | 1.12264 | 1.43493 | 1.93371 | 2.15925 | 2.73361 |
| 0.580 | 0.71605 | 1.13487 | 1.44923 | 1.94013 | 2.16684 | 2.73770 |
| 0.600 | 0.72679 | 1.14715 | 1.46365 | 1.94707 | 2.17455 | 2.74192 |
| 0.620 | 0.73798 | 1.15949 | 1.47814 | 1.95457 | 2.18238 | 2.74634 |
| 0.640 | 0.74959 | 1.17187 | 1.49268 | 1.96264 | 2.19031 | 2.75090 |
| 0.660 | 0.76161 | 1.18431 | 1.50720 | 1.97128 | 2.19836 | 2.75562 |
| 0.680 | 0.77401 | 1.19682 | 1.52170 | 1.98051 | 2.20652 | 2.76052 |
| 0.700 | 0.78677 | 1.20938 | 1.53614 | 1.99035 | 2.21476 | 2.76562 |
| 0.720 | 0.79988 | 1.22203 | 1.55049 | 2.00079 | 2.22312 | 2.77087 |
| 0.740 | 0.81331 | 1.23473 | 1.56473 | 2.01182 | 2.23159 | 2.77632 |
| 0.760 | 0.82705 | 1.24753 | 1.57883 | 2.02346 | 2.24014 | 2.78196 |
| 0.780 | 0.84107 | 1.26040 | 1.59279 | 2.03570 | 2.24881 | 2.78781 |
| 0.800 | 0.85537 | 1.27334 | 1.60657 | 2.04851 | 2.25757 | 2.79385 |
| 0.820 | 0.86993 | 1.28639 | 1.62019 | 2.06190 | 2.26644 | 2.80007 |
| 0.840 | 0.88473 | 1.29952 | 1.63364 | 2.07585 | 2.27540 | 2.80654 |
| 0.860 | 0.89976 | 1.31272 | 1.64691 | 2.09034 | 2.28446 | 2.81321 |
| 0.880 | 0.91500 | 1.32604 | 1.66000 | 2.10534 | 2.29362 | 2.82010 |
| 0.900 | 0.93044 | 1.33943 | 1.67292 | 2.12083 | 2.30289 | 2.82724 |
| 0.920 | 0.94608 | 1.35291 | 1.68568 | 2.13677 | 2.31225 | 2.83460 |
| 0.940 | 0.96189 | 1.36650 | 1.69830 | 2.15315 | 2.32171 | 2.84221 |
| 0.960 | 0.97787 | 1.38017 | 1.71076 | 2.16991 | 2.33129 | 2.85007 |
| 0.980 | 0.99401 | 1.39394 | 1.72311 | 2.18703 | 2.34095 | 2.85818 |
| 1.000 | 1.01029 | 1.40781 | 1.73535 | 2.20447 | 2.35073 | 2.86657 |

$H/R = 1.700 \qquad n = 2.$

| $H/L$ | $\Omega_1$ | $\Omega_2$ | $\Omega_3$ | $\Omega_4$ | $\Omega_5$ | $\Omega_6$ |
|---|---|---|---|---|---|---|
| 0. | 0.66186 | 0.89321 | 1.26260 | 1.95902 | 2.15757 | 2.76682 |
| 0.010 | 0.66182 | 0.89336 | 1.26266 | 1.95903 | 2.15762 | 2.76683 |
| 0.050 | 0.66099 | 0.89704 | 1.26445 | 1.95913 | 2.15869 | 2.76744 |
| 0.090 | 0.65932 | 0.90534 | 1.26860 | 1.95940 | 2.16117 | 2.76887 |
| 0.100 | 0.65883 | 0.90807 | 1.27001 | 1.95949 | 2.16200 | 2.76936 |
| 0.120 | 0.65782 | 0.91425 | 1.27328 | 1.95974 | 2.16392 | 2.77045 |
| 0.140 | 0.65687 | 0.92132 | 1.27714 | 1.96005 | 2.16616 | 2.77174 |
| 0.160 | 0.65603 | 0.92916 | 1.28160 | 1.96045 | 2.16873 | 2.77321 |
| 0.180 | 0.65539 | 0.93770 | 1.28665 | 1.96094 | 2.17160 | 2.77484 |
| 0.200 | 0.65503 | 0.94684 | 1.29230 | 1.96155 | 2.17477 | 2.77666 |
| 0.220 | 0.65500 | 0.95652 | 1.29855 | 1.96228 | 2.17822 | 2.77863 |
| 0.240 | 0.65537 | 0.96665 | 1.30539 | 1.96315 | 2.18194 | 2.78076 |
| 0.260 | 0.65618 | 0.97718 | 1.31281 | 1.96419 | 2.18593 | 2.78304 |
| 0.280 | 0.65746 | 0.98802 | 1.32081 | 1.96541 | 2.19018 | 2.78548 |
| 0.300 | 0.65927 | 0.99915 | 1.32939 | 1.96682 | 2.19465 | 2.78804 |
| 0.320 | 0.66161 | 1.01050 | 1.33852 | 1.96844 | 2.19935 | 2.79072 |
| 0.340 | 0.66452 | 1.02204 | 1.34822 | 1.97029 | 2.20428 | 2.79355 |
| 0.360 | 0.66800 | 1.03373 | 1.35844 | 1.97238 | 2.20942 | 2.79650 |
| 0.380 | 0.67206 | 1.04554 | 1.36919 | 1.97474 | 2.21475 | 2.79955 |
| 0.400 | 0.67671 | 1.05744 | 1.38044 | 1.97739 | 2.22028 | 2.80271 |
| 0.420 | 0.68195 | 1.06942 | 1.39216 | 1.98034 | 2.22598 | 2.80600 |
| 0.440 | 0.68777 | 1.08145 | 1.40434 | 1.98359 | 2.23187 | 2.80940 |
| 0.460 | 0.69417 | 1.09353 | 1.41694 | 1.98720 | 2.23792 | 2.81291 |
| 0.480 | 0.70113 | 1.10564 | 1.42994 | 1.99114 | 2.24413 | 2.81653 |
| 0.500 | 0.70864 | 1.11778 | 1.44331 | 1.99547 | 2.25050 | 2.82026 |
| 0.520 | 0.71670 | 1.12994 | 1.45699 | 2.00017 | 2.25702 | 2.82410 |
| 0.540 | 0.72528 | 1.14212 | 1.47098 | 2.00529 | 2.26369 | 2.82806 |
| 0.560 | 0.73436 | 1.15433 | 1.48523 | 2.01082 | 2.27049 | 2.83215 |
| 0.580 | 0.74394 | 1.16656 | 1.49971 | 2.01680 | 2.27743 | 2.83633 |
| 0.600 | 0.75398 | 1.17882 | 1.51436 | 2.02322 | 2.28450 | 2.84065 |
| 0.620 | 0.76448 | 1.19110 | 1.52917 | 2.03013 | 2.29172 | 2.84510 |
| 0.640 | 0.77541 | 1.20343 | 1.54409 | 2.03751 | 2.29905 | 2.84968 |
| 0.660 | 0.78675 | 1.21579 | 1.55910 | 2.04540 | 2.30650 | 2.85439 |
| 0.680 | 0.79849 | 1.22820 | 1.57414 | 2.05380 | 2.31408 | 2.85924 |
| 0.700 | 0.81059 | 1.24065 | 1.58920 | 2.06272 | 2.32180 | 2.86426 |
| 0.720 | 0.82306 | 1.25316 | 1.60423 | 2.07217 | 2.32961 | 2.86940 |
| 0.740 | 0.83586 | 1.26572 | 1.61922 | 2.08217 | 2.33755 | 2.87469 |
| 0.760 | 0.84899 | 1.27835 | 1.63413 | 2.09270 | 2.34561 | 2.88014 |
| 0.780 | 0.86242 | 1.29105 | 1.64895 | 2.10379 | 2.35378 | 2.88577 |
| 0.800 | 0.87614 | 1.30381 | 1.66365 | 2.11542 | 2.36207 | 2.89158 |
| 0.820 | 0.89014 | 1.31664 | 1.67820 | 2.12759 | 2.37046 | 2.89755 |
| 0.840 | 0.90438 | 1.32956 | 1.69262 | 2.14030 | 2.37898 | 2.90370 |
| 0.860 | 0.91888 | 1.34254 | 1.70688 | 2.15353 | 2.38759 | 2.91005 |
| 0.880 | 0.93361 | 1.35562 | 1.72096 | 2.16728 | 2.39634 | 2.91659 |
| 0.900 | 0.94854 | 1.36877 | 1.73488 | 2.18152 | 2.40519 | 2.92332 |
| 0.920 | 0.96369 | 1.38202 | 1.74863 | 2.19624 | 2.41416 | 2.93027 |
| 0.940 | 0.97903 | 1.39534 | 1.76221 | 2.21142 | 2.42324 | 2.93741 |
| 0.960 | 0.99455 | 1.40875 | 1.77562 | 2.22702 | 2.43244 | 2.94481 |
| 0.980 | 1.01025 | 1.42225 | 1.78889 | 2.24302 | 2.44176 | 2.95240 |
| 1.000 | 1.02610 | 1.43584 | 1.80200 | 2.25938 | 2.45120 | 2.96024 |

$H/R = 1.800 \qquad n = 2.$

| $H/L$ | $\Omega_1$ | $\Omega_2$ | $\Omega_3$ | $\Omega_4$ | $\Omega_5$ | $\Omega_6$ |
|---|---|---|---|---|---|---|
| 0. | 0.69711 | 0.92100 | 1.31376 | 2.02181 | 2.30247 | 2.85257 |
| 0.010 | 0.69707 | 0.92115 | 1.31382 | 2.02182 | 2.30250 | 2.85261 |
| 0.050 | 0.69614 | 0.92486 | 1.31553 | 2.02211 | 2.30333 | 2.85325 |
| 0.100 | 0.69365 | 0.93595 | 1.32088 | 2.02302 | 2.30590 | 2.85525 |
| 0.120 | 0.69247 | 0.94216 | 1.32403 | 2.02358 | 2.30740 | 2.85644 |
| 0.140 | 0.69130 | 0.94925 | 1.32775 | 2.02425 | 2.30916 | 2.85779 |
| 0.160 | 0.69023 | 0.95712 | 1.33205 | 2.02504 | 2.31119 | 2.85935 |
| 0.180 | 0.68934 | 0.96569 | 1.33692 | 2.02596 | 2.31347 | 2.86112 |
| 0.200 | 0.68869 | 0.97487 | 1.34237 | 2.02701 | 2.31599 | 2.86306 |
| 0.220 | 0.68835 | 0.98458 | 1.34839 | 2.02821 | 2.31877 | 2.86517 |
| 0.240 | 0.68838 | 0.99475 | 1.35498 | 2.02957 | 2.32178 | 2.86745 |
| 0.260 | 0.68883 | 1.00531 | 1.36215 | 2.03109 | 2.32502 | 2.86988 |
| 0.280 | 0.68973 | 1.01620 | 1.36989 | 2.03278 | 2.32850 | 2.87248 |
| 0.300 | 0.69113 | 1.02738 | 1.37819 | 2.03467 | 2.33218 | 2.87524 |
| 0.320 | 0.69304 | 1.03878 | 1.38705 | 2.03675 | 2.33609 | 2.87812 |
| 0.340 | 0.69548 | 1.05037 | 1.39646 | 2.03904 | 2.34021 | 2.88115 |
| 0.360 | 0.69848 | 1.06212 | 1.40641 | 2.04155 | 2.34453 | 2.88431 |
| 0.380 | 0.70204 | 1.07400 | 1.41688 | 2.04430 | 2.34905 | 2.88759 |
| 0.400 | 0.70618 | 1.08596 | 1.42787 | 2.04729 | 2.35377 | 2.89099 |
| 0.420 | 0.71088 | 1.09800 | 1.43934 | 2.05056 | 2.35865 | 2.89451 |
| 0.440 | 0.71615 | 1.11009 | 1.45129 | 2.05409 | 2.36374 | 2.89815 |
| 0.460 | 0.72199 | 1.12223 | 1.46368 | 2.05791 | 2.36899 | 2.90191 |
| 0.480 | 0.72838 | 1.13439 | 1.47651 | 2.06203 | 2.37441 | 2.90575 |
| 0.500 | 0.73532 | 1.14657 | 1.48973 | 2.06648 | 2.38002 | 2.90972 |
| 0.520 | 0.74279 | 1.15877 | 1.50332 | 2.07124 | 2.38577 | 2.91379 |
| 0.540 | 0.75078 | 1.17099 | 1.51725 | 2.07635 | 2.39170 | 2.91798 |
| 0.560 | 0.75927 | 1.18321 | 1.53150 | 2.08182 | 2.39777 | 2.92227 |
| 0.580 | 0.76826 | 1.19544 | 1.54602 | 2.08767 | 2.40400 | 2.92667 |
| 0.600 | 0.77772 | 1.20770 | 1.56079 | 2.09389 | 2.41038 | 2.93120 |
| 0.620 | 0.78763 | 1.21996 | 1.57576 | 2.10053 | 2.41690 | 2.93584 |
| 0.640 | 0.79797 | 1.23225 | 1.59091 | 2.10758 | 2.42357 | 2.94058 |
| 0.660 | 0.80875 | 1.24456 | 1.60621 | 2.11504 | 2.43039 | 2.94544 |
| 0.680 | 0.81992 | 1.25691 | 1.62162 | 2.12295 | 2.43734 | 2.95045 |
| 0.700 | 0.83147 | 1.26928 | 1.63710 | 2.13132 | 2.44445 | 2.95557 |
| 0.720 | 0.84339 | 1.28170 | 1.65264 | 2.14014 | 2.45168 | 2.96082 |
| 0.740 | 0.85566 | 1.29416 | 1.66817 | 2.14943 | 2.45906 | 2.96623 |
| 0.760 | 0.86826 | 1.30667 | 1.68371 | 2.15920 | 2.46656 | 2.97175 |
| 0.780 | 0.88118 | 1.31924 | 1.69921 | 2.16945 | 2.47420 | 2.97744 |
| 0.800 | 0.89439 | 1.33186 | 1.71464 | 2.18020 | 2.48198 | 2.98328 |
| 0.820 | 0.90790 | 1.34455 | 1.72999 | 2.19144 | 2.48989 | 2.98925 |
| 0.840 | 0.92167 | 1.35729 | 1.74523 | 2.20317 | 2.49793 | 2.99539 |
| 0.860 | 0.93570 | 1.37011 | 1.76035 | 2.21538 | 2.50612 | 3.00169 |
| 0.880 | 0.94997 | 1.38300 | 1.77533 | 2.22808 | 2.51444 | 3.00818 |
| 0.900 | 0.96447 | 1.39596 | 1.79017 | 2.24124 | 2.52288 | 3.01483 |
| 0.920 | 0.97919 | 1.40900 | 1.80484 | 2.25488 | 2.53146 | 3.02166 |
| 0.940 | 0.99411 | 1.42213 | 1.81937 | 2.26896 | 2.54019 | 3.02870 |
| 0.960 | 1.00923 | 1.43534 | 1.83373 | 2.28348 | 2.54905 | 3.03591 |
| 0.980 | 1.02454 | 1.44862 | 1.84792 | 2.29841 | 2.55804 | 3.04333 |
| 1.000 | 1.04001 | 1.46199 | 1.86195 | 2.31372 | 2.56718 | 3.05094 |

$H/R = 1.900 \qquad n = 2.$

| $H/L$ | $\Omega_1$ | $\Omega_2$ | $\Omega_3$ | $\Omega_4$ | $\Omega_5$ | $\Omega_6$ |
|---|---|---|---|---|---|---|
| 0. | 0.72554 | 0.94727 | 1.35870 | 2.07987 | 2.42738 | 2.93410 |
| 0.010 | 0.72550 | 0.94742 | 1.35876 | 2.07989 | 2.42741 | 2.93414 |
| 0.050 | 0.72451 | 0.95110 | 1.36042 | 2.08026 | 2.42811 | 2.93479 |
| 0.100 | 0.72186 | 0.96213 | 1.36560 | 2.08141 | 2.43031 | 2.93680 |
| 0.120 | 0.72057 | 0.96830 | 1.36864 | 2.08210 | 2.43159 | 2.93798 |
| 0.140 | 0.71929 | 0.97536 | 1.37225 | 2.08293 | 2.43310 | 2.93934 |
| 0.160 | 0.71808 | 0.98319 | 1.37640 | 2.08390 | 2.43484 | 2.94094 |
| 0.180 | 0.71702 | 0.99172 | 1.38111 | 2.08500 | 2.43680 | 2.94269 |
| 0.200 | 0.71618 | 1.00086 | 1.38639 | 2.08625 | 2.43898 | 2.94465 |
| 0.220 | 0.71564 | 1.01053 | 1.39222 | 2.08767 | 2.44137 | 2.94678 |
| 0.240 | 0.71544 | 1.02067 | 1.39861 | 2.08925 | 2.44400 | 2.94912 |
| 0.260 | 0.71562 | 1.03120 | 1.40557 | 2.09100 | 2.44682 | 2.95160 |
| 0.280 | 0.71625 | 1.04208 | 1.41308 | 2.09291 | 2.44984 | 2.95425 |
| 0.300 | 0.71734 | 1.05324 | 1.42114 | 2.09503 | 2.45308 | 2.95706 |
| 0.320 | 0.71892 | 1.06463 | 1.42973 | 2.09733 | 2.45650 | 2.96000 |
| 0.340 | 0.72102 | 1.07623 | 1.43889 | 2.09983 | 2.46012 | 2.96309 |
| 0.360 | 0.72366 | 1.08798 | 1.44857 | 2.10255 | 2.46394 | 2.96635 |
| 0.380 | 0.72683 | 1.09986 | 1.45878 | 2.10549 | 2.46796 | 2.96971 |
| 0.400 | 0.73056 | 1.11185 | 1.46950 | 2.10866 | 2.47216 | 2.97322 |
| 0.420 | 0.73484 | 1.12391 | 1.48071 | 2.11207 | 2.47652 | 2.97685 |
| 0.440 | 0.73969 | 1.13603 | 1.49240 | 2.11574 | 2.48106 | 2.98059 |
| 0.460 | 0.74507 | 1.14818 | 1.50456 | 2.11967 | 2.48580 | 2.98447 |
| 0.480 | 0.75101 | 1.16037 | 1.51715 | 2.12387 | 2.49070 | 2.98845 |
| 0.500 | 0.75748 | 1.17258 | 1.53016 | 2.12836 | 2.49578 | 2.99255 |
| 0.520 | 0.76448 | 1.18480 | 1.54357 | 2.13315 | 2.50101 | 2.99675 |
| 0.540 | 0.77199 | 1.19704 | 1.55735 | 2.13824 | 2.50639 | 3.00109 |
| 0.560 | 0.78001 | 1.20927 | 1.57147 | 2.14365 | 2.51195 | 3.00552 |
| 0.580 | 0.78851 | 1.22152 | 1.58589 | 2.14940 | 2.51767 | 3.01007 |
| 0.600 | 0.79749 | 1.23376 | 1.60060 | 2.15548 | 2.52357 | 3.01475 |
| 0.620 | 0.80692 | 1.24602 | 1.61556 | 2.16193 | 2.52960 | 3.01951 |
| 0.640 | 0.81680 | 1.25829 | 1.63073 | 2.16876 | 2.53576 | 3.02440 |
| 0.660 | 0.82709 | 1.27058 | 1.64609 | 2.17594 | 2.54212 | 3.02941 |
| 0.680 | 0.83780 | 1.28288 | 1.66162 | 2.18353 | 2.54859 | 3.03456 |
| 0.700 | 0.84889 | 1.29522 | 1.67727 | 2.19151 | 2.55522 | 3.03982 |
| 0.720 | 0.86036 | 1.30758 | 1.69300 | 2.19992 | 2.56202 | 3.04518 |
| 0.740 | 0.87218 | 1.31997 | 1.70882 | 2.20874 | 2.56893 | 3.05071 |
| 0.760 | 0.88434 | 1.33240 | 1.72466 | 2.21799 | 2.57601 | 3.05635 |
| 0.780 | 0.89683 | 1.34488 | 1.74051 | 2.22767 | 2.58323 | 3.06213 |
| 0.800 | 0.90963 | 1.35741 | 1.75635 | 2.23781 | 2.59061 | 3.06804 |
| 0.820 | 0.92272 | 1.36999 | 1.77215 | 2.24840 | 2.59812 | 3.07408 |
| 0.840 | 0.93609 | 1.38263 | 1.78789 | 2.25942 | 2.60578 | 3.08030 |
| 0.860 | 0.94973 | 1.39532 | 1.80354 | 2.27091 | 2.61358 | 3.08666 |
| 0.880 | 0.96362 | 1.40808 | 1.81908 | 2.28285 | 2.62154 | 3.09316 |
| 0.900 | 0.97775 | 1.42092 | 1.83452 | 2.29524 | 2.62965 | 3.09983 |
| 0.920 | 0.99211 | 1.43381 | 1.84982 | 2.30807 | 2.63789 | 3.10666 |
| 0.940 | 1.00668 | 1.44678 | 1.86497 | 2.32133 | 2.64630 | 3.11367 |
| 0.960 | 1.02145 | 1.45984 | 1.87998 | 2.33502 | 2.65484 | 3.12084 |
| 0.980 | 1.03643 | 1.47296 | 1.89482 | 2.34913 | 2.66355 | 3.12822 |
| 1.000 | 1.05158 | 1.48616 | 1.90952 | 2.36363 | 2.67241 | 3.13575 |

$H/R = 2.000 \qquad n = 2.$

| $H/L$ | $\Omega_1$ | $\Omega_2$ | $\Omega_3$ | $\Omega_4$ | $\Omega_5$ | $\Omega_6$ |
|---|---|---|---|---|---|---|
| 0. | 0.74738 | 0.97220 | 1.39689 | 2.13463 | 2.51103 | 3.01143 |
| 0.010 | 0.74733 | 0.97235 | 1.39696 | 2.13465 | 2.51104 | 3.01146 |
| 0.050 | 0.74634 | 0.97596 | 1.39856 | 2.13502 | 2.51172 | 3.01210 |
| 0.100 | 0.74366 | 0.98680 | 1.40360 | 2.13622 | 2.51377 | 3.01407 |
| 0.120 | 0.74236 | 0.99287 | 1.40656 | 2.13695 | 2.51496 | 3.01525 |
| 0.140 | 0.74103 | 0.99982 | 1.41006 | 2.13779 | 2.51639 | 3.01658 |
| 0.160 | 0.73976 | 1.00755 | 1.41411 | 2.13878 | 2.51802 | 3.01813 |
| 0.180 | 0.73863 | 1.01597 | 1.41869 | 2.13993 | 2.51985 | 3.01989 |
| 0.200 | 0.73770 | 1.02501 | 1.42383 | 2.14120 | 2.52190 | 3.02182 |
| 0.220 | 0.73703 | 1.03458 | 1.42950 | 2.14264 | 2.52415 | 3.02392 |
| 0.240 | 0.73669 | 1.04462 | 1.43572 | 2.14425 | 2.52661 | 3.02621 |
| 0.260 | 0.73671 | 1.05507 | 1.44249 | 2.14602 | 2.52926 | 3.02866 |
| 0.280 | 0.73715 | 1.06586 | 1.44980 | 2.14796 | 2.53213 | 3.03129 |
| 0.300 | 0.73803 | 1.07695 | 1.45764 | 2.15008 | 2.53517 | 3.03406 |
| 0.320 | 0.73938 | 1.08828 | 1.46603 | 2.15239 | 2.53841 | 3.03699 |
| 0.340 | 0.74123 | 1.09983 | 1.47494 | 2.15490 | 2.54185 | 3.04006 |
| 0.360 | 0.74359 | 1.11153 | 1.48439 | 2.15761 | 2.54546 | 3.04328 |
| 0.380 | 0.74649 | 1.12338 | 1.49434 | 2.16053 | 2.54924 | 3.04662 |
| 0.400 | 0.74990 | 1.13534 | 1.50481 | 2.16367 | 2.55324 | 3.05010 |
| 0.420 | 0.75387 | 1.14738 | 1.51577 | 2.16705 | 2.55740 | 3.05372 |
| 0.440 | 0.75837 | 1.15948 | 1.52720 | 2.17067 | 2.56172 | 3.05747 |
| 0.460 | 0.76342 | 1.17164 | 1.53910 | 2.17453 | 2.56622 | 3.06132 |
| 0.480 | 0.76900 | 1.18382 | 1.55144 | 2.17865 | 2.57088 | 3.06528 |
| 0.500 | 0.77510 | 1.19604 | 1.56421 | 2.18304 | 2.57573 | 3.06937 |
| 0.520 | 0.78173 | 1.20826 | 1.57738 | 2.18771 | 2.58072 | 3.07358 |
| 0.540 | 0.78886 | 1.22049 | 1.59093 | 2.19267 | 2.58587 | 3.07790 |
| 0.560 | 0.79650 | 1.23274 | 1.60484 | 2.19792 | 2.59120 | 3.08232 |
| 0.580 | 0.80462 | 1.24499 | 1.61907 | 2.20349 | 2.59669 | 3.08687 |
| 0.600 | 0.81321 | 1.25723 | 1.63360 | 2.20937 | 2.60231 | 3.09150 |
| 0.620 | 0.82225 | 1.26949 | 1.64841 | 2.21560 | 2.60812 | 3.09627 |
| 0.640 | 0.83175 | 1.28175 | 1.66345 | 2.22216 | 2.61404 | 3.10114 |
| 0.660 | 0.84166 | 1.29402 | 1.67872 | 2.22908 | 2.62013 | 3.10613 |
| 0.680 | 0.85198 | 1.30630 | 1.69416 | 2.23637 | 2.62640 | 3.11125 |
| 0.700 | 0.86270 | 1.31860 | 1.70977 | 2.24402 | 2.63278 | 3.11649 |
| 0.720 | 0.87380 | 1.33094 | 1.72549 | 2.25207 | 2.63932 | 3.12183 |
| 0.740 | 0.88526 | 1.34328 | 1.74132 | 2.26050 | 2.64601 | 3.12730 |
| 0.760 | 0.89707 | 1.35568 | 1.75722 | 2.26934 | 2.65285 | 3.13290 |
| 0.780 | 0.90921 | 1.36810 | 1.77314 | 2.27860 | 2.65982 | 3.13864 |
| 0.800 | 0.92166 | 1.38056 | 1.78910 | 2.28827 | 2.66695 | 3.14450 |
| 0.820 | 0.93442 | 1.39308 | 1.80504 | 2.29837 | 2.67424 | 3.15050 |
| 0.840 | 0.94747 | 1.40565 | 1.82095 | 2.30890 | 2.68167 | 3.15664 |
| 0.860 | 0.96079 | 1.41826 | 1.83680 | 2.31985 | 2.68923 | 3.16292 |
| 0.880 | 0.97437 | 1.43094 | 1.85258 | 2.33125 | 2.69693 | 3.16934 |
| 0.900 | 0.98820 | 1.44368 | 1.86826 | 2.34307 | 2.70481 | 3.17592 |
| 0.920 | 1.00226 | 1.45649 | 1.88383 | 2.35533 | 2.71282 | 3.18263 |
| 0.940 | 1.01655 | 1.46936 | 1.89929 | 2.36802 | 2.72099 | 3.18951 |
| 0.960 | 1.03105 | 1.48230 | 1.91461 | 2.38112 | 2.72930 | 3.19655 |
| 0.980 | 1.04575 | 1.49532 | 1.92979 | 2.39464 | 2.73777 | 3.20377 |
| 1.000 | 1.06065 | 1.50840 | 1.94482 | 2.40856 | 2.74638 | 3.21116 |

$H/R = 0.010 \qquad n = 3.$

| $H/L$ | $\Omega_1$ | $\Omega_2$ | $\Omega_3$ | $\Omega_4$ | $\Omega_5$ | $\Omega_6$ |
|---|---|---|---|---|---|---|
| 0. | 0.000118 | 0.00955 | 0.01701 | 1.00005 | 1.00018 | 1.87061 |
| 0.001 | 0.000130 | 0.00962 | 0.01709 | 1.00006 | 1.00018 | 1.87061 |
| 0.005 | 0.00105 | 0.01107 | 0.01880 | 1.00007 | 1.00022 | 1.87054 |
| 0.010 | 0.00260 | 0.01419 | 0.02363 | 1.00011 | 1.00035 | 1.87035 |
| 0.020 | 0.00421 | 0.02230 | 0.03753 | 1.00027 | 1.00086 | 1.86958 |
| 0.030 | 0.00488 | 0.03154 | 0.05325 | 1.00052 | 1.00171 | 1.86833 |
| 0.040 | 0.00549 | 0.04115 | 0.06953 | 1.00087 | 1.00291 | 1.86663 |
| 0.050 | 0.00633 | 0.05092 | 0.08604 | 1.00132 | 1.00445 | 1.86453 |
| 0.060 | 0.00752 | 0.06076 | 0.10268 | 1.00187 | 1.00633 | 1.86207 |
| 0.070 | 0.00910 | 0.07065 | 0.11938 | 1.00252 | 1.00853 | 1.85930 |
| 0.080 | 0.01105 | 0.08057 | 0.13613 | 1.00327 | 1.01107 | 1.85627 |
| 0.090 | 0.01337 | 0.09051 | 0.15290 | 1.00411 | 1.01393 | 1.85305 |
| 0.100 | 0.01602 | 0.10046 | 0.16968 | 1.00505 | 1.01711 | 1.84963 |
| 0.120 | 0.02226 | 0.12038 | 0.20326 | 1.00724 | 1.02439 | 1.84243 |
| 0.140 | 0.02964 | 0.14033 | 0.23684 | 1.00982 | 1.03289 | 1.83491 |
| 0.160 | 0.03808 | 0.16029 | 0.27040 | 1.01279 | 1.04255 | 1.82725 |
| 0.180 | 0.04750 | 0.18025 | 0.30392 | 1.01614 | 1.05330 | 1.81957 |
| 0.200 | 0.05785 | 0.20023 | 0.33739 | 1.01987 | 1.06509 | 1.81198 |
| 0.220 | 0.06904 | 0.22021 | 0.37080 | 1.02398 | 1.07786 | 1.80454 |
| 0.240 | 0.08103 | 0.24019 | 0.40414 | 1.02846 | 1.09154 | 1.79733 |
| 0.260 | 0.09375 | 0.26018 | 0.43740 | 1.03331 | 1.10610 | 1.79037 |
| 0.280 | 0.10715 | 0.28016 | 0.47057 | 1.03852 | 1.12146 | 1.78373 |
| 0.300 | 0.12117 | 0.30015 | 0.50363 | 1.04409 | 1.13756 | 1.77743 |
| 0.320 | 0.13577 | 0.32015 | 0.53657 | 1.05002 | 1.15436 | 1.77150 |
| 0.340 | 0.15089 | 0.34014 | 0.56939 | 1.05629 | 1.17181 | 1.76596 |
| 0.360 | 0.16650 | 0.36013 | 0.60206 | 1.06289 | 1.18986 | 1.76086 |
| 0.380 | 0.18254 | 0.38012 | 0.63458 | 1.06983 | 1.20846 | 1.75621 |
| 0.400 | 0.19899 | 0.40012 | 0.66692 | 1.07710 | 1.22759 | 1.75202 |
| 0.420 | 0.21581 | 0.42011 | 0.69907 | 1.08469 | 1.24718 | 1.74834 |
| 0.440 | 0.23296 | 0.44011 | 0.73101 | 1.09259 | 1.26720 | 1.74518 |
| 0.460 | 0.25041 | 0.46010 | 0.76272 | 1.10079 | 1.28763 | 1.74258 |
| 0.480 | 0.26814 | 0.48010 | 0.79418 | 1.10930 | 1.30844 | 1.74056 |
| 0.500 | 0.28613 | 0.50009 | 0.82536 | 1.11810 | 1.32958 | 1.73913 |
| 0.520 | 0.30434 | 0.52009 | 0.85624 | 1.12719 | 1.35102 | 1.73833 |
| 0.540 | 0.32277 | 0.54009 | 0.88678 | 1.13654 | 1.37276 | 1.73821 |
| 0.560 | 0.34138 | 0.56008 | 0.91695 | 1.14618 | 1.39476 | 1.73880 |
| 0.580 | 0.36016 | 0.58008 | 0.94672 | 1.15608 | 1.41700 | 1.74012 |
| 0.600 | 0.37909 | 0.60008 | 0.97606 | 1.16625 | 1.43945 | 1.74220 |
| 0.620 | 0.39817 | 0.62008 | 1.00491 | 1.17667 | 1.46210 | 1.74511 |
| 0.640 | 0.41737 | 0.64008 | 1.03324 | 1.18732 | 1.48492 | 1.74888 |
| 0.660 | 0.43669 | 0.66007 | 1.06101 | 1.19822 | 1.50790 | 1.75354 |
| 0.680 | 0.45611 | 0.68007 | 1.08817 | 1.20935 | 1.53102 | 1.75916 |
| 0.700 | 0.47562 | 0.70007 | 1.11466 | 1.22072 | 1.55425 | 1.76576 |
| 0.720 | 0.49521 | 0.72007 | 1.14045 | 1.23229 | 1.57760 | 1.77342 |
| 0.740 | 0.51488 | 0.74006 | 1.16550 | 1.24408 | 1.60103 | 1.78215 |
| 0.760 | 0.53462 | 0.76006 | 1.18975 | 1.25608 | 1.62454 | 1.79200 |
| 0.780 | 0.55441 | 0.78006 | 1.21317 | 1.26829 | 1.64810 | 1.80302 |
| 0.800 | 0.57426 | 0.80006 | 1.23573 | 1.28068 | 1.67171 | 1.81521 |
| 0.820 | 0.59415 | 0.82006 | 1.25741 | 1.29327 | 1.69535 | 1.82861 |
| 0.840 | 0.61409 | 0.84006 | 1.27820 | 1.30604 | 1.71900 | 1.84322 |
| 0.860 | 0.63406 | 0.86005 | 1.29808 | 1.31900 | 1.74266 | 1.85906 |
| 0.880 | 0.65407 | 0.88006 | 1.31707 | 1.33212 | 1.76631 | 1.87608 |
| 0.900 | 0.67410 | 0.90006 | 1.33520 | 1.34541 | 1.78994 | 1.89427 |
| 0.920 | 0.69416 | 0.92005 | 1.35250 | 1.35888 | 1.81354 | 1.91361 |
| 0.940 | 0.71424 | 0.94005 | 1.36900 | 1.37249 | 1.83709 | 1.93404 |
| 0.960 | 0.73433 | 0.96005 | 1.38476 | 1.38627 | 1.86058 | 1.95551 |
| 0.980 | 0.75444 | 0.98005 | 1.39982 | 1.40019 | 1.88399 | 1.97794 |
| 1.000 | 0.77456 | 1.00005 | 1.41423 | 1.41429 | 1.90732 | 2.00129 |

$H/R = 0.020$ $n = 3.$ $H/R = 0.030$

| $H/L$ | $\Omega_1$ | $\Omega_2$ | $\Omega_3$ | $\Omega_1$ | $\Omega_2$ | $\Omega_3$ | $H/L$ |
|---|---|---|---|---|---|---|---|
| 0. | 0.000471 | 0.01910 | 0.03402 | 0.00106 | 0.02865 | 0.05103 | 0. |
| 0.001 | 0.000473 | 0.01914 | 0.03406 | 0.00106 | 0.02867 | 0.05105 | 0.001 |
| 0.005 | 0.000800 | 0.01996 | 0.03493 | 0.00117 | 0.02924 | 0.05163 | 0.005 |
| 0.010 | 0.00217 | 0.02213 | 0.03760 | 0.00197 | 0.03086 | 0.05344 | 0.010 |
| 0.020 | 0.00528 | 0.02838 | 0.04726 | 0.00506 | 0.03602 | 0.06039 | 0.020 |
| 0.030 | 0.00747 | 0.03604 | 0.06036 | 0.00812 | 0.04257 | 0.07088 | 0.030 |
| 0.040 | 0.00879 | 0.04461 | 0.07506 | 0.01033 | 0.05005 | 0.08361 | 0.040 |
| 0.050 | 0.00989 | 0.05370 | 0.09055 | 0.01241 | 0.05823 | 0.09768 | 0.050 |
| 0.060 | 0.01105 | 0.06308 | 0.10647 | 0.01405 | 0.06692 | 0.11256 | 0.060 |
| 0.070 | 0.01240 | 0.07263 | 0.12266 | 0.01563 | 0.07594 | 0.12796 | 0.070 |
| 0.080 | 0.01406 | 0.08230 | 0.13900 | 0.01734 | 0.08520 | 0.14369 | 0.080 |
| 0.090 | 0.01607 | 0.09204 | 0.15546 | 0.01928 | 0.09462 | 0.15965 | 0.090 |
| 0.100 | 0.01843 | 0.10184 | 0.17199 | 0.02149 | 0.10415 | 0.17578 | 0.100 |
| 0.120 | 0.02419 | 0.12153 | 0.20519 | 0.02688 | 0.12345 | 0.20837 | 0.120 |
| 0.140 | 0.03121 | 0.14131 | 0.23849 | 0.03354 | 0.14295 | 0.24123 | 0.140 |
| 0.160 | 0.03940 | 0.16114 | 0.27185 | 0.04142 | 0.16258 | 0.27424 | 0.160 |
| 0.180 | 0.04864 | 0.18102 | 0.30520 | 0.05041 | 0.18229 | 0.30734 | 0.180 |
| 0.200 | 0.05884 | 0.20091 | 0.33855 | 0.06041 | 0.20206 | 0.34046 | 0.200 |
| 0.220 | 0.06992 | 0.22083 | 0.37185 | 0.07134 | 0.22187 | 0.37359 | 0.220 |
| 0.240 | 0.08182 | 0.24076 | 0.40510 | 0.08310 | 0.24172 | 0.40670 | 0.240 |
| 0.260 | 0.09447 | 0.26070 | 0.43828 | 0.09565 | 0.26158 | 0.43975 | 0.260 |
| 0.280 | 0.10781 | 0.28065 | 0.47138 | 0.10889 | 0.28147 | 0.47274 | 0.280 |
| 0.300 | 0.12179 | 0.30061 | 0.50439 | 0.12279 | 0.30137 | 0.50565 | 0.300 |
| 0.320 | 0.13634 | 0.32057 | 0.53728 | 0.13728 | 0.32129 | 0.53847 | 0.320 |
| 0.340 | 0.15143 | 0.34054 | 0.57005 | 0.15231 | 0.34121 | 0.57116 | 0.340 |
| 0.360 | 0.16700 | 0.36051 | 0.60269 | 0.16783 | 0.36114 | 0.60373 | 0.360 |
| 0.380 | 0.18302 | 0.38048 | 0.63517 | 0.18380 | 0.38108 | 0.63615 | 0.380 |
| 0.400 | 0.19944 | 0.40046 | 0.66748 | 0.20019 | 0.40103 | 0.66841 | 0.400 |
| 0.420 | 0.21623 | 0.42044 | 0.69960 | 0.21694 | 0.42098 | 0.70048 | 0.420 |
| 0.440 | 0.23337 | 0.44042 | 0.73151 | 0.23404 | 0.44093 | 0.73236 | 0.440 |
| 0.460 | 0.25080 | 0.46040 | 0.76320 | 0.25145 | 0.46089 | 0.76400 | 0.460 |
| 0.480 | 0.26852 | 0.48038 | 0.79463 | 0.26914 | 0.48086 | 0.79539 | 0.480 |
| 0.500 | 0.28649 | 0.50037 | 0.82580 | 0.28708 | 0.50082 | 0.82651 | 0.500 |
| 0.520 | 0.30469 | 0.52035 | 0.85665 | 0.30526 | 0.52079 | 0.85733 | 0.520 |
| 0.540 | 0.32310 | 0.54034 | 0.88717 | 0.32364 | 0.54076 | 0.88783 | 0.540 |
| 0.560 | 0.34170 | 0.56033 | 0.91733 | 0.34222 | 0.56073 | 0.91795 | 0.560 |
| 0.580 | 0.36046 | 0.58032 | 0.94708 | 0.36097 | 0.58071 | 0.94768 | 0.580 |
| 0.600 | 0.37939 | 0.60031 | 0.97640 | 0.37988 | 0.60069 | 0.97696 | 0.600 |
| 0.620 | 0.39845 | 0.62030 | 1.00523 | 0.39893 | 0.62067 | 1.00578 | 0.620 |
| 0.640 | 0.41765 | 0.64029 | 1.03355 | 0.41811 | 0.64064 | 1.03407 | 0.640 |
| 0.660 | 0.43695 | 0.66028 | 1.06130 | 0.43740 | 0.66062 | 1.06180 | 0.660 |
| 0.680 | 0.45636 | 0.68027 | 1.08845 | 0.45680 | 0.68061 | 1.08892 | 0.680 |
| 0.700 | 0.47587 | 0.70026 | 1.11493 | 0.47629 | 0.70059 | 1.11538 | 0.700 |
| 0.720 | 0.49545 | 0.72026 | 1.14071 | 0.49586 | 0.72057 | 1.14113 | 0.720 |
| 0.740 | 0.51511 | 0.74025 | 1.16574 | 0.51551 | 0.74056 | 1.16614 | 0.740 |
| 0.760 | 0.53484 | 0.76024 | 1.18998 | 0.53523 | 0.76054 | 1.19035 | 0.760 |
| 0.780 | 0.55463 | 0.78024 | 1.21338 | 0.55501 | 0.78053 | 1.21374 | 0.780 |
| 0.800 | 0.57447 | 0.80023 | 1.23594 | 0.57484 | 0.80052 | 1.23627 | 0.800 |
| 0.820 | 0.59437 | 0.82023 | 1.25761 | 0.59472 | 0.82051 | 1.25792 | 0.820 |
| 0.840 | 0.61430 | 0.84022 | 1.27838 | 0.61464 | 0.84049 | 1.27867 | 0.840 |
| 0.860 | 0.63426 | 0.86021 | 1.29824 | 0.63460 | 0.86048 | 1.29853 | 0.860 |
| 0.880 | 0.65427 | 0.88021 | 1.31724 | 0.65459 | 0.88047 | 1.31749 | 0.880 |
| 0.900 | 0.67429 | 0.90020 | 1.33535 | 0.67461 | 0.90046 | 1.33560 | 0.900 |
| 0.920 | 0.69435 | 0.92020 | 1.35264 | 0.69466 | 0.92045 | 1.35286 | 0.920 |
| 0.940 | 0.71442 | 0.94020 | 1.36913 | 0.71473 | 0.94044 | 1.36934 | 0.940 |
| 0.960 | 0.73451 | 0.96019 | 1.38487 | 0.73481 | 0.96043 | 1.38504 | 0.960 |
| 0.980 | 0.75461 | 0.98019 | 1.39990 | 0.75491 | 0.98042 | 1.40001 | 0.980 |
| 1.000 | 0.77473 | 1.00018 | 1.41427 | 0.77502 | 1.00042 | 1.41432 | 1.000 |

$H/R = 0.040$ $n = 3.$ $H/R = 0.050$

| $H/L$ | $\Omega_1$ | $\Omega_2$ | $\Omega_3$ | $\Omega_1$ | $\Omega_2$ | $\Omega_3$ | $H/L$ |
|---|---|---|---|---|---|---|---|
| 0. | 0.00188 | 0.03820 | 0.06802 | 0.00294 | 0.04775 | 0.08500 | 0. |
| 0.001 | 0.00188 | 0.03822 | 0.06804 | 0.00294 | 0.04777 | 0.08502 | 0.001 |
| 0.005 | 0.00194 | 0.03865 | 0.06848 | 0.00298 | 0.04811 | 0.08536 | 0.005 |
| 0.010 | 0.00236 | 0.03992 | 0.06984 | 0.00322 | 0.04915 | 0.08646 | 0.010 |
| 0.020 | 0.00481 | 0.04427 | 0.07518 | 0.00498 | 0.05288 | 0.09077 | 0.020 |
| 0.030 | 0.00809 | 0.05007 | 0.08363 | 0.00803 | 0.05808 | 0.09775 | 0.030 |
| 0.040 | 0.01117 | 0.05676 | 0.09449 | 0.01136 | 0.06419 | 0.10702 | 0.040 |
| 0.050 | 0.01402 | 0.06414 | 0.10701 | 0.01451 | 0.07095 | 0.11808 | 0.050 |
| 0.060 | 0.01611 | 0.07209 | 0.12067 | 0.01730 | 0.07827 | 0.13048 | 0.060 |
| 0.070 | 0.01818 | 0.08049 | 0.13509 | 0.02000 | 0.08606 | 0.14384 | 0.070 |
| 0.080 | 0.02020 | 0.08922 | 0.15005 | 0.02244 | 0.09425 | 0.15791 | 0.080 |
| 0.090 | 0.02230 | 0.09821 | 0.16537 | 0.02488 | 0.10276 | 0.17251 | 0.090 |
| 0.100 | 0.02458 | 0.10740 | 0.18098 | 0.02738 | 0.11153 | 0.18749 | 0.100 |
| 0.120 | 0.02988 | 0.12616 | 0.21275 | 0.03286 | 0.12963 | 0.21828 | 0.120 |
| 0.140 | 0.03632 | 0.14527 | 0.24501 | 0.03927 | 0.14825 | 0.24981 | 0.140 |
| 0.160 | 0.04394 | 0.16460 | 0.27757 | 0.04675 | 0.16721 | 0.28179 | 0.160 |
| 0.180 | 0.05269 | 0.18408 | 0.31030 | 0.05531 | 0.18640 | 0.31407 | 0.180 |
| 0.200 | 0.06247 | 0.20367 | 0.34313 | 0.06491 | 0.20575 | 0.34654 | 0.200 |
| 0.220 | 0.07321 | 0.22334 | 0.37602 | 0.07547 | 0.22522 | 0.37912 | 0.220 |
| 0.240 | 0.08483 | 0.24305 | 0.40892 | 0.08692 | 0.24479 | 0.41176 | 0.240 |
| 0.260 | 0.09723 | 0.26282 | 0.44180 | 0.09919 | 0.26441 | 0.44442 | 0.260 |
| 0.280 | 0.11037 | 0.28262 | 0.47464 | 0.11220 | 0.28410 | 0.47707 | 0.280 |
| 0.300 | 0.12417 | 0.30244 | 0.50742 | 0.12588 | 0.30382 | 0.50968 | 0.300 |
| 0.320 | 0.13857 | 0.32229 | 0.54012 | 0.14018 | 0.32358 | 0.54224 | 0.320 |
| 0.340 | 0.15352 | 0.34215 | 0.57272 | 0.15505 | 0.34337 | 0.57470 | 0.340 |
| 0.360 | 0.16898 | 0.36203 | 0.60519 | 0.17042 | 0.36318 | 0.60706 | 0.360 |
| 0.380 | 0.18489 | 0.38193 | 0.63753 | 0.18626 | 0.38301 | 0.63930 | 0.380 |
| 0.400 | 0.20122 | 0.40183 | 0.66971 | 0.20252 | 0.40286 | 0.67138 | 0.400 |
| 0.420 | 0.21793 | 0.42174 | 0.70172 | 0.21917 | 0.42273 | 0.70330 | 0.420 |
| 0.440 | 0.23498 | 0.44166 | 0.73352 | 0.23617 | 0.44260 | 0.73503 | 0.440 |
| 0.460 | 0.25235 | 0.46159 | 0.76511 | 0.25349 | 0.46249 | 0.76653 | 0.460 |
| 0.480 | 0.27000 | 0.48152 | 0.79645 | 0.27109 | 0.48238 | 0.79781 | 0.480 |
| 0.500 | 0.28791 | 0.50146 | 0.82752 | 0.28896 | 0.50229 | 0.82881 | 0.500 |
| 0.520 | 0.30605 | 0.52141 | 0.85830 | 0.30706 | 0.52220 | 0.85953 | 0.520 |
| 0.540 | 0.32441 | 0.54136 | 0.88874 | 0.32538 | 0.54212 | 0.88991 | 0.540 |
| 0.560 | 0.34296 | 0.56131 | 0.91882 | 0.34390 | 0.56204 | 0.91994 | 0.560 |
| 0.580 | 0.36168 | 0.58126 | 0.94851 | 0.36259 | 0.58197 | 0.94957 | 0.580 |
| 0.600 | 0.38057 | 0.60122 | 0.97776 | 0.38144 | 0.60191 | 0.97878 | 0.600 |
| 0.620 | 0.39959 | 0.62118 | 1.00653 | 0.40044 | 0.62184 | 1.00750 | 0.620 |
| 0.640 | 0.41875 | 0.64114 | 1.03479 | 0.41957 | 0.64179 | 1.03572 | 0.640 |
| 0.660 | 0.43802 | 0.66111 | 1.06248 | 0.43881 | 0.66173 | 1.06337 | 0.660 |
| 0.680 | 0.45740 | 0.68107 | 1.08957 | 0.45817 | 0.68168 | 1.09040 | 0.680 |
| 0.700 | 0.47687 | 0.70105 | 1.11599 | 0.47762 | 0.70163 | 1.11679 | 0.700 |
| 0.720 | 0.49643 | 0.72102 | 1.14172 | 0.49715 | 0.72159 | 1.14247 | 0.720 |
| 0.740 | 0.51606 | 0.74099 | 1.16669 | 0.51677 | 0.74155 | 1.16741 | 0.740 |
| 0.760 | 0.53576 | 0.76097 | 1.19088 | 0.53645 | 0.76151 | 1.19155 | 0.760 |
| 0.780 | 0.55553 | 0.78094 | 1.21424 | 0.55620 | 0.78147 | 1.21487 | 0.780 |
| 0.800 | 0.57534 | 0.80091 | 1.23674 | 0.57600 | 0.80143 | 1.23733 | 0.800 |
| 0.820 | 0.59521 | 0.82089 | 1.25836 | 0.59584 | 0.82139 | 1.25892 | 0.820 |
| 0.840 | 0.61512 | 0.84087 | 1.27908 | 0.61574 | 0.84136 | 1.27961 | 0.840 |
| 0.860 | 0.63507 | 0.86085 | 1.29892 | 0.63567 | 0.86133 | 1.29940 | 0.860 |
| 0.880 | 0.65505 | 0.88083 | 1.31785 | 0.65563 | 0.88130 | 1.31831 | 0.880 |
| 0.900 | 0.67506 | 0.90081 | 1.33593 | 0.67563 | 0.90127 | 1.33634 | 0.900 |
| 0.920 | 0.69509 | 0.92080 | 1.35316 | 0.69565 | 0.92124 | 1.35354 | 0.920 |
| 0.940 | 0.71515 | 0.94078 | 1.36960 | 0.71570 | 0.94122 | 1.36990 | 0.940 |
| 0.960 | 0.73522 | 0.96076 | 1.38524 | 0.73576 | 0.96119 | 1.38545 | 0.960 |
| 0.980 | 0.75532 | 0.98075 | 1.40011 | 0.75583 | 0.98117 | 1.40023 | 0.980 |
| 1.000 | 0.77541 | 1.00073 | 1.41440 | 0.77593 | 1.00115 | 1.41450 | 1.000 |

$H/R = 0.060$ $n = 3.$ $H/R = 0.070$

| $H/L$ | $\Omega_1$ | $\Omega_2$ | $\Omega_3$ | $\Omega_1$ | $\Omega_2$ | $\Omega_3$ | $H/L$ |
|---|---|---|---|---|---|---|---|
| 0. | 0.00422 | 0.05730 | 0.10196 | 0.00574 | 0.06686 | 0.11890 | 0. |
| 0.001 | 0.00422 | 0.05732 | 0.10197 | 0.00574 | 0.06687 | 0.11891 | 0.001 |
| 0.005 | 0.00426 | 0.05760 | 0.10227 | 0.00577 | 0.06712 | 0.11916 | 0.005 |
| 0.010 | 0.00443 | 0.05848 | 0.10318 | 0.00590 | 0.06787 | 0.11994 | 0.010 |
| 0.020 | 0.00569 | 0.06172 | 0.10679 | 0.00683 | 0.07072 | 0.12305 | 0.020 |
| 0.030 | 0.00826 | 0.06641 | 0.11270 | 0.00895 | 0.07497 | 0.12817 | 0.030 |
| 0.040 | 0.01153 | 0.07204 | 0.12070 | 0.01193 | 0.08019 | 0.13516 | 0.040 |
| 0.050 | 0.01490 | 0.07833 | 0.13047 | 0.01531 | 0.08608 | 0.14383 | 0.050 |
| 0.060 | 0.01816 | 0.08515 | 0.14166 | 0.01877 | 0.09249 | 0.15394 | 0.060 |
| 0.070 | 0.02105 | 0.09242 | 0.15396 | 0.02216 | 0.09934 | 0.16522 | 0.070 |
| 0.080 | 0.02409 | 0.10010 | 0.16711 | 0.02544 | 0.10659 | 0.17745 | 0.080 |
| 0.090 | 0.02698 | 0.10814 | 0.18090 | 0.02878 | 0.11419 | 0.19043 | 0.090 |
| 0.100 | 0.02976 | 0.11647 | 0.19520 | 0.03180 | 0.12210 | 0.20400 | 0.100 |
| 0.120 | 0.03565 | 0.13384 | 0.22489 | 0.03816 | 0.13873 | 0.23249 | 0.120 |
| 0.140 | 0.04221 | 0.15189 | 0.25557 | 0.04501 | 0.15616 | 0.26223 | 0.140 |
| 0.160 | 0.04968 | 0.17040 | 0.28688 | 0.05260 | 0.17417 | 0.29280 | 0.160 |
| 0.180 | 0.05815 | 0.18924 | 0.31863 | 0.06107 | 0.19259 | 0.32394 | 0.180 |
| 0.200 | 0.06761 | 0.20831 | 0.35066 | 0.07048 | 0.21133 | 0.35547 | 0.200 |
| 0.220 | 0.07803 | 0.22754 | 0.38287 | 0.08079 | 0.23029 | 0.38727 | 0.220 |
| 0.240 | 0.08933 | 0.24691 | 0.41520 | 0.09198 | 0.24942 | 0.41924 | 0.240 |
| 0.260 | 0.10146 | 0.26637 | 0.44760 | 0.10398 | 0.26869 | 0.45134 | 0.260 |
| 0.280 | 0.11434 | 0.28591 | 0.48003 | 0.11674 | 0.28806 | 0.48349 | 0.280 |
| 0.300 | 0.12790 | 0.30551 | 0.51244 | 0.13019 | 0.30752 | 0.51568 | 0.300 |
| 0.320 | 0.14209 | 0.32517 | 0.54481 | 0.14427 | 0.32704 | 0.54784 | 0.320 |
| 0.340 | 0.15686 | 0.34486 | 0.57712 | 0.15893 | 0.34663 | 0.57996 | 0.340 |
| 0.360 | 0.17214 | 0.36459 | 0.60934 | 0.17412 | 0.36625 | 0.61201 | 0.360 |
| 0.380 | 0.18790 | 0.38434 | 0.64144 | 0.18979 | 0.38592 | 0.64397 | 0.380 |
| 0.400 | 0.20409 | 0.40413 | 0.67341 | 0.20590 | 0.40562 | 0.67581 | 0.400 |
| 0.420 | 0.22067 | 0.42393 | 0.70522 | 0.22240 | 0.42535 | 0.70749 | 0.420 |
| 0.440 | 0.23760 | 0.44375 | 0.73685 | 0.23927 | 0.44511 | 0.73900 | 0.440 |
| 0.460 | 0.25486 | 0.46358 | 0.76827 | 0.25646 | 0.46488 | 0.77032 | 0.460 |
| 0.480 | 0.27241 | 0.48343 | 0.79946 | 0.27395 | 0.48468 | 0.80141 | 0.480 |
| 0.500 | 0.29023 | 0.50330 | 0.83039 | 0.29171 | 0.50449 | 0.83224 | 0.500 |
| 0.520 | 0.30828 | 0.52317 | 0.86102 | 0.30971 | 0.52432 | 0.86279 | 0.520 |
| 0.540 | 0.32656 | 0.54305 | 0.89134 | 0.32793 | 0.54416 | 0.89302 | 0.540 |
| 0.560 | 0.34503 | 0.56294 | 0.92130 | 0.34636 | 0.56401 | 0.92291 | 0.560 |
| 0.580 | 0.36369 | 0.58284 | 0.95087 | 0.36497 | 0.58387 | 0.95240 | 0.580 |
| 0.600 | 0.38250 | 0.60274 | 0.98001 | 0.38374 | 0.60374 | 0.98147 | 0.600 |
| 0.620 | 0.40146 | 0.62266 | 1.00869 | 0.40267 | 0.62362 | 1.01008 | 0.620 |
| 0.640 | 0.42056 | 0.64258 | 1.03684 | 0.42173 | 0.64351 | 1.03816 | 0.640 |
| 0.660 | 0.43978 | 0.66250 | 1.06443 | 0.44090 | 0.66340 | 1.06569 | 0.660 |
| 0.680 | 0.45910 | 0.68242 | 1.09142 | 0.46020 | 0.68330 | 1.09261 | 0.680 |
| 0.700 | 0.47852 | 0.70235 | 1.11775 | 0.47959 | 0.70320 | 1.11888 | 0.700 |
| 0.720 | 0.49804 | 0.72229 | 1.14339 | 0.49907 | 0.72311 | 1.14445 | 0.720 |
| 0.740 | 0.51762 | 0.74222 | 1.16827 | 0.51863 | 0.74303 | 1.16928 | 0.740 |
| 0.760 | 0.53728 | 0.76217 | 1.19236 | 0.53826 | 0.76295 | 1.19333 | 0.760 |
| 0.780 | 0.55701 | 0.78211 | 1.21564 | 0.55796 | 0.78287 | 1.21654 | 0.780 |
| 0.800 | 0.57679 | 0.80206 | 1.23806 | 0.57771 | 0.80280 | 1.23890 | 0.800 |
| 0.820 | 0.59662 | 0.82201 | 1.25959 | 0.59752 | 0.82273 | 1.26037 | 0.820 |
| 0.840 | 0.61649 | 0.84196 | 1.28024 | 0.61737 | 0.84267 | 1.28096 | 0.840 |
| 0.860 | 0.63640 | 0.86192 | 1.29998 | 0.63726 | 0.86261 | 1.30065 | 0.860 |
| 0.880 | 0.65635 | 0.88187 | 1.31884 | 0.65719 | 0.88255 | 1.31944 | 0.880 |
| 0.900 | 0.67633 | 0.90183 | 1.33682 | 0.67715 | 0.90249 | 1.33736 | 0.900 |
| 0.920 | 0.69633 | 0.92179 | 1.35395 | 0.69713 | 0.92243 | 1.35439 | 0.920 |
| 0.940 | 0.71636 | 0.94175 | 1.37023 | 0.71714 | 0.94239 | 1.37056 | 0.940 |
| 0.960 | 0.73641 | 0.96172 | 1.38566 | 0.73718 | 0.96234 | 1.38588 | 0.960 |
| 0.980 | 0.75647 | 0.98168 | 1.40037 | 0.75722 | 0.98229 | 1.40054 | 0.980 |
| 1.000 | 0.77655 | 1.00165 | 1.41463 | 0.77728 | 1.00224 | 1.41479 | 1.000 |

$H/R = 0.080$ $n = 3.$ $H/R = 0.090$

| $H/L$ | $\Omega_1$ | $\Omega_2$ | $\Omega_3$ | $\Omega_1$ | $\Omega_2$ | $\Omega_3$ | $H/L$ |
|---|---|---|---|---|---|---|---|
| 0. | 0.00748 | 0.07641 | 0.13582 | 0.00945 | 0.08597 | 0.15271 | 0. |
| 0.001 | 0.00748 | 0.07642 | 0.13583 | 0.00945 | 0.08598 | 0.15272 | 0.001 |
| 0.005 | 0.00751 | 0.07664 | 0.13605 | 0.00948 | 0.08617 | 0.15291 | 0.005 |
| 0.010 | 0.00763 | 0.07731 | 0.13673 | 0.00958 | 0.08677 | 0.15352 | 0.010 |
| 0.020 | 0.00833 | 0.07985 | 0.13946 | 0.01017 | 0.08906 | 0.15594 | 0.020 |
| 0.030 | 0.01005 | 0.08372 | 0.14396 | 0.01157 | 0.09260 | 0.15996 | 0.030 |
| 0.040 | 0.01269 | 0.08856 | 0.15015 | 0.01386 | 0.09710 | 0.16551 | 0.040 |
| 0.050 | 0.01592 | 0.09410 | 0.15791 | 0.01685 | 0.10232 | 0.17251 | 0.050 |
| 0.060 | 0.01942 | 0.10017 | 0.16706 | 0.02026 | 0.10809 | 0.18084 | 0.060 |
| 0.070 | 0.02299 | 0.10666 | 0.17741 | 0.02388 | 0.11428 | 0.19034 | 0.070 |
| 0.080 | 0.02654 | 0.11354 | 0.18875 | 0.02758 | 0.12084 | 0.20087 | 0.080 |
| 0.090 | 0.03003 | 0.12076 | 0.20092 | 0.03129 | 0.12774 | 0.21226 | 0.090 |
| 0.100 | 0.03366 | 0.12830 | 0.21377 | 0.03499 | 0.13495 | 0.22438 | 0.100 |
| 0.120 | 0.04043 | 0.14420 | 0.24101 | 0.04239 | 0.15018 | 0.25036 | 0.120 |
| 0.140 | 0.04763 | 0.16100 | 0.26975 | 0.05008 | 0.16636 | 0.27806 | 0.140 |
| 0.160 | 0.05545 | 0.17847 | 0.29951 | 0.05817 | 0.18328 | 0.30696 | 0.160 |
| 0.180 | 0.06402 | 0.19645 | 0.32998 | 0.06691 | 0.20079 | 0.33671 | 0.180 |
| 0.200 | 0.07342 | 0.21481 | 0.36096 | 0.07640 | 0.21875 | 0.36708 | 0.200 |
| 0.220 | 0.08370 | 0.23346 | 0.39228 | 0.08667 | 0.23705 | 0.39790 | 0.220 |
| 0.240 | 0.09480 | 0.25233 | 0.42386 | 0.09774 | 0.25563 | 0.42904 | 0.240 |
| 0.260 | 0.10671 | 0.27137 | 0.45561 | 0.10958 | 0.27442 | 0.46041 | 0.260 |
| 0.280 | 0.11936 | 0.29055 | 0.48747 | 0.12215 | 0.29338 | 0.49193 | 0.280 |
| 0.300 | 0.13270 | 0.30984 | 0.51938 | 0.13540 | 0.31248 | 0.52355 | 0.300 |
| 0.320 | 0.14668 | 0.32922 | 0.55132 | 0.14929 | 0.33169 | 0.55522 | 0.320 |
| 0.340 | 0.16124 | 0.34867 | 0.58323 | 0.16375 | 0.35099 | 0.58690 | 0.340 |
| 0.360 | 0.17634 | 0.36818 | 0.61509 | 0.17876 | 0.37038 | 0.61855 | 0.360 |
| 0.380 | 0.19191 | 0.38775 | 0.64688 | 0.19424 | 0.38982 | 0.65014 | 0.380 |
| 0.400 | 0.20793 | 0.40736 | 0.67855 | 0.21018 | 0.40933 | 0.68165 | 0.400 |
| 0.420 | 0.22436 | 0.42700 | 0.71010 | 0.22652 | 0.42888 | 0.71303 | 0.420 |
| 0.440 | 0.24115 | 0.44668 | 0.74148 | 0.24323 | 0.44847 | 0.74426 | 0.440 |
| 0.460 | 0.25827 | 0.46639 | 0.77267 | 0.26027 | 0.46810 | 0.77532 | 0.460 |
| 0.480 | 0.27569 | 0.48612 | 0.80364 | 0.27763 | 0.48776 | 0.80616 | 0.480 |
| 0.500 | 0.29339 | 0.50587 | 0.83437 | 0.29526 | 0.50745 | 0.83677 | 0.500 |
| 0.520 | 0.31133 | 0.52564 | 0.86482 | 0.31314 | 0.52716 | 0.86711 | 0.520 |
| 0.540 | 0.32950 | 0.54543 | 0.89496 | 0.33125 | 0.54689 | 0.89714 | 0.540 |
| 0.560 | 0.34787 | 0.56524 | 0.92475 | 0.34956 | 0.56664 | 0.92683 | 0.560 |
| 0.580 | 0.36643 | 0.58506 | 0.95416 | 0.36807 | 0.58641 | 0.95614 | 0.580 |
| 0.600 | 0.38516 | 0.60489 | 0.98315 | 0.38675 | 0.60619 | 0.98504 | 0.600 |
| 0.620 | 0.40404 | 0.62473 | 1.01167 | 0.40558 | 0.62600 | 1.01346 | 0.620 |
| 0.640 | 0.42306 | 0.64458 | 1.03967 | 0.42455 | 0.64581 | 1.04139 | 0.640 |
| 0.660 | 0.44220 | 0.66444 | 1.06713 | 0.44365 | 0.66563 | 1.06876 | 0.660 |
| 0.680 | 0.46145 | 0.68431 | 1.09398 | 0.46286 | 0.68546 | 1.09553 | 0.680 |
| 0.700 | 0.48080 | 0.70419 | 1.12018 | 0.48217 | 0.70531 | 1.12164 | 0.700 |
| 0.720 | 0.50025 | 0.72407 | 1.14569 | 0.50158 | 0.72516 | 1.14706 | 0.720 |
| 0.740 | 0.51978 | 0.74396 | 1.17044 | 0.52107 | 0.74502 | 1.17174 | 0.740 |
| 0.760 | 0.53938 | 0.76386 | 1.19442 | 0.54064 | 0.76489 | 1.19563 | 0.760 |
| 0.780 | 0.55905 | 0.78376 | 1.21756 | 0.56028 | 0.78476 | 1.21870 | 0.780 |
| 0.800 | 0.57878 | 0.80366 | 1.23985 | 0.57997 | 0.80464 | 1.24091 | 0.800 |
| 0.820 | 0.59855 | 0.82357 | 1.26126 | 0.59972 | 0.82452 | 1.26224 | 0.820 |
| 0.840 | 0.61838 | 0.84349 | 1.28178 | 0.61952 | 0.84442 | 1.28268 | 0.840 |
| 0.860 | 0.63825 | 0.86341 | 1.30140 | 0.63936 | 0.86431 | 1.30221 | 0.860 |
| 0.880 | 0.65815 | 0.88333 | 1.32011 | 0.65924 | 0.88422 | 1.32083 | 0.880 |
| 0.900 | 0.67809 | 0.90325 | 1.33794 | 0.67915 | 0.90412 | 1.33853 | 0.900 |
| 0.920 | 0.69806 | 0.92319 | 1.35485 | 0.69909 | 0.92403 | 1.35533 | 0.920 |
| 0.940 | 0.71805 | 0.94312 | 1.37089 | 0.71906 | 0.94395 | 1.37124 | 0.940 |
| 0.960 | 0.73806 | 0.96305 | 1.38611 | 0.73905 | 0.96387 | 1.38636 | 0.960 |
| 0.980 | 0.75808 | 0.98300 | 1.40072 | 0.75905 | 0.98379 | 1.40093 | 0.980 |
| 1.000 | 0.77812 | 1.00293 | 1.41496 | 0.77908 | 1.00371 | 1.41516 | 1.000 |

$H/R = 0.100 \qquad n = 3.$

| $H/L$ | $\Omega_1$ | $\Omega_2$ | $\Omega_3$ | $\Omega_4$ | $\Omega_5$ | $\Omega_6$ |
|---|---|---|---|---|---|---|
| 0. | 0.01163 | 0.09553 | 0.16956 | 1.00495 | 1.01754 | 1.85211 |
| 0.001 | 0.01163 | 0.09554 | 0.16957 | 1.00495 | 1.01753 | 1.85211 |
| 0.005 | 0.01166 | 0.09571 | 0.16975 | 1.00497 | 1.01757 | 1.85206 |
| 0.010 | 0.01176 | 0.09625 | 0.17029 | 1.00502 | 1.01769 | 1.85193 |
| 0.020 | 0.01230 | 0.09833 | 0.17248 | 1.00521 | 1.01815 | 1.85140 |
| 0.030 | 0.01344 | 0.10159 | 0.17611 | 1.00552 | 1.01891 | 1.85051 |
| 0.040 | 0.01543 | 0.10579 | 0.18113 | 1.00596 | 1.02000 | 1.84930 |
| 0.050 | 0.01814 | 0.11072 | 0.18749 | 1.00650 | 1.02140 | 1.84778 |
| 0.060 | 0.02138 | 0.11620 | 0.19511 | 1.00715 | 1.02312 | 1.84597 |
| 0.070 | 0.02496 | 0.12213 | 0.20387 | 1.00790 | 1.02517 | 1.84390 |
| 0.080 | 0.02871 | 0.12842 | 0.21364 | 1.00874 | 1.02754 | 1.84160 |
| 0.090 | 0.03255 | 0.13503 | 0.22431 | 1.00967 | 1.03023 | 1.83909 |
| 0.100 | 0.03643 | 0.14195 | 0.23573 | 1.01070 | 1.03325 | 1.83640 |
| 0.120 | 0.04424 | 0.15658 | 0.26046 | 1.01302 | 1.04022 | 1.83062 |
| 0.140 | 0.05226 | 0.17217 | 0.28709 | 1.01570 | 1.04839 | 1.82439 |
| 0.160 | 0.06079 | 0.18855 | 0.31509 | 1.01873 | 1.05772 | 1.81791 |
| 0.180 | 0.06973 | 0.20558 | 0.34409 | 1.02213 | 1.06816 | 1.81130 |
| 0.200 | 0.07935 | 0.22311 | 0.37382 | 1.02590 | 1.07963 | 1.80466 |
| 0.220 | 0.08968 | 0.24105 | 0.40409 | 1.03003 | 1.09208 | 1.79808 |
| 0.240 | 0.10076 | 0.25931 | 0.43476 | 1.03452 | 1.10546 | 1.79163 |
| 0.260 | 0.11256 | 0.27782 | 0.46571 | 1.03937 | 1.11970 | 1.78538 |
| 0.280 | 0.12508 | 0.29654 | 0.49687 | 1.04458 | 1.13474 | 1.77936 |
| 0.300 | 0.13826 | 0.31543 | 0.52817 | 1.05014 | 1.15054 | 1.77363 |
| 0.320 | 0.15206 | 0.33446 | 0.55956 | 1.05604 | 1.16704 | 1.76822 |
| 0.340 | 0.16644 | 0.35360 | 0.59098 | 1.06229 | 1.18419 | 1.76317 |
| 0.360 | 0.18136 | 0.37283 | 0.62240 | 1.06887 | 1.20194 | 1.75850 |
| 0.380 | 0.19676 | 0.39215 | 0.65378 | 1.07578 | 1.22026 | 1.75426 |
| 0.400 | 0.21261 | 0.41153 | 0.68509 | 1.08302 | 1.23909 | 1.75047 |
| 0.420 | 0.22886 | 0.43098 | 0.71629 | 1.09057 | 1.25841 | 1.74715 |
| 0.440 | 0.24549 | 0.45047 | 0.74736 | 1.09844 | 1.27818 | 1.74433 |
| 0.460 | 0.26246 | 0.47001 | 0.77826 | 1.10661 | 1.29834 | 1.74205 |
| 0.480 | 0.27974 | 0.48959 | 0.80897 | 1.11508 | 1.31890 | 1.74033 |
| 0.500 | 0.29730 | 0.50920 | 0.83944 | 1.12384 | 1.33979 | 1.73920 |
| 0.520 | 0.31512 | 0.52885 | 0.86965 | 1.13289 | 1.36101 | 1.73869 |
| 0.540 | 0.33317 | 0.54852 | 0.89956 | 1.14222 | 1.38251 | 1.73885 |
| 0.560 | 0.35142 | 0.56821 | 0.92914 | 1.15181 | 1.40428 | 1.73971 |
| 0.580 | 0.36987 | 0.58793 | 0.95834 | 1.16168 | 1.42631 | 1.74129 |
| 0.600 | 0.38849 | 0.60766 | 0.98713 | 1.17180 | 1.44855 | 1.74363 |
| 0.620 | 0.40727 | 0.62741 | 1.01545 | 1.18218 | 1.47099 | 1.74680 |
| 0.640 | 0.42620 | 0.64717 | 1.04328 | 1.19280 | 1.49361 | 1.75082 |
| 0.660 | 0.44525 | 0.66696 | 1.07055 | 1.20366 | 1.51640 | 1.75574 |
| 0.680 | 0.46441 | 0.68675 | 1.09723 | 1.21476 | 1.53932 | 1.76161 |
| 0.700 | 0.48369 | 0.70656 | 1.12325 | 1.22609 | 1.56237 | 1.76847 |
| 0.720 | 0.50306 | 0.72637 | 1.14858 | 1.23764 | 1.58554 | 1.77637 |
| 0.740 | 0.52251 | 0.74620 | 1.17317 | 1.24940 | 1.60879 | 1.78535 |
| 0.760 | 0.54204 | 0.76604 | 1.19697 | 1.26138 | 1.63212 | 1.79545 |
| 0.780 | 0.56164 | 0.78588 | 1.21995 | 1.27357 | 1.65552 | 1.80670 |
| 0.800 | 0.58130 | 0.80573 | 1.24208 | 1.28598 | 1.67896 | 1.81914 |
| 0.820 | 0.60102 | 0.82559 | 1.26332 | 1.29857 | 1.70243 | 1.70244 |
| 0.840 | 0.62079 | 0.84546 | 1.28365 | 1.31138 | 1.72593 | 1.84760 |
| 0.860 | 0.64060 | 0.86533 | 1.30307 | 1.32441 | 1.74943 | 1.86364 |
| 0.880 | 0.66044 | 0.88521 | 1.32158 | 1.33767 | 1.77292 | 1.88086 |
| 0.900 | 0.68033 | 0.90509 | 1.33916 | 1.35117 | 1.79639 | 1.89924 |
| 0.920 | 0.70024 | 0.92498 | 1.35580 | 1.36495 | 1.81984 | 1.91874 |
| 0.940 | 0.72019 | 0.94487 | 1.37158 | 1.37900 | 1.84323 | 1.93931 |
| 0.960 | 0.74015 | 0.96477 | 1.38663 | 1.39320 | 1.86656 | 1.96090 |
| 0.980 | 0.76013 | 0.98467 | 1.40116 | 1.40740 | 1.88983 | 1.98345 |
| 1.000 | 0.78013 | 1.00458 | 1.41537 | 1.42146 | 1.91302 | 2.00688 |

$H/R = 0.120$ $\quad n = 3. \quad$ $H/R = 0.140$

| $H/L$ | $\Omega_1$ | $\Omega_2$ | $\Omega_3$ | $\Omega_1$ | $\Omega_2$ | $\Omega_3$ | $H/L$ |
|---|---|---|---|---|---|---|---|
| 0. | 0.01666 | 0.11466 | 0.20316 | 0.02252 | 0.13379 | 0.23659 | 0. |
| 0.001 | 0.01666 | 0.11466 | 0.20317 | 0.02252 | 0.13380 | 0.23659 | 0.001 |
| 0.005 | 0.01668 | 0.11481 | 0.20332 | 0.02255 | 0.13392 | 0.23672 | 0.005 |
| 0.010 | 0.01677 | 0.11526 | 0.20377 | 0.02263 | 0.13431 | 0.23711 | 0.010 |
| 0.020 | 0.01720 | 0.11701 | 0.20560 | 0.02304 | 0.13583 | 0.23868 | 0.020 |
| 0.030 | 0.01811 | 0.11981 | 0.20863 | 0.02377 | 0.13827 | 0.24129 | 0.030 |
| 0.040 | 0.01962 | 0.12350 | 0.21285 | 0.02502 | 0.14154 | 0.24492 | 0.040 |
| 0.050 | 0.02182 | 0.12790 | 0.21822 | 0.02683 | 0.14550 | 0.24956 | 0.050 |
| 0.060 | 0.02462 | 0.13289 | 0.22470 | 0.02921 | 0.15005 | 0.25519 | 0.060 |
| 0.070 | 0.02791 | 0.13833 | 0.23223 | 0.03212 | 0.15507 | 0.26175 | 0.070 |
| 0.080 | 0.03155 | 0.14416 | 0.24071 | 0.03548 | 0.16049 | 0.26920 | 0.080 |
| 0.090 | 0.03543 | 0.15030 | 0.25008 | 0.03919 | 0.16624 | 0.27749 | 0.090 |
| 0.100 | 0.03948 | 0.15673 | 0.26024 | 0.04318 | 0.17227 | 0.28656 | 0.100 |
| 0.120 | 0.04789 | 0.17037 | 0.28257 | 0.05174 | 0.18509 | 0.30674 | 0.120 |
| 0.140 | 0.05656 | 0.18492 | 0.30708 | 0.06085 | 0.19880 | 0.32923 | 0.140 |
| 0.160 | 0.06571 | 0.20029 | 0.33324 | 0.07033 | 0.21329 | 0.35357 | 0.160 |
| 0.180 | 0.07512 | 0.21636 | 0.36066 | 0.08034 | 0.22850 | 0.37935 | 0.180 |
| 0.200 | 0.08512 | 0.23303 | 0.38901 | 0.09069 | 0.24433 | 0.40627 | 0.200 |
| 0.220 | 0.09568 | 0.25019 | 0.41809 | 0.10162 | 0.26072 | 0.43408 | 0.220 |
| 0.240 | 0.10688 | 0.26776 | 0.44772 | 0.11300 | 0.27758 | 0.46259 | 0.240 |
| 0.260 | 0.11872 | 0.28567 | 0.47777 | 0.12497 | 0.29483 | 0.49164 | 0.260 |
| 0.280 | 0.13120 | 0.30385 | 0.50813 | 0.13752 | 0.31243 | 0.52110 | 0.280 |
| 0.300 | 0.14430 | 0.32227 | 0.53871 | 0.15062 | 0.33032 | 0.55088 | 0.300 |
| 0.320 | 0.15799 | 0.34087 | 0.56945 | 0.16427 | 0.34844 | 0.58090 | 0.320 |
| 0.340 | 0.17224 | 0.35964 | 0.60030 | 0.17845 | 0.36678 | 0.61110 | 0.340 |
| 0.360 | 0.18701 | 0.37854 | 0.63119 | 0.19311 | 0.38529 | 0.64140 | 0.360 |
| 0.380 | 0.20226 | 0.39755 | 0.66210 | 0.20824 | 0.40396 | 0.67176 | 0.380 |
| 0.400 | 0.21795 | 0.41667 | 0.69297 | 0.22381 | 0.42275 | 0.70212 | 0.400 |
| 0.420 | 0.23405 | 0.43586 | 0.72377 | 0.23977 | 0.44166 | 0.73246 | 0.420 |
| 0.440 | 0.25053 | 0.45513 | 0.75446 | 0.25611 | 0.46066 | 0.76272 | 0.440 |
| 0.460 | 0.26734 | 0.47447 | 0.78501 | 0.27279 | 0.47976 | 0.79288 | 0.460 |
| 0.480 | 0.28447 | 0.49386 | 0.81539 | 0.28978 | 0.49893 | 0.82288 | 0.480 |
| 0.500 | 0.30189 | 0.51330 | 0.84556 | 0.30705 | 0.51816 | 0.85269 | 0.500 |
| 0.520 | 0.31957 | 0.53278 | 0.87548 | 0.32459 | 0.53745 | 0.88227 | 0.520 |
| 0.540 | 0.33748 | 0.55230 | 0.90512 | 0.34237 | 0.55680 | 0.91159 | 0.540 |
| 0.560 | 0.35561 | 0.57186 | 0.93444 | 0.36037 | 0.57619 | 0.94059 | 0.560 |
| 0.580 | 0.37394 | 0.59145 | 0.96338 | 0.37858 | 0.59563 | 0.96924 | 0.580 |
| 0.600 | 0.39245 | 0.61106 | 0.99193 | 0.39696 | 0.61510 | 0.99751 | 0.600 |
| 0.620 | 0.41112 | 0.63070 | 1.02002 | 0.41551 | 0.63461 | 1.02531 | 0.620 |
| 0.640 | 0.42993 | 0.65036 | 1.04761 | 0.43421 | 0.65415 | 1.05264 | 0.640 |
| 0.660 | 0.44888 | 0.67005 | 1.07467 | 0.45305 | 0.67371 | 1.07942 | 0.660 |
| 0.680 | 0.46795 | 0.68975 | 1.10112 | 0.47201 | 0.69330 | 1.10561 | 0.680 |
| 0.700 | 0.48713 | 0.70947 | 1.12693 | 0.49109 | 0.71292 | 1.13117 | 0.700 |
| 0.720 | 0.50641 | 0.72920 | 1.15205 | 0.51027 | 0.73256 | 1.15602 | 0.720 |
| 0.740 | 0.52578 | 0.74895 | 1.17642 | 0.52954 | 0.75221 | 1.18014 | 0.740 |
| 0.760 | 0.54522 | 0.76871 | 1.20002 | 0.54890 | 0.77189 | 1.20346 | 0.760 |
| 0.780 | 0.56475 | 0.78849 | 1.22277 | 0.56833 | 0.79158 | 1.22595 | 0.780 |
| 0.800 | 0.58433 | 0.80827 | 1.24467 | 0.58784 | 0.81129 | 1.24757 | 0.800 |
| 0.820 | 0.60397 | 0.82807 | 1.26569 | 0.60740 | 0.83101 | 1.26829 | 0.820 |
| 0.840 | 0.62367 | 0.84788 | 1.28578 | 0.62702 | 0.85074 | 1.28807 | 0.840 |
| 0.860 | 0.64342 | 0.86769 | 1.30494 | 0.64669 | 0.87050 | 1.30691 | 0.860 |
| 0.880 | 0.66320 | 0.88752 | 1.32316 | 0.66640 | 0.89025 | 1.32477 | 0.880 |
| 0.900 | 0.68302 | 0.90735 | 1.34042 | 0.68615 | 0.91002 | 1.34171 | 0.900 |
| 0.920 | 0.70288 | 0.92719 | 1.35677 | 0.70594 | 0.92980 | 1.35777 | 0.920 |
| 0.940 | 0.72276 | 0.94703 | 1.37232 | 0.72576 | 0.94959 | 1.37312 | 0.940 |
| 0.960 | 0.74267 | 0.96689 | 1.38723 | 0.74560 | 0.96939 | 1.38793 | 0.960 |
| 0.980 | 0.76260 | 0.98675 | 1.40171 | 0.76548 | 0.98920 | 1.40236 | 0.980 |
| 1.000 | 0.78255 | 1.00661 | 1.41589 | 0.78537 | 1.00902 | 1.41651 | 1.000 |

$H/R = 0.160$ $\quad n = 3. \quad$ $H/R = 0.180$

| $H/L$ | $\Omega_1$ | $\Omega_2$ | $\Omega_3$ | $\Omega_1$ | $\Omega_2$ | $\Omega_3$ | $H/L$ |
|---|---|---|---|---|---|---|---|
| 0. | 0.02919 | 0.15294 | 0.26979 | 0.03664 | 0.17209 | 0.30276 | 0. |
| 0.001 | 0.02920 | 0.15294 | 0.26980 | 0.03664 | 0.17210 | 0.30276 | 0.001 |
| 0.005 | 0.02922 | 0.15305 | 0.26991 | 0.03667 | 0.17219 | 0.30286 | 0.005 |
| 0.010 | 0.02931 | 0.15339 | 0.27025 | 0.03675 | 0.17250 | 0.30317 | 0.010 |
| 0.020 | 0.02966 | 0.15473 | 0.27163 | 0.03709 | 0.17369 | 0.30439 | 0.020 |
| 0.030 | 0.03026 | 0.15690 | 0.27392 | 0.03771 | 0.17564 | 0.30643 | 0.030 |
| 0.040 | 0.03141 | 0.15982 | 0.27711 | 0.03869 | 0.17828 | 0.30928 | 0.040 |
| 0.050 | 0.03294 | 0.16341 | 0.28120 | 0.04002 | 0.18156 | 0.31293 | 0.050 |
| 0.060 | 0.03500 | 0.16757 | 0.28616 | 0.04182 | 0.18538 | 0.31736 | 0.060 |
| 0.070 | 0.03755 | 0.17221 | 0.29196 | 0.04409 | 0.18969 | 0.32257 | 0.070 |
| 0.080 | 0.04058 | 0.17727 | 0.29859 | 0.04681 | 0.19440 | 0.32852 | 0.080 |
| 0.090 | 0.04402 | 0.18266 | 0.30600 | 0.04996 | 0.19948 | 0.33520 | 0.090 |
| 0.100 | 0.04782 | 0.18835 | 0.31414 | 0.05350 | 0.20486 | 0.34257 | 0.100 |
| 0.120 | 0.05623 | 0.20049 | 0.33244 | 0.06156 | 0.21639 | 0.35924 | 0.120 |
| 0.140 | 0.06544 | 0.21349 | 0.35308 | 0.07064 | 0.22878 | 0.37823 | 0.140 |
| 0.160 | 0.07523 | 0.22725 | 0.37567 | 0.08048 | 0.24192 | 0.39920 | 0.160 |
| 0.180 | 0.08551 | 0.24170 | 0.39985 | 0.09093 | 0.25572 | 0.42185 | 0.180 |
| 0.200 | 0.09627 | 0.25678 | 0.42532 | 0.10188 | 0.27013 | 0.44589 | 0.200 |
| 0.220 | 0.10743 | 0.27243 | 0.45182 | 0.11330 | 0.28512 | 0.47108 | 0.220 |
| 0.240 | 0.11906 | 0.28859 | 0.47915 | 0.12522 | 0.30062 | 0.49721 | 0.240 |
| 0.260 | 0.13125 | 0.30519 | 0.50713 | 0.13756 | 0.31658 | 0.52410 | 0.260 |
| 0.280 | 0.14392 | 0.32218 | 0.53565 | 0.15036 | 0.33297 | 0.55161 | 0.280 |
| 0.300 | 0.15709 | 0.33950 | 0.56457 | 0.16365 | 0.34973 | 0.57962 | 0.300 |
| 0.320 | 0.17076 | 0.35712 | 0.59381 | 0.17739 | 0.36681 | 0.60802 | 0.320 |
| 0.340 | 0.18492 | 0.37498 | 0.62328 | 0.19155 | 0.38419 | 0.63673 | 0.340 |
| 0.360 | 0.19953 | 0.39307 | 0.65292 | 0.20616 | 0.40182 | 0.66567 | 0.360 |
| 0.380 | 0.21459 | 0.41134 | 0.68268 | 0.22118 | 0.41967 | 0.69477 | 0.380 |
| 0.400 | 0.23006 | 0.42978 | 0.71249 | 0.23659 | 0.43772 | 0.72398 | 0.400 |
| 0.420 | 0.24591 | 0.44836 | 0.74231 | 0.25237 | 0.45595 | 0.75322 | 0.420 |
| 0.440 | 0.26213 | 0.46707 | 0.77209 | 0.26849 | 0.47432 | 0.78247 | 0.440 |
| 0.460 | 0.27868 | 0.48588 | 0.80179 | 0.28494 | 0.49283 | 0.81167 | 0.460 |
| 0.480 | 0.29555 | 0.50479 | 0.83136 | 0.30169 | 0.51146 | 0.84077 | 0.480 |
| 0.500 | 0.31269 | 0.52379 | 0.86076 | 0.31872 | 0.53020 | 0.86973 | 0.500 |
| 0.520 | 0.33010 | 0.54287 | 0.88996 | 0.33601 | 0.54902 | 0.89850 | 0.520 |
| 0.540 | 0.34775 | 0.56201 | 0.91891 | 0.35354 | 0.56794 | 0.92704 | 0.540 |
| 0.560 | 0.36562 | 0.58122 | 0.94757 | 0.37129 | 0.58693 | 0.95529 | 0.560 |
| 0.580 | 0.38370 | 0.60047 | 0.97588 | 0.38924 | 0.60600 | 0.98323 | 0.580 |
| 0.600 | 0.40196 | 0.61978 | 1.00381 | 0.40738 | 0.62512 | 1.01078 | 0.600 |
| 0.620 | 0.42039 | 0.63914 | 1.03129 | 0.42569 | 0.64430 | 1.03790 | 0.620 |
| 0.640 | 0.43897 | 0.65854 | 1.05830 | 0.44416 | 0.66353 | 1.06455 | 0.640 |
| 0.660 | 0.45769 | 0.67797 | 1.08478 | 0.46277 | 0.68281 | 1.09066 | 0.660 |
| 0.680 | 0.47655 | 0.69743 | 1.11066 | 0.48151 | 0.70213 | 1.11618 | 0.680 |
| 0.700 | 0.49552 | 0.71693 | 1.13590 | 0.50037 | 0.72149 | 1.14107 | 0.700 |
| 0.720 | 0.51459 | 0.73645 | 1.16045 | 0.51934 | 0.74088 | 1.16525 | 0.720 |
| 0.740 | 0.53377 | 0.75600 | 1.18424 | 0.53841 | 0.76031 | 1.18868 | 0.740 |
| 0.760 | 0.55303 | 0.77557 | 1.20725 | 0.55757 | 0.77977 | 1.21131 | 0.760 |
| 0.780 | 0.57237 | 0.79517 | 1.22941 | 0.57681 | 0.79926 | 1.23308 | 0.780 |
| 0.800 | 0.59178 | 0.81479 | 1.25069 | 0.59613 | 0.81877 | 1.25396 | 0.800 |
| 0.820 | 0.61125 | 0.83442 | 1.27105 | 0.61551 | 0.83830 | 1.27390 | 0.820 |
| 0.840 | 0.63079 | 0.85408 | 1.29047 | 0.63496 | 0.85786 | 1.29290 | 0.840 |
| 0.860 | 0.65038 | 0.87374 | 1.30891 | 0.65446 | 0.87744 | 1.31093 | 0.860 |
| 0.880 | 0.67001 | 0.89343 | 1.32641 | 0.67401 | 0.89704 | 1.32806 | 0.880 |
| 0.900 | 0.68969 | 0.91313 | 1.34301 | 0.69361 | 0.91666 | 1.34437 | 0.900 |
| 0.920 | 0.70941 | 0.93283 | 1.35883 | 0.71325 | 0.93629 | 1.35998 | 0.920 |
| 0.940 | 0.72915 | 0.95256 | 1.37402 | 0.73292 | 0.95594 | 1.37503 | 0.940 |
| 0.960 | 0.74893 | 0.97230 | 1.38875 | 0.75263 | 0.97560 | 1.38967 | 0.960 |
| 0.980 | 0.76874 | 0.99204 | 1.40312 | 0.77236 | 0.99528 | 1.40399 | 0.980 |
| 1.000 | 0.78856 | 1.01180 | 1.41722 | 0.79212 | 1.01497 | 1.41804 | 1.000 |

$H/R = 0.200 \qquad n = 3.$

| $H/L$ | $\Omega_1$ | $\Omega_2$ | $\Omega_3$ | $\Omega_4$ | $\Omega_5$ | $\Omega_6$ |
|---|---|---|---|---|---|---|
| 0. | 0.04483 | 0.19126 | 0.33543 | 1.01995 | 1.06956 | 1.81804 |
| 0.001 | 0.04483 | 0.19126 | 0.33543 | 1.01995 | 1.06956 | 1.81804 |
| 0.005 | 0.04486 | 0.19135 | 0.33552 | 1.01996 | 1.06960 | 1.81802 |
| 0.010 | 0.04494 | 0.19162 | 0.33580 | 1.02002 | 1.06970 | 1.81794 |
| 0.020 | 0.04526 | 0.19270 | 0.33691 | 1.02020 | 1.07012 | 1.81764 |
| 0.030 | 0.04584 | 0.19447 | 0.33875 | 1.02052 | 1.07082 | 1.81714 |
| 0.040 | 0.04672 | 0.19688 | 0.34132 | 1.02097 | 1.07180 | 1.81646 |
| 0.050 | 0.04795 | 0.19988 | 0.34461 | 1.02154 | 1.07306 | 1.81557 |
| 0.060 | 0.04956 | 0.20342 | 0.34863 | 1.02223 | 1.07460 | 1.81453 |
| 0.070 | 0.05159 | 0.20742 | 0.35334 | 1.02303 | 1.07642 | 1.81331 |
| 0.080 | 0.05405 | 0.21184 | 0.35875 | 1.02395 | 1.07852 | 1.81191 |
| 0.090 | 0.05693 | 0.21661 | 0.36482 | 1.02498 | 1.08091 | 1.81039 |
| 0.100 | 0.06020 | 0.22171 | 0.37154 | 1.02611 | 1.08359 | 1.80872 |
| 0.120 | 0.06781 | 0.23268 | 0.38682 | 1.02868 | 1.08977 | 1.80503 |
| 0.140 | 0.07660 | 0.24454 | 0.40433 | 1.03164 | 1.09706 | 1.80095 |
| 0.160 | 0.08632 | 0.25713 | 0.42383 | 1.03496 | 1.10544 | 1.79656 |
| 0.180 | 0.09677 | 0.27037 | 0.44504 | 1.03863 | 1.11487 | 1.79196 |
| 0.200 | 0.10782 | 0.28420 | 0.46772 | 1.04266 | 1.12532 | 1.78723 |
| 0.220 | 0.11939 | 0.29859 | 0.49162 | 1.04701 | 1.13674 | 1.78247 |
| 0.240 | 0.13143 | 0.31350 | 0.51655 | 1.05169 | 1.14907 | 1.77771 |
| 0.260 | 0.14397 | 0.32887 | 0.54234 | 1.05671 | 1.16228 | 1.77304 |
| 0.280 | 0.15691 | 0.34468 | 0.56884 | 1.06205 | 1.17630 | 1.76851 |
| 0.300 | 0.17029 | 0.36089 | 0.59591 | 1.06772 | 1.19110 | 1.76416 |
| 0.320 | 0.18405 | 0.37745 | 0.62343 | 1.07370 | 1.20661 | 1.76006 |
| 0.340 | 0.19832 | 0.39433 | 0.65134 | 1.08001 | 1.22280 | 1.75623 |
| 0.360 | 0.21296 | 0.41149 | 0.67953 | 1.08665 | 1.23961 | 1.75273 |
| 0.380 | 0.22794 | 0.42890 | 0.70793 | 1.09359 | 1.25701 | 1.74957 |
| 0.400 | 0.24332 | 0.44654 | 0.73649 | 1.10083 | 1.27496 | 1.74680 |
| 0.420 | 0.25904 | 0.46438 | 0.76513 | 1.10839 | 1.29341 | 1.74447 |
| 0.440 | 0.27510 | 0.48240 | 0.79380 | 1.11625 | 1.31232 | 1.74259 |
| 0.460 | 0.29147 | 0.50058 | 0.82246 | 1.12441 | 1.33167 | 1.74122 |
| 0.480 | 0.30813 | 0.51891 | 0.85104 | 1.13286 | 1.35142 | 1.74037 |
| 0.500 | 0.32506 | 0.53735 | 0.87951 | 1.14160 | 1.37154 | 1.74009 |
| 0.520 | 0.34225 | 0.55592 | 0.90781 | 1.15063 | 1.39200 | 1.74042 |
| 0.540 | 0.35967 | 0.57458 | 0.93589 | 1.15993 | 1.41278 | 1.74137 |
| 0.560 | 0.37730 | 0.59334 | 0.96371 | 1.16951 | 1.43384 | 1.74302 |
| 0.580 | 0.39514 | 0.61218 | 0.99122 | 1.17936 | 1.45518 | 1.74539 |
| 0.600 | 0.41317 | 0.63110 | 1.01835 | 1.18947 | 1.47675 | 1.74851 |
| 0.620 | 0.43136 | 0.65009 | 1.04506 | 1.19986 | 1.49854 | 1.75244 |
| 0.640 | 0.44971 | 0.66914 | 1.07130 | 1.21051 | 1.52052 | 1.75723 |
| 0.660 | 0.46821 | 0.68825 | 1.09701 | 1.22141 | 1.54268 | 1.76292 |
| 0.680 | 0.48684 | 0.70740 | 1.12213 | 1.23257 | 1.56499 | 1.76955 |
| 0.700 | 0.50560 | 0.72661 | 1.14660 | 1.24401 | 1.58745 | 1.77719 |
| 0.720 | 0.52446 | 0.74586 | 1.17037 | 1.25570 | 1.61003 | 1.78584 |
| 0.740 | 0.54342 | 0.76515 | 1.19337 | 1.26766 | 1.63271 | 1.79559 |
| 0.760 | 0.56248 | 0.78448 | 1.21556 | 1.27991 | 1.65546 | 1.80644 |
| 0.780 | 0.58163 | 0.80385 | 1.23689 | 1.29244 | 1.67829 | 1.81845 |
| 0.800 | 0.60084 | 0.82324 | 1.25731 | 1.30527 | 1.70117 | 1.83161 |
| 0.820 | 0.62014 | 0.84266 | 1.27680 | 1.31839 | 1.72408 | 1.84595 |
| 0.840 | 0.63949 | 0.86212 | 1.29535 | 1.33183 | 1.74702 | 1.86147 |
| 0.860 | 0.65890 | 0.88159 | 1.31298 | 1.34554 | 1.76998 | 1.87816 |
| 0.880 | 0.67837 | 0.90110 | 1.32976 | 1.35945 | 1.79293 | 1.89599 |
| 0.900 | 0.69788 | 0.92062 | 1.34580 | 1.37350 | 1.81586 | 1.91493 |
| 0.920 | 0.71744 | 0.94017 | 1.36121 | 1.38759 | 1.83877 | 1.93494 |
| 0.940 | 0.73703 | 0.95973 | 1.37615 | 1.40161 | 1.86165 | 1.95597 |
| 0.960 | 0.75666 | 0.97931 | 1.39071 | 1.41551 | 1.88448 | 1.97795 |
| 0.980 | 0.77632 | 0.99891 | 1.40497 | 1.42926 | 1.90727 | 2.00083 |
| 1.000 | 0.79600 | 1.01854 | 1.41898 | 1.44285 | 1.92999 | 2.02453 |

$H/R = 0.250 \qquad n = 3.$

| $H/L$ | $\Omega_1$ | $\Omega_2$ | $\Omega_3$ |
|---|---|---|---|
| 0. | 0.06830 | 0.23922 | 0.41559 |
| 0.001 | 0.06830 | 0.23922 | 0.41559 |
| 0.005 | 0.06832 | 0.23929 | 0.41566 |
| 0.010 | 0.06839 | 0.23951 | 0.41588 |
| 0.020 | 0.06869 | 0.24038 | 0.41677 |
| 0.030 | 0.06920 | 0.24181 | 0.41826 |
| 0.040 | 0.06996 | 0.24379 | 0.42033 |
| 0.050 | 0.07098 | 0.24627 | 0.42300 |
| 0.060 | 0.07230 | 0.24922 | 0.42624 |
| 0.070 | 0.07392 | 0.25261 | 0.43006 |
| 0.080 | 0.07591 | 0.25640 | 0.43445 |
| 0.090 | 0.07824 | 0.26054 | 0.43940 |
| 0.100 | 0.08094 | 0.26500 | 0.44489 |
| 0.120 | 0.08738 | 0.27475 | 0.45747 |
| 0.140 | 0.09515 | 0.28544 | 0.47205 |
| 0.160 | 0.10407 | 0.29688 | 0.48848 |
| 0.180 | 0.11399 | 0.30897 | 0.50659 |
| 0.200 | 0.12475 | 0.32164 | 0.52620 |
| 0.220 | 0.13621 | 0.33482 | 0.54714 |
| 0.240 | 0.14828 | 0.34849 | 0.56924 |
| 0.260 | 0.16089 | 0.36262 | 0.59233 |
| 0.280 | 0.17396 | 0.37717 | 0.61629 |
| 0.300 | 0.18749 | 0.39213 | 0.64096 |
| 0.320 | 0.20143 | 0.40747 | 0.66625 |
| 0.340 | 0.21575 | 0.42315 | 0.69205 |
| 0.360 | 0.23043 | 0.43917 | 0.71826 |
| 0.380 | 0.24544 | 0.45549 | 0.74480 |
| 0.400 | 0.26078 | 0.47208 | 0.77160 |
| 0.420 | 0.27645 | 0.48893 | 0.79857 |
| 0.440 | 0.29241 | 0.50601 | 0.82566 |
| 0.460 | 0.30864 | 0.52330 | 0.85281 |
| 0.480 | 0.32514 | 0.54077 | 0.87995 |
| 0.500 | 0.34189 | 0.55843 | 0.90702 |
| 0.520 | 0.35888 | 0.57625 | 0.93398 |
| 0.540 | 0.37607 | 0.59421 | 0.96076 |
| 0.560 | 0.39347 | 0.61231 | 0.98730 |
| 0.580 | 0.41107 | 0.63053 | 1.01356 |
| 0.600 | 0.42884 | 0.64885 | 1.03945 |
| 0.620 | 0.44678 | 0.66728 | 1.06494 |
| 0.640 | 0.46488 | 0.68581 | 1.08996 |
| 0.660 | 0.48312 | 0.70442 | 1.11445 |
| 0.680 | 0.50149 | 0.72310 | 1.13833 |
| 0.700 | 0.51998 | 0.74186 | 1.16154 |
| 0.720 | 0.53859 | 0.76069 | 1.18404 |
| 0.740 | 0.55730 | 0.77958 | 1.20576 |
| 0.760 | 0.57611 | 0.79853 | 1.22666 |
| 0.780 | 0.59501 | 0.81753 | 1.24668 |
| 0.800 | 0.61399 | 0.83658 | 1.26583 |
| 0.820 | 0.63305 | 0.85568 | 1.28412 |
| 0.840 | 0.65217 | 0.87481 | 1.30157 |
| 0.860 | 0.67136 | 0.89399 | 1.31828 |
| 0.880 | 0.69060 | 0.91321 | 1.33432 |
| 0.900 | 0.70989 | 0.93245 | 1.34981 |
| 0.920 | 0.72924 | 0.95174 | 1.36484 |
| 0.940 | 0.74863 | 0.97105 | 1.37949 |
| 0.960 | 0.76805 | 0.99040 | 1.39384 |
| 0.980 | 0.78752 | 1.00977 | 1.40793 |
| 1.000 | 0.80701 | 1.02915 | 1.42178 |

$H/R = 0.300 \qquad n = 3.$

| $H/L$ | $\Omega_1$ | $\Omega_2$ | $\Omega_3$ | $\Omega_4$ | $\Omega_5$ | $\Omega_6$ |
|---|---|---|---|---|---|---|
| 0. | 0.09556 | 0.28721 | 0.49302 | 1.04543 | 1.15517 | 1.78652 |
| 0.001 | 0.09556 | 0.28721 | 0.49302 | 1.04543 | 1.15517 | 1.78651 |
| 0.005 | 0.09558 | 0.28727 | 0.49309 | 1.04544 | 1.15520 | 1.78650 |
| 0.010 | 0.09565 | 0.28746 | 0.49327 | 1.04550 | 1.15529 | 1.78646 |
| 0.020 | 0.09593 | 0.28819 | 0.49402 | 1.04569 | 1.15566 | 1.78629 |
| 0.030 | 0.09639 | 0.28939 | 0.49527 | 1.04601 | 1.15627 | 1.78601 |
| 0.040 | 0.09706 | 0.29106 | 0.49701 | 1.04645 | 1.15712 | 1.78562 |
| 0.050 | 0.09795 | 0.29317 | 0.49925 | 1.04702 | 1.15822 | 1.78512 |
| 0.060 | 0.09908 | 0.29571 | 0.50198 | 1.04772 | 1.15956 | 1.78452 |
| 0.070 | 0.10048 | 0.29863 | 0.50520 | 1.04854 | 1.16115 | 1.78383 |
| 0.080 | 0.10215 | 0.30193 | 0.50889 | 1.04948 | 1.16298 | 1.78303 |
| 0.090 | 0.10411 | 0.30556 | 0.51307 | 1.05053 | 1.16505 | 1.78214 |
| 0.100 | 0.10638 | 0.30951 | 0.51772 | 1.05170 | 1.16737 | 1.78116 |
| 0.120 | 0.11185 | 0.31824 | 0.52839 | 1.05438 | 1.17272 | 1.77899 |
| 0.140 | 0.11857 | 0.32792 | 0.54083 | 1.05749 | 1.17903 | 1.77654 |
| 0.160 | 0.12648 | 0.33840 | 0.55494 | 1.06102 | 1.18628 | 1.77388 |
| 0.180 | 0.13548 | 0.34955 | 0.57060 | 1.06494 | 1.19447 | 1.77106 |
| 0.200 | 0.14546 | 0.36128 | 0.58771 | 1.06924 | 1.20356 | 1.76813 |
| 0.220 | 0.15630 | 0.37352 | 0.60613 | 1.07388 | 1.21354 | 1.76514 |
| 0.240 | 0.16789 | 0.38623 | 0.62572 | 1.07888 | 1.22437 | 1.76215 |
| 0.260 | 0.18014 | 0.39938 | 0.64637 | 1.08421 | 1.23603 | 1.75921 |
| 0.280 | 0.19297 | 0.41294 | 0.66795 | 1.08987 | 1.24847 | 1.75638 |
| 0.300 | 0.20631 | 0.42689 | 0.69033 | 1.09584 | 1.26167 | 1.75367 |
| 0.320 | 0.22011 | 0.44120 | 0.71341 | 1.10213 | 1.27560 | 1.75117 |
| 0.340 | 0.23432 | 0.45587 | 0.73709 | 1.10871 | 1.29020 | 1.74890 |
| 0.360 | 0.24893 | 0.47086 | 0.76127 | 1.11561 | 1.30543 | 1.74689 |
| 0.380 | 0.26388 | 0.48618 | 0.78586 | 1.12281 | 1.32128 | 1.74520 |
| 0.400 | 0.27915 | 0.50179 | 0.81078 | 1.13030 | 1.33769 | 1.74386 |
| 0.420 | 0.29472 | 0.51767 | 0.83596 | 1.13809 | 1.35464 | 1.74291 |
| 0.440 | 0.31059 | 0.53382 | 0.86131 | 1.14619 | 1.37209 | 1.74239 |
| 0.460 | 0.32670 | 0.55021 | 0.88678 | 1.15456 | 1.39000 | 1.74235 |
| 0.480 | 0.34307 | 0.56682 | 0.91229 | 1.16324 | 1.40834 | 1.74281 |
| 0.500 | 0.35967 | 0.58365 | 0.93779 | 1.17220 | 1.42709 | 1.74383 |
| 0.520 | 0.37650 | 0.60066 | 0.96319 | 1.18146 | 1.44620 | 1.74542 |
| 0.540 | 0.39353 | 0.61787 | 0.98845 | 1.19102 | 1.46565 | 1.74766 |
| 0.560 | 0.41075 | 0.63523 | 1.01349 | 1.20086 | 1.48542 | 1.75058 |
| 0.580 | 0.42815 | 0.65276 | 1.03825 | 1.21101 | 1.50548 | 1.75421 |
| 0.600 | 0.44573 | 0.67042 | 1.06266 | 1.22146 | 1.52580 | 1.75863 |
| 0.620 | 0.46346 | 0.68822 | 1.08667 | 1.23223 | 1.54634 | 1.76384 |
| 0.640 | 0.48134 | 0.70614 | 1.11020 | 1.24329 | 1.56711 | 1.76994 |
| 0.660 | 0.49936 | 0.72417 | 1.13317 | 1.25468 | 1.58805 | 1.77695 |
| 0.680 | 0.51750 | 0.74231 | 1.15554 | 1.26639 | 1.60916 | 1.78491 |
| 0.700 | 0.53577 | 0.76055 | 1.17725 | 1.27843 | 1.63042 | 1.79389 |
| 0.720 | 0.55416 | 0.77888 | 1.19823 | 1.29078 | 1.65178 | 1.80392 |
| 0.740 | 0.57264 | 0.79729 | 1.21846 | 1.30345 | 1.67325 | 1.81504 |
| 0.760 | 0.59122 | 0.81579 | 1.23791 | 1.31642 | 1.69481 | 1.82727 |
| 0.780 | 0.60988 | 0.83435 | 1.25657 | 1.32965 | 1.71642 | 1.84062 |
| 0.800 | 0.62863 | 0.85299 | 1.27448 | 1.34310 | 1.73809 | 1.85514 |
| 0.820 | 0.64747 | 0.87169 | 1.29166 | 1.35671 | 1.75981 | 1.87078 |
| 0.840 | 0.66637 | 0.89045 | 1.30819 | 1.37042 | 1.78155 | 1.88755 |
| 0.860 | 0.68533 | 0.90926 | 1.32414 | 1.38415 | 1.80333 | 1.90540 |
| 0.880 | 0.70436 | 0.92813 | 1.33960 | 1.39784 | 1.82514 | 1.92432 |
| 0.900 | 0.72344 | 0.94704 | 1.35463 | 1.41143 | 1.84699 | 1.94423 |
| 0.920 | 0.74258 | 0.96600 | 1.36930 | 1.42491 | 1.86886 | 1.96510 |
| 0.940 | 0.76176 | 0.98501 | 1.38367 | 1.43823 | 1.89075 | 1.98685 |
| 0.960 | 0.78099 | 1.00406 | 1.39778 | 1.45142 | 1.91266 | 2.00941 |
| 0.980 | 0.80025 | 1.02314 | 1.41168 | 1.46447 | 1.93458 | 2.03273 |
| 1.000 | 0.81956 | 1.04226 | 1.42537 | 1.47741 | 1.95649 | 2.05677 |

$H/R = 0.350$ $n = 3.$

| $H/L$ | $\Omega_1$ | $\Omega_2$ | $\Omega_3$ |
|---|---|---|---|
| 0. | 0.12608 | 0.33519 | 0.56707 |
| 0.001 | 0.12608 | 0.33519 | 0.56707 |
| 0.005 | 0.12610 | 0.33524 | 0.56712 |
| 0.010 | 0.12616 | 0.33540 | 0.56728 |
| 0.020 | 0.12641 | 0.33603 | 0.56793 |
| 0.030 | 0.12683 | 0.33707 | 0.56901 |
| 0.040 | 0.12744 | 0.33852 | 0.57052 |
| 0.050 | 0.12823 | 0.34036 | 0.57245 |
| 0.060 | 0.12923 | 0.34258 | 0.57481 |
| 0.070 | 0.13045 | 0.34515 | 0.57759 |
| 0.080 | 0.13190 | 0.34806 | 0.58080 |
| 0.090 | 0.13360 | 0.35129 | 0.58442 |
| 0.100 | 0.13555 | 0.35482 | 0.58845 |
| 0.120 | 0.14026 | 0.36269 | 0.59771 |
| 0.140 | 0.14608 | 0.37151 | 0.60854 |
| 0.160 | 0.15302 | 0.38114 | 0.62088 |
| 0.180 | 0.16104 | 0.39147 | 0.63463 |
| 0.200 | 0.17006 | 0.40240 | 0.64972 |
| 0.220 | 0.18002 | 0.41385 | 0.66605 |
| 0.240 | 0.19082 | 0.42577 | 0.68352 |
| 0.260 | 0.20239 | 0.43813 | 0.70202 |
| 0.280 | 0.21462 | 0.45088 | 0.72146 |
| 0.300 | 0.22746 | 0.46400 | 0.74174 |
| 0.320 | 0.24084 | 0.47748 | 0.76275 |
| 0.340 | 0.25471 | 0.49130 | 0.78439 |
| 0.360 | 0.26900 | 0.50544 | 0.80658 |
| 0.380 | 0.28368 | 0.51989 | 0.82923 |
| 0.400 | 0.29872 | 0.53463 | 0.85226 |
| 0.420 | 0.31409 | 0.54966 | 0.87559 |
| 0.440 | 0.32975 | 0.56496 | 0.89913 |
| 0.460 | 0.34569 | 0.58051 | 0.92282 |
| 0.480 | 0.36188 | 0.59631 | 0.94658 |
| 0.500 | 0.37831 | 0.61232 | 0.97036 |
| 0.520 | 0.39495 | 0.62856 | 0.99405 |
| 0.540 | 0.41180 | 0.64500 | 1.01762 |
| 0.560 | 0.42884 | 0.66163 | 1.04098 |
| 0.580 | 0.44606 | 0.67844 | 1.06405 |
| 0.600 | 0.46344 | 0.69541 | 1.08678 |
| 0.620 | 0.48098 | 0.71255 | 1.10910 |
| 0.640 | 0.49866 | 0.72983 | 1.13094 |
| 0.660 | 0.51648 | 0.74724 | 1.15225 |
| 0.680 | 0.53443 | 0.76479 | 1.17296 |
| 0.700 | 0.55249 | 0.78246 | 1.19304 |
| 0.720 | 0.57066 | 0.80024 | 1.21245 |
| 0.740 | 0.58894 | 0.81813 | 1.23119 |
| 0.760 | 0.60731 | 0.83612 | 1.24926 |
| 0.780 | 0.62578 | 0.85420 | 1.26668 |
| 0.800 | 0.64432 | 0.87236 | 1.28350 |
| 0.820 | 0.66295 | 0.89061 | 1.29976 |
| 0.840 | 0.68164 | 0.90894 | 1.31551 |
| 0.860 | 0.70040 | 0.92734 | 1.33084 |
| 0.880 | 0.71923 | 0.94580 | 1.34578 |
| 0.900 | 0.73811 | 0.96434 | 1.36039 |
| 0.920 | 0.75705 | 0.98293 | 1.37470 |
| 0.940 | 0.77604 | 1.00158 | 1.38878 |
| 0.960 | 0.79507 | 1.02029 | 1.40263 |
| 0.980 | 0.81414 | 1.03904 | 1.41629 |
| 1.000 | 0.83326 | 1.05784 | 1.42977 |

$H/R = 0.400 \qquad n = 3.$

| $H/L$ | $\Omega_1$ | $\Omega_2$ | $\Omega_3$ | $\Omega_4$ | $\Omega_5$ | $\Omega_6$ |
|---|---|---|---|---|---|---|
| 0. | 0.15935 | 0.38306 | 0.63714 | 1.08227 | 1.27357 | 1.76321 |
| 0.001 | 0.15935 | 0.38307 | 0.63715 | 1.08227 | 1.27358 | 1.76321 |
| 0.005 | 0.15937 | 0.38311 | 0.63719 | 1.08228 | 1.27360 | 1.76321 |
| 0.010 | 0.15943 | 0.38325 | 0.63734 | 1.08234 | 1.27368 | 1.76318 |
| 0.020 | 0.15966 | 0.38380 | 0.63791 | 1.08252 | 1.27399 | 1.76310 |
| 0.030 | 0.16004 | 0.38473 | 0.63886 | 1.08284 | 1.27451 | 1.76295 |
| 0.040 | 0.16058 | 0.38601 | 0.64019 | 1.08328 | 1.27523 | 1.76276 |
| 0.050 | 0.16129 | 0.38764 | 0.64189 | 1.08385 | 1.27616 | 1.76249 |
| 0.060 | 0.16218 | 0.38961 | 0.64397 | 1.08455 | 1.27729 | 1.76218 |
| 0.070 | 0.16326 | 0.39191 | 0.64643 | 1.08537 | 1.27865 | 1.76183 |
| 0.080 | 0.16454 | 0.39452 | 0.64926 | 1.08630 | 1.28020 | 1.76141 |
| 0.090 | 0.16603 | 0.39742 | 0.65245 | 1.08737 | 1.28195 | 1.76096 |
| 0.100 | 0.16774 | 0.40061 | 0.65601 | 1.08854 | 1.28391 | 1.76045 |
| 0.120 | 0.17185 | 0.40776 | 0.66419 | 1.09126 | 1.28843 | 1.75934 |
| 0.140 | 0.17693 | 0.41584 | 0.67378 | 1.09444 | 1.29377 | 1.75808 |
| 0.160 | 0.18302 | 0.42473 | 0.68472 | 1.09806 | 1.29991 | 1.75672 |
| 0.180 | 0.19011 | 0.43433 | 0.69694 | 1.10212 | 1.30683 | 1.75530 |
| 0.200 | 0.19819 | 0.44455 | 0.71039 | 1.10659 | 1.31453 | 1.75384 |
| 0.220 | 0.20720 | 0.45531 | 0.72500 | 1.11146 | 1.32299 | 1.75239 |
| 0.240 | 0.21708 | 0.46655 | 0.74067 | 1.11673 | 1.33220 | 1.75098 |
| 0.260 | 0.22778 | 0.47821 | 0.75734 | 1.12236 | 1.34213 | 1.74965 |
| 0.280 | 0.23922 | 0.49028 | 0.77490 | 1.12836 | 1.35276 | 1.74844 |
| 0.300 | 0.25134 | 0.50271 | 0.79328 | 1.13471 | 1.36409 | 1.74738 |
| 0.320 | 0.26406 | 0.51549 | 0.81239 | 1.14141 | 1.37606 | 1.74652 |
| 0.340 | 0.27733 | 0.52860 | 0.83214 | 1.14844 | 1.38866 | 1.74591 |
| 0.360 | 0.29110 | 0.54202 | 0.85244 | 1.15579 | 1.40187 | 1.74556 |
| 0.380 | 0.30532 | 0.55574 | 0.87323 | 1.16348 | 1.41566 | 1.74554 |
| 0.400 | 0.31994 | 0.56975 | 0.89440 | 1.17149 | 1.43000 | 1.74587 |
| 0.420 | 0.33493 | 0.58404 | 0.91588 | 1.17981 | 1.44484 | 1.74662 |
| 0.440 | 0.35025 | 0.59859 | 0.93760 | 1.18845 | 1.46018 | 1.74779 |
| 0.460 | 0.36587 | 0.61339 | 0.95948 | 1.19742 | 1.47597 | 1.74945 |
| 0.480 | 0.38178 | 0.62845 | 0.98145 | 1.20670 | 1.49219 | 1.75164 |
| 0.500 | 0.39794 | 0.64374 | 1.00342 | 1.21630 | 1.50880 | 1.75441 |
| 0.520 | 0.41433 | 0.65925 | 1.02532 | 1.22624 | 1.52577 | 1.75781 |
| 0.540 | 0.43094 | 0.67497 | 1.04710 | 1.23651 | 1.54307 | 1.76188 |
| 0.560 | 0.44775 | 0.69089 | 1.06866 | 1.24710 | 1.56065 | 1.76667 |
| 0.580 | 0.46474 | 0.70701 | 1.08996 | 1.25802 | 1.57852 | 1.77225 |
| 0.600 | 0.48191 | 0.72332 | 1.11091 | 1.26928 | 1.59660 | 1.77865 |
| 0.620 | 0.49923 | 0.73979 | 1.13147 | 1.28086 | 1.61490 | 1.78594 |
| 0.640 | 0.51671 | 0.75643 | 1.15157 | 1.29276 | 1.63336 | 1.79416 |
| 0.660 | 0.53431 | 0.77323 | 1.17118 | 1.30498 | 1.65197 | 1.80337 |
| 0.680 | 0.55206 | 0.79017 | 1.19025 | 1.31746 | 1.67070 | 1.81359 |
| 0.700 | 0.56992 | 0.80725 | 1.20877 | 1.33022 | 1.68955 | 1.82488 |
| 0.720 | 0.58788 | 0.82446 | 1.22672 | 1.34317 | 1.70848 | 1.83725 |
| 0.740 | 0.60596 | 0.84180 | 1.24412 | 1.35630 | 1.72752 | 1.85070 |
| 0.760 | 0.62414 | 0.85925 | 1.26099 | 1.36953 | 1.74664 | 1.86524 |
| 0.780 | 0.64241 | 0.87682 | 1.27735 | 1.38282 | 1.76588 | 1.88084 |
| 0.800 | 0.66075 | 0.89448 | 1.29324 | 1.39613 | 1.78525 | 1.89747 |
| 0.820 | 0.67918 | 0.91225 | 1.30870 | 1.40938 | 1.80478 | 1.91506 |
| 0.840 | 0.69769 | 0.93011 | 1.32379 | 1.42256 | 1.82447 | 1.93359 |
| 0.860 | 0.71625 | 0.94806 | 1.33854 | 1.43563 | 1.84434 | 1.95298 |
| 0.880 | 0.73489 | 0.96609 | 1.35299 | 1.44857 | 1.86441 | 1.97320 |
| 0.900 | 0.75358 | 0.98420 | 1.36718 | 1.46137 | 1.88468 | 1.99414 |
| 0.920 | 0.77233 | 1.00238 | 1.38114 | 1.47403 | 1.90513 | 2.01579 |
| 0.940 | 0.79113 | 1.02063 | 1.39491 | 1.48657 | 1.92574 | 2.03809 |
| 0.960 | 0.80998 | 1.03896 | 1.40847 | 1.49900 | 1.94652 | 2.06099 |
| 0.980 | 0.82888 | 1.05734 | 1.42187 | 1.51137 | 1.96740 | 2.08444 |
| 1.000 | 0.84781 | 1.07579 | 1.43510 | 1.52367 | 1.98837 | 2.10841 |

$H/R = 0.450$ $n = 3.$

| $H/L$ | $\Omega_1$ | $\Omega_2$ | $\Omega_3$ |
|---|---|---|---|
| 0. | 0.19495 | 0.43071 | 0.70286 |
| 0.001 | 0.19495 | 0.43071 | 0.70286 |
| 0.005 | 0.19497 | 0.43075 | 0.70291 |
| 0.010 | 0.19502 | 0.43088 | 0.70303 |
| 0.020 | 0.19523 | 0.43137 | 0.70354 |
| 0.030 | 0.19557 | 0.43220 | 0.70439 |
| 0.040 | 0.19606 | 0.43335 | 0.70558 |
| 0.050 | 0.19670 | 0.43482 | 0.70711 |
| 0.060 | 0.19750 | 0.43660 | 0.70897 |
| 0.070 | 0.19846 | 0.43868 | 0.71116 |
| 0.080 | 0.19959 | 0.44104 | 0.71369 |
| 0.090 | 0.20091 | 0.44369 | 0.71655 |
| 0.100 | 0.20241 | 0.44659 | 0.71973 |
| 0.120 | 0.20603 | 0.45314 | 0.72706 |
| 0.140 | 0.21050 | 0.46060 | 0.73565 |
| 0.160 | 0.21586 | 0.46885 | 0.74546 |
| 0.180 | 0.22214 | 0.47781 | 0.75644 |
| 0.200 | 0.22933 | 0.48739 | 0.76854 |
| 0.220 | 0.23742 | 0.49752 | 0.78170 |
| 0.240 | 0.24637 | 0.50814 | 0.79585 |
| 0.260 | 0.25615 | 0.51921 | 0.81092 |
| 0.280 | 0.26670 | 0.53067 | 0.82685 |
| 0.300 | 0.27796 | 0.54250 | 0.84355 |
| 0.320 | 0.28988 | 0.55467 | 0.86095 |
| 0.340 | 0.30241 | 0.56717 | 0.87896 |
| 0.360 | 0.31549 | 0.57997 | 0.89751 |
| 0.380 | 0.32907 | 0.59307 | 0.91653 |
| 0.400 | 0.34311 | 0.60644 | 0.93593 |
| 0.420 | 0.35756 | 0.62009 | 0.95564 |
| 0.440 | 0.37239 | 0.63399 | 0.97558 |
| 0.460 | 0.38756 | 0.64815 | 0.99566 |
| 0.480 | 0.40305 | 0.66255 | 1.01584 |
| 0.500 | 0.41883 | 0.67718 | 1.03601 |
| 0.520 | 0.43486 | 0.69204 | 1.05613 |
| 0.540 | 0.45114 | 0.70711 | 1.07611 |
| 0.560 | 0.46764 | 0.72239 | 1.09590 |
| 0.580 | 0.48434 | 0.73787 | 1.11543 |
| 0.600 | 0.50122 | 0.75354 | 1.13464 |
| 0.620 | 0.51828 | 0.76939 | 1.15349 |
| 0.640 | 0.53550 | 0.78542 | 1.17195 |
| 0.660 | 0.55287 | 0.80162 | 1.18997 |
| 0.680 | 0.57037 | 0.81797 | 1.20755 |
| 0.700 | 0.58801 | 0.83447 | 1.22468 |
| 0.720 | 0.60576 | 0.85112 | 1.24136 |
| 0.740 | 0.62362 | 0.86791 | 1.25759 |
| 0.760 | 0.64158 | 0.88482 | 1.27342 |
| 0.780 | 0.65963 | 0.90187 | 1.28886 |
| 0.800 | 0.67778 | 0.91902 | 1.30395 |
| 0.820 | 0.69601 | 0.93629 | 1.31871 |
| 0.840 | 0.71432 | 0.95367 | 1.33318 |
| 0.860 | 0.73269 | 0.97115 | 1.34739 |
| 0.880 | 0.75114 | 0.98872 | 1.36137 |
| 0.900 | 0.76965 | 1.00639 | 1.37515 |
| 0.920 | 0.78821 | 1.02414 | 1.38872 |
| 0.940 | 0.80683 | 1.04198 | 1.40213 |
| 0.960 | 0.82550 | 1.05989 | 1.41540 |
| 0.980 | 0.84422 | 1.07788 | 1.42850 |
| 1.000 | 0.86298 | 1.09594 | 1.44149 |

$H/R = 0.500 \qquad n = 3.$

| $H/L$ | $\Omega_1$ | $\Omega_2$ | $\Omega_3$ | $\Omega_4$ | $\Omega_5$ | $\Omega_6$ |
|---|---|---|---|---|---|---|
| 0. | 0.23251 | 0.47797 | 0.76409 | 1.13183 | 1.42106 | 1.75283 |
| 0.010 | 0.23257 | 0.47812 | 0.76424 | 1.13189 | 1.42115 | 1.75281 |
| 0.050 | 0.23408 | 0.48171 | 0.76792 | 1.13339 | 1.42318 | 1.75263 |
| 0.100 | 0.23916 | 0.49251 | 0.77933 | 1.13803 | 1.42952 | 1.75210 |
| 0.120 | 0.24236 | 0.49856 | 0.78595 | 1.14074 | 1.43322 | 1.75184 |
| 0.140 | 0.24631 | 0.50548 | 0.79372 | 1.14392 | 1.43757 | 1.75158 |
| 0.160 | 0.25104 | 0.51318 | 0.80259 | 1.14756 | 1.44258 | 1.75134 |
| 0.180 | 0.25660 | 0.52157 | 0.81254 | 1.15167 | 1.44823 | 1.75115 |
| 0.200 | 0.26300 | 0.53059 | 0.82350 | 1.15622 | 1.45450 | 1.75102 |
| 0.220 | 0.27024 | 0.54016 | 0.83543 | 1.16121 | 1.46139 | 1.75101 |
| 0.240 | 0.27831 | 0.55022 | 0.84827 | 1.16663 | 1.46889 | 1.75113 |
| 0.260 | 0.28718 | 0.56074 | 0.86197 | 1.17246 | 1.47695 | 1.75142 |
| 0.280 | 0.29682 | 0.57166 | 0.87646 | 1.17871 | 1.48560 | 1.75191 |
| 0.300 | 0.30719 | 0.58295 | 0.89167 | 1.18537 | 1.49478 | 1.75263 |
| 0.320 | 0.31825 | 0.59459 | 0.90753 | 1.19242 | 1.50448 | 1.75364 |
| 0.340 | 0.32993 | 0.60654 | 0.92398 | 1.19986 | 1.51469 | 1.75498 |
| 0.360 | 0.34222 | 0.61881 | 0.94093 | 1.20769 | 1.52537 | 1.75667 |
| 0.380 | 0.35505 | 0.63135 | 0.95832 | 1.21590 | 1.53649 | 1.75878 |
| 0.400 | 0.36838 | 0.64417 | 0.97607 | 1.22449 | 1.54804 | 1.76134 |
| 0.420 | 0.38217 | 0.65726 | 0.99411 | 1.23344 | 1.55996 | 1.76441 |
| 0.440 | 0.39638 | 0.67060 | 1.01236 | 1.24277 | 1.57223 | 1.76803 |
| 0.460 | 0.41098 | 0.68418 | 1.03075 | 1.25247 | 1.58481 | 1.77227 |
| 0.480 | 0.42594 | 0.69801 | 1.04922 | 1.26253 | 1.59766 | 1.77717 |
| 0.500 | 0.44121 | 0.71206 | 1.06769 | 1.27295 | 1.61076 | 1.78280 |
| 0.520 | 0.45678 | 0.72633 | 1.08610 | 1.28374 | 1.62407 | 1.78920 |
| 0.540 | 0.47262 | 0.74083 | 1.10440 | 1.29485 | 1.63756 | 1.79642 |
| 0.560 | 0.48872 | 0.75553 | 1.12251 | 1.30630 | 1.65120 | 1.80452 |
| 0.580 | 0.50504 | 0.77043 | 1.14039 | 1.31807 | 1.66500 | 1.81353 |
| 0.600 | 0.52157 | 0.78551 | 1.15801 | 1.33013 | 1.67891 | 1.82348 |
| 0.620 | 0.53829 | 0.80080 | 1.17532 | 1.34245 | 1.69296 | 1.83438 |
| 0.640 | 0.55519 | 0.81625 | 1.19229 | 1.35500 | 1.70718 | 1.84624 |
| 0.660 | 0.57226 | 0.83188 | 1.20892 | 1.36774 | 1.72158 | 1.85905 |
| 0.680 | 0.58947 | 0.84768 | 1.22520 | 1.38062 | 1.73620 | 1.87277 |
| 0.700 | 0.60683 | 0.86363 | 1.24111 | 1.39359 | 1.75109 | 1.88737 |
| 0.720 | 0.62432 | 0.87974 | 1.25669 | 1.40662 | 1.76630 | 1.90280 |
| 0.740 | 0.64193 | 0.89600 | 1.27192 | 1.41965 | 1.78188 | 1.91903 |
| 0.760 | 0.65965 | 0.91239 | 1.28684 | 1.43264 | 1.79786 | 1.93598 |
| 0.780 | 0.67747 | 0.92891 | 1.30147 | 1.44556 | 1.81426 | 1.95363 |
| 0.800 | 0.69540 | 0.94557 | 1.31584 | 1.45837 | 1.83112 | 1.97190 |
| 0.820 | 0.71340 | 0.96235 | 1.32994 | 1.47106 | 1.84843 | 1.99078 |
| 0.840 | 0.73150 | 0.97924 | 1.34383 | 1.48362 | 1.86620 | 2.01020 |
| 0.860 | 0.74967 | 0.99625 | 1.35752 | 1.49604 | 1.88441 | 2.03013 |
| 0.880 | 0.76791 | 1.01337 | 1.37102 | 1.50832 | 1.90302 | 2.05055 |
| 0.900 | 0.78622 | 1.03058 | 1.38435 | 1.52049 | 1.92202 | 2.07142 |
| 0.920 | 0.80459 | 1.04790 | 1.39754 | 1.53254 | 1.94135 | 2.09272 |
| 0.940 | 0.82303 | 1.06530 | 1.41059 | 1.54452 | 1.96098 | 2.11442 |
| 0.960 | 0.84152 | 1.08280 | 1.42352 | 1.55644 | 1.98085 | 2.13649 |
| 0.980 | 0.86005 | 1.10039 | 1.43632 | 1.56832 | 2.00092 | 2.15891 |
| 1.000 | 0.87864 | 1.11805 | 1.44901 | 1.58019 | 2.02114 | 2.18164 |

$H/R = 0.600 \qquad n = 3.$

| $H/L$ | $\Omega_1$ | $\Omega_2$ | $\Omega_3$ | $\Omega_4$ | $\Omega_5$ | $\Omega_6$ |
|---|---|---|---|---|---|---|
| 0. | 0.31233 | 0.57048 | 0.87379 | 1.19595 | 1.57883 | 1.77088 |
| 0.010 | 0.31237 | 0.57061 | 0.87392 | 1.19601 | 1.57890 | 1.77090 |
| 0.050 | 0.31358 | 0.57369 | 0.87699 | 1.19746 | 1.58031 | 1.77131 |
| 0.100 | 0.31759 | 0.58302 | 0.88652 | 1.20197 | 1.58469 | 1.77267 |
| 0.120 | 0.32009 | 0.58831 | 0.89204 | 1.20461 | 1.58723 | 1.77351 |
| 0.140 | 0.32318 | 0.59438 | 0.89853 | 1.20772 | 1.59021 | 1.77456 |
| 0.160 | 0.32690 | 0.60119 | 0.90594 | 1.21130 | 1.59360 | 1.77583 |
| 0.180 | 0.33127 | 0.60867 | 0.91424 | 1.21535 | 1.59741 | 1.77736 |
| 0.200 | 0.33634 | 0.61675 | 0.92340 | 1.21985 | 1.60159 | 1.77917 |
| 0.220 | 0.34211 | 0.62537 | 0.93338 | 1.22482 | 1.60615 | 1.78127 |
| 0.240 | 0.34861 | 0.63450 | 0.94412 | 1.23023 | 1.61106 | 1.78373 |
| 0.260 | 0.35583 | 0.64408 | 0.95560 | 1.23610 | 1.61628 | 1.78657 |
| 0.280 | 0.36376 | 0.65407 | 0.96774 | 1.24242 | 1.62182 | 1.78984 |
| 0.300 | 0.37239 | 0.66444 | 0.98049 | 1.24916 | 1.62762 | 1.79355 |
| 0.320 | 0.38170 | 0.67515 | 0.99379 | 1.25637 | 1.63369 | 1.79777 |
| 0.340 | 0.39167 | 0.68619 | 1.00760 | 1.26400 | 1.63998 | 1.80251 |
| 0.360 | 0.40225 | 0.69753 | 1.02184 | 1.27206 | 1.64649 | 1.80782 |
| 0.380 | 0.41342 | 0.70915 | 1.03645 | 1.28054 | 1.65320 | 1.81375 |
| 0.400 | 0.42515 | 0.72103 | 1.05137 | 1.28947 | 1.66010 | 1.82029 |
| 0.420 | 0.43741 | 0.73318 | 1.06652 | 1.29879 | 1.66718 | 1.82749 |
| 0.440 | 0.45014 | 0.74557 | 1.08187 | 1.30854 | 1.67444 | 1.83535 |
| 0.460 | 0.46334 | 0.75821 | 1.09734 | 1.31869 | 1.68191 | 1.84390 |
| 0.480 | 0.47696 | 0.77107 | 1.11288 | 1.32922 | 1.68959 | 1.85312 |
| 0.500 | 0.49097 | 0.78414 | 1.12843 | 1.34013 | 1.69752 | 1.86302 |
| 0.520 | 0.50534 | 0.79744 | 1.14393 | 1.35141 | 1.70572 | 1.87358 |
| 0.540 | 0.52006 | 0.81094 | 1.15937 | 1.36301 | 1.71422 | 1.88478 |
| 0.560 | 0.53508 | 0.82465 | 1.17469 | 1.37491 | 1.72308 | 1.89663 |
| 0.580 | 0.55039 | 0.83855 | 1.18985 | 1.38709 | 1.73234 | 1.90907 |
| 0.600 | 0.56597 | 0.85264 | 1.20484 | 1.39950 | 1.74203 | 1.92210 |
| 0.620 | 0.62001 | 0.86693 | 1.21964 | 1.41210 | 1.75221 | 1.93569 |
| 0.640 | 0.59785 | 0.88139 | 1.23424 | 1.42484 | 1.76291 | 1.94979 |
| 0.660 | 0.61411 | 0.89602 | 1.24861 | 1.43769 | 1.77418 | 1.96441 |
| 0.680 | 0.63057 | 0.91083 | 1.26278 | 1.45059 | 1.78604 | 1.97950 |
| 0.700 | 0.64720 | 0.92580 | 1.27674 | 1.46350 | 1.79853 | 1.99505 |
| 0.720 | 0.66400 | 0.94093 | 1.29049 | 1.47639 | 1.81165 | 2.01102 |
| 0.740 | 0.68096 | 0.95621 | 1.30406 | 1.48921 | 1.82544 | 2.02739 |
| 0.760 | 0.69805 | 0.97165 | 1.31743 | 1.50194 | 1.83986 | 2.04413 |
| 0.780 | 0.71528 | 0.98723 | 1.33064 | 1.51456 | 1.85493 | 2.06123 |
| 0.800 | 0.73263 | 1.00295 | 1.34369 | 1.52705 | 1.87062 | 2.07867 |
| 0.820 | 0.75010 | 1.01880 | 1.35660 | 1.53941 | 1.88691 | 2.09641 |
| 0.840 | 0.76767 | 1.03479 | 1.36937 | 1.55164 | 1.90376 | 2.11445 |
| 0.860 | 0.78534 | 1.05089 | 1.38202 | 1.56374 | 1.92114 | 2.13274 |
| 0.880 | 0.80311 | 1.06712 | 1.39458 | 1.57572 | 1.93899 | 2.15126 |
| 0.900 | 0.82095 | 1.08347 | 1.40703 | 1.58762 | 1.95728 | 2.17000 |
| 0.920 | 0.83888 | 1.09993 | 1.41939 | 1.59943 | 1.97592 | 2.18893 |
| 0.940 | 0.85688 | 1.11649 | 1.43167 | 1.61119 | 1.99489 | 2.20798 |
| 0.960 | 0.87494 | 1.13316 | 1.44388 | 1.62290 | 2.01411 | 2.22713 |
| 0.980 | 0.89307 | 1.14993 | 1.45603 | 1.63461 | 2.03355 | 2.24633 |
| 1.000 | 0.91127 | 1.16680 | 1.46812 | 1.64631 | 2.05314 | 2.26550 |

$H/R = 0.700 \qquad n = 3.$

| $H/L$ | $\Omega_1$ | $\Omega_2$ | $\Omega_3$ | $\Omega_4$ | $\Omega_5$ | $\Omega_6$ |
|---|---|---|---|---|---|---|
| 0. | 0.39677 | 0.65872 | 0.96946 | 1.27664 | 1.67773 | 1.87882 |
| 0.010 | 0.39680 | 0.65883 | 0.96957 | 1.27669 | 1.67775 | 1.87887 |
| 0.050 | 0.39774 | 0.66157 | 0.97220 | 1.27805 | 1.67835 | 1.88001 |
| 0.100 | 0.40085 | 0.66991 | 0.98035 | 1.28230 | 1.68021 | 1.88364 |
| 0.120 | 0.40279 | 0.67465 | 0.98508 | 1.28480 | 1.68130 | 1.88578 |
| 0.140 | 0.40519 | 0.68014 | 0.99062 | 1.28775 | 1.68258 | 1.88832 |
| 0.160 | 0.40809 | 0.68630 | 0.99696 | 1.29114 | 1.68408 | 1.89127 |
| 0.180 | 0.41152 | 0.69310 | 1.00407 | 1.29499 | 1.68576 | 1.89463 |
| 0.200 | 0.41551 | 0.70048 | 1.01190 | 1.29929 | 1.68765 | 1.89842 |
| 0.220 | 0.42010 | 0.70839 | 1.02043 | 1.30405 | 1.68974 | 1.90262 |
| 0.240 | 0.42529 | 0.71679 | 1.02963 | 1.30926 | 1.69205 | 1.90725 |
| 0.260 | 0.43111 | 0.72563 | 1.03944 | 1.31492 | 1.69456 | 1.91231 |
| 0.280 | 0.43756 | 0.73488 | 1.04983 | 1.32104 | 1.69731 | 1.91782 |
| 0.300 | 0.44465 | 0.74451 | 1.06075 | 1.32761 | 1.70029 | 1.92376 |
| 0.320 | 0.45237 | 0.75448 | 1.07214 | 1.33462 | 1.70351 | 1.93013 |
| 0.340 | 0.46070 | 0.76478 | 1.08396 | 1.34210 | 1.70698 | 1.93695 |
| 0.360 | 0.46964 | 0.77537 | 1.09615 | 1.35002 | 1.71074 | 1.94420 |
| 0.380 | 0.47917 | 0.78624 | 1.10867 | 1.35839 | 1.71478 | 1.95188 |
| 0.400 | 0.48926 | 0.79738 | 1.12145 | 1.36721 | 1.71912 | 1.95999 |
| 0.420 | 0.49990 | 0.80877 | 1.13444 | 1.37646 | 1.72378 | 1.96851 |
| 0.440 | 0.51105 | 0.82040 | 1.14761 | 1.38614 | 1.72880 | 1.97745 |
| 0.460 | 0.52270 | 0.83226 | 1.16089 | 1.39623 | 1.73420 | 1.98679 |
| 0.480 | 0.53481 | 0.84434 | 1.17424 | 1.40673 | 1.73999 | 1.99651 |
| 0.500 | 0.54736 | 0.85664 | 1.18762 | 1.41761 | 1.74621 | 2.00663 |
| 0.520 | 0.56033 | 0.86915 | 1.20100 | 1.42883 | 1.75288 | 2.01712 |
| 0.540 | 0.57368 | 0.88185 | 1.21434 | 1.44040 | 1.76005 | 2.02796 |
| 0.560 | 0.58740 | 0.89476 | 1.22761 | 1.45226 | 1.76771 | 2.03914 |
| 0.580 | 0.60146 | 0.90785 | 1.24079 | 1.46437 | 1.77594 | 2.05066 |
| 0.600 | 0.61585 | 0.92113 | 1.25386 | 1.47670 | 1.78473 | 2.06247 |
| 0.620 | 0.63053 | 0.93460 | 1.26681 | 1.48921 | 1.79411 | 2.07460 |
| 0.640 | 0.64549 | 0.94823 | 1.27964 | 1.50185 | 1.80412 | 2.08700 |
| 0.660 | 0.66073 | 0.96204 | 1.29233 | 1.51459 | 1.81477 | 2.09966 |
| 0.680 | 0.67616 | 0.97603 | 1.30490 | 1.52736 | 1.82609 | 2.11257 |
| 0.700 | 0.69185 | 0.99017 | 1.31733 | 1.54014 | 1.83808 | 2.12571 |
| 0.720 | 0.70774 | 1.00448 | 1.32965 | 1.55289 | 1.85073 | 2.13905 |
| 0.740 | 0.72383 | 1.01893 | 1.34184 | 1.56559 | 1.86406 | 2.15260 |
| 0.760 | 0.74010 | 1.03355 | 1.35392 | 1.57822 | 1.87804 | 2.16631 |
| 0.780 | 0.75655 | 1.04830 | 1.36591 | 1.59073 | 1.89266 | 2.18017 |
| 0.800 | 0.77315 | 1.06320 | 1.37781 | 1.60315 | 1.90790 | 2.19418 |
| 0.820 | 0.78990 | 1.07825 | 1.38962 | 1.61545 | 1.92373 | 2.20831 |
| 0.840 | 0.80678 | 1.09342 | 1.40136 | 1.62764 | 1.94010 | 2.22254 |
| 0.860 | 0.82379 | 1.10873 | 1.41303 | 1.63973 | 1.95697 | 2.23685 |
| 0.880 | 0.84092 | 1.12416 | 1.42466 | 1.65172 | 1.97428 | 2.25124 |
| 0.900 | 0.85817 | 1.13973 | 1.43624 | 1.66365 | 1.99202 | 2.26567 |
| 0.920 | 0.87551 | 1.15541 | 1.44778 | 1.67549 | 2.01009 | 2.28014 |
| 0.940 | 0.89295 | 1.17120 | 1.45929 | 1.68731 | 2.02848 | 2.29464 |
| 0.960 | 0.91048 | 1.18711 | 1.47079 | 1.69909 | 2.04710 | 2.30915 |
| 0.980 | 0.92810 | 1.20313 | 1.48227 | 1.71085 | 2.06592 | 2.32367 |
| 1.000 | 0.94579 | 1.21926 | 1.49376 | 1.72262 | 2.08488 | 2.33819 |

$H/R = 0.800 \qquad n = 3.$

| $H/L$ | $\Omega_1$ | $\Omega_2$ | $\Omega_3$ | $\Omega_4$ | $\Omega_5$ | $\Omega_6$ |
|---|---|---|---|---|---|---|
| 0. | 0.48404 | 0.74086 | 1.05558 | 1.37510 | 1.71597 | 2.06735 |
| 0.010 | 0.48407 | 0.74096 | 1.05567 | 1.37515 | 1.71600 | 2.06739 |
| 0.050 | 0.48476 | 0.74348 | 1.05798 | 1.37639 | 1.71655 | 2.06826 |
| 0.100 | 0.48707 | 0.75118 | 1.06512 | 1.38023 | 1.71828 | 2.07096 |
| 0.120 | 0.48853 | 0.75557 | 1.06926 | 1.38249 | 1.71929 | 2.07255 |
| 0.140 | 0.49035 | 0.76066 | 1.07412 | 1.38517 | 1.72052 | 2.07445 |
| 0.160 | 0.49256 | 0.76639 | 1.07968 | 1.38826 | 1.72194 | 2.07662 |
| 0.180 | 0.49520 | 0.77272 | 1.08590 | 1.39177 | 1.72356 | 2.07910 |
| 0.200 | 0.49830 | 0.77961 | 1.09277 | 1.39571 | 1.72540 | 2.08187 |
| 0.220 | 0.50189 | 0.78701 | 1.10026 | 1.40007 | 1.72747 | 2.08494 |
| 0.240 | 0.50599 | 0.79488 | 1.10832 | 1.40487 | 1.72975 | 2.08829 |
| 0.260 | 0.51063 | 0.80318 | 1.11693 | 1.41009 | 1.73228 | 2.09195 |
| 0.280 | 0.51582 | 0.81188 | 1.12606 | 1.41575 | 1.73505 | 2.09590 |
| 0.300 | 0.52158 | 0.82095 | 1.13564 | 1.42185 | 1.73807 | 2.10015 |
| 0.320 | 0.52790 | 0.83037 | 1.14566 | 1.42839 | 1.74138 | 2.10467 |
| 0.340 | 0.53479 | 0.84009 | 1.15606 | 1.43538 | 1.74496 | 2.10950 |
| 0.360 | 0.54224 | 0.85011 | 1.16679 | 1.44280 | 1.74885 | 2.11461 |
| 0.380 | 0.55026 | 0.86040 | 1.17782 | 1.45068 | 1.75304 | 2.11999 |
| 0.400 | 0.55882 | 0.87095 | 1.18908 | 1.45902 | 1.75755 | 2.12566 |
| 0.420 | 0.56791 | 0.88175 | 1.20056 | 1.46777 | 1.76242 | 2.13159 |
| 0.440 | 0.57752 | 0.89278 | 1.21219 | 1.47697 | 1.76764 | 2.13780 |
| 0.460 | 0.58764 | 0.90403 | 1.22394 | 1.48660 | 1.77325 | 2.14427 |
| 0.480 | 0.59824 | 0.91549 | 1.23577 | 1.49663 | 1.77927 | 2.15100 |
| 0.500 | 0.60930 | 0.92716 | 1.24763 | 1.50706 | 1.78571 | 2.15797 |
| 0.520 | 0.62080 | 0.93904 | 1.25951 | 1.51786 | 1.79259 | 2.16521 |
| 0.540 | 0.63272 | 0.95110 | 1.27138 | 1.52901 | 1.79993 | 2.17267 |
| 0.560 | 0.64504 | 0.96336 | 1.28321 | 1.54047 | 1.80778 | 2.18039 |
| 0.580 | 0.65775 | 0.97580 | 1.29498 | 1.55223 | 1.81613 | 2.18832 |
| 0.600 | 0.67081 | 0.98842 | 1.30668 | 1.56421 | 1.82503 | 2.19648 |
| 0.620 | 0.68421 | 1.00121 | 1.31830 | 1.57642 | 1.83448 | 2.20486 |
| 0.640 | 0.69793 | 1.01418 | 1.32983 | 1.58880 | 1.84450 | 2.21347 |
| 0.660 | 0.71196 | 1.02731 | 1.34128 | 1.60130 | 1.85512 | 2.22227 |
| 0.680 | 0.72626 | 1.04061 | 1.35262 | 1.61389 | 1.86635 | 2.23130 |
| 0.700 | 0.740834 | 1.05407 | 1.36390 | 1.62652 | 1.87819 | 2.24052 |
| 0.720 | 0.75567 | 1.06769 | 1.37507 | 1.63917 | 1.89064 | 2.24995 |
| 0.740 | 0.77073 | 1.08146 | 1.38618 | 1.65182 | 1.90371 | 2.25957 |
| 0.760 | 0.78602 | 1.09537 | 1.39723 | 1.66442 | 1.91737 | 2.26940 |
| 0.780 | 0.80151 | 1.10944 | 1.40820 | 1.67695 | 1.93163 | 2.27942 |
| 0.800 | 0.81719 | 1.12365 | 1.41912 | 1.68942 | 1.94645 | 2.28964 |
| 0.820 | 0.83306 | 1.13799 | 1.43001 | 1.70180 | 1.96181 | 2.30005 |
| 0.840 | 0.84910 | 1.15248 | 1.44086 | 1.71411 | 1.97767 | 2.31066 |
| 0.860 | 0.86530 | 1.16710 | 1.45170 | 1.72633 | 1.99398 | 2.32146 |
| 0.880 | 0.88166 | 1.18185 | 1.46251 | 1.73848 | 2.01073 | 2.33247 |
| 0.900 | 0.89814 | 1.19672 | 1.47332 | 1.75056 | 2.02785 | 2.34367 |
| 0.920 | 0.91478 | 1.21172 | 1.48413 | 1.76259 | 2.04530 | 2.35507 |
| 0.940 | 0.93153 | 1.22684 | 1.49495 | 1.77456 | 2.06302 | 2.36667 |
| 0.960 | 0.94839 | 1.24207 | 1.50579 | 1.78652 | 2.08099 | 2.37848 |
| 0.980 | 0.96537 | 1.25742 | 1.51666 | 1.79845 | 2.09913 | 2.39050 |
| 1.000 | 0.98245 | 1.27289 | 1.52755 | 1.81038 | 2.11741 | 2.40272 |

$H/R = 0.900 \qquad n = 3.$

| $H/L$ | $\Omega_1$ | $\Omega_2$ | $\Omega_3$ | $\Omega_4$ | $\Omega_5$ | $\Omega_6$ |
|---|---|---|---|---|---|---|
| 0. | 0.57214 | 0.81584 | 1.13675 | 1.49046 | 1.75138 | 2.25918 |
| 0.010 | 0.57216 | 0.81594 | 1.13684 | 1.49051 | 1.75141 | 2.25877 |
| 0.050 | 0.57261 | 0.81834 | 1.13890 | 1.49156 | 1.75210 | 2.25373 |
| 0.100 | 0.57419 | 0.82569 | 1.14528 | 1.49486 | 1.75427 | 2.24780 |
| 0.120 | 0.57520 | 0.82988 | 1.14899 | 1.49681 | 1.75555 | 2.24596 |
| 0.140 | 0.57650 | 0.83473 | 1.15334 | 1.49912 | 1.75706 | 2.24445 |
| 0.160 | 0.57810 | 0.84021 | 1.15832 | 1.50180 | 1.75881 | 2.24327 |
| 0.180 | 0.58005 | 0.84627 | 1.16390 | 1.50485 | 1.76081 | 2.24243 |
| 0.200 | 0.58237 | 0.85286 | 1.17006 | 1.50828 | 1.76306 | 2.24194 |
| 0.220 | 0.58510 | 0.85995 | 1.17678 | 1.51210 | 1.76556 | 2.24179 |
| 0.240 | 0.58826 | 0.86749 | 1.18403 | 1.51629 | 1.76832 | 2.24197 |
| 0.260 | 0.59189 | 0.87545 | 1.19177 | 1.52089 | 1.77133 | 2.24247 |
| 0.280 | 0.59599 | 0.88379 | 1.19998 | 1.52589 | 1.77462 | 2.24333 |
| 0.300 | 0.60060 | 0.89250 | 1.20863 | 1.53131 | 1.77817 | 2.24451 |
| 0.320 | 0.60571 | 0.90153 | 1.21767 | 1.53714 | 1.78202 | 2.24602 |
| 0.340 | 0.61133 | 0.91086 | 1.22706 | 1.54340 | 1.78615 | 2.24786 |
| 0.360 | 0.61748 | 0.92048 | 1.23678 | 1.55009 | 1.79058 | 2.25002 |
| 0.380 | 0.62415 | 0.93036 | 1.24677 | 1.55721 | 1.79531 | 2.25250 |
| 0.400 | 0.63134 | 0.94049 | 1.25700 | 1.56476 | 1.80036 | 2.25530 |
| 0.420 | 0.63903 | 0.95085 | 1.26742 | 1.57277 | 1.80574 | 2.25840 |
| 0.440 | 0.64724 | 0.96144 | 1.27801 | 1.58121 | 1.81146 | 2.26182 |
| 0.460 | 0.65593 | 0.97224 | 1.28871 | 1.59009 | 1.81752 | 2.26555 |
| 0.480 | 0.66510 | 0.98324 | 1.29951 | 1.59940 | 1.82395 | 2.26957 |
| 0.500 | 0.67474 | 0.99443 | 1.31036 | 1.60913 | 1.83076 | 2.27389 |
| 0.520 | 0.68483 | 1.00582 | 1.32125 | 1.61926 | 1.83795 | 2.27852 |
| 0.540 | 0.69535 | 1.01738 | 1.33213 | 1.62978 | 1.84556 | 2.28342 |
| 0.560 | 0.70630 | 1.02913 | 1.34300 | 1.64066 | 1.85360 | 2.28863 |
| 0.580 | 0.71764 | 1.04105 | 1.35384 | 1.65189 | 1.86208 | 2.29411 |
| 0.600 | 0.72937 | 1.05315 | 1.36462 | 1.66341 | 1.87102 | 2.29989 |
| 0.620 | 0.74147 | 1.06541 | 1.37535 | 1.67520 | 1.88044 | 2.30594 |
| 0.640 | 0.75391 | 1.07783 | 1.38600 | 1.68724 | 1.89035 | 2.31227 |
| 0.660 | 0.76669 | 1.09041 | 1.39659 | 1.69946 | 1.90078 | 2.31888 |
| 0.680 | 0.779[illegible]3 | 1.10316 | 1.40711 | 1.71184 | 1.91172 | 2.32575 |
| 0.700 | 0.7931[illegible] | 1.11605 | 1.41756 | 1.72433 | 1.92321 | 2.33290 |
| 0.720 | 0.80686 | 1.12910 | 1.42795 | 1.73690 | 1.93523 | 2.34034 |
| 0.740 | 0.82081 | 1.14229 | 1.43828 | 1.74952 | 1.94781 | 2.34802 |
| 0.760 | 0.83501 | 1.15563 | 1.44858 | 1.76215 | 1.96092 | 2.35599 |
| 0.780 | 0.84946 | 1.16911 | 1.45882 | 1.77476 | 1.97456 | 2.36423 |
| 0.800 | 0.86413 | 1.18273 | 1.46904 | 1.78734 | 1.98872 | 2.37272 |
| 0.820 | 0.87903 | 1.19649 | 1.47923 | 1.79987 | 2.00338 | 2.38150 |
| 0.840 | 0.89412 | 1.21038 | 1.48941 | 1.81233 | 2.01852 | 2.39054 |
| 0.860 | 0.90941 | 1.22442 | 1.49959 | 1.82474 | 2.03409 | 2.39985 |
| 0.880 | 0.92488 | 1.23858 | 1.50978 | 1.83707 | 2.05007 | 2.40945 |
| 0.900 | 0.94051 | 1.25285 | 1.51999 | 1.84933 | 2.06643 | 2.41931 |
| 0.920 | 0.95632 | 1.26726 | 1.53022 | 1.86155 | 2.08310 | 2.42945 |
| 0.940 | 0.97226 | 1.28178 | 1.54049 | 1.87370 | 2.10006 | 2.43987 |
| 0.960 | 0.98836 | 1.29642 | 1.55080 | 1.88582 | 2.11725 | 2.45057 |
| 0.980 | 1.00459 | 1.31118 | 1.56116 | 1.89791 | 2.13464 | 2.46153 |
| 1.000 | 1.02095 | 1.32606 | 1.57158 | 1.90996 | 2.15217 | 2.47279 |

$H/R = 1.000 \qquad n = 3.$

| $H/L$ | $\Omega_1$ | $\Omega_2$ | $\Omega_3$ | $\Omega_4$ | $\Omega_5$ | $\Omega_6$ |
|---|---|---|---|---|---|---|
| 0. | 0.65849 | 0.88365 | 1.21733 | 1.61842 | 1.79897 | 2.35095 |
| 0.010 | 0.65850 | 0.88375 | 1.21741 | 1.61845 | 1.79900 | 2.34734 |
| 0.050 | 0.65872 | 0.88612 | 1.21928 | 1.61927 | 1.79988 | 2.34718 |
| 0.100 | 0.65957 | 0.89336 | 1.22510 | 1.62188 | 1.80258 | 2.34683 |
| 0.120 | 0.66018 | 0.89748 | 1.22848 | 1.62341 | 1.80415 | 2.34671 |
| 0.140 | 0.66098 | 0.90225 | 1.23244 | 1.62525 | 1.80602 | 2.34665 |
| 0.160 | 0.66202 | 0.90764 | 1.23700 | 1.62738 | 1.80818 | 2.34666 |
| 0.180 | 0.66334 | 0.91359 | 1.24209 | 1.62983 | 1.81062 | 2.34679 |
| 0.200 | 0.66496 | 0.92006 | 1.24773 | 1.63260 | 1.81334 | 2.34705 |
| 0.220 | 0.66693 | 0.92701 | 1.25388 | 1.63569 | 1.81634 | 2.34745 |
| 0.240 | 0.66927 | 0.93440 | 1.26054 | 1.63913 | 1.81963 | 2.34803 |
| 0.260 | 0.67200 | 0.94220 | 1.26766 | 1.64293 | 1.82319 | 2.34880 |
| 0.280 | 0.67516 | 0.95037 | 1.27522 | 1.64707 | 1.82704 | 2.34977 |
| 0.300 | 0.67876 | 0.95889 | 1.28318 | 1.65160 | 1.83117 | 2.35096 |
| 0.320 | 0.68282 | 0.96771 | 1.29154 | 1.65652 | 1.83557 | 2.35236 |
| 0.340 | 0.68735 | 0.97683 | 1.30023 | 1.66184 | 1.84026 | 2.35400 |
| 0.360 | 0.69236 | 0.98621 | 1.30925 | 1.66757 | 1.84524 | 2.35588 |
| 0.380 | 0.69785 | 0.99585 | 1.31853 | 1.67372 | 1.85049 | 2.35800 |
| 0.400 | 0.70383 | 1.00571 | 1.32807 | 1.68031 | 1.85604 | 2.36039 |
| 0.420 | 0.71029 | 1.01580 | 1.33781 | 1.68735 | 1.86186 | 2.36301 |
| 0.440 | 0.71724 | 1.02609 | 1.34772 | 1.69483 | 1.86798 | 2.36591 |
| 0.460 | 0.72466 | 1.03658 | 1.35777 | 1.70278 | 1.87439 | 2.36905 |
| 0.480 | 0.73255 | 1.04726 | 1.36793 | 1.71118 | 1.88112 | 2.37247 |
| 0.500 | 0.74090 | 1.05812 | 1.37817 | 1.72004 | 1.88813 | 2.37614 |
| 0.520 | 0.74969 | 1.06916 | 1.38845 | 1.72935 | 1.89549 | 2.38008 |
| 0.540 | 0.75893 | 1.08036 | 1.39877 | 1.73909 | 1.90315 | 2.38429 |
| 0.560 | 0.76859 | 1.09173 | 1.40908 | 1.74928 | 1.91113 | 2.38876 |
| 0.580 | 0.77866 | 1.10326 | 1.41938 | 1.75987 | 1.91948 | 2.39350 |
| 0.600 | 0.78912 | 1.11495 | 1.42965 | 1.77084 | 1.92819 | 2.39852 |
| 0.620 | 0.79997 | 1.12679 | 1.43987 | 1.78218 | 1.93726 | 2.40380 |
| 0.640 | 0.81119 | 1.13878 | 1.45004 | 1.79384 | 1.94671 | 2.40935 |
| 0.660 | 0.82276 | 1.15093 | 1.46016 | 1.80579 | 1.95656 | 2.41516 |
| 0.680 | 0.83466 | 1.16322 | 1.47022 | 1.81801 | 1.96685 | 2.42126 |
| 0.700 | 0.84689 | 1.17565 | 1.48022 | 1.83042 | 1.97755 | 2.42762 |
| 0.720 | 0.85943 | 1.18823 | 1.49017 | 1.84301 | 1.98872 | 2.43426 |
| 0.740 | 0.87226 | 1.20095 | 1.50008 | 1.85571 | 2.00034 | 2.44118 |
| 0.760 | 0.88538 | 1.21380 | 1.50993 | 1.86849 | 2.01244 | 2.44834 |
| 0.780 | 0.89877 | 1.22680 | 1.51977 | 1.88133 | 2.02502 | 2.45582 |
| 0.800 | 0.91240 | 1.23992 | 1.52957 | 1.89417 | 2.03808 | 2.46356 |
| 0.820 | 0.92628 | 1.25317 | 1.53936 | 1.90698 | 2.05162 | 2.47158 |
| 0.840 | 0.94039 | 1.26655 | 1.54914 | 1.91974 | 2.06562 | 2.47987 |
| 0.860 | 0.95472 | 1.28007 | 1.55893 | 1.93245 | 2.08007 | 2.48846 |
| 0.880 | 0.96926 | 1.29370 | 1.56872 | 1.94507 | 2.09494 | 2.49735 |
| 0.900 | 0.98400 | 1.30747 | 1.57854 | 1.95761 | 2.11023 | 2.50650 |
| 0.920 | 0.99892 | 1.32135 | 1.58840 | 1.97007 | 2.12586 | 2.51596 |
| 0.940 | 1.01402 | 1.33535 | 1.59830 | 1.98245 | 2.14182 | 2.52572 |
| 0.960 | 1.02929 | 1.34947 | 1.60825 | 1.99475 | 2.15807 | 2.53577 |
| 0.980 | 1.04472 | 1.36370 | 1.61825 | 2.00698 | 2.17456 | 2.54611 |
| 1.000 | 1.06030 | 1.37805 | 1.62832 | 2.01918 | 2.19125 | 2.55677 |

$H/R = 1.100 \qquad n = 3.$

| $H/L$ | $\Omega_1$ | $\Omega_2$ | $\Omega_3$ | $\Omega_4$ | $\Omega_5$ | $\Omega_6$ |
|---|---|---|---|---|---|---|
| 0. | 0.73997 | 0.94506 | 1.30064 | 1.75088 | 1.86536 | 2.46241 |
| 0.010 | 0.73997 | 0.94516 | 1.30071 | 1.75090 | 1.86541 | 2.46242 |
| 0.050 | 0.73997 | 0.94756 | 1.30245 | 1.75140 | 1.86650 | 2.46249 |
| 0.100 | 0.74012 | 0.95487 | 1.30785 | 1.75303 | 1.86989 | 2.46277 |
| 0.120 | 0.74034 | 0.95903 | 1.31100 | 1.75402 | 1.87185 | 2.46298 |
| 0.140 | 0.74069 | 0.96384 | 1.31468 | 1.75520 | 1.87416 | 2.46324 |
| 0.160 | 0.74121 | 0.96925 | 1.31892 | 1.75662 | 1.87680 | 2.46358 |
| 0.180 | 0.74195 | 0.97522 | 1.32368 | 1.75828 | 1.87976 | 2.46403 |
| 0.200 | 0.74294 | 0.98170 | 1.32894 | 1.76018 | 1.88303 | 2.46458 |
| 0.220 | 0.74423 | 0.98864 | 1.33470 | 1.76235 | 1.88661 | 2.46524 |
| 0.240 | 0.74584 | 0.99602 | 1.34093 | 1.76481 | 1.89049 | 2.46606 |
| 0.260 | 0.74780 | 1.00380 | 1.34762 | 1.76757 | 1.89465 | 2.46700 |
| 0.280 | 0.75013 | 1.01193 | 1.35474 | 1.77066 | 1.89909 | 2.46813 |
| 0.300 | 0.75288 | 1.02039 | 1.36226 | 1.77410 | 1.90379 | 2.46938 |
| 0.320 | 0.75603 | 1.02915 | 1.37016 | 1.77789 | 1.90877 | 2.47084 |
| 0.340 | 0.75962 | 1.03818 | 1.37842 | 1.78207 | 1.91399 | 2.47250 |
| 0.360 | 0.76365 | 1.04746 | 1.38699 | 1.78664 | 1.91948 | 2.47433 |
| 0.380 | 0.76814 | 1.05698 | 1.39586 | 1.79163 | 1.92520 | 2.47636 |
| 0.400 | 0.77308 | 1.06672 | 1.40499 | 1.79706 | 1.93115 | 2.47861 |
| 0.420 | 0.77849 | 1.07665 | 1.41435 | 1.80292 | 1.93735 | 2.48109 |
| 0.440 | 0.78434 | 1.08678 | 1.42391 | 1.80926 | 1.94377 | 2.48375 |
| 0.460 | 0.79067 | 1.09709 | 1.43363 | 1.81607 | 1.95042 | 2.48666 |
| 0.480 | 0.79744 | 1.10756 | 1.44348 | 1.82337 | 1.95730 | 2.48979 |
| 0.500 | 0.80467 | 1.11821 | 1.45344 | 1.83116 | 1.96440 | 2.49316 |
| 0.520 | 0.81233 | 1.12901 | 1.46349 | 1.83944 | 1.97173 | 2.49675 |
| 0.540 | 0.82043 | 1.13996 | 1.47358 | 1.84823 | 1.97927 | 2.50058 |
| 0.560 | 0.82895 | 1.15106 | 1.48371 | 1.85751 | 1.98706 | 2.50467 |
| 0.580 | 0.83788 | 1.16231 | 1.49385 | 1.86730 | 1.99507 | 2.50898 |
| 0.600 | 0.84722 | 1.17369 | 1.50397 | 1.87756 | 2.00333 | 2.51355 |
| 0.620 | 0.85693 | 1.18522 | 1.51408 | 1.88828 | 2.01182 | 2.51835 |
| 0.640 | 0.86703 | 1.19689 | 1.52415 | 1.89944 | 2.02056 | 2.52341 |
| 0.660 | 0.87749 | 1.20868 | 1.53418 | 1.91103 | 2.02956 | 2.52875 |
| 0.680 | 0.88830 | 1.22061 | 1.54416 | 1.92301 | 2.03884 | 2.53432 |
| 0.700 | 0.89945 | 1.23268 | 1.55408 | 1.93535 | 2.04842 | 2.54014 |
| 0.720 | 0.91092 | 1.24487 | 1.56396 | 1.94799 | 2.05830 | 2.54623 |
| 0.740 | 0.92271 | 1.25720 | 1.57379 | 1.96091 | 2.06850 | 2.55259 |
| 0.760 | 0.93479 | 1.26964 | 1.58357 | 1.97405 | 2.07903 | 2.55920 |
| 0.780 | 0.94717 | 1.28222 | 1.59331 | 1.98736 | 2.08995 | 2.56609 |
| 0.800 | 0.95981 | 1.29491 | 1.60303 | 2.00079 | 2.10125 | 2.57324 |
| 0.820 | 0.97272 | 1.30773 | 1.61272 | 2.01427 | 2.11297 | 2.58068 |
| 0.840 | 0.98589 | 1.32068 | 1.62239 | 2.02776 | 2.12512 | 2.58839 |
| 0.860 | 0.99929 | 1.33374 | 1.63206 | 2.04122 | 2.13773 | 2.59639 |
| 0.880 | 1.01292 | 1.34692 | 1.64175 | 2.05459 | 2.15079 | 2.60467 |
| 0.900 | 1.02677 | 1.36022 | 1.65143 | 2.06785 | 2.16430 | 2.61322 |
| 0.920 | 1.04083 | 1.37364 | 1.66114 | 2.08097 | 2.17827 | 2.62209 |
| 0.940 | 1.05508 | 1.38717 | 1.67090 | 2.09393 | 2.19268 | 2.63124 |
| 0.960 | 1.06953 | 1.40081 | 1.68068 | 2.10673 | 2.20749 | 2.64068 |
| 0.980 | 1.08415 | 1.41456 | 1.69052 | 2.11938 | 2.22268 | 2.65043 |
| 1.000 | 1.09895 | 1.42843 | 1.70042 | 2.13188 | 2.23817 | 2.66048 |

$H/R = 1.200 \qquad n = 3.$

| $H/L$ | $\Omega_1$ | $\Omega_2$ | $\Omega_3$ | $\Omega_4$ | $\Omega_5$ | $\Omega_6$ |
|---|---|---|---|---|---|---|
| 0. | 0.81364 | 1.00114 | 1.38739 | 1.87799 | 1.95743 | 2.60706 |
| 0.010 | 0.81363 | 1.00124 | 1.38745 | 1.87799 | 1.95749 | 2.60706 |
| 0.050 | 0.81342 | 1.00370 | 1.38910 | 1.87814 | 1.95881 | 2.60712 |
| 0.100 | 0.81295 | 1.01117 | 1.39422 | 1.87870 | 1.96287 | 2.60732 |
| 0.120 | 0.81281 | 1.01540 | 1.39722 | 1.87910 | 1.96520 | 2.60745 |
| 0.140 | 0.81275 | 1.02029 | 1.40072 | 1.87963 | 1.96789 | 2.60765 |
| 0.160 | 0.81282 | 1.02578 | 1.40475 | 1.88030 | 1.97096 | 2.60791 |
| 0.180 | 0.81306 | 1.03183 | 1.40929 | 1.88117 | 1.97434 | 2.60824 |
| 0.200 | 0.81352 | 1.03837 | 1.41433 | 1.88223 | 1.97805 | 2.60866 |
| 0.220 | 0.81423 | 1.04538 | 1.41985 | 1.88351 | 1.98206 | 2.60920 |
| 0.240 | 0.81522 | 1.05281 | 1.42583 | 1.88504 | 1.98635 | 2.60985 |
| 0.260 | 0.81653 | 1.06062 | 1.43227 | 1.88685 | 1.99090 | 2.61062 |
| 0.280 | 0.81818 | 1.06878 | 1.43913 | 1.88895 | 1.99572 | 2.61154 |
| 0.300 | 0.82020 | 1.07724 | 1.44640 | 1.89136 | 2.00077 | 2.61257 |
| 0.320 | 0.82260 | 1.08600 | 1.45407 | 1.89412 | 2.00606 | 2.61377 |
| 0.340 | 0.82542 | 1.09502 | 1.46210 | 1.89722 | 2.01157 | 2.61515 |
| 0.360 | 0.82864 | 1.10427 | 1.47048 | 1.90072 | 2.01729 | 2.61670 |
| 0.380 | 0.83229 | 1.11375 | 1.47917 | 1.90460 | 2.02322 | 2.61842 |
| 0.400 | 0.83639 | 1.12342 | 1.48815 | 1.90889 | 2.02933 | 2.62031 |
| 0.420 | 0.84091 | 1.13328 | 1.49739 | 1.91362 | 2.03564 | 2.62242 |
| 0.440 | 0.84587 | 1.14332 | 1.50687 | 1.91880 | 2.04213 | 2.62471 |
| 0.460 | 0.85128 | 1.15352 | 1.51655 | 1.92443 | 2.04879 | 2.62720 |
| 0.480 | 0.85713 | 1.16388 | 1.52640 | 1.93056 | 2.05563 | 2.62991 |
| 0.500 | 0.86341 | 1.17438 | 1.53641 | 1.93715 | 2.06264 | 2.63282 |
| 0.520 | 0.87012 | 1.18502 | 1.54654 | 1.94427 | 2.06982 | 2.63595 |
| 0.540 | 0.87726 | 1.19580 | 1.55675 | 1.95189 | 2.07717 | 2.63929 |
| 0.560 | 0.88481 | 1.20671 | 1.56704 | 1.96001 | 2.08467 | 2.64286 |
| 0.580 | 0.89277 | 1.21775 | 1.57737 | 1.96867 | 2.09233 | 2.64667 |
| 0.600 | 0.90113 | 1.22891 | 1.58772 | 1.97784 | 2.10015 | 2.65070 |
| 0.620 | 0.90987 | 1.24020 | 1.59809 | 1.98753 | 2.10814 | 2.65495 |
| 0.640 | 0.91901 | 1.25161 | 1.60843 | 1.99774 | 2.11628 | 2.65943 |
| 0.660 | 0.92850 | 1.26314 | 1.61875 | 2.00844 | 2.12459 | 2.66418 |
| 0.680 | 0.93835 | 1.27478 | 1.62904 | 2.01965 | 2.13305 | 2.66915 |
| 0.700 | 0.94855 | 1.28656 | 1.63928 | 2.03131 | 2.14170 | 2.67436 |
| 0.720 | 0.95909 | 1.29843 | 1.64948 | 2.04343 | 2.15051 | 2.67983 |
| 0.740 | 0.96994 | 1.31043 | 1.65963 | 2.05598 | 2.15950 | 2.68554 |
| 0.760 | 0.98111 | 1.32254 | 1.66971 | 2.06893 | 2.16868 | 2.69150 |
| 0.780 | 0.99257 | 1.33476 | 1.67976 | 2.08224 | 2.17805 | 2.69772 |
| 0.800 | 1.00433 | 1.34710 | 1.68974 | 2.09587 | 2.18765 | 2.70422 |
| 0.820 | 1.01636 | 1.35955 | 1.69969 | 2.10979 | 2.19746 | 2.71096 |
| 0.840 | 1.02866 | 1.37212 | 1.70961 | 2.12393 | 2.20753 | 2.71797 |
| 0.860 | 1.04122 | 1.38480 | 1.71950 | 2.13824 | 2.21789 | 2.72528 |
| 0.880 | 1.05401 | 1.39759 | 1.72938 | 2.15269 | 2.22853 | 2.73284 |
| 0.900 | 1.06705 | 1.41048 | 1.73923 | 2.16718 | 2.23952 | 2.74069 |
| 0.920 | 1.08030 | 1.42349 | 1.74909 | 2.18165 | 2.25088 | 2.74884 |
| 0.940 | 1.09378 | 1.43660 | 1.75896 | 2.19604 | 2.26266 | 2.75726 |
| 0.960 | 1.10746 | 1.44982 | 1.76884 | 2.21027 | 2.27489 | 2.76597 |
| 0.980 | 1.12134 | 1.46315 | 1.77875 | 2.22428 | 2.28761 | 2.77498 |
| 1.000 | 1.13540 | 1.47658 | 1.78870 | 2.23803 | 2.30084 | 2.78430 |

$H/R = 1.300 \qquad n = 3.$

| $H/L$ | $\Omega_1$ | $\Omega_2$ | $\Omega_3$ | $\Omega_4$ | $\Omega_5$ | $\Omega_6$ |
|---|---|---|---|---|---|---|
| 0. | 0.87780 | 1.05287 | 1.47451 | 1.99278 | 2.08198 | 2.77221 |
| 0.010 | 0.87778 | 1.05297 | 1.47457 | 1.99279 | 2.08203 | 2.77219 |
| 0.050 | 0.87740 | 1.05548 | 1.47616 | 1.99292 | 2.08324 | 2.77203 |
| 0.100 | 0.87643 | 1.06308 | 1.48111 | 1.99340 | 2.08697 | 2.77158 |
| 0.120 | 0.87601 | 1.06739 | 1.48401 | 1.99373 | 2.08911 | 2.77140 |
| 0.140 | 0.87563 | 1.07234 | 1.48741 | 1.99418 | 2.09160 | 2.77121 |
| 0.160 | 0.87534 | 1.07789 | 1.49133 | 1.99474 | 2.09444 | 2.77104 |
| 0.180 | 0.87519 | 1.08399 | 1.49574 | 1.99546 | 2.09759 | 2.77092 |
| 0.200 | 0.87522 | 1.09059 | 1.50064 | 1.99633 | 2.10103 | 2.77084 |
| 0.220 | 0.87547 | 1.09765 | 1.50601 | 1.99739 | 2.10477 | 2.77087 |
| 0.240 | 0.87598 | 1.10511 | 1.51186 | 1.99867 | 2.10878 | 2.77098 |
| 0.260 | 0.87678 | 1.11295 | 1.51816 | 2.00017 | 2.11304 | 2.77121 |
| 0.280 | 0.87789 | 1.12112 | 1.52491 | 2.00192 | 2.11756 | 2.77156 |
| 0.300 | 0.87935 | 1.12959 | 1.53206 | 2.00393 | 2.12230 | 2.77202 |
| 0.320 | 0.88117 | 1.13835 | 1.53963 | 2.00623 | 2.12727 | 2.77265 |
| 0.340 | 0.88337 | 1.14735 | 1.54760 | 2.00882 | 2.13246 | 2.77344 |
| 0.360 | 0.88596 | 1.15658 | 1.55593 | 2.01174 | 2.13784 | 2.77435 |
| 0.380 | 0.88895 | 1.16602 | 1.56460 | 2.01499 | 2.14342 | 2.77547 |
| 0.400 | 0.89236 | 1.17565 | 1.57360 | 2.01861 | 2.14919 | 2.77675 |
| 0.420 | 0.89619 | 1.18546 | 1.58290 | 2.02258 | 2.15514 | 2.77823 |
| 0.440 | 0.90043 | 1.19542 | 1.59247 | 2.02696 | 2.16126 | 2.77986 |
| 0.460 | 0.90511 | 1.20554 | 1.60230 | 2.03173 | 2.16755 | 2.78173 |
| 0.480 | 0.91021 | 1.21580 | 1.61235 | 2.03692 | 2.17400 | 2.78376 |
| 0.500 | 0.91572 | 1.22621 | 1.62261 | 2.04254 | 2.18062 | 2.78599 |
| 0.520 | 0.92166 | 1.23673 | 1.63303 | 2.04861 | 2.18738 | 2.78844 |
| 0.540 | 0.92801 | 1.24738 | 1.64360 | 2.05513 | 2.19430 | 2.79109 |
| 0.560 | 0.93477 | 1.25815 | 1.65429 | 2.06212 | 2.20137 | 2.79396 |
| 0.580 | 0.94193 | 1.26903 | 1.66507 | 2.06959 | 2.20858 | 2.79704 |
| 0.600 | 0.94949 | 1.28002 | 1.67593 | 2.07754 | 2.21593 | 2.80034 |
| 0.620 | 0.95743 | 1.29112 | 1.68684 | 2.08599 | 2.22343 | 2.80386 |
| 0.640 | 0.96575 | 1.30233 | 1.69777 | 2.09492 | 2.23105 | 2.80760 |
| 0.660 | 0.97444 | 1.31365 | 1.70871 | 2.10435 | 2.23883 | 2.81159 |
| 0.680 | 0.98348 | 1.32507 | 1.71964 | 2.11426 | 2.24672 | 2.81578 |
| 0.700 | 0.99288 | 1.33660 | 1.73056 | 2.12468 | 2.25477 | 2.82022 |
| 0.720 | 1.00261 | 1.34823 | 1.74143 | 2.13556 | 2.26295 | 2.82490 |
| 0.740 | 1.01268 | 1.35997 | 1.75227 | 2.14692 | 2.27127 | 2.82980 |
| 0.760 | 1.02306 | 1.37182 | 1.76305 | 2.15873 | 2.27972 | 2.83496 |
| 0.780 | 1.03375 | 1.38376 | 1.77378 | 2.17100 | 2.28831 | 2.84037 |
| 0.800 | 1.04473 | 1.39580 | 1.78444 | 2.18366 | 2.29704 | 2.84603 |
| 0.820 | 1.05600 | 1.40795 | 1.79505 | 2.19673 | 2.30592 | 2.85193 |
| 0.840 | 1.06755 | 1.42021 | 1.80560 | 2.21017 | 2.31496 | 2.85811 |
| 0.860 | 1.07936 | 1.43256 | 1.81610 | 2.22394 | 2.32415 | 2.86454 |
| 0.880 | 1.09143 | 1.44502 | 1.82654 | 2.23802 | 2.33350 | 2.87123 |
| 0.900 | 1.10375 | 1.45758 | 1.83694 | 2.25238 | 2.34303 | 2.87822 |
| 0.920 | 1.11630 | 1.47025 | 1.84731 | 2.26697 | 2.35275 | 2.88546 |
| 0.940 | 1.12909 | 1.48301 | 1.85765 | 2.28174 | 2.36266 | 2.89299 |
| 0.960 | 1.14209 | 1.49588 | 1.86796 | 2.29664 | 2.37282 | 2.90081 |
| 0.980 | 1.15530 | 1.50885 | 1.87826 | 2.31165 | 2.38321 | 2.90894 |
| 1.000 | 1.16871 | 1.52191 | 1.88857 | 2.32666 | 2.39389 | 2.91734 |

$H/R = 1.400 \qquad n = 3.$

| $H/L$ | $\Omega_1$ | $\Omega_2$ | $\Omega_3$ | $\Omega_4$ | $\Omega_5$ | $\Omega_6$ |
|---|---|---|---|---|---|---|
| 0. | 0.93256 | 1.10103 | 1.55696 | 2.09412 | 2.23924 | 2.93610 |
| 0.010 | 0.93253 | 1.10113 | 1.55703 | 2.09413 | 2.23927 | 2.93609 |
| 0.050 | 0.93204 | 1.10365 | 1.55857 | 2.09445 | 2.24017 | 2.93545 |
| 0.100 | 0.93073 | 1.11129 | 1.56339 | 2.09551 | 2.24299 | 2.93373 |
| 0.120 | 0.93010 | 1.11561 | 1.56621 | 2.09614 | 2.24462 | 2.93287 |
| 0.140 | 0.92950 | 1.12057 | 1.56954 | 2.09690 | 2.24653 | 2.93198 |
| 0.160 | 0.92897 | 1.12613 | 1.57335 | 2.09781 | 2.24874 | 2.93105 |
| 0.180 | 0.92854 | 1.13223 | 1.57768 | 2.09887 | 2.25120 | 2.93020 |
| 0.200 | 0.92827 | 1.13883 | 1.58247 | 2.10010 | 2.25393 | 2.92939 |
| 0.220 | 0.92820 | 1.14588 | 1.58774 | 2.10149 | 2.25692 | 2.92867 |
| 0.240 | 0.92836 | 1.15332 | 1.59349 | 2.10307 | 2.26017 | 2.92803 |
| 0.260 | 0.92878 | 1.16114 | 1.59968 | 2.10483 | 2.26365 | 2.92751 |
| 0.280 | 0.92950 | 1.16928 | 1.60634 | 2.10681 | 2.26738 | 2.92716 |
| 0.300 | 0.93054 | 1.17773 | 1.61342 | 2.10902 | 2.27133 | 2.92692 |
| 0.320 | 0.93192 | 1.18644 | 1.62092 | 2.11144 | 2.27551 | 2.92685 |
| 0.340 | 0.93365 | 1.19540 | 1.62883 | 2.11411 | 2.27989 | 2.92693 |
| 0.360 | 0.93577 | 1.20458 | 1.63713 | 2.11706 | 2.28450 | 2.92719 |
| 0.380 | 0.93826 | 1.21396 | 1.64580 | 2.12026 | 2.28930 | 2.92762 |
| 0.400 | 0.94114 | 1.22353 | 1.65482 | 2.12375 | 2.29429 | 2.92825 |
| 0.420 | 0.94442 | 1.23327 | 1.66419 | 2.12755 | 2.29948 | 2.92904 |
| 0.440 | 0.94811 | 1.24317 | 1.67386 | 2.13166 | 2.30486 | 2.93005 |
| 0.460 | 0.95221 | 1.25321 | 1.68383 | 2.13608 | 2.31042 | 2.93124 |
| 0.480 | 0.95671 | 1.26337 | 1.69407 | 2.14085 | 2.31614 | 2.93262 |
| 0.500 | 0.96163 | 1.27368 | 1.70455 | 2.14599 | 2.32204 | 2.93417 |
| 0.520 | 0.96694 | 1.28410 | 1.71526 | 2.15147 | 2.32812 | 2.93595 |
| 0.540 | 0.97267 | 1.29464 | 1.72617 | 2.15733 | 2.33434 | 2.93795 |
| 0.560 | 0.97879 | 1.30529 | 1.73726 | 2.16359 | 2.34073 | 2.94014 |
| 0.580 | 0.98530 | 1.31604 | 1.74850 | 2.17025 | 2.34728 | 2.94252 |
| 0.600 | 0.99221 | 1.32689 | 1.75987 | 2.17731 | 2.35398 | 2.94513 |
| 0.620 | 0.99949 | 1.33785 | 1.77134 | 2.18480 | 2.36083 | 2.94796 |
| 0.640 | 1.00715 | 1.34891 | 1.78289 | 2.19272 | 2.36783 | 2.95098 |
| 0.660 | 1.01517 | 1.36006 | 1.79450 | 2.20105 | 2.37499 | 2.95425 |
| 0.680 | 1.02355 | 1.37131 | 1.80617 | 2.20984 | 2.38228 | 2.95771 |
| 0.700 | 1.03228 | 1.38265 | 1.81784 | 2.21905 | 2.38971 | 2.96141 |
| 0.720 | 1.04135 | 1.39409 | 1.82951 | 2.22872 | 2.39729 | 2.96533 |
| 0.740 | 1.05074 | 1.40562 | 1.84118 | 2.23883 | 2.40502 | 2.96949 |
| 0.760 | 1.06047 | 1.41725 | 1.85282 | 2.24936 | 2.41288 | 2.97386 |
| 0.780 | 1.07050 | 1.42897 | 1.86441 | 2.26032 | 2.42089 | 2.97848 |
| 0.800 | 1.08083 | 1.44080 | 1.87596 | 2.27171 | 2.42904 | 2.98333 |
| 0.820 | 1.09146 | 1.45271 | 1.88745 | 2.28351 | 2.43733 | 2.98845 |
| 0.840 | 1.10237 | 1.46472 | 1.89888 | 2.29570 | 2.44577 | 2.99379 |
| 0.860 | 1.11355 | 1.47683 | 1.91024 | 2.30826 | 2.45435 | 2.99936 |
| 0.880 | 1.12500 | 1.48902 | 1.92154 | 2.32119 | 2.46310 | 3.00521 |
| 0.900 | 1.13670 | 1.50132 | 1.93277 | 2.33445 | 2.47197 | 3.01131 |
| 0.920 | 1.14865 | 1.51371 | 1.94394 | 2.34803 | 2.48103 | 3.01769 |
| 0.940 | 1.16084 | 1.52619 | 1.95504 | 2.36187 | 2.49024 | 3.02432 |
| 0.960 | 1.17325 | 1.53878 | 1.96610 | 2.37599 | 2.49960 | 3.03123 |
| 0.980 | 1.18588 | 1.55145 | 1.97710 | 2.39034 | 2.50915 | 3.03840 |
| 1.000 | 1.19873 | 1.56423 | 1.98805 | 2.40486 | 2.51887 | 3.04585 |

$H/R = 1.500 \qquad n = 3.$

| $H/L$ | $\Omega_1$ | $\Omega_2$ | $\Omega_3$ | $\Omega_4$ | $\Omega_5$ | $\Omega_6$ |
|---|---|---|---|---|---|---|
| 0. | 0.97935 | 1.14616 | 1.63150 | 2.18459 | 2.41221 | 3.08126 |
| 0.010 | 0.97932 | 1.14627 | 1.63156 | 2.18460 | 2.41224 | 3.08120 |
| 0.050 | 0.97877 | 1.14877 | 1.63306 | 2.18504 | 2.41296 | 3.08024 |
| 0.100 | 0.97726 | 1.15633 | 1.63775 | 2.18641 | 2.41519 | 3.07768 |
| 0.120 | 0.97654 | 1.16061 | 1.64051 | 2.18724 | 2.41649 | 3.07644 |
| 0.140 | 0.97581 | 1.16553 | 1.64374 | 2.18820 | 2.41803 | 3.07515 |
| 0.160 | 0.97512 | 1.17104 | 1.64747 | 2.18934 | 2.41980 | 3.07386 |
| 0.180 | 0.97452 | 1.17708 | 1.65167 | 2.19064 | 2.42179 | 3.07264 |
| 0.200 | 0.97406 | 1.18362 | 1.65636 | 2.19212 | 2.42400 | 3.07148 |
| 0.220 | 0.97377 | 1.19060 | 1.66151 | 2.19376 | 2.42644 | 3.07040 |
| 0.240 | 0.97369 | 1.19798 | 1.66713 | 2.19559 | 2.42911 | 3.06946 |
| 0.260 | 0.97386 | 1.20572 | 1.67320 | 2.19761 | 2.43199 | 3.06865 |
| 0.280 | 0.97430 | 1.21380 | 1.67973 | 2.19982 | 2.43508 | 3.06798 |
| 0.300 | 0.97504 | 1.22216 | 1.68669 | 2.20226 | 2.43838 | 3.06748 |
| 0.320 | 0.97609 | 1.23081 | 1.69407 | 2.20488 | 2.44189 | 3.06712 |
| 0.340 | 0.97748 | 1.23968 | 1.70187 | 2.20774 | 2.44560 | 3.06692 |
| 0.360 | 0.97923 | 1.24879 | 1.71007 | 2.21083 | 2.44951 | 3.06691 |
| 0.380 | 0.98134 | 1.25809 | 1.71865 | 2.21415 | 2.45362 | 3.06708 |
| 0.400 | 0.98382 | 1.26758 | 1.72761 | 2.21772 | 2.45792 | 3.06743 |
| 0.420 | 0.98669 | 1.27723 | 1.73692 | 2.22155 | 2.46238 | 3.06796 |
| 0.440 | 0.98994 | 1.28704 | 1.74657 | 2.22564 | 2.46706 | 3.06866 |
| 0.460 | 0.99359 | 1.29698 | 1.75654 | 2.23002 | 2.47193 | 3.06957 |
| 0.480 | 0.99763 | 1.30708 | 1.76681 | 2.23467 | 2.47694 | 3.07066 |
| 0.500 | 1.00205 | 1.31728 | 1.77736 | 2.23963 | 2.48218 | 3.07192 |
| 0.520 | 1.00688 | 1.32761 | 1.78818 | 2.24490 | 2.48756 | 3.07342 |
| 0.540 | 1.01210 | 1.33804 | 1.79923 | 2.25047 | 2.49311 | 3.07508 |
| 0.560 | 1.01770 | 1.34859 | 1.81050 | 2.25638 | 2.49885 | 3.07694 |
| 0.580 | 1.02369 | 1.35923 | 1.82198 | 2.26263 | 2.50473 | 3.07899 |
| 0.600 | 1.03006 | 1.36997 | 1.83362 | 2.26921 | 2.51081 | 3.08125 |
| 0.620 | 1.03680 | 1.38080 | 1.84541 | 2.27617 | 2.51702 | 3.08371 |
| 0.640 | 1.04391 | 1.39173 | 1.85735 | 2.28347 | 2.52343 | 3.08638 |
| 0.660 | 1.05139 | 1.40275 | 1.86939 | 2.29114 | 2.52998 | 3.08924 |
| 0.680 | 1.05922 | 1.41387 | 1.88151 | 2.29920 | 2.53667 | 3.09232 |
| 0.700 | 1.06739 | 1.42506 | 1.89370 | 2.30765 | 2.54353 | 3.09559 |
| 0.720 | 1.07591 | 1.43635 | 1.90594 | 2.31647 | 2.55056 | 3.09909 |
| 0.740 | 1.08475 | 1.44773 | 1.91822 | 2.32569 | 2.55774 | 3.10277 |
| 0.760 | 1.09391 | 1.45919 | 1.93050 | 2.33529 | 2.56507 | 3.10668 |
| 0.780 | 1.10339 | 1.47074 | 1.94278 | 2.34529 | 2.57255 | 3.11083 |
| 0.800 | 1.11318 | 1.48238 | 1.95504 | 2.35568 | 2.58020 | 3.11518 |
| 0.820 | 1.12326 | 1.49411 | 1.96727 | 2.36645 | 2.58803 | 3.11973 |
| 0.840 | 1.13363 | 1.50593 | 1.97946 | 2.37761 | 2.59598 | 3.12453 |
| 0.860 | 1.14427 | 1.51785 | 1.99159 | 2.38912 | 2.60410 | 3.12955 |
| 0.880 | 1.15519 | 1.52985 | 2.00367 | 2.40101 | 2.61237 | 3.13480 |
| 0.900 | 1.16636 | 1.54193 | 2.01567 | 2.41323 | 2.62082 | 3.14029 |
| 0.920 | 1.17779 | 1.55411 | 2.02762 | 2.42579 | 2.62941 | 3.14601 |
| 0.940 | 1.18946 | 1.56638 | 2.03949 | 2.43867 | 2.63817 | 3.15198 |
| 0.960 | 1.20137 | 1.57874 | 2.05129 | 2.45184 | 2.64710 | 3.15820 |
| 0.980 | 1.21351 | 1.59119 | 2.06303 | 2.46527 | 2.65620 | 3.16467 |
| 1.000 | 1.22587 | 1.60373 | 2.07470 | 2.47897 | 2.66546 | 3.17138 |

$H/R = 1.600 \qquad n = 3.$

| $H/L$ | $\Omega_1$ | $\Omega_2$ | $\Omega_3$ | $\Omega_4$ | $\Omega_5$ | $\Omega_6$ |
|---|---|---|---|---|---|---|
| 0. | 1.02008 | 1.18868 | 1.69805 | 2.26729 | 2.56911 | 3.20638 |
| 0.010 | 1.02006 | 1.18878 | 1.69811 | 2.26731 | 2.56916 | 3.20635 |
| 0.050 | 1.01948 | 1.19123 | 1.69957 | 2.26779 | 2.56978 | 3.20657 |
| 0.100 | 1.01790 | 1.19865 | 1.70412 | 2.26926 | 2.57172 | 3.20349 |
| 0.120 | 1.01712 | 1.20285 | 1.70679 | 2.27014 | 2.57287 | 3.20244 |
| 0.140 | 1.01632 | 1.20769 | 1.70994 | 2.27118 | 2.57420 | 3.20136 |
| 0.160 | 1.01555 | 1.21310 | 1.71355 | 2.27239 | 2.57575 | 3.20028 |
| 0.180 | 1.01486 | 1.21905 | 1.71764 | 2.27377 | 2.57750 | 3.19922 |
| 0.200 | 1.01428 | 1.22549 | 1.72219 | 2.27532 | 2.57946 | 3.19824 |
| 0.220 | 1.01385 | 1.23236 | 1.72721 | 2.27706 | 2.58161 | 3.19733 |
| 0.240 | 1.01362 | 1.23964 | 1.73270 | 2.27897 | 2.58397 | 3.19653 |
| 0.260 | 1.01362 | 1.24727 | 1.73861 | 2.28107 | 2.58653 | 3.19584 |
| 0.280 | 1.01386 | 1.25525 | 1.74497 | 2.28336 | 2.58926 | 3.19531 |
| 0.300 | 1.01438 | 1.26351 | 1.75176 | 2.28585 | 2.59221 | 3.19490 |
| 0.320 | 1.01520 | 1.27204 | 1.75898 | 2.28855 | 2.59533 | 3.19463 |
| 0.340 | 1.01634 | 1.28082 | 1.76661 | 2.29144 | 2.59866 | 3.19455 |
| 0.360 | 1.01781 | 1.28982 | 1.77465 | 2.29456 | 2.60217 | 3.19462 |
| 0.380 | 1.01962 | 1.29903 | 1.78307 | 2.29791 | 2.60586 | 3.19484 |
| 0.400 | 1.02180 | 1.30842 | 1.79186 | 2.30148 | 2.60972 | 3.19524 |
| 0.420 | 1.02433 | 1.31798 | 1.80102 | 2.30529 | 2.61380 | 3.19582 |
| 0.440 | 1.02724 | 1.32769 | 1.81053 | 2.30934 | 2.61805 | 3.19658 |
| 0.460 | 1.03052 | 1.33755 | 1.82038 | 2.31364 | 2.62246 | 3.19750 |
| 0.480 | 1.03419 | 1.34754 | 1.83054 | 2.31819 | 2.62707 | 3.19861 |
| 0.500 | 1.03822 | 1.35766 | 1.84100 | 2.32302 | 2.63182 | 3.19990 |
| 0.520 | 1.04265 | 1.36788 | 1.85174 | 2.32812 | 2.63678 | 3.20137 |
| 0.540 | 1.04745 | 1.37824 | 1.86275 | 2.33350 | 2.64189 | 3.20301 |
| 0.560 | 1.05263 | 1.38868 | 1.87400 | 2.33918 | 2.64718 | 3.20486 |
| 0.580 | 1.05819 | 1.39922 | 1.88546 | 2.34516 | 2.65264 | 3.20688 |
| 0.600 | 1.06412 | 1.40986 | 1.89714 | 2.35144 | 2.65829 | 3.20909 |
| 0.620 | 1.07041 | 1.42059 | 1.90901 | 2.35804 | 2.66408 | 3.21151 |
| 0.640 | 1.07707 | 1.43142 | 1.92103 | 2.36497 | 2.67003 | 3.21409 |
| 0.660 | 1.08407 | 1.44233 | 1.93320 | 2.37222 | 2.67616 | 3.21687 |
| 0.680 | 1.09144 | 1.45332 | 1.94550 | 2.37980 | 2.68245 | 3.21985 |
| 0.700 | 1.09914 | 1.46441 | 1.95789 | 2.38773 | 2.68894 | 3.22301 |
| 0.720 | 1.10719 | 1.47557 | 1.97038 | 2.39601 | 2.69554 | 3.22637 |
| 0.740 | 1.11556 | 1.48683 | 1.98293 | 2.40464 | 2.70234 | 3.22992 |
| 0.760 | 1.12425 | 1.49816 | 1.99551 | 2.41363 | 2.70930 | 3.23368 |
| 0.780 | 1.13326 | 1.50958 | 2.00813 | 2.42298 | 2.71640 | 3.23761 |
| 0.800 | 1.14257 | 1.52109 | 2.02077 | 2.43268 | 2.72368 | 3.24176 |
| 0.820 | 1.15218 | 1.53267 | 2.03340 | 2.44274 | 2.73114 | 3.24612 |
| 0.840 | 1.16208 | 1.54434 | 2.04602 | 2.45317 | 2.73873 | 3.25068 |
| 0.860 | 1.17226 | 1.55609 | 2.05861 | 2.46395 | 2.74650 | 3.25543 |
| 0.880 | 1.18272 | 1.56794 | 2.07116 | 2.47509 | 2.75445 | 3.26042 |
| 0.900 | 1.19344 | 1.57987 | 2.08366 | 2.48654 | 2.76258 | 3.26559 |
| 0.920 | 1.20441 | 1.59188 | 2.09610 | 2.49834 | 2.77084 | 3.27098 |
| 0.940 | 1.21564 | 1.60397 | 2.10849 | 2.51045 | 2.77929 | 3.27660 |
| 0.960 | 1.22710 | 1.61616 | 2.12081 | 2.52290 | 2.78790 | 3.28243 |
| 0.980 | 1.23881 | 1.62843 | 2.13305 | 2.53563 | 2.79671 | 3.28849 |
| 1.000 | 1.25073 | 1.64079 | 2.14524 | 2.54862 | 2.80569 | 3.29477 |

$H/R = 1.700 \qquad n = 3.$

| $H/L$ | $\Omega_1$ | $\Omega_2$ | $\Omega_3$ | $\Omega_4$ | $\Omega_5$ | $\Omega_6$ |
|---|---|---|---|---|---|---|
| 0. | 1.05648 | 1.22886 | 1.75817 | 2.34437 | 2.69214 | 3.31803 |
| 0.010 | 1.05646 | 1.22896 | 1.75823 | 2.34440 | 2.69217 | 3.31802 |
| 0.050 | 1.05588 | 1.23135 | 1.75965 | 2.34487 | 2.69276 | 3.31739 |
| 0.100 | 1.05429 | 1.23858 | 1.76406 | 2.34636 | 2.69455 | 3.31574 |
| 0.120 | 1.05350 | 1.24268 | 1.76665 | 2.34724 | 2.69562 | 3.31492 |
| 0.140 | 1.05267 | 1.24742 | 1.76970 | 2.34829 | 2.69687 | 3.31404 |
| 0.160 | 1.05186 | 1.25272 | 1.77322 | 2.34950 | 2.69829 | 3.31318 |
| 0.180 | 1.05111 | 1.25855 | 1.77719 | 2.35089 | 2.69992 | 3.31232 |
| 0.200 | 1.05046 | 1.26486 | 1.78161 | 2.35244 | 2.70174 | 3.31150 |
| 0.220 | 1.04996 | 1.27161 | 1.78649 | 2.35416 | 2.70376 | 3.31076 |
| 0.240 | 1.04962 | 1.27877 | 1.79180 | 2.35607 | 2.70594 | 3.31013 |
| 0.260 | 1.04949 | 1.28629 | 1.79756 | 2.35816 | 2.70832 | 3.30958 |
| 0.280 | 1.04959 | 1.29414 | 1.80376 | 2.36044 | 2.71089 | 3.30917 |
| 0.300 | 1.04995 | 1.30229 | 1.81037 | 2.36290 | 2.71362 | 3.30889 |
| 0.320 | 1.05060 | 1.31071 | 1.81740 | 2.36556 | 2.71655 | 3.30873 |
| 0.340 | 1.05154 | 1.31937 | 1.82483 | 2.36843 | 2.71966 | 3.30874 |
| 0.360 | 1.05280 | 1.32826 | 1.83267 | 2.37149 | 2.72295 | 3.30887 |
| 0.380 | 1.05438 | 1.33736 | 1.84089 | 2.37477 | 2.72642 | 3.30920 |
| 0.400 | 1.05630 | 1.34665 | 1.84949 | 2.37827 | 2.73004 | 3.30967 |
| 0.420 | 1.05857 | 1.35611 | 1.85846 | 2.38199 | 2.73386 | 3.31029 |
| 0.440 | 1.06119 | 1.36572 | 1.86778 | 2.38593 | 2.73785 | 3.31108 |
| 0.460 | 1.06418 | 1.37548 | 1.87743 | 2.39011 | 2.74203 | 3.31207 |
| 0.480 | 1.06753 | 1.38539 | 1.88739 | 2.39454 | 2.74636 | 3.31320 |
| 0.500 | 1.07125 | 1.39541 | 1.89768 | 2.39921 | 2.75088 | 3.31450 |
| 0.520 | 1.07533 | 1.40555 | 1.90825 | 2.40414 | 2.75554 | 3.31601 |
| 0.540 | 1.07979 | 1.41580 | 1.91910 | 2.40934 | 2.76040 | 3.31767 |
| 0.560 | 1.08461 | 1.42615 | 1.93021 | 2.41479 | 2.76542 | 3.31951 |
| 0.580 | 1.08980 | 1.43662 | 1.94155 | 2.42053 | 2.77060 | 3.32151 |
| 0.600 | 1.09535 | 1.44716 | 1.95312 | 2.42656 | 2.77597 | 3.32371 |
| 0.620 | 1.10126 | 1.45781 | 1.96489 | 2.43287 | 2.78147 | 3.32609 |
| 0.640 | 1.10753 | 1.46853 | 1.97685 | 2.43948 | 2.78715 | 3.32865 |
| 0.660 | 1.11415 | 1.47936 | 1.98897 | 2.44640 | 2.79300 | 3.33137 |
| 0.680 | 1.12111 | 1.49025 | 2.00124 | 2.45363 | 2.79903 | 3.33429 |
| 0.700 | 1.12841 | 1.50124 | 2.01363 | 2.46118 | 2.80519 | 3.33741 |
| 0.720 | 1.13605 | 1.51231 | 2.02614 | 2.46905 | 2.81152 | 3.34068 |
| 0.740 | 1.14401 | 1.52345 | 2.03874 | 2.47725 | 2.81804 | 3.34415 |
| 0.760 | 1.15229 | 1.53468 | 2.05140 | 2.48577 | 2.82470 | 3.34781 |
| 0.780 | 1.16088 | 1.54599 | 2.06412 | 2.49463 | 2.83153 | 3.35164 |
| 0.800 | 1.16978 | 1.55738 | 2.07689 | 2.50383 | 2.83854 | 3.35567 |
| 0.820 | 1.17898 | 1.56885 | 2.08966 | 2.51338 | 2.84570 | 3.35990 |
| 0.840 | 1.18847 | 1.58039 | 2.10246 | 2.52328 | 2.85303 | 3.36430 |
| 0.860 | 1.19824 | 1.59203 | 2.11525 | 2.53350 | 2.86052 | 3.36891 |
| 0.880 | 1.20829 | 1.60374 | 2.12802 | 2.54405 | 2.86818 | 3.37369 |
| 0.900 | 1.21860 | 1.61553 | 2.14075 | 2.55495 | 2.87603 | 3.37868 |
| 0.920 | 1.22918 | 1.62740 | 2.15345 | 2.56619 | 2.88402 | 3.38387 |
| 0.940 | 1.24001 | 1.63936 | 2.16610 | 2.57773 | 2.89220 | 3.38926 |
| 0.960 | 1.25108 | 1.65141 | 2.17868 | 2.58959 | 2.90054 | 3.39484 |
| 0.980 | 1.26238 | 1.66353 | 2.19122 | 2.60176 | 2.90907 | 3.40064 |
| 1.000 | 1.27393 | 1.67574 | 2.20368 | 2.61420 | 2.91775 | 3.40663 |

$H/R = 1.800 \qquad n = 3.$

| $H/L$ | $\Omega_1$ | $\Omega_2$ | $\Omega_3$ | $\Omega_4$ | $\Omega_5$ | $\Omega_6$ |
|---|---|---|---|---|---|---|
| 0. | 1.08985 | 1.26691 | 1.81355 | 2.41704 | 2.78844 | 3.42140 |
| 0.010 | 1.08983 | 1.26700 | 1.81361 | 2.41707 | 2.78847 | 3.42137 |
| 0.050 | 1.08926 | 1.26933 | 1.81498 | 2.41754 | 2.78904 | 3.42084 |
| 0.100 | 1.08768 | 1.27638 | 1.81927 | 2.41900 | 2.79074 | 3.41937 |
| 0.120 | 1.08689 | 1.28039 | 1.82178 | 2.41987 | 2.79176 | 3.41863 |
| 0.140 | 1.08606 | 1.28500 | 1.82475 | 2.42089 | 2.79295 | 3.41782 |
| 0.160 | 1.08524 | 1.29018 | 1.82816 | 2.42207 | 2.79433 | 3.41702 |
| 0.180 | 1.08445 | 1.29589 | 1.83202 | 2.42343 | 2.79590 | 3.41625 |
| 0.200 | 1.08377 | 1.30208 | 1.83632 | 2.42495 | 2.79762 | 3.41550 |
| 0.220 | 1.08320 | 1.30871 | 1.84105 | 2.42664 | 2.79955 | 3.41482 |
| 0.240 | 1.08279 | 1.31574 | 1.84623 | 2.42850 | 2.80166 | 3.41422 |
| 0.260 | 1.08257 | 1.32313 | 1.85182 | 2.43054 | 2.80392 | 3.41372 |
| 0.280 | 1.08257 | 1.33085 | 1.85785 | 2.43276 | 2.80638 | 3.41333 |
| 0.300 | 1.08281 | 1.33889 | 1.86429 | 2.43517 | 2.80901 | 3.41306 |
| 0.320 | 1.08331 | 1.34719 | 1.87113 | 2.43776 | 2.81182 | 3.41293 |
| 0.340 | 1.08409 | 1.35575 | 1.87838 | 2.44055 | 2.81479 | 3.41292 |
| 0.360 | 1.08517 | 1.36452 | 1.88602 | 2.44353 | 2.81795 | 3.41306 |
| 0.380 | 1.08656 | 1.37351 | 1.89404 | 2.44671 | 2.82129 | 3.41337 |
| 0.400 | 1.08827 | 1.38270 | 1.90243 | 2.45012 | 2.82477 | 3.41382 |
| 0.420 | 1.09033 | 1.39205 | 1.91119 | 2.45371 | 2.82843 | 3.41441 |
| 0.440 | 1.09271 | 1.40158 | 1.92029 | 2.45754 | 2.83228 | 3.41516 |
| 0.460 | 1.09545 | 1.41124 | 1.92973 | 2.46158 | 2.83628 | 3.41610 |
| 0.480 | 1.09854 | 1.42105 | 1.93949 | 2.46585 | 2.84045 | 3.41717 |
| 0.500 | 1.10198 | 1.43099 | 1.94957 | 2.47038 | 2.84481 | 3.41843 |
| 0.520 | 1.10577 | 1.44104 | 1.95994 | 2.47512 | 2.84930 | 3.41985 |
| 0.540 | 1.10993 | 1.45120 | 1.97059 | 2.48013 | 2.85397 | 3.42144 |
| 0.560 | 1.11445 | 1.46148 | 1.98152 | 2.48539 | 2.85882 | 3.42321 |
| 0.580 | 1.11932 | 1.47185 | 1.99267 | 2.49091 | 2.86382 | 3.42514 |
| 0.600 | 1.12453 | 1.48231 | 2.00408 | 2.49667 | 2.86898 | 3.42724 |
| 0.620 | 1.13011 | 1.49287 | 2.01569 | 2.50274 | 2.87429 | 3.42953 |
| 0.640 | 1.13604 | 1.50352 | 2.02750 | 2.50908 | 2.87977 | 3.43199 |
| 0.660 | 1.14231 | 1.51426 | 2.03949 | 2.51570 | 2.88542 | 3.43461 |
| 0.680 | 1.14892 | 1.52507 | 2.05165 | 2.52263 | 2.89125 | 3.43741 |
| 0.700 | 1.15586 | 1.53596 | 2.06394 | 2.52985 | 2.89720 | 3.44038 |
| 0.720 | 1.16313 | 1.54695 | 2.07637 | 2.53737 | 2.90333 | 3.44353 |
| 0.740 | 1.17073 | 1.55800 | 2.08891 | 2.54523 | 2.90962 | 3.44687 |
| 0.760 | 1.17864 | 1.56914 | 2.10154 | 2.55336 | 2.91609 | 3.45036 |
| 0.780 | 1.18687 | 1.58035 | 2.11424 | 2.56184 | 2.92269 | 3.45406 |
| 0.800 | 1.19540 | 1.59165 | 2.12700 | 2.57064 | 2.92947 | 3.45793 |
| 0.820 | 1.20423 | 1.60301 | 2.13980 | 2.57976 | 2.93642 | 3.46197 |
| 0.840 | 1.21335 | 1.61446 | 2.15262 | 2.58921 | 2.94350 | 3.46619 |
| 0.860 | 1.22276 | 1.62598 | 2.16546 | 2.59897 | 2.95077 | 3.47061 |
| 0.880 | 1.23244 | 1.63758 | 2.17831 | 2.60908 | 2.95819 | 3.47520 |
| 0.900 | 1.24239 | 1.64926 | 2.19113 | 2.61952 | 2.96577 | 3.47997 |
| 0.920 | 1.25259 | 1.66103 | 2.20394 | 2.63027 | 2.97352 | 3.48494 |
| 0.940 | 1.26307 | 1.67287 | 2.21671 | 2.64133 | 2.98146 | 3.49009 |
| 0.960 | 1.27378 | 1.68479 | 2.22944 | 2.65272 | 2.98955 | 3.49545 |
| 0.980 | 1.28473 | 1.69678 | 2.24211 | 2.66442 | 2.99780 | 3.50099 |
| 1.000 | 1.29593 | 1.70887 | 2.25473 | 2.67640 | 3.00623 | 3.50673 |

$H/R = 1.900 \qquad n = 3.$

| $H/L$ | $\Omega_1$ | $\Omega_2$ | $\Omega_3$ | $\Omega_4$ | $\Omega_5$ | $\Omega_6$ |
|---|---|---|---|---|---|---|
| 0. | 1.12104 | 1.30300 | 1.86541 | 2.48593 | 2.87063 | 3.51893 |
| 0.010 | 1.12101 | 1.30310 | 1.86548 | 2.48595 | 2.87066 | 3.51891 |
| 0.050 | 1.12046 | 1.30535 | 1.86681 | 2.48641 | 2.87119 | 3.51840 |
| 0.100 | 1.11891 | 1.31223 | 1.87098 | 2.48782 | 2.87287 | 3.51697 |
| 0.120 | 1.11813 | 1.31614 | 1.87343 | 2.48865 | 2.87385 | 3.51627 |
| 0.140 | 1.11730 | 1.32064 | 1.87631 | 2.48966 | 2.87500 | 3.51549 |
| 0.160 | 1.11647 | 1.32572 | 1.87963 | 2.49083 | 2.87634 | 3.51471 |
| 0.180 | 1.11567 | 1.33131 | 1.88338 | 2.49213 | 2.87784 | 3.51394 |
| 0.200 | 1.11495 | 1.33738 | 1.88757 | 2.49361 | 2.87954 | 3.51321 |
| 0.220 | 1.11434 | 1.34389 | 1.89218 | 2.49527 | 2.88139 | 3.51251 |
| 0.240 | 1.11388 | 1.35080 | 1.89722 | 2.49707 | 2.88343 | 3.51191 |
| 0.260 | 1.11359 | 1.35806 | 1.90267 | 2.49907 | 2.88563 | 3.51140 |
| 0.280 | 1.11350 | 1.36566 | 1.90854 | 2.50120 | 2.88801 | 3.51097 |
| 0.300 | 1.11363 | 1.37359 | 1.91481 | 2.50355 | 2.89057 | 3.51069 |
| 0.320 | 1.11402 | 1.38177 | 1.92148 | 2.50607 | 2.89329 | 3.51049 |
| 0.340 | 1.11467 | 1.39021 | 1.92854 | 2.50876 | 2.89620 | 3.51045 |
| 0.360 | 1.11561 | 1.39889 | 1.93600 | 2.51165 | 2.89926 | 3.51055 |
| 0.380 | 1.11683 | 1.40778 | 1.94383 | 2.51474 | 2.90248 | 3.51077 |
| 0.400 | 1.11837 | 1.41686 | 1.95203 | 2.51803 | 2.90589 | 3.51116 |
| 0.420 | 1.12023 | 1.42613 | 1.96058 | 2.52150 | 2.90943 | 3.51167 |
| 0.440 | 1.12241 | 1.43555 | 1.96947 | 2.52520 | 2.91318 | 3.51237 |
| 0.460 | 1.12493 | 1.44513 | 1.97872 | 2.52912 | 2.91707 | 3.51318 |
| 0.480 | 1.12778 | 1.45485 | 1.98827 | 2.53326 | 2.92112 | 3.51420 |
| 0.500 | 1.13098 | 1.46469 | 1.99814 | 2.53761 | 2.92534 | 3.51536 |
| 0.520 | 1.13453 | 1.47467 | 2.00831 | 2.54219 | 2.92973 | 3.51665 |
| 0.540 | 1.13842 | 1.48475 | 2.01876 | 2.54702 | 2.93427 | 3.51814 |
| 0.560 | 1.14266 | 1.49494 | 2.02948 | 2.55207 | 2.93895 | 3.51980 |
| 0.580 | 1.14725 | 1.50523 | 2.04045 | 2.55738 | 2.94383 | 3.52160 |
| 0.600 | 1.15218 | 1.51562 | 2.05167 | 2.56296 | 2.94884 | 3.52359 |
| 0.620 | 1.15746 | 1.52611 | 2.06311 | 2.56879 | 2.95401 | 3.52573 |
| 0.640 | 1.16308 | 1.53668 | 2.07475 | 2.57489 | 2.95935 | 3.52804 |
| 0.660 | 1.16904 | 1.54733 | 2.08659 | 2.58126 | 2.96485 | 3.53053 |
| 0.680 | 1.17533 | 1.55807 | 2.09860 | 2.58792 | 2.97050 | 3.53319 |
| 0.700 | 1.18195 | 1.56889 | 2.11078 | 2.59485 | 2.97631 | 3.53599 |
| 0.720 | 1.18890 | 1.57979 | 2.12308 | 2.60207 | 2.98227 | 3.53899 |
| 0.740 | 1.19617 | 1.59076 | 2.13551 | 2.60960 | 2.98840 | 3.54215 |
| 0.760 | 1.20375 | 1.60182 | 2.14805 | 2.61743 | 2.99466 | 3.54548 |
| 0.780 | 1.21164 | 1.61295 | 2.16068 | 2.62554 | 3.00109 | 3.54901 |
| 0.800 | 1.21984 | 1.62416 | 2.17338 | 2.63399 | 3.00769 | 3.55270 |
| 0.820 | 1.22834 | 1.63543 | 2.18615 | 2.64274 | 3.01443 | 3.55654 |
| 0.840 | 1.23711 | 1.64680 | 2.19896 | 2.65181 | 3.02135 | 3.56058 |
| 0.860 | 1.24618 | 1.65822 | 2.21179 | 2.66119 | 3.02842 | 3.56479 |
| 0.880 | 1.25552 | 1.66973 | 2.22464 | 2.67090 | 3.03563 | 3.56917 |
| 0.900 | 1.26514 | 1.68132 | 2.23750 | 2.68092 | 3.04302 | 3.57372 |
| 0.920 | 1.27502 | 1.69298 | 2.25034 | 2.69127 | 3.05056 | 3.57848 |
| 0.940 | 1.28516 | 1.70472 | 2.26316 | 2.70194 | 3.05825 | 3.58339 |
| 0.960 | 1.29555 | 1.71654 | 2.27595 | 2.71290 | 3.06612 | 3.58852 |
| 0.980 | 1.30617 | 1.72843 | 2.28871 | 2.72417 | 3.07416 | 3.59380 |
| 1.000 | 1.31705 | 1.74040 | 2.30141 | 2.73575 | 3.08235 | 3.59928 |

$H/R = 2.000 \qquad n = 3.$

| $H/L$ | $\Omega_1$ | $\Omega_2$ | $\Omega_3$ | $\Omega_4$ | $\Omega_5$ | $\Omega_6$ |
|---|---|---|---|---|---|---|
| 0. | 1.15054 | 1.33729 | 1.91452 | 2.55134 | 2.94635 | 3.61152 |
| 0.010 | 1.15052 | 1.33738 | 1.91457 | 2.55136 | 2.94637 | 3.61152 |
| 0.050 | 1.14998 | 1.33958 | 1.91588 | 2.55181 | 2.94689 | 3.61102 |
| 0.100 | 1.14847 | 1.34629 | 1.91995 | 2.55319 | 2.94852 | 3.60964 |
| 0.120 | 1.14769 | 1.35011 | 1.92233 | 2.55401 | 2.94948 | 3.60892 |
| 0.140 | 1.14686 | 1.35452 | 1.92513 | 2.55498 | 2.95060 | 3.60816 |
| 0.160 | 1.14603 | 1.35949 | 1.92837 | 2.55612 | 2.95191 | 3.60737 |
| 0.180 | 1.14522 | 1.36498 | 1.93203 | 2.55739 | 2.95337 | 3.60661 |
| 0.200 | 1.14448 | 1.37093 | 1.93611 | 2.55883 | 2.95501 | 3.60585 |
| 0.220 | 1.14384 | 1.37732 | 1.94061 | 2.56045 | 2.95682 | 3.60515 |
| 0.240 | 1.14333 | 1.38412 | 1.94552 | 2.56218 | 2.95881 | 3.60452 |
| 0.260 | 1.14298 | 1.39127 | 1.95084 | 2.56411 | 2.96098 | 3.60399 |
| 0.280 | 1.14282 | 1.39876 | 1.95657 | 2.56623 | 2.96329 | 3.60353 |
| 0.300 | 1.14287 | 1.40658 | 1.96268 | 2.56848 | 2.96579 | 3.60318 |
| 0.320 | 1.14315 | 1.41465 | 1.96921 | 2.57093 | 2.96843 | 3.60297 |
| 0.340 | 1.14368 | 1.42300 | 1.97610 | 2.57355 | 2.97124 | 3.60285 |
| 0.360 | 1.14449 | 1.43156 | 1.98339 | 2.57637 | 2.97424 | 3.60289 |
| 0.380 | 1.14558 | 1.44035 | 1.99104 | 2.57936 | 2.97740 | 3.60305 |
| 0.400 | 1.14696 | 1.44934 | 1.99904 | 2.58255 | 2.98071 | 3.60336 |
| 0.420 | 1.14865 | 1.45851 | 2.00742 | 2.58592 | 2.98418 | 3.60380 |
| 0.440 | 1.15065 | 1.46785 | 2.01612 | 2.58950 | 2.98780 | 3.60440 |
| 0.460 | 1.15297 | 1.47734 | 2.02517 | 2.59330 | 2.99161 | 3.60515 |
| 0.480 | 1.15562 | 1.48697 | 2.03453 | 2.59729 | 2.99557 | 3.60604 |
| 0.500 | 1.15861 | 1.49674 | 2.04421 | 2.60150 | 2.99968 | 3.60709 |
| 0.520 | 1.16192 | 1.50664 | 2.05418 | 2.60593 | 3.00396 | 3.60832 |
| 0.540 | 1.16558 | 1.51665 | 2.06444 | 2.61060 | 3.00839 | 3.60970 |
| 0.560 | 1.16957 | 1.52677 | 2.07497 | 2.61549 | 3.01297 | 3.61122 |
| 0.580 | 1.17391 | 1.53698 | 2.08576 | 2.62062 | 3.01773 | 3.61291 |
| 0.600 | 1.17858 | 1.54730 | 2.09680 | 2.62599 | 3.02261 | 3.61476 |
| 0.620 | 1.18359 | 1.55771 | 2.10807 | 2.63162 | 3.02768 | 3.61678 |
| 0.640 | 1.18893 | 1.56821 | 2.11955 | 2.63751 | 3.03289 | 3.61896 |
| 0.660 | 1.19461 | 1.57879 | 2.13122 | 2.64364 | 3.03823 | 3.62131 |
| 0.680 | 1.20062 | 1.58946 | 2.14309 | 2.65006 | 3.04375 | 3.62380 |
| 0.700 | 1.20694 | 1.60021 | 2.15512 | 2.65674 | 3.04943 | 3.62648 |
| 0.720 | 1.21360 | 1.61104 | 2.16730 | 2.66370 | 3.05524 | 3.62931 |
| 0.740 | 1.22056 | 1.62194 | 2.17962 | 2.67095 | 3.06122 | 3.63233 |
| 0.760 | 1.22785 | 1.63292 | 2.19206 | 2.67848 | 3.06734 | 3.63549 |
| 0.780 | 1.23543 | 1.64398 | 2.20460 | 2.68630 | 3.07362 | 3.63882 |
| 0.800 | 1.24333 | 1.65510 | 2.21723 | 2.69445 | 3.08006 | 3.64235 |
| 0.820 | 1.25151 | 1.66630 | 2.22993 | 2.70287 | 3.08665 | 3.64601 |
| 0.840 | 1.25998 | 1.67758 | 2.24269 | 2.71160 | 3.09337 | 3.64986 |
| 0.860 | 1.26874 | 1.68894 | 2.25549 | 2.72064 | 3.10027 | 3.65387 |
| 0.880 | 1.27778 | 1.70036 | 2.26833 | 2.72999 | 3.10733 | 3.65807 |
| 0.900 | 1.28708 | 1.71185 | 2.28117 | 2.73966 | 3.11450 | 3.66242 |
| 0.920 | 1.29665 | 1.72342 | 2.29403 | 2.74961 | 3.12186 | 3.66697 |
| 0.940 | 1.30647 | 1.73507 | 2.30688 | 2.75990 | 3.12937 | 3.67167 |
| 0.960 | 1.31656 | 1.74680 | 2.31971 | 2.77048 | 3.13705 | 3.67657 |
| 0.980 | 1.32688 | 1.75859 | 2.33250 | 2.78137 | 3.14489 | 3.68162 |
| 1.000 | 1.33744 | 1.77045 | 2.34527 | 2.79258 | 3.15287 | 3.68687 |

$H/R = 0.010 \qquad n = 4.$

| $H/L$ | $\Omega_1$ | $\Omega_2$ | $\Omega_3$ | $\Omega_4$ | $\Omega_5$ | $\Omega_6$ |
|---|---|---|---|---|---|---|
| 0. | 0.000226 | 0.01273 | 0.02218 | 1.00009 | 1.00030 | 1.87042 |
| 0.001 | 0.000229 | 0.01278 | 0.02224 | 1.00009 | 1.00030 | 1.87042 |
| 0.005 | 0.00071 | 0.01384 | 0.02364 | 1.00010 | 1.00034 | 1.87036 |
| 0.010 | 0.00194 | 0.01648 | 0.02765 | 1.00014 | 1.00046 | 1.87017 |
| 0.020 | 0.00371 | 0.02389 | 0.04017 | 1.00030 | 1.00097 | 1.86941 |
| 0.030 | 0.00461 | 0.03267 | 0.05513 | 1.00055 | 1.00183 | 1.86816 |
| 0.040 | 0.00537 | 0.04202 | 0.07097 | 1.00090 | 1.00304 | 1.86647 |
| 0.050 | 0.00629 | 0.05162 | 0.08721 | 1.00135 | 1.00457 | 1.86437 |
| 0.060 | 0.00754 | 0.06135 | 0.10366 | 1.00190 | 1.00644 | 1.86191 |
| 0.070 | 0.00914 | 0.07116 | 0.12023 | 1.00255 | 1.00865 | 1.85916 |
| 0.080 | 0.01112 | 0.08101 | 0.13687 | 1.00330 | 1.01119 | 1.85614 |
| 0.090 | 0.01345 | 0.09090 | 0.15356 | 1.00415 | 1.01405 | 1.85292 |
| 0.100 | 0.01611 | 0.10081 | 0.17027 | 1.00509 | 1.01722 | 1.84951 |
| 0.120 | 0.02235 | 0.12068 | 0.20376 | 1.00728 | 1.02451 | 1.84233 |
| 0.140 | 0.02973 | 0.14058 | 0.23727 | 1.00986 | 1.03301 | 1.83481 |
| 0.160 | 0.03818 | 0.16051 | 0.27077 | 1.01282 | 1.04266 | 1.82716 |
| 0.180 | 0.04760 | 0.18045 | 0.30425 | 1.01617 | 1.05341 | 1.81949 |
| 0.200 | 0.05794 | 0.20041 | 0.33769 | 1.01991 | 1.06520 | 1.81191 |
| 0.220 | 0.06914 | 0.22037 | 0.37107 | 1.02402 | 1.07796 | 1.80448 |
| 0.240 | 0.08112 | 0.24034 | 0.40439 | 1.02850 | 1.09165 | 1.79727 |
| 0.260 | 0.09384 | 0.26031 | 0.43763 | 1.03335 | 1.10620 | 1.79033 |
| 0.280 | 0.10724 | 0.28029 | 0.47077 | 1.03856 | 1.12155 | 1.78369 |
| 0.300 | 0.12126 | 0.30027 | 0.50382 | 1.04413 | 1.13766 | 1.77739 |
| 0.320 | 0.13585 | 0.32026 | 0.53675 | 1.05005 | 1.15446 | 1.77147 |
| 0.340 | 0.15097 | 0.34024 | 0.56956 | 1.05632 | 1.17190 | 1.76593 |
| 0.360 | 0.16657 | 0.36023 | 0.60222 | 1.06293 | 1.18995 | 1.76084 |
| 0.380 | 0.18262 | 0.38022 | 0.63473 | 1.06986 | 1.20855 | 1.75619 |
| 0.400 | 0.19906 | 0.40020 | 0.66706 | 1.07713 | 1.22767 | 1.75201 |
| 0.420 | 0.21588 | 0.42019 | 0.69921 | 1.08472 | 1.24725 | 1.74833 |
| 0.440 | 0.23303 | 0.44019 | 0.73114 | 1.09262 | 1.26728 | 1.74517 |
| 0.460 | 0.25048 | 0.46018 | 0.76285 | 1.10082 | 1.28771 | 1.74256 |
| 0.480 | 0.26821 | 0.48017 | 0.79430 | 1.10932 | 1.30851 | 1.74055 |
| 0.500 | 0.28619 | 0.50016 | 0.82547 | 1.11813 | 1.32965 | 1.73913 |
| 0.520 | 0.30441 | 0.52016 | 0.85634 | 1.12722 | 1.35110 | 1.73834 |
| 0.540 | 0.32283 | 0.54015 | 0.88688 | 1.13657 | 1.37284 | 1.73822 |
| 0.560 | 0.34144 | 0.56015 | 0.91704 | 1.14621 | 1.39483 | 1.73880 |
| 0.580 | 0.36022 | 0.58014 | 0.94681 | 1.15612 | 1.41707 | 1.74012 |
| 0.600 | 0.37915 | 0.60014 | 0.97615 | 1.16629 | 1.43952 | 1.74221 |
| 0.620 | 0.39822 | 0.62013 | 1.00499 | 1.17669 | 1.46217 | 1.74513 |
| 0.640 | 0.41742 | 0.64013 | 1.03332 | 1.18736 | 1.48499 | 1.74890 |
| 0.660 | 0.43674 | 0.66012 | 1.06108 | 1.19826 | 1.50796 | 1.75356 |
| 0.680 | 0.45615 | 0.68012 | 1.08824 | 1.20938 | 1.53108 | 1.75917 |
| 0.700 | 0.47566 | 0.70012 | 1.11473 | 1.22075 | 1.55432 | 1.76579 |
| 0.720 | 0.49526 | 0.72012 | 1.14052 | 1.23232 | 1.57766 | 1.77344 |
| 0.740 | 0.51493 | 0.74011 | 1.16556 | 1.24412 | 1.60109 | 1.78218 |
| 0.760 | 0.53466 | 0.76011 | 1.18980 | 1.25612 | 1.62459 | 1.79203 |
| 0.780 | 0.55445 | 0.78010 | 1.21322 | 1.26831 | 1.64815 | 1.80304 |
| 0.800 | 0.57430 | 0.80010 | 1.23579 | 1.28071 | 1.67176 | 1.81523 |
| 0.820 | 0.59419 | 0.82010 | 1.25746 | 1.29329 | 1.69541 | 1.82864 |
| 0.840 | 0.61413 | 0.84010 | 1.27824 | 1.30608 | 1.71905 | 1.84326 |
| 0.860 | 0.63411 | 0.86010 | 1.29812 | 1.31902 | 1.74272 | 1.85908 |
| 0.880 | 0.65411 | 0.88010 | 1.31712 | 1.33214 | 1.76637 | 1.87611 |
| 0.900 | 0.67414 | 0.90009 | 1.33524 | 1.34544 | 1.78999 | 1.89431 |
| 0.920 | 0.69419 | 0.92009 | 1.35253 | 1.35891 | 1.81359 | 1.91365 |
| 0.940 | 0.71428 | 0.94009 | 1.36903 | 1.37252 | 1.83713 | 1.93408 |
| 0.960 | 0.73437 | 0.96009 | 1.38478 | 1.38630 | 1.86062 | 1.95554 |
| 0.980 | 0.75448 | 0.98009 | 1.39985 | 1.40022 | 1.88403 | 1.97798 |
| 1.000 | 0.77460 | 1.00009 | 1.41424 | 1.41432 | 1.90736 | 2.00133 |

$H/R = 0.020$ $n = 4.$ $H/R = 0.030$

| $H/L$ | $\Omega_1$ | $\Omega_2$ | $\Omega_3$ | $\Omega_1$ | $\Omega_2$ | $\Omega_3$ | $H/L$ |
|---|---|---|---|---|---|---|---|
| 0. | 0.000903 | 0.02547 | 0.04436 | 0.00203 | 0.03820 | 0.06653 | 0. |
| 0.001 | 0.000905 | 0.02549 | 0.04439 | 0.00203 | 0.03821 | 0.06655 | 0.001 |
| 0.005 | 0.00101 | 0.02605 | 0.04510 | 0.00208 | 0.03860 | 0.06702 | 0.005 |
| 0.010 | 0.00168 | 0.02768 | 0.04727 | 0.00237 | 0.03974 | 0.06849 | 0.010 |
| 0.020 | 0.00409 | 0.03296 | 0.05530 | 0.00412 | 0.04379 | 0.07416 | 0.020 |
| 0.030 | 0.00629 | 0.03985 | 0.06682 | 0.00660 | 0.04943 | 0.08294 | 0.030 |
| 0.040 | 0.00796 | 0.04777 | 0.08032 | 0.00902 | 0.05613 | 0.09403 | 0.040 |
| 0.050 | 0.00937 | 0.05635 | 0.09494 | 0.01119 | 0.06360 | 0.10671 | 0.050 |
| 0.060 | 0.01076 | 0.06534 | 0.11023 | 0.01316 | 0.07166 | 0.12045 | 0.060 |
| 0.070 | 0.01231 | 0.07460 | 0.12592 | 0.01510 | 0.08016 | 0.13493 | 0.070 |
| 0.080 | 0.01411 | 0.08404 | 0.14189 | 0.01709 | 0.08897 | 0.14992 | 0.080 |
| 0.090 | 0.01621 | 0.09359 | 0.15804 | 0.01927 | 0.09802 | 0.16527 | 0.090 |
| 0.100 | 0.01864 | 0.10324 | 0.17432 | 0.02169 | 0.10724 | 0.18089 | 0.100 |
| 0.120 | 0.02448 | 0.12270 | 0.20715 | 0.02734 | 0.12606 | 0.21269 | 0.120 |
| 0.140 | 0.03155 | 0.14232 | 0.24018 | 0.03417 | 0.14520 | 0.24496 | 0.140 |
| 0.160 | 0.03976 | 0.16203 | 0.27332 | 0.04213 | 0.16456 | 0.27752 | 0.160 |
| 0.180 | 0.04900 | 0.18180 | 0.30652 | 0.05116 | 0.18405 | 0.31026 | 0.180 |
| 0.200 | 0.05920 | 0.20162 | 0.33973 | 0.06118 | 0.20365 | 0.34310 | 0.200 |
| 0.220 | 0.07028 | 0.22147 | 0.37292 | 0.07211 | 0.22332 | 0.37599 | 0.220 |
| 0.240 | 0.08218 | 0.24135 | 0.40608 | 0.08388 | 0.24304 | 0.40889 | 0.240 |
| 0.260 | 0.09482 | 0.26125 | 0.43919 | 0.09641 | 0.26281 | 0.44177 | 0.260 |
| 0.280 | 0.10815 | 0.28116 | 0.47222 | 0.10965 | 0.28261 | 0.47462 | 0.280 |
| 0.300 | 0.12212 | 0.30108 | 0.50516 | 0.12353 | 0.30243 | 0.50740 | 0.300 |
| 0.320 | 0.13667 | 0.32102 | 0.53801 | 0.13800 | 0.32228 | 0.54010 | 0.320 |
| 0.340 | 0.15174 | 0.34095 | 0.57074 | 0.15301 | 0.34215 | 0.57269 | 0.340 |
| 0.360 | 0.16731 | 0.36090 | 0.60333 | 0.16852 | 0.36203 | 0.60517 | 0.360 |
| 0.380 | 0.18332 | 0.38085 | 0.63577 | 0.18447 | 0.38192 | 0.63751 | 0.380 |
| 0.400 | 0.19973 | 0.40081 | 0.66805 | 0.20084 | 0.40183 | 0.66969 | 0.400 |
| 0.420 | 0.21652 | 0.42077 | 0.70014 | 0.21758 | 0.42174 | 0.70170 | 0.420 |
| 0.440 | 0.23364 | 0.44074 | 0.73203 | 0.23466 | 0.44166 | 0.73350 | 0.440 |
| 0.460 | 0.25107 | 0.46071 | 0.76369 | 0.25205 | 0.46159 | 0.76509 | 0.460 |
| 0.480 | 0.26878 | 0.48068 | 0.79510 | 0.26973 | 0.48152 | 0.79643 | 0.480 |
| 0.500 | 0.28674 | 0.50065 | 0.82623 | 0.28765 | 0.50146 | 0.82751 | 0.500 |
| 0.520 | 0.30494 | 0.52063 | 0.85707 | 0.30582 | 0.52140 | 0.85828 | 0.520 |
| 0.540 | 0.32334 | 0.54060 | 0.88757 | 0.32419 | 0.54135 | 0.88872 | 0.540 |
| 0.560 | 0.34193 | 0.56058 | 0.91771 | 0.34275 | 0.56131 | 0.91880 | 0.560 |
| 0.580 | 0.36070 | 0.58056 | 0.94744 | 0.36149 | 0.58126 | 0.94849 | 0.580 |
| 0.600 | 0.37961 | 0.60054 | 0.97674 | 0.38038 | 0.60122 | 0.97774 | 0.600 |
| 0.620 | 0.39867 | 0.62053 | 1.00556 | 0.39942 | 0.62118 | 1.00651 | 0.620 |
| 0.640 | 0.41786 | 0.64051 | 1.03386 | 0.41858 | 0.64115 | 1.03477 | 0.640 |
| 0.660 | 0.43716 | 0.66050 | 1.06160 | 0.43787 | 0.66111 | 1.06246 | 0.660 |
| 0.680 | 0.45657 | 0.68048 | 1.08873 | 0.45725 | 0.68107 | 1.08954 | 0.680 |
| 0.700 | 0.47606 | 0.70047 | 1.11520 | 0.47673 | 0.70105 | 1.11597 | 0.700 |
| 0.720 | 0.49565 | 0.72045 | 1.14096 | 0.49629 | 0.72102 | 1.14169 | 0.720 |
| 0.740 | 0.51530 | 0.74044 | 1.16597 | 0.51593 | 0.74099 | 1.16667 | 0.740 |
| 0.760 | 0.53503 | 0.76043 | 1.19020 | 0.53564 | 0.76096 | 1.19085 | 0.760 |
| 0.780 | 0.55481 | 0.78042 | 1.21359 | 0.55541 | 0.78094 | 1.21421 | 0.780 |
| 0.800 | 0.57465 | 0.80041 | 1.23613 | 0.57523 | 0.80091 | 1.23671 | 0.800 |
| 0.820 | 0.59454 | 0.82040 | 1.25778 | 0.59510 | 0.82089 | 1.25832 | 0.820 |
| 0.840 | 0.61446 | 0.84039 | 1.27855 | 0.61502 | 0.84087 | 1.27906 | 0.840 |
| 0.860 | 0.63443 | 0.86038 | 1.29840 | 0.63497 | 0.86085 | 1.29888 | 0.860 |
| 0.880 | 0.65442 | 0.88037 | 1.31738 | 0.65496 | 0.88083 | 1.31783 | 0.880 |
| 0.900 | 0.67445 | 0.90036 | 1.33549 | 0.67497 | 0.90081 | 1.33590 | 0.900 |
| 0.920 | 0.69450 | 0.92036 | 1.35277 | 0.69501 | 0.92080 | 1.35315 | 0.920 |
| 0.940 | 0.71457 | 0.94035 | 1.36925 | 0.71506 | 0.94078 | 1.36959 | 0.940 |
| 0.960 | 0.73466 | 0.96034 | 1.38498 | 0.73514 | 0.96076 | 1.38526 | 0.960 |
| 0.980 | 0.75476 | 0.98034 | 1.39998 | 0.75524 | 0.98074 | 1.40016 | 0.980 |
| 1.000 | 0.77488 | 1.00032 | 1.41432 | 0.77534 | 1.00073 | 1.41445 | 1.000 |

$H/R = 0.040$ $n = 4.$ $H/R = 0.050$

| $H/L$ | $\Omega_1$ | $\Omega_2$ | $\Omega_3$ | $\Omega_1$ | $\Omega_2$ | $\Omega_3$ | $H/L$ |
|---|---|---|---|---|---|---|---|
| 0. | 0.00360 | 0.05093 | 0.08868 | 0.00561 | 0.06367 | 0.11081 | 0. |
| 0.001 | 0.00360 | 0.05095 | 0.08869 | 0.00561 | 0.06368 | 0.11082 | 0.001 |
| 0.005 | 0.00364 | 0.05123 | 0.08905 | 0.00565 | 0.06391 | 0.11111 | 0.005 |
| 0.010 | 0.00381 | 0.05211 | 0.09015 | 0.00578 | 0.06462 | 0.11199 | 0.010 |
| 0.020 | 0.00490 | 0.05536 | 0.09449 | 0.00656 | 0.06731 | 0.11549 | 0.020 |
| 0.030 | 0.00705 | 0.06011 | 0.10143 | 0.00819 | 0.07139 | 0.12118 | 0.030 |
| 0.040 | 0.00964 | 0.06591 | 0.11056 | 0.01056 | 0.07650 | 0.12884 | 0.040 |
| 0.050 | 0.01210 | 0.07250 | 0.12142 | 0.01324 | 0.08240 | 0.13816 | 0.050 |
| 0.060 | 0.01476 | 0.07971 | 0.13357 | 0.01615 | 0.08892 | 0.14887 | 0.060 |
| 0.070 | 0.01716 | 0.08743 | 0.14670 | 0.01876 | 0.09595 | 0.16067 | 0.070 |
| 0.080 | 0.01957 | 0.09555 | 0.16055 | 0.02163 | 0.10343 | 0.17335 | 0.080 |
| 0.090 | 0.02203 | 0.10400 | 0.17494 | 0.02439 | 0.11128 | 0.18671 | 0.090 |
| 0.100 | 0.02463 | 0.11271 | 0.18975 | 0.02727 | 0.11944 | 0.20060 | 0.100 |
| 0.120 | 0.03042 | 0.13069 | 0.22023 | 0.03343 | 0.13650 | 0.22960 | 0.120 |
| 0.140 | 0.03719 | 0.14921 | 0.25152 | 0.04035 | 0.15428 | 0.25973 | 0.140 |
| 0.160 | 0.04502 | 0.16808 | 0.28331 | 0.04819 | 0.17256 | 0.29059 | 0.160 |
| 0.180 | 0.05389 | 0.18719 | 0.31543 | 0.05698 | 0.19120 | 0.32196 | 0.180 |
| 0.200 | 0.06374 | 0.20648 | 0.34777 | 0.06673 | 0.21010 | 0.35369 | 0.200 |
| 0.220 | 0.07452 | 0.22589 | 0.38024 | 0.07738 | 0.22919 | 0.38564 | 0.220 |
| 0.240 | 0.08614 | 0.24540 | 0.41279 | 0.08888 | 0.24843 | 0.41775 | 0.240 |
| 0.260 | 0.09855 | 0.26499 | 0.44537 | 0.10116 | 0.26779 | 0.44996 | 0.260 |
| 0.280 | 0.11167 | 0.28463 | 0.47795 | 0.11416 | 0.28724 | 0.48221 | 0.280 |
| 0.300 | 0.12545 | 0.30432 | 0.51051 | 0.12783 | 0.30676 | 0.51448 | 0.300 |
| 0.320 | 0.13983 | 0.32406 | 0.54300 | 0.14211 | 0.32634 | 0.54672 | 0.320 |
| 0.340 | 0.15476 | 0.34382 | 0.57542 | 0.15694 | 0.34596 | 0.57891 | 0.340 |
| 0.360 | 0.17018 | 0.36361 | 0.60774 | 0.17227 | 0.36563 | 0.61102 | 0.360 |
| 0.380 | 0.18607 | 0.38342 | 0.63994 | 0.18808 | 0.38534 | 0.64303 | 0.380 |
| 0.400 | 0.20237 | 0.40324 | 0.67198 | 0.20430 | 0.40507 | 0.67491 | 0.400 |
| 0.420 | 0.21905 | 0.42309 | 0.70387 | 0.22090 | 0.42483 | 0.70664 | 0.420 |
| 0.440 | 0.23607 | 0.44295 | 0.73556 | 0.23786 | 0.44461 | 0.73820 | 0.440 |
| 0.460 | 0.25341 | 0.46282 | 0.76705 | 0.25514 | 0.46441 | 0.76955 | 0.460 |
| 0.480 | 0.27103 | 0.48270 | 0.79829 | 0.27270 | 0.48423 | 0.80068 | 0.480 |
| 0.500 | 0.28892 | 0.50260 | 0.82928 | 0.29053 | 0.50406 | 0.83154 | 0.500 |
| 0.520 | 0.30704 | 0.52250 | 0.85996 | 0.30859 | 0.52390 | 0.86212 | 0.520 |
| 0.540 | 0.32537 | 0.54241 | 0.89033 | 0.32687 | 0.54376 | 0.89238 | 0.540 |
| 0.560 | 0.34390 | 0.56232 | 0.92033 | 0.34535 | 0.56363 | 0.92229 | 0.560 |
| 0.580 | 0.36260 | 0.58224 | 0.94995 | 0.36401 | 0.58350 | 0.95182 | 0.580 |
| 0.600 | 0.38146 | 0.60217 | 0.97913 | 0.38282 | 0.60338 | 0.98091 | 0.600 |
| 0.620 | 0.40046 | 0.62209 | 1.00784 | 0.40179 | 0.62327 | 1.00953 | 0.620 |
| 0.640 | 0.41959 | 0.64203 | 1.03602 | 0.42088 | 0.64317 | 1.03764 | 0.640 |
| 0.660 | 0.43884 | 0.66197 | 1.06365 | 0.44010 | 0.66307 | 1.06519 | 0.660 |
| 0.680 | 0.45820 | 0.68191 | 1.09068 | 0.45942 | 0.68298 | 1.09214 | 0.680 |
| 0.700 | 0.47766 | 0.70185 | 1.11705 | 0.47884 | 0.70290 | 1.11843 | 0.700 |
| 0.720 | 0.49719 | 0.72181 | 1.14271 | 0.49835 | 0.72282 | 1.14402 | 0.720 |
| 0.740 | 0.51681 | 0.74176 | 1.16763 | 0.51793 | 0.74274 | 1.16887 | 0.740 |
| 0.760 | 0.53649 | 0.76171 | 1.19176 | 0.53759 | 0.76267 | 1.19293 | 0.760 |
| 0.780 | 0.55624 | 0.78167 | 1.21507 | 0.55731 | 0.78260 | 1.21616 | 0.780 |
| 0.800 | 0.57605 | 0.80162 | 1.23751 | 0.57708 | 0.80254 | 1.23854 | 0.800 |
| 0.820 | 0.59590 | 0.82158 | 1.25909 | 0.59691 | 0.82247 | 1.26004 | 0.820 |
| 0.840 | 0.61579 | 0.84155 | 1.27976 | 0.61678 | 0.84241 | 1.28065 | 0.840 |
| 0.860 | 0.63572 | 0.86151 | 1.29954 | 0.63669 | 0.86236 | 1.30036 | 0.860 |
| 0.880 | 0.65569 | 0.88148 | 1.31843 | 0.65664 | 0.88231 | 1.31920 | 0.880 |
| 0.900 | 0.67569 | 0.90144 | 1.33646 | 0.67661 | 0.90225 | 1.33716 | 0.900 |
| 0.920 | 0.69571 | 0.92141 | 1.35366 | 0.69662 | 0.92220 | 1.35426 | 0.920 |
| 0.940 | 0.71575 | 0.94138 | 1.37002 | 0.71664 | 0.94216 | 1.37052 | 0.940 |
| 0.960 | 0.73581 | 0.96135 | 1.38559 | 0.73668 | 0.96212 | 1.38595 | 0.960 |
| 0.980 | 0.75589 | 0.98132 | 1.40037 | 0.75674 | 0.98208 | 1.40063 | 0.980 |
| 1.000 | 0.77599 | 1.00130 | 1.41464 | 0.77682 | 1.00203 | 1.41488 | 1.000 |

$H/R = 0.060$ $n = 4.$ $H/R = 0.070$

| $H/L$ | $\Omega_1$ | $\Omega_2$ | $\Omega_3$ | $\Omega_1$ | $\Omega_2$ | $\Omega_3$ | $H/L$ |
|---|---|---|---|---|---|---|---|
| 0. | 0.00806 | 0.07641 | 0.13291 | 0.01094 | 0.08914 | 0.15498 | 0. |
| 0.001 | 0.00806 | 0.07641 | 0.13292 | 0.01094 | 0.08915 | 0.15499 | 0.001 |
| 0.005 | 0.00810 | 0.07661 | 0.13316 | 0.01097 | 0.08932 | 0.15519 | 0.005 |
| 0.010 | 0.00821 | 0.07720 | 0.13390 | 0.01108 | 0.08983 | 0.15583 | 0.010 |
| 0.020 | 0.00876 | 0.07949 | 0.13683 | 0.01157 | 0.09182 | 0.15835 | 0.020 |
| 0.030 | 0.01010 | 0.08305 | 0.14164 | 0.01264 | 0.09496 | 0.16250 | 0.030 |
| 0.040 | 0.01210 | 0.08760 | 0.14818 | 0.01435 | 0.09905 | 0.16819 | 0.040 |
| 0.050 | 0.01463 | 0.09293 | 0.15628 | 0.01662 | 0.10390 | 0.17532 | 0.050 |
| 0.060 | 0.01745 | 0.09888 | 0.16573 | 0.01932 | 0.10938 | 0.18373 | 0.060 |
| 0.070 | 0.02022 | 0.10535 | 0.17633 | 0.02229 | 0.11537 | 0.19327 | 0.070 |
| 0.080 | 0.02343 | 0.11227 | 0.18788 | 0.02544 | 0.12181 | 0.20379 | 0.080 |
| 0.090 | 0.02658 | 0.11958 | 0.20021 | 0.02895 | 0.12864 | 0.21515 | 0.090 |
| 0.100 | 0.02968 | 0.12722 | 0.21318 | 0.03204 | 0.13581 | 0.22721 | 0.100 |
| 0.120 | 0.03631 | 0.14334 | 0.24060 | 0.03909 | 0.15104 | 0.25302 | 0.120 |
| 0.140 | 0.04351 | 0.16033 | 0.26944 | 0.04664 | 0.16724 | 0.28052 | 0.140 |
| 0.160 | 0.05149 | 0.17795 | 0.29927 | 0.05482 | 0.18417 | 0.30923 | 0.160 |
| 0.180 | 0.06031 | 0.19605 | 0.32979 | 0.06377 | 0.20168 | 0.33881 | 0.180 |
| 0.200 | 0.07002 | 0.21449 | 0.36079 | 0.07351 | 0.21962 | 0.36902 | 0.200 |
| 0.220 | 0.08059 | 0.23321 | 0.39214 | 0.08407 | 0.23791 | 0.39969 | 0.220 |
| 0.240 | 0.09199 | 0.25213 | 0.42374 | 0.09541 | 0.25646 | 0.43071 | 0.240 |
| 0.260 | 0.10417 | 0.27120 | 0.45550 | 0.10750 | 0.27522 | 0.46196 | 0.260 |
| 0.280 | 0.11706 | 0.29041 | 0.48736 | 0.12031 | 0.29415 | 0.49338 | 0.280 |
| 0.300 | 0.13062 | 0.30972 | 0.51929 | 0.13376 | 0.31322 | 0.52491 | 0.300 |
| 0.320 | 0.14479 | 0.32912 | 0.55123 | 0.14783 | 0.33240 | 0.55650 | 0.320 |
| 0.340 | 0.15952 | 0.34858 | 0.58314 | 0.16246 | 0.35168 | 0.58810 | 0.340 |
| 0.360 | 0.17476 | 0.36811 | 0.61501 | 0.17760 | 0.37104 | 0.61969 | 0.360 |
| 0.380 | 0.19047 | 0.38768 | 0.64680 | 0.19322 | 0.39046 | 0.65122 | 0.380 |
| 0.400 | 0.20660 | 0.40730 | 0.67848 | 0.20926 | 0.40994 | 0.68266 | 0.400 |
| 0.420 | 0.22313 | 0.42695 | 0.71002 | 0.22570 | 0.42947 | 0.71399 | 0.420 |
| 0.440 | 0.24001 | 0.44664 | 0.74140 | 0.24249 | 0.44904 | 0.74517 | 0.440 |
| 0.460 | 0.25721 | 0.46635 | 0.77259 | 0.25961 | 0.46864 | 0.77618 | 0.460 |
| 0.480 | 0.27470 | 0.48609 | 0.80357 | 0.27703 | 0.48828 | 0.80697 | 0.480 |
| 0.500 | 0.29246 | 0.50585 | 0.83430 | 0.29472 | 0.50796 | 0.83754 | 0.500 |
| 0.520 | 0.31047 | 0.52562 | 0.86475 | 0.31265 | 0.52765 | 0.86784 | 0.520 |
| 0.540 | 0.32869 | 0.54541 | 0.89489 | 0.33080 | 0.54737 | 0.89783 | 0.540 |
| 0.560 | 0.34711 | 0.56522 | 0.92468 | 0.34916 | 0.56710 | 0.92748 | 0.560 |
| 0.580 | 0.36571 | 0.58504 | 0.95409 | 0.36771 | 0.58686 | 0.95676 | 0.580 |
| 0.600 | 0.38448 | 0.60487 | 0.98308 | 0.38641 | 0.60663 | 0.98562 | 0.600 |
| 0.620 | 0.40339 | 0.62471 | 1.01159 | 0.40527 | 0.62642 | 1.01401 | 0.620 |
| 0.640 | 0.42245 | 0.64457 | 1.03960 | 0.42427 | 0.64622 | 1.04191 | 0.640 |
| 0.660 | 0.44162 | 0.66443 | 1.06706 | 0.44339 | 0.66603 | 1.06924 | 0.660 |
| 0.680 | 0.46090 | 0.68430 | 1.09390 | 0.46263 | 0.68585 | 1.09597 | 0.680 |
| 0.700 | 0.48028 | 0.70418 | 1.12010 | 0.48196 | 0.70569 | 1.12207 | 0.700 |
| 0.720 | 0.49975 | 0.72406 | 1.14561 | 0.50139 | 0.72553 | 1.14745 | 0.720 |
| 0.740 | 0.51930 | 0.74395 | 1.17036 | 0.52090 | 0.74538 | 1.17211 | 0.740 |
| 0.760 | 0.53892 | 0.76385 | 1.19433 | 0.54048 | 0.76524 | 1.19597 | 0.760 |
| 0.780 | 0.55861 | 0.78375 | 1.21748 | 0.56013 | 0.78510 | 1.21902 | 0.780 |
| 0.800 | 0.57835 | 0.80366 | 1.23977 | 0.57983 | 0.80497 | 1.24121 | 0.800 |
| 0.820 | 0.59814 | 0.82356 | 1.26119 | 0.59959 | 0.82485 | 1.26253 | 0.820 |
| 0.840 | 0.61799 | 0.84348 | 1.28172 | 0.61940 | 0.84474 | 1.28296 | 0.840 |
| 0.860 | 0.63787 | 0.86340 | 1.30135 | 0.63925 | 0.86463 | 1.30248 | 0.860 |
| 0.880 | 0.65779 | 0.88332 | 1.32009 | 0.65914 | 0.88452 | 1.32111 | 0.880 |
| 0.900 | 0.67774 | 0.90325 | 1.33795 | 0.67906 | 0.90442 | 1.33884 | 0.900 |
| 0.920 | 0.69771 | 0.92318 | 1.35494 | 0.69901 | 0.92432 | 1.35568 | 0.920 |
| 0.940 | 0.71771 | 0.94311 | 1.37106 | 0.71898 | 0.94423 | 1.37163 | 0.940 |
| 0.960 | 0.73773 | 0.96304 | 1.38634 | 0.73897 | 0.96414 | 1.38677 | 0.960 |
| 0.980 | 0.75777 | 0.98298 | 1.40094 | 0.75898 | 0.98407 | 1.40131 | 0.980 |
| 1.000 | 0.77783 | 1.00293 | 1.41517 | 0.77901 | 1.00398 | 1.41552 | 1.000 |

$H/R = 0.080$ $n = 4.$ $H/R = 0.090$

| $H/L$ | $\Omega_1$ | $\Omega_2$ | $\Omega_3$ | $\Omega_1$ | $\Omega_2$ | $\Omega_3$ | $H/L$ |
|---|---|---|---|---|---|---|---|
| 0. | 0.01423 | 0.10189 | 0.17701 | 0.01794 | 0.11463 | 0.19900 | 0. |
| 0.001 | 0.01423 | 0.10189 | 0.17702 | 0.01794 | 0.11463 | 0.19900 | 0.001 |
| 0.005 | 0.01427 | 0.10204 | 0.17720 | 0.01797 | 0.11476 | 0.19916 | 0.005 |
| 0.010 | 0.01437 | 0.10249 | 0.17775 | 0.01807 | 0.11516 | 0.19966 | 0.010 |
| 0.020 | 0.01484 | 0.10424 | 0.17996 | 0.01858 | 0.11673 | 0.20163 | 0.020 |
| 0.030 | 0.01575 | 0.10705 | 0.18362 | 0.01931 | 0.11926 | 0.20488 | 0.030 |
| 0.040 | 0.01720 | 0.11074 | 0.18865 | 0.02061 | 0.12263 | 0.20939 | 0.040 |
| 0.050 | 0.01922 | 0.11519 | 0.19499 | 0.02243 | 0.12673 | 0.21509 | 0.050 |
| 0.060 | 0.02173 | 0.12025 | 0.20253 | 0.02472 | 0.13143 | 0.22190 | 0.060 |
| 0.070 | 0.02460 | 0.12584 | 0.21116 | 0.02744 | 0.13664 | 0.22976 | 0.070 |
| 0.080 | 0.02774 | 0.13186 | 0.22077 | 0.03049 | 0.14231 | 0.23856 | 0.080 |
| 0.090 | 0.03108 | 0.13828 | 0.23123 | 0.03380 | 0.14836 | 0.24822 | 0.090 |
| 0.100 | 0.03451 | 0.14503 | 0.24243 | 0.03733 | 0.15475 | 0.25863 | 0.100 |
| 0.120 | 0.04186 | 0.15946 | 0.26667 | 0.04499 | 0.16844 | 0.28137 | 0.120 |
| 0.140 | 0.04975 | 0.17489 | 0.29281 | 0.05300 | 0.18317 | 0.30617 | 0.140 |
| 0.160 | 0.05820 | 0.19114 | 0.32036 | 0.06163 | 0.19876 | 0.33254 | 0.160 |
| 0.180 | 0.06731 | 0.20804 | 0.34894 | 0.07093 | 0.21505 | 0.36009 | 0.180 |
| 0.200 | 0.07715 | 0.22545 | 0.37830 | 0.08089 | 0.23191 | 0.38856 | 0.200 |
| 0.220 | 0.08773 | 0.24327 | 0.40824 | 0.09154 | 0.24925 | 0.41771 | 0.220 |
| 0.240 | 0.09906 | 0.26142 | 0.43861 | 0.10289 | 0.26696 | 0.44740 | 0.240 |
| 0.260 | 0.11110 | 0.27982 | 0.46930 | 0.11492 | 0.28499 | 0.47749 | 0.260 |
| 0.280 | 0.12384 | 0.29844 | 0.50023 | 0.12761 | 0.30327 | 0.50788 | 0.280 |
| 0.300 | 0.13721 | 0.31724 | 0.53132 | 0.14092 | 0.32177 | 0.53849 | 0.300 |
| 0.320 | 0.15119 | 0.33618 | 0.56252 | 0.15483 | 0.34044 | 0.56925 | 0.320 |
| 0.340 | 0.16573 | 0.35524 | 0.59377 | 0.16928 | 0.35927 | 0.60011 | 0.340 |
| 0.360 | 0.18077 | 0.37440 | 0.62503 | 0.18423 | 0.37821 | 0.63102 | 0.360 |
| 0.380 | 0.19629 | 0.39365 | 0.65627 | 0.19966 | 0.39727 | 0.66193 | 0.380 |
| 0.400 | 0.21224 | 0.41298 | 0.68744 | 0.21552 | 0.41641 | 0.69281 | 0.400 |
| 0.420 | 0.22859 | 0.43236 | 0.71853 | 0.23177 | 0.43564 | 0.72362 | 0.420 |
| 0.440 | 0.24529 | 0.45180 | 0.74948 | 0.24839 | 0.45493 | 0.75433 | 0.440 |
| 0.460 | 0.26233 | 0.47129 | 0.78027 | 0.26533 | 0.47429 | 0.78488 | 0.460 |
| 0.480 | 0.27966 | 0.49082 | 0.81088 | 0.28258 | 0.49370 | 0.81527 | 0.480 |
| 0.500 | 0.29727 | 0.51039 | 0.84126 | 0.30011 | 0.51315 | 0.84543 | 0.500 |
| 0.520 | 0.31513 | 0.52999 | 0.87138 | 0.31788 | 0.53265 | 0.87536 | 0.520 |
| 0.540 | 0.33321 | 0.54962 | 0.90120 | 0.33589 | 0.55218 | 0.90500 | 0.540 |
| 0.560 | 0.35150 | 0.56928 | 0.93070 | 0.35410 | 0.57175 | 0.93431 | 0.560 |
| 0.580 | 0.36997 | 0.58896 | 0.95982 | 0.37251 | 0.59134 | 0.96326 | 0.580 |
| 0.600 | 0.38862 | 0.60866 | 0.98853 | 0.39108 | 0.61096 | 0.99181 | 0.600 |
| 0.620 | 0.40742 | 0.62838 | 1.01679 | 0.40982 | 0.63062 | 1.01991 | 0.620 |
| 0.640 | 0.42636 | 0.64812 | 1.04454 | 0.42869 | 0.65028 | 1.04750 | 0.640 |
| 0.660 | 0.44542 | 0.66787 | 1.07174 | 0.44770 | 0.66997 | 1.07455 | 0.660 |
| 0.680 | 0.46460 | 0.68764 | 1.09835 | 0.46682 | 0.68968 | 1.10101 | 0.680 |
| 0.700 | 0.48389 | 0.70742 | 1.12430 | 0.48604 | 0.70940 | 1.12683 | 0.700 |
| 0.720 | 0.50326 | 0.72722 | 1.14957 | 0.50537 | 0.72914 | 1.15195 | 0.720 |
| 0.740 | 0.52272 | 0.74703 | 1.17410 | 0.52478 | 0.74889 | 1.17633 | 0.740 |
| 0.760 | 0.54226 | 0.76684 | 1.19784 | 0.54427 | 0.76866 | 1.19993 | 0.760 |
| 0.780 | 0.56187 | 0.78666 | 1.22077 | 0.56382 | 0.78844 | 1.22271 | 0.780 |
| 0.800 | 0.58153 | 0.80650 | 1.24284 | 0.58344 | 0.80823 | 1.24465 | 0.800 |
| 0.820 | 0.60125 | 0.82634 | 1.26403 | 0.60312 | 0.82803 | 1.26569 | 0.820 |
| 0.840 | 0.62102 | 0.84619 | 1.28434 | 0.62285 | 0.84783 | 1.28585 | 0.840 |
| 0.860 | 0.64084 | 0.86605 | 1.30373 | 0.64262 | 0.86765 | 1.30508 | 0.860 |
| 0.880 | 0.66069 | 0.88591 | 1.32222 | 0.66243 | 0.88748 | 1.32340 | 0.880 |
| 0.900 | 0.68057 | 0.90578 | 1.33978 | 0.68228 | 0.90731 | 1.34078 | 0.900 |
| 0.920 | 0.70049 | 0.92565 | 1.35645 | 0.70216 | 0.92716 | 1.35724 | 0.920 |
| 0.940 | 0.72043 | 0.94553 | 1.37222 | 0.72206 | 0.94700 | 1.37284 | 0.940 |
| 0.960 | 0.74039 | 0.96542 | 1.38724 | 0.74199 | 0.96686 | 1.38776 | 0.960 |
| 0.980 | 0.76038 | 0.98531 | 1.40174 | 0.76194 | 0.98672 | 1.40222 | 0.980 |
| 1.000 | 0.78037 | 1.00520 | 1.41591 | 0.78191 | 1.00658 | 1.41637 | 1.000 |

$H/R = 0.100 \qquad n = 4.$

| $H/L$ | $\Omega_1$ | $\Omega_2$ | $\Omega_3$ | $\Omega_4$ | $\Omega_5$ | $\Omega_6$ |
|---|---|---|---|---|---|---|
| 0. | 0.02204 | 0.12737 | 0.22093 | 1.00849 | 1.02940 | 1.84056 |
| 0.001 | 0.02205 | 0.12738 | 0.22094 | 1.00850 | 1.02940 | 1.84056 |
| 0.005 | 0.02208 | 0.12749 | 0.22108 | 1.00852 | 1.02944 | 1.84053 |
| 0.010 | 0.02217 | 0.12786 | 0.22153 | 1.00855 | 1.02955 | 1.84042 |
| 0.020 | 0.02256 | 0.12928 | 0.22330 | 1.00873 | 1.03002 | 1.83997 |
| 0.030 | 0.02334 | 0.13158 | 0.22624 | 1.00902 | 1.03080 | 1.83925 |
| 0.040 | 0.02452 | 0.13467 | 0.23031 | 1.00942 | 1.03188 | 1.83823 |
| 0.050 | 0.02615 | 0.13845 | 0.23548 | 1.00993 | 1.03329 | 1.83696 |
| 0.060 | 0.02827 | 0.14283 | 0.24170 | 1.01055 | 1.03500 | 1.83545 |
| 0.070 | 0.03082 | 0.14773 | 0.24889 | 1.01127 | 1.03703 | 1.83370 |
| 0.080 | 0.03373 | 0.15306 | 0.25699 | 1.01209 | 1.03936 | 1.83175 |
| 0.090 | 0.03697 | 0.15879 | 0.26592 | 1.01301 | 1.04202 | 1.82961 |
| 0.100 | 0.04046 | 0.16485 | 0.27561 | 1.01402 | 1.04497 | 1.82731 |
| 0.120 | 0.04803 | 0.17789 | 0.29695 | 1.01633 | 1.05180 | 1.82227 |
| 0.140 | 0.05623 | 0.19197 | 0.32045 | 1.01901 | 1.05982 | 1.81680 |
| 0.160 | 0.06521 | 0.20693 | 0.34565 | 1.02205 | 1.06896 | 1.81102 |
| 0.180 | 0.07465 | 0.22263 | 0.37217 | 1.02547 | 1.07918 | 1.80507 |
| 0.200 | 0.08475 | 0.23896 | 0.39971 | 1.02923 | 1.09044 | 1.79905 |
| 0.220 | 0.09549 | 0.25580 | 0.42806 | 1.03337 | 1.10268 | 1.79303 |
| 0.240 | 0.10688 | 0.27307 | 0.45702 | 1.03787 | 1.11582 | 1.78711 |
| 0.260 | 0.11892 | 0.29069 | 0.48646 | 1.04271 | 1.12982 | 1.78134 |
| 0.280 | 0.13159 | 0.30862 | 0.51628 | 1.04791 | 1.14464 | 1.77576 |
| 0.300 | 0.14485 | 0.32680 | 0.54638 | 1.05347 | 1.16020 | 1.77043 |
| 0.320 | 0.15870 | 0.34518 | 0.57668 | 1.05936 | 1.17647 | 1.76541 |
| 0.340 | 0.17308 | 0.36374 | 0.60711 | 1.06560 | 1.19338 | 1.76071 |
| 0.360 | 0.18795 | 0.38245 | 0.63764 | 1.07217 | 1.21091 | 1.75638 |
| 0.380 | 0.20329 | 0.40129 | 0.66820 | 1.07907 | 1.22902 | 1.75245 |
| 0.400 | 0.21906 | 0.42025 | 0.69876 | 1.08629 | 1.24764 | 1.74894 |
| 0.420 | 0.23522 | 0.43930 | 0.72926 | 1.09383 | 1.26674 | 1.74589 |
| 0.440 | 0.25175 | 0.45843 | 0.75969 | 1.10168 | 1.28629 | 1.74334 |
| 0.460 | 0.26861 | 0.47764 | 0.78998 | 1.10983 | 1.30626 | 1.74132 |
| 0.480 | 0.28577 | 0.49691 | 0.82013 | 1.11829 | 1.32662 | 1.73983 |
| 0.500 | 0.30321 | 0.51623 | 0.85006 | 1.12703 | 1.34732 | 1.73894 |
| 0.520 | 0.32090 | 0.53561 | 0.87977 | 1.13606 | 1.36835 | 1.73867 |
| 0.540 | 0.33883 | 0.55504 | 0.90920 | 1.14537 | 1.38967 | 1.73905 |
| 0.560 | 0.35696 | 0.57450 | 0.93831 | 1.15495 | 1.41127 | 1.74012 |
| 0.580 | 0.37529 | 0.59401 | 0.96708 | 1.16479 | 1.43313 | 1.74191 |
| 0.600 | 0.39379 | 0.61354 | 0.99544 | 1.17489 | 1.45520 | 1.74448 |
| 0.620 | 0.41246 | 0.63311 | 1.02335 | 1.18525 | 1.47748 | 1.74785 |
| 0.640 | 0.43126 | 0.65270 | 1.05078 | 1.19586 | 1.49994 | 1.75209 |
| 0.660 | 0.45020 | 0.67231 | 1.07766 | 1.20670 | 1.52257 | 1.75721 |
| 0.680 | 0.46926 | 0.69195 | 1.10394 | 1.21778 | 1.54535 | 1.76330 |
| 0.700 | 0.48843 | 0.71161 | 1.12960 | 1.22910 | 1.56826 | 1.77037 |
| 0.720 | 0.50770 | 0.73129 | 1.15455 | 1.24063 | 1.59129 | 1.77848 |
| 0.740 | 0.52705 | 0.75098 | 1.17878 | 1.25239 | 1.61441 | 1.78767 |
| 0.760 | 0.54648 | 0.77070 | 1.20222 | 1.26437 | 1.63761 | 1.79797 |
| 0.780 | 0.56599 | 0.79042 | 1.22484 | 1.27656 | 1.66086 | 1.80943 |
| 0.800 | 0.58556 | 0.81016 | 1.24661 | 1.28897 | 1.68418 | 1.82205 |
| 0.820 | 0.60519 | 0.82991 | 1.26750 | 1.30160 | 1.70753 | 1.83587 |
| 0.840 | 0.62487 | 0.84968 | 1.28747 | 1.31446 | 1.73090 | 1.85089 |
| 0.860 | 0.64459 | 0.86945 | 1.30652 | 1.32756 | 1.75429 | 1.86710 |
| 0.880 | 0.66437 | 0.88924 | 1.32464 | 1.34092 | 1.77766 | 1.88448 |
| 0.900 | 0.68417 | 0.90903 | 1.34181 | 1.35456 | 1.80101 | 1.90300 |
| 0.920 | 0.70401 | 0.92883 | 1.35807 | 1.36847 | 1.82434 | 1.92264 |
| 0.940 | 0.72388 | 0.94865 | 1.37352 | 1.38262 | 1.84763 | 1.94332 |
| 0.960 | 0.74377 | 0.96847 | 1.38834 | 1.39682 | 1.87086 | 1.96502 |
| 0.980 | 0.76368 | 0.98829 | 1.40275 | 1.41095 | 1.89403 | 1.98765 |
| 1.000 | 0.78362 | 1.00813 | 1.41688 | 1.42492 | 1.91711 | 2.01116 |

$H/R = 0.120$ $n = 4.$ $H/R = 0.140$

| $H/L$ | $\Omega_1$ | $\Omega_2$ | $\Omega_3$ | $\Omega_1$ | $\Omega_2$ | $\Omega_3$ | $H/L$ |
|---|---|---|---|---|---|---|---|
| 0. | 0.03141 | 0.15287 | 0.26462 | 0.04225 | 0.17838 | 0.30802 | 0. |
| 0.001 | 0.03142 | 0.15288 | 0.26463 | 0.04225 | 0.17839 | 0.30803 | 0.001 |
| 0.005 | 0.03144 | 0.15297 | 0.26474 | 0.04228 | 0.17847 | 0.30813 | 0.005 |
| 0.010 | 0.03154 | 0.15327 | 0.26511 | 0.04237 | 0.17873 | 0.30845 | 0.010 |
| 0.020 | 0.03191 | 0.15447 | 0.26660 | 0.04273 | 0.17975 | 0.30972 | 0.020 |
| 0.030 | 0.03258 | 0.15641 | 0.26906 | 0.04337 | 0.18144 | 0.31183 | 0.030 |
| 0.040 | 0.03363 | 0.15906 | 0.27248 | 0.04429 | 0.18374 | 0.31478 | 0.040 |
| 0.050 | 0.03504 | 0.16233 | 0.27683 | 0.04557 | 0.18663 | 0.31854 | 0.050 |
| 0.060 | 0.03686 | 0.16618 | 0.28209 | 0.04720 | 0.19003 | 0.32309 | 0.060 |
| 0.070 | 0.03909 | 0.17051 | 0.28822 | 0.04919 | 0.19391 | 0.32842 | 0.070 |
| 0.080 | 0.04171 | 0.17529 | 0.29518 | 0.05155 | 0.19822 | 0.33449 | 0.080 |
| 0.090 | 0.04471 | 0.18045 | 0.30290 | 0.05428 | 0.20291 | 0.34126 | 0.090 |
| 0.100 | 0.04801 | 0.18595 | 0.31135 | 0.05734 | 0.20794 | 0.34872 | 0.100 |
| 0.120 | 0.05545 | 0.19786 | 0.33020 | 0.06441 | 0.21890 | 0.36549 | 0.120 |
| 0.140 | 0.06376 | 0.21081 | 0.35131 | 0.07252 | 0.23089 | 0.38449 | 0.140 |
| 0.160 | 0.07296 | 0.22465 | 0.37426 | 0.08152 | 0.24378 | 0.40540 | 0.160 |
| 0.180 | 0.08249 | 0.23927 | 0.39872 | 0.09127 | 0.25744 | 0.42792 | 0.180 |
| 0.200 | 0.09288 | 0.25455 | 0.42441 | 0.10169 | 0.27179 | 0.45179 | 0.200 |
| 0.220 | 0.10377 | 0.27042 | 0.45107 | 0.11266 | 0.28675 | 0.47677 | 0.220 |
| 0.240 | 0.11528 | 0.28679 | 0.47853 | 0.12431 | 0.30226 | 0.50269 | 0.240 |
| 0.260 | 0.12738 | 0.30359 | 0.50661 | 0.13644 | 0.31824 | 0.52935 | 0.260 |
| 0.280 | 0.14005 | 0.32076 | 0.53520 | 0.14913 | 0.33465 | 0.55664 | 0.280 |
| 0.300 | 0.15328 | 0.33825 | 0.56418 | 0.16236 | 0.35142 | 0.58443 | 0.300 |
| 0.320 | 0.16704 | 0.35601 | 0.59347 | 0.17607 | 0.36851 | 0.61262 | 0.320 |
| 0.340 | 0.18131 | 0.37400 | 0.62298 | 0.19026 | 0.38589 | 0.64113 | 0.340 |
| 0.360 | 0.19606 | 0.39219 | 0.65266 | 0.20491 | 0.40352 | 0.66987 | 0.360 |
| 0.380 | 0.21125 | 0.41056 | 0.68244 | 0.21998 | 0.42136 | 0.69878 | 0.380 |
| 0.400 | 0.22685 | 0.42908 | 0.71227 | 0.23545 | 0.43940 | 0.72781 | 0.400 |
| 0.420 | 0.24285 | 0.44773 | 0.74211 | 0.25130 | 0.45760 | 0.75689 | 0.420 |
| 0.440 | 0.25920 | 0.46649 | 0.77190 | 0.26749 | 0.47596 | 0.78598 | 0.440 |
| 0.460 | 0.27589 | 0.48536 | 0.80161 | 0.28401 | 0.49444 | 0.81502 | 0.460 |
| 0.480 | 0.29287 | 0.50433 | 0.83120 | 0.30083 | 0.51305 | 0.84398 | 0.480 |
| 0.500 | 0.31014 | 0.52336 | 0.86061 | 0.31793 | 0.53176 | 0.87280 | 0.500 |
| 0.520 | 0.32766 | 0.54248 | 0.88983 | 0.33528 | 0.55056 | 0.90143 | 0.520 |
| 0.540 | 0.34542 | 0.56165 | 0.91878 | 0.35287 | 0.56945 | 0.92984 | 0.540 |
| 0.560 | 0.36339 | 0.58089 | 0.94745 | 0.37067 | 0.58842 | 0.95798 | 0.560 |
| 0.580 | 0.38156 | 0.60017 | 0.97577 | 0.38868 | 0.60745 | 0.98579 | 0.580 |
| 0.600 | 0.39991 | 0.61950 | 1.00371 | 0.40686 | 0.62655 | 1.01323 | 0.600 |
| 0.620 | 0.41842 | 0.63888 | 1.03121 | 0.42522 | 0.64570 | 1.04025 | 0.620 |
| 0.640 | 0.43708 | 0.65829 | 1.05823 | 0.44372 | 0.66491 | 1.06679 | 0.640 |
| 0.660 | 0.45588 | 0.67774 | 1.08472 | 0.46237 | 0.68416 | 1.09281 | 0.660 |
| 0.680 | 0.47480 | 0.69722 | 1.11061 | 0.48115 | 0.70345 | 1.11825 | 0.680 |
| 0.700 | 0.49384 | 0.71673 | 1.13589 | 0.50004 | 0.72279 | 1.14305 | 0.700 |
| 0.720 | 0.51298 | 0.73627 | 1.16046 | 0.51904 | 0.74216 | 1.16717 | 0.720 |
| 0.740 | 0.53221 | 0.75583 | 1.18431 | 0.53814 | 0.76156 | 1.19054 | 0.740 |
| 0.760 | 0.55152 | 0.77541 | 1.20736 | 0.55732 | 0.78099 | 1.21313 | 0.760 |
| 0.780 | 0.57091 | 0.79502 | 1.22959 | 0.57659 | 0.80046 | 1.23487 | 0.780 |
| 0.800 | 0.59037 | 0.81464 | 1.25096 | 0.59593 | 0.81995 | 1.25574 | 0.800 |
| 0.820 | 0.60990 | 0.83429 | 1.27144 | 0.61533 | 0.83947 | 1.27570 | 0.820 |
| 0.840 | 0.62947 | 0.85395 | 1.29098 | 0.63480 | 0.85900 | 1.29471 | 0.840 |
| 0.860 | 0.64910 | 0.87362 | 1.30958 | 0.65432 | 0.87856 | 1.31278 | 0.860 |
| 0.880 | 0.66878 | 0.89331 | 1.32722 | 0.67388 | 0.89814 | 1.32991 | 0.880 |
| 0.900 | 0.68849 | 0.91302 | 1.34394 | 0.69349 | 0.91774 | 1.34618 | 0.900 |
| 0.920 | 0.70824 | 0.93273 | 1.35981 | 0.71314 | 0.93735 | 1.36172 | 0.920 |
| 0.940 | 0.72802 | 0.95246 | 1.37500 | 0.73283 | 0.95698 | 1.37669 | 0.940 |
| 0.960 | 0.74783 | 0.97220 | 1.38967 | 0.75255 | 0.97663 | 1.39124 | 0.960 |
| 0.980 | 0.76766 | 0.99196 | 1.40400 | 0.77229 | 0.99629 | 1.40548 | 0.980 |
| 1.000 | 0.78752 | 1.01172 | 1.41807 | 0.79206 | 1.01596 | 1.41947 | 1.000 |

$H/R = 0.160$ $n = 4.$ $H/R = 0.180$

| $H/L$ | $\Omega_1$ | $\Omega_2$ | $\Omega_3$ | $\Omega_1$ | $\Omega_2$ | $\Omega_3$ | $H/L$ |
|---|---|---|---|---|---|---|---|
| 0. | 0.05445 | 0.20390 | 0.35108 | 0.06792 | 0.22943 | 0.39372 | 0. |
| 0.001 | 0.05445 | 0.20390 | 0.35108 | 0.06792 | 0.22943 | 0.39373 | 0.001 |
| 0.005 | 0.05448 | 0.20398 | 0.35117 | 0.06795 | 0.22950 | 0.39381 | 0.005 |
| 0.010 | 0.05457 | 0.20420 | 0.35145 | 0.06803 | 0.22970 | 0.39406 | 0.010 |
| 0.020 | 0.05492 | 0.20511 | 0.35257 | 0.06837 | 0.23050 | 0.39505 | 0.020 |
| 0.030 | 0.05552 | 0.20659 | 0.35442 | 0.06894 | 0.23183 | 0.39670 | 0.030 |
| 0.040 | 0.05639 | 0.20863 | 0.35700 | 0.06977 | 0.23366 | 0.39900 | 0.040 |
| 0.050 | 0.05755 | 0.21120 | 0.36031 | 0.07086 | 0.23597 | 0.40195 | 0.050 |
| 0.060 | 0.05911 | 0.21425 | 0.36432 | 0.07224 | 0.23874 | 0.40554 | 0.060 |
| 0.070 | 0.06086 | 0.21775 | 0.36903 | 0.07391 | 0.24192 | 0.40974 | 0.070 |
| 0.080 | 0.06302 | 0.22167 | 0.37440 | 0.07593 | 0.24550 | 0.41456 | 0.080 |
| 0.090 | 0.06551 | 0.22595 | 0.38042 | 0.07821 | 0.24943 | 0.41997 | 0.090 |
| 0.100 | 0.06834 | 0.23057 | 0.38707 | 0.08083 | 0.25370 | 0.42596 | 0.100 |
| 0.120 | 0.07495 | 0.24071 | 0.40212 | 0.08700 | 0.26312 | 0.43957 | 0.120 |
| 0.140 | 0.08271 | 0.25188 | 0.41931 | 0.09434 | 0.27357 | 0.45522 | 0.140 |
| 0.160 | 0.09146 | 0.26394 | 0.43840 | 0.10273 | 0.28490 | 0.47271 | 0.160 |
| 0.180 | 0.10106 | 0.27678 | 0.45913 | 0.11205 | 0.29701 | 0.49184 | 0.180 |
| 0.200 | 0.11138 | 0.29030 | 0.48129 | 0.12218 | 0.30981 | 0.51242 | 0.200 |
| 0.220 | 0.12247 | 0.30445 | 0.50465 | 0.13302 | 0.32323 | 0.53425 | 0.220 |
| 0.240 | 0.13400 | 0.31916 | 0.52904 | 0.14453 | 0.33721 | 0.55718 | 0.240 |
| 0.260 | 0.14615 | 0.33437 | 0.55429 | 0.15663 | 0.35171 | 0.58103 | 0.260 |
| 0.280 | 0.15885 | 0.35003 | 0.58025 | 0.16925 | 0.36668 | 0.60568 | 0.280 |
| 0.300 | 0.17206 | 0.36610 | 0.60681 | 0.18232 | 0.38207 | 0.63099 | 0.300 |
| 0.320 | 0.18572 | 0.38253 | 0.63385 | 0.19598 | 0.39785 | 0.65687 | 0.320 |
| 0.340 | 0.19984 | 0.39927 | 0.66129 | 0.21003 | 0.41398 | 0.68321 | 0.340 |
| 0.360 | 0.21441 | 0.41631 | 0.68904 | 0.22450 | 0.43043 | 0.70992 | 0.360 |
| 0.380 | 0.22937 | 0.43360 | 0.71702 | 0.23937 | 0.44716 | 0.73692 | 0.380 |
| 0.400 | 0.24472 | 0.45113 | 0.74517 | 0.25460 | 0.46415 | 0.76415 | 0.400 |
| 0.420 | 0.26044 | 0.46885 | 0.77344 | 0.27019 | 0.48137 | 0.79154 | 0.420 |
| 0.440 | 0.27649 | 0.48675 | 0.80175 | 0.28610 | 0.49881 | 0.81902 | 0.440 |
| 0.460 | 0.29286 | 0.50482 | 0.83005 | 0.30233 | 0.51643 | 0.84653 | 0.460 |
| 0.480 | 0.30952 | 0.52303 | 0.85831 | 0.31884 | 0.53423 | 0.87403 | 0.480 |
| 0.500 | 0.32645 | 0.54138 | 0.88646 | 0.33562 | 0.55218 | 0.90145 | 0.500 |
| 0.520 | 0.34364 | 0.55984 | 0.91446 | 0.35265 | 0.57027 | 0.92874 | 0.520 |
| 0.540 | 0.36106 | 0.57841 | 0.94225 | 0.36991 | 0.58848 | 0.95586 | 0.540 |
| 0.560 | 0.37870 | 0.59707 | 0.96979 | 0.38739 | 0.60682 | 0.98273 | 0.560 |
| 0.580 | 0.39654 | 0.61582 | 0.99702 | 0.40506 | 0.62526 | 1.00932 | 0.580 |
| 0.600 | 0.41457 | 0.63465 | 1.02389 | 0.42293 | 0.64380 | 1.03555 | 0.600 |
| 0.620 | 0.43276 | 0.65355 | 1.05035 | 0.44096 | 0.66242 | 1.06138 | 0.620 |
| 0.640 | 0.45111 | 0.67252 | 1.07635 | 0.45915 | 0.68113 | 1.08676 | 0.640 |
| 0.660 | 0.46960 | 0.69155 | 1.10183 | 0.47748 | 0.69991 | 1.11160 | 0.660 |
| 0.680 | 0.48823 | 0.71063 | 1.12672 | 0.49596 | 0.71876 | 1.13587 | 0.680 |
| 0.700 | 0.50697 | 0.72976 | 1.15097 | 0.51454 | 0.73767 | 1.15950 | 0.700 |
| 0.720 | 0.52582 | 0.74895 | 1.17454 | 0.53325 | 0.75664 | 1.18243 | 0.720 |
| 0.740 | 0.54478 | 0.76818 | 1.19735 | 0.55206 | 0.77566 | 1.20460 | 0.740 |
| 0.760 | 0.56383 | 0.78744 | 1.21938 | 0.57097 | 0.79473 | 1.22596 | 0.760 |
| 0.780 | 0.58296 | 0.80674 | 1.24055 | 0.58996 | 0.81385 | 1.24646 | 0.780 |
| 0.800 | 0.60217 | 0.82607 | 1.26082 | 0.60904 | 0.83301 | 1.26607 | 0.800 |
| 0.820 | 0.62145 | 0.84544 | 1.28018 | 0.62818 | 0.85221 | 1.28476 | 0.820 |
| 0.840 | 0.64079 | 0.86484 | 1.29859 | 0.64740 | 0.87145 | 1.30255 | 0.840 |
| 0.860 | 0.66019 | 0.88426 | 1.31608 | 0.66667 | 0.89073 | 1.31949 | 0.860 |
| 0.880 | 0.67964 | 0.90371 | 1.33271 | 0.68600 | 0.91003 | 1.33567 | 0.880 |
| 0.900 | 0.69914 | 0.92318 | 1.34859 | 0.70538 | 0.92937 | 1.35120 | 0.900 |
| 0.920 | 0.71868 | 0.94268 | 1.36385 | 0.72481 | 0.94873 | 1.36621 | 0.920 |
| 0.940 | 0.73827 | 0.96220 | 1.37863 | 0.74428 | 0.96812 | 1.38083 | 0.940 |
| 0.960 | 0.75788 | 0.98174 | 1.39305 | 0.76378 | 0.98754 | 1.39512 | 0.960 |
| 0.980 | 0.77752 | 1.00130 | 1.40720 | 0.78332 | 1.00698 | 1.40917 | 0.980 |
| 1.000 | 0.79720 | 1.02087 | 1.42112 | 0.80289 | 1.02644 | 1.42298 | 1.000 |

$H/R = 0.200 \qquad n = 4.$

| $H/L$ | $\Omega_1$ | $\Omega_2$ | $\Omega_3$ | $\Omega_4$ | $\Omega_5$ | $\Omega_6$ |
|---|---|---|---|---|---|---|
| 0. | 0.08256 | 0.25497 | 0.43589 | 1.03405 | 1.11379 | 1.79451 |
| 0.010 | 0.08266 | 0.25521 | 0.43619 | 1.03412 | 1.11393 | 1.79444 |
| 0.020 | 0.08299 | 0.25593 | 0.43708 | 1.03428 | 1.11433 | 1.79425 |
| 0.030 | 0.08354 | 0.25713 | 0.43857 | 1.03458 | 1.11500 | 1.79391 |
| 0.040 | 0.08432 | 0.25879 | 0.44065 | 1.03498 | 1.11594 | 1.79343 |
| 0.050 | 0.08536 | 0.26090 | 0.44331 | 1.03550 | 1.11715 | 1.79284 |
| 0.060 | 0.08665 | 0.26342 | 0.44655 | 1.03613 | 1.11863 | 1.79211 |
| 0.070 | 0.08821 | 0.26634 | 0.45036 | 1.03688 | 1.12037 | 1.79127 |
| 0.080 | 0.09006 | 0.26962 | 0.45472 | 1.03773 | 1.12238 | 1.79031 |
| 0.090 | 0.09216 | 0.27326 | 0.45963 | 1.03868 | 1.12466 | 1.78926 |
| 0.100 | 0.09465 | 0.27722 | 0.46507 | 1.03975 | 1.12720 | 1.78808 |
| 0.120 | 0.10041 | 0.28600 | 0.47747 | 1.04219 | 1.13306 | 1.78548 |
| 0.140 | 0.10732 | 0.29580 | 0.49180 | 1.04503 | 1.13995 | 1.78257 |
| 0.160 | 0.11531 | 0.30649 | 0.50790 | 1.04826 | 1.14785 | 1.77940 |
| 0.180 | 0.12427 | 0.31795 | 0.52560 | 1.05185 | 1.15674 | 1.77604 |
| 0.200 | 0.13409 | 0.33010 | 0.54473 | 1.05581 | 1.16658 | 1.77256 |
| 0.220 | 0.14469 | 0.34286 | 0.56515 | 1.06012 | 1.17732 | 1.76902 |
| 0.240 | 0.15598 | 0.35620 | 0.58669 | 1.06478 | 1.18894 | 1.76548 |
| 0.260 | 0.16791 | 0.37005 | 0.60921 | 1.06977 | 1.20140 | 1.76199 |
| 0.280 | 0.18045 | 0.38437 | 0.63257 | 1.07511 | 1.21466 | 1.75860 |
| 0.300 | 0.19341 | 0.39914 | 0.65666 | 1.08076 | 1.22866 | 1.75537 |
| 0.320 | 0.20690 | 0.41430 | 0.68137 | 1.08674 | 1.24338 | 1.75233 |
| 0.340 | 0.22083 | 0.42984 | 0.70660 | 1.09305 | 1.25876 | 1.74953 |
| 0.360 | 0.23515 | 0.44571 | 0.73225 | 1.09967 | 1.27478 | 1.74702 |
| 0.380 | 0.24994 | 0.46189 | 0.75825 | 1.10660 | 1.29138 | 1.74483 |
| 0.400 | 0.26505 | 0.47835 | 0.78452 | 1.11384 | 1.30854 | 1.74301 |
| 0.420 | 0.28051 | 0.49507 | 0.81099 | 1.12139 | 1.32620 | 1.74158 |
| 0.440 | 0.29628 | 0.51202 | 0.83760 | 1.12924 | 1.34436 | 1.74059 |
| 0.460 | 0.31236 | 0.52919 | 0.86427 | 1.13739 | 1.36295 | 1.74008 |
| 0.480 | 0.32873 | 0.54655 | 0.89095 | 1.14582 | 1.38197 | 1.74008 |
| 0.500 | 0.34536 | 0.56409 | 0.91759 | 1.15455 | 1.40137 | 1.74065 |
| 0.520 | 0.36223 | 0.58179 | 0.94413 | 1.16357 | 1.42112 | 1.74180 |
| 0.540 | 0.37934 | 0.59964 | 0.97049 | 1.17288 | 1.44120 | 1.74358 |
| 0.560 | 0.39666 | 0.61762 | 0.99664 | 1.18246 | 1.46159 | 1.74604 |
| 0.580 | 0.41417 | 0.63573 | 1.02252 | 1.19232 | 1.48225 | 1.74921 |
| 0.600 | 0.43187 | 0.65395 | 1.04804 | 1.20246 | 1.50317 | 1.75316 |
| 0.620 | 0.44974 | 0.67228 | 1.07318 | 1.21288 | 1.52433 | 1.75791 |
| 0.640 | 0.46778 | 0.69070 | 1.09785 | 1.22357 | 1.54569 | 1.76353 |
| 0.660 | 0.48596 | 0.70921 | 1.12200 | 1.23456 | 1.56723 | 1.77005 |
| 0.680 | 0.50427 | 0.72780 | 1.14556 | 1.24582 | 1.58894 | 1.77752 |
| 0.700 | 0.52271 | 0.74647 | 1.16848 | 1.25737 | 1.61080 | 1.78597 |
| 0.720 | 0.54126 | 0.76521 | 1.19068 | 1.26922 | 1.63278 | 1.79546 |
| 0.740 | 0.55992 | 0.78402 | 1.21213 | 1.28138 | 1.65489 | 1.80604 |
| 0.760 | 0.57868 | 0.80287 | 1.23275 | 1.29385 | 1.67707 | 1.81772 |
| 0.780 | 0.59754 | 0.82179 | 1.25251 | 1.30664 | 1.69935 | 1.83052 |
| 0.800 | 0.61647 | 0.84076 | 1.27140 | 1.31974 | 1.72167 | 1.84446 |
| 0.820 | 0.63548 | 0.85979 | 1.28942 | 1.33313 | 1.74404 | 1.85955 |
| 0.840 | 0.65456 | 0.87885 | 1.30661 | 1.34676 | 1.76645 | 1.87580 |
| 0.860 | 0.67371 | 0.89795 | 1.32304 | 1.36057 | 1.78887 | 1.89315 |
| 0.880 | 0.69291 | 0.91710 | 1.33882 | 1.37446 | 1.81132 | 1.91158 |
| 0.900 | 0.71217 | 0.93628 | 1.35404 | 1.38836 | 1.83377 | 1.93107 |
| 0.920 | 0.73147 | 0.95549 | 1.36884 | 1.40217 | 1.85619 | 1.95156 |
| 0.940 | 0.75083 | 0.97474 | 1.38328 | 1.41586 | 1.87861 | 1.97300 |
| 0.960 | 0.77022 | 0.99402 | 1.39744 | 1.42939 | 1.90102 | 1.99533 |
| 0.980 | 0.78965 | 1.01333 | 1.41137 | 1.44277 | 1.92338 | 2.01849 |
| 1.000 | 0.80911 | 1.03266 | 1.42510 | 1.45602 | 1.94570 | 2.04241 |

$H/R = 0.250$ $\quad n = 4.$

| $H/L$ | $\Omega_1$ | $\Omega_2$ | $\Omega_3$ |
|---|---|---|---|
| 0. | 0.12362 | 0.31881 | 0.53873 |
| 0.001 | 0.12362 | 0.31881 | 0.53874 |
| 0.005 | 0.12364 | 0.31886 | 0.53879 |
| 0.010 | 0.12372 | 0.31901 | 0.53898 |
| 0.020 | 0.12401 | 0.31959 | 0.53969 |
| 0.030 | 0.12450 | 0.32056 | 0.54089 |
| 0.040 | 0.12520 | 0.32190 | 0.54256 |
| 0.050 | 0.12610 | 0.32362 | 0.54469 |
| 0.060 | 0.12723 | 0.32568 | 0.54730 |
| 0.070 | 0.12859 | 0.32809 | 0.55037 |
| 0.080 | 0.13018 | 0.33082 | 0.55390 |
| 0.090 | 0.13201 | 0.33387 | 0.55787 |
| 0.100 | 0.13409 | 0.33720 | 0.56228 |
| 0.120 | 0.13901 | 0.34468 | 0.57241 |
| 0.140 | 0.14495 | 0.35314 | 0.58417 |
| 0.160 | 0.15192 | 0.36247 | 0.59749 |
| 0.180 | 0.15984 | 0.37257 | 0.61227 |
| 0.200 | 0.16869 | 0.38335 | 0.62840 |
| 0.220 | 0.17839 | 0.39475 | 0.64576 |
| 0.240 | 0.18886 | 0.40671 | 0.66425 |
| 0.260 | 0.20005 | 0.41918 | 0.68374 |
| 0.280 | 0.21188 | 0.43213 | 0.70414 |
| 0.300 | 0.22430 | 0.44553 | 0.72533 |
| 0.320 | 0.23727 | 0.45934 | 0.74723 |
| 0.340 | 0.25072 | 0.47353 | 0.76973 |
| 0.360 | 0.26462 | 0.48809 | 0.79274 |
| 0.380 | 0.27895 | 0.50298 | 0.81618 |
| 0.400 | 0.29363 | 0.51818 | 0.83998 |
| 0.420 | 0.30868 | 0.53369 | 0.86406 |
| 0.440 | 0.32405 | 0.54946 | 0.88834 |
| 0.460 | 0.33973 | 0.56549 | 0.91276 |
| 0.480 | 0.35569 | 0.58175 | 0.93725 |
| 0.500 | 0.37192 | 0.59824 | 0.96173 |
| 0.520 | 0.38836 | 0.61493 | 0.98615 |
| 0.540 | 0.40510 | 0.63181 | 1.01044 |
| 0.560 | 0.42201 | 0.64887 | 1.03453 |
| 0.580 | 0.43912 | 0.66609 | 1.05835 |
| 0.600 | 0.45642 | 0.68346 | 1.08184 |
| 0.620 | 0.47390 | 0.70098 | 1.10492 |
| 0.640 | 0.49153 | 0.71863 | 1.12754 |
| 0.660 | 0.50932 | 0.73640 | 1.14963 |
| 0.680 | 0.52725 | 0.75429 | 1.17111 |
| 0.700 | 0.54530 | 0.77227 | 1.19194 |
| 0.720 | 0.56348 | 0.79036 | 1.21206 |
| 0.740 | 0.58176 | 0.80854 | 1.23144 |
| 0.760 | 0.60016 | 0.82681 | 1.25005 |
| 0.780 | 0.61864 | 0.84516 | 1.26792 |
| 0.800 | 0.63722 | 0.86359 | 1.28506 |
| 0.820 | 0.65588 | 0.88208 | 1.30155 |
| 0.840 | 0.67462 | 0.90064 | 1.31743 |
| 0.860 | 0.69343 | 0.91927 | 1.33280 |
| 0.880 | 0.71231 | 0.93795 | 1.34774 |
| 0.900 | 0.73124 | 0.95669 | 1.36232 |
| 0.920 | 0.75024 | 0.97548 | 1.37659 |
| 0.940 | 0.76928 | 0.99431 | 1.39061 |
| 0.960 | 0.78837 | 1.01320 | 1.40442 |
| 0.980 | 0.80751 | 1.03213 | 1.41802 |
| 1.000 | 0.82669 | 1.05110 | 1.43145 |

$H/R = 0.300 \qquad n = 4.$

| $H/L$ | $\Omega_1$ | $\Omega_2$ | $\Omega_3$ | $\Omega_4$ | $\Omega_5$ | $\Omega_6$ |
|---|---|---|---|---|---|---|
| 0. | 0.16997 | 0.38258 | 0.63697 | 1.07706 | 1.24635 | 1.75991 |
| 0.001 | 0.16998 | 0.38258 | 0.63697 | 1.07705 | 1.24636 | 1.75991 |
| 0.005 | 0.17000 | 0.38262 | 0.63702 | 1.07707 | 1.24638 | 1.75990 |
| 0.010 | 0.17006 | 0.38275 | 0.63717 | 1.07711 | 1.24646 | 1.75988 |
| 0.020 | 0.17032 | 0.38324 | 0.63776 | 1.07729 | 1.24680 | 1.75980 |
| 0.030 | 0.17077 | 0.38405 | 0.63876 | 1.07758 | 1.24734 | 1.75965 |
| 0.040 | 0.17139 | 0.38518 | 0.64015 | 1.07798 | 1.24812 | 1.75945 |
| 0.050 | 0.17219 | 0.38663 | 0.64194 | 1.07850 | 1.24911 | 1.75920 |
| 0.060 | 0.17319 | 0.38838 | 0.64411 | 1.07913 | 1.25033 | 1.75889 |
| 0.070 | 0.17438 | 0.39043 | 0.64668 | 1.07987 | 1.25177 | 1.75854 |
| 0.080 | 0.17577 | 0.39276 | 0.64963 | 1.08072 | 1.25343 | 1.75812 |
| 0.090 | 0.17736 | 0.39537 | 0.65296 | 1.08169 | 1.25530 | 1.75766 |
| 0.100 | 0.17917 | 0.39825 | 0.65666 | 1.08277 | 1.25739 | 1.75717 |
| 0.120 | 0.18343 | 0.40475 | 0.66516 | 1.08525 | 1.26222 | 1.75606 |
| 0.140 | 0.18859 | 0.41217 | 0.67509 | 1.08815 | 1.26790 | 1.75481 |
| 0.160 | 0.19464 | 0.42042 | 0.68637 | 1.09147 | 1.27443 | 1.75346 |
| 0.180 | 0.20159 | 0.42942 | 0.69895 | 1.09519 | 1.28177 | 1.75204 |
| 0.200 | 0.20941 | 0.43909 | 0.71274 | 1.09930 | 1.28994 | 1.75058 |
| 0.220 | 0.21812 | 0.44938 | 0.72767 | 1.10378 | 1.29889 | 1.74913 |
| 0.240 | 0.22754 | 0.46023 | 0.74365 | 1.10864 | 1.30861 | 1.74772 |
| 0.260 | 0.23774 | 0.47159 | 0.76059 | 1.11386 | 1.31906 | 1.74640 |
| 0.280 | 0.24865 | 0.48343 | 0.77841 | 1.11943 | 1.33025 | 1.74519 |
| 0.300 | 0.26019 | 0.49570 | 0.79703 | 1.12535 | 1.34212 | 1.74414 |
| 0.320 | 0.27233 | 0.50839 | 0.81635 | 1.13159 | 1.35467 | 1.74327 |
| 0.340 | 0.28501 | 0.52146 | 0.83630 | 1.13817 | 1.36784 | 1.74265 |
| 0.360 | 0.29819 | 0.53490 | 0.85679 | 1.14508 | 1.38162 | 1.74230 |
| 0.380 | 0.31184 | 0.54868 | 0.87774 | 1.15230 | 1.39599 | 1.74226 |
| 0.400 | 0.32590 | 0.56279 | 0.89907 | 1.15984 | 1.41089 | 1.74258 |
| 0.420 | 0.34036 | 0.57721 | 0.92071 | 1.16771 | 1.42633 | 1.74331 |
| 0.440 | 0.35517 | 0.59191 | 0.94257 | 1.17589 | 1.44223 | 1.74447 |
| 0.460 | 0.37031 | 0.60688 | 0.96460 | 1.18438 | 1.45861 | 1.74610 |
| 0.480 | 0.38577 | 0.62212 | 0.98672 | 1.19320 | 1.47540 | 1.74825 |
| 0.500 | 0.40149 | 0.63760 | 1.00885 | 1.20233 | 1.49260 | 1.75099 |
| 0.520 | 0.41748 | 0.65331 | 1.03092 | 1.21178 | 1.51017 | 1.75432 |
| 0.540 | 0.43371 | 0.66924 | 1.05286 | 1.22155 | 1.52806 | 1.75831 |
| 0.560 | 0.45017 | 0.68536 | 1.07460 | 1.23166 | 1.54628 | 1.76301 |
| 0.580 | 0.46684 | 0.70168 | 1.09607 | 1.24210 | 1.56477 | 1.76848 |
| 0.600 | 0.48370 | 0.71819 | 1.11719 | 1.25287 | 1.58353 | 1.77474 |
| 0.620 | 0.50075 | 0.73486 | 1.13792 | 1.26398 | 1.60251 | 1.78186 |
| 0.640 | 0.51796 | 0.75169 | 1.15817 | 1.27543 | 1.62169 | 1.78988 |
| 0.660 | 0.53534 | 0.76867 | 1.17790 | 1.28721 | 1.64106 | 1.79886 |
| 0.680 | 0.55285 | 0.78579 | 1.19706 | 1.29932 | 1.66057 | 1.80880 |
| 0.700 | 0.57051 | 0.80305 | 1.21563 | 1.31172 | 1.68024 | 1.81978 |
| 0.720 | 0.58830 | 0.82043 | 1.23358 | 1.32439 | 1.70003 | 1.83179 |
| 0.740 | 0.60618 | 0.83793 | 1.25093 | 1.33729 | 1.71992 | 1.84487 |
| 0.760 | 0.62420 | 0.85553 | 1.26769 | 1.35036 | 1.73992 | 1.85903 |
| 0.780 | 0.64233 | 0.87324 | 1.28389 | 1.36356 | 1.76002 | 1.87424 |
| 0.800 | 0.66055 | 0.89106 | 1.29959 | 1.37681 | 1.78023 | 1.89048 |
| 0.820 | 0.67885 | 0.90896 | 1.31483 | 1.39006 | 1.80055 | 1.90774 |
| 0.840 | 0.69724 | 0.92695 | 1.32969 | 1.40327 | 1.82099 | 1.92597 |
| 0.860 | 0.71570 | 0.94503 | 1.34421 | 1.41639 | 1.84155 | 1.94511 |
| 0.880 | 0.73425 | 0.96317 | 1.35843 | 1.42938 | 1.86222 | 1.96512 |
| 0.900 | 0.75286 | 0.98140 | 1.37240 | 1.44224 | 1.88301 | 1.98594 |
| 0.920 | 0.77152 | 0.99969 | 1.38616 | 1.45498 | 1.90391 | 2.00752 |
| 0.940 | 0.79025 | 1.01805 | 1.39973 | 1.46759 | 1.92490 | 2.02981 |
| 0.960 | 0.80903 | 1.03647 | 1.41312 | 1.48010 | 1.94598 | 2.05277 |
| 0.980 | 0.82786 | 1.05496 | 1.42637 | 1.49253 | 1.96711 | 2.07633 |
| 1.000 | 0.84673 | 1.07349 | 1.43945 | 1.50493 | 1.98827 | 2.10048 |

$H/R = 0.350$ $n = 4.$

| $H/L$ | $\Omega_1$ | $\Omega_2$ | $\Omega_3$ |
|---|---|---|---|
| 0. | 0.22046 | 0.44613 | 0.72951 |
| 0.001 | 0.22046 | 0.44613 | 0.72951 |
| 0.005 | 0.22048 | 0.44616 | 0.72955 |
| 0.010 | 0.22054 | 0.44627 | 0.72968 |
| 0.020 | 0.22078 | 0.44669 | 0.73019 |
| 0.030 | 0.22117 | 0.44740 | 0.73104 |
| 0.040 | 0.22172 | 0.44838 | 0.73223 |
| 0.050 | 0.22244 | 0.44963 | 0.73376 |
| 0.060 | 0.22332 | 0.45115 | 0.73561 |
| 0.070 | 0.22437 | 0.45294 | 0.73781 |
| 0.080 | 0.22559 | 0.45498 | 0.74033 |
| 0.090 | 0.22700 | 0.45726 | 0.74318 |
| 0.100 | 0.22858 | 0.45979 | 0.74635 |
| 0.120 | 0.23231 | 0.46553 | 0.75364 |
| 0.140 | 0.23681 | 0.47213 | 0.76217 |
| 0.160 | 0.24210 | 0.47951 | 0.77189 |
| 0.180 | 0.24819 | 0.48762 | 0.78275 |
| 0.200 | 0.25508 | 0.49638 | 0.79468 |
| 0.220 | 0.26276 | 0.50575 | 0.80763 |
| 0.240 | 0.27120 | 0.51567 | 0.82154 |
| 0.260 | 0.28037 | 0.52610 | 0.83634 |
| 0.280 | 0.29025 | 0.53700 | 0.85195 |
| 0.300 | 0.30065 | 0.54834 | 0.86829 |
| 0.320 | 0.31196 | 0.56009 | 0.88531 |
| 0.340 | 0.32370 | 0.57222 | 0.90292 |
| 0.360 | 0.33598 | 0.58472 | 0.92105 |
| 0.380 | 0.34876 | 0.59756 | 0.93962 |
| 0.400 | 0.36201 | 0.61072 | 0.95856 |
| 0.420 | 0.37569 | 0.62419 | 0.97780 |
| 0.440 | 0.38976 | 0.63796 | 0.99725 |
| 0.460 | 0.40421 | 0.65201 | 1.01685 |
| 0.480 | 0.41899 | 0.66632 | 1.03653 |
| 0.500 | 0.43409 | 0.68089 | 1.05620 |
| 0.520 | 0.44948 | 0.69570 | 1.07581 |
| 0.540 | 0.46513 | 0.71073 | 1.09527 |
| 0.560 | 0.48104 | 0.72599 | 1.11453 |
| 0.580 | 0.49719 | 0.74146 | 1.13352 |
| 0.600 | 0.51353 | 0.75712 | 1.15218 |
| 0.620 | 0.53008 | 0.77297 | 1.17047 |
| 0.640 | 0.54682 | 0.78900 | 1.18834 |
| 0.660 | 0.56372 | 0.80520 | 1.20577 |
| 0.680 | 0.58079 | 0.82156 | 1.22273 |
| 0.700 | 0.59801 | 0.83806 | 1.23924 |
| 0.720 | 0.61537 | 0.85472 | 1.25529 |
| 0.740 | 0.63286 | 0.87152 | 1.27092 |
| 0.760 | 0.65047 | 0.88844 | 1.28615 |
| 0.780 | 0.66819 | 0.90549 | 1.30101 |
| 0.800 | 0.68602 | 0.92265 | 1.31553 |
| 0.820 | 0.70395 | 0.93993 | 1.32978 |
| 0.840 | 0.72197 | 0.95731 | 1.34376 |
| 0.860 | 0.74008 | 0.97479 | 1.35751 |
| 0.880 | 0.75827 | 0.99237 | 1.37107 |
| 0.900 | 0.77653 | 1.01004 | 1.38444 |
| 0.920 | 0.79486 | 1.02779 | 1.39767 |
| 0.940 | 0.81326 | 1.04563 | 1.41074 |
| 0.960 | 0.83171 | 1.06354 | 1.42369 |
| 0.980 | 0.85023 | 1.08153 | 1.43651 |
| 1.000 | 0.86880 | 1.09959 | 1.44922 |

$H/R = 0.400 \qquad n = 4.$

| $H/L$ | $\Omega_1$ | $\Omega_2$ | $\Omega_3$ | $\Omega_4$ | $\Omega_5$ | $\Omega_6$ |
|---|---|---|---|---|---|---|
| 0. | 0.27416 | 0.50922 | 0.81561 | 1.13847 | 1.41926 | 1.74701 |
| 0.001 | 0.27416 | 0.50922 | 0.81561 | 1.13847 | 1.41927 | 1.74701 |
| 0.005 | 0.27417 | 0.50925 | 0.81565 | 1.13848 | 1.41928 | 1.74701 |
| 0.010 | 0.27423 | 0.50935 | 0.81576 | 1.13853 | 1.41935 | 1.74701 |
| 0.020 | 0.27444 | 0.50972 | 0.81620 | 1.13870 | 1.41961 | 1.74701 |
| 0.030 | 0.27478 | 0.51034 | 0.81694 | 1.13899 | 1.42004 | 1.74699 |
| 0.040 | 0.27528 | 0.51121 | 0.81797 | 1.13939 | 1.42066 | 1.74697 |
| 0.050 | 0.27591 | 0.51231 | 0.81929 | 1.13990 | 1.42144 | 1.74695 |
| 0.060 | 0.27669 | 0.51367 | 0.82090 | 1.14053 | 1.42239 | 1.74693 |
| 0.070 | 0.27762 | 0.51525 | 0.82280 | 1.14127 | 1.42353 | 1.74691 |
| 0.080 | 0.27870 | 0.51706 | 0.82499 | 1.14214 | 1.42483 | 1.74688 |
| 0.090 | 0.27994 | 0.51910 | 0.82747 | 1.14310 | 1.42631 | 1.74685 |
| 0.100 | 0.28134 | 0.52136 | 0.83022 | 1.14419 | 1.42795 | 1.74684 |
| 0.120 | 0.28461 | 0.52650 | 0.83655 | 1.14668 | 1.43174 | 1.74680 |
| 0.140 | 0.28857 | 0.53244 | 0.84396 | 1.14962 | 1.43622 | 1.74682 |
| 0.160 | 0.29321 | 0.53913 | 0.85242 | 1.15299 | 1.44134 | 1.74689 |
| 0.180 | 0.29856 | 0.54650 | 0.86187 | 1.15679 | 1.44713 | 1.74704 |
| 0.200 | 0.30463 | 0.55450 | 0.87228 | 1.16101 | 1.45355 | 1.74729 |
| 0.220 | 0.31142 | 0.56310 | 0.88359 | 1.16565 | 1.46060 | 1.74768 |
| 0.240 | 0.31891 | 0.57224 | 0.89575 | 1.17070 | 1.46826 | 1.74822 |
| 0.260 | 0.32711 | 0.58187 | 0.90870 | 1.17614 | 1.47651 | 1.74896 |
| 0.280 | 0.33598 | 0.59198 | 0.92238 | 1.18199 | 1.48533 | 1.74992 |
| 0.300 | 0.34550 | 0.60252 | 0.93673 | 1.18822 | 1.49469 | 1.75115 |
| 0.320 | 0.35565 | 0.61347 | 0.95169 | 1.19485 | 1.50460 | 1.75267 |
| 0.340 | 0.36638 | 0.62480 | 0.96717 | 1.20185 | 1.51502 | 1.75453 |
| 0.360 | 0.37768 | 0.63648 | 0.98313 | 1.20923 | 1.52591 | 1.75677 |
| 0.380 | 0.38951 | 0.64851 | 0.99948 | 1.21699 | 1.53724 | 1.75942 |
| 0.400 | 0.40182 | 0.66086 | 1.01615 | 1.22513 | 1.54902 | 1.76255 |
| 0.420 | 0.41462 | 0.67353 | 1.03308 | 1.23363 | 1.56118 | 1.76618 |
| 0.440 | 0.42784 | 0.68648 | 1.05020 | 1.24251 | 1.57373 | 1.77036 |
| 0.460 | 0.44146 | 0.69971 | 1.06744 | 1.25177 | 1.58659 | 1.77516 |
| 0.480 | 0.45545 | 0.71321 | 1.08473 | 1.26139 | 1.59977 | 1.78062 |
| 0.500 | 0.46980 | 0.72697 | 1.10200 | 1.27137 | 1.61323 | 1.78678 |
| 0.520 | 0.48445 | 0.74098 | 1.11920 | 1.28171 | 1.62695 | 1.79369 |
| 0.540 | 0.49942 | 0.75522 | 1.13626 | 1.29239 | 1.64089 | 1.80139 |
| 0.560 | 0.51466 | 0.76968 | 1.15313 | 1.30342 | 1.65504 | 1.80994 |
| 0.580 | 0.53017 | 0.78437 | 1.16977 | 1.31476 | 1.66939 | 1.81935 |
| 0.600 | 0.54591 | 0.79926 | 1.18614 | 1.32638 | 1.68394 | 1.82965 |
| 0.620 | 0.56188 | 0.81434 | 1.20222 | 1.33827 | 1.69869 | 1.84085 |
| 0.640 | 0.57805 | 0.82961 | 1.21798 | 1.35038 | 1.71364 | 1.85295 |
| 0.660 | 0.59442 | 0.84507 | 1.23341 | 1.36266 | 1.72884 | 1.86594 |
| 0.680 | 0.61099 | 0.86070 | 1.24853 | 1.37507 | 1.74429 | 1.87982 |
| 0.700 | 0.62770 | 0.87650 | 1.26334 | 1.38758 | 1.76004 | 1.89454 |
| 0.720 | 0.64457 | 0.89245 | 1.27784 | 1.40013 | 1.77610 | 1.91007 |
| 0.740 | 0.66160 | 0.90855 | 1.29207 | 1.41268 | 1.79251 | 1.92637 |
| 0.760 | 0.67875 | 0.92480 | 1.30605 | 1.42518 | 1.80928 | 1.94339 |
| 0.780 | 0.69604 | 0.94119 | 1.31979 | 1.43762 | 1.82645 | 1.96109 |
| 0.800 | 0.71345 | 0.95771 | 1.33332 | 1.44997 | 1.84400 | 1.97945 |
| 0.820 | 0.73096 | 0.97435 | 1.34667 | 1.46221 | 1.86194 | 1.99841 |
| 0.840 | 0.74858 | 0.99111 | 1.35985 | 1.47434 | 1.88025 | 2.01793 |
| 0.860 | 0.76630 | 1.00800 | 1.37288 | 1.48636 | 1.89891 | 2.03801 |
| 0.880 | 0.78411 | 1.02498 | 1.38577 | 1.49828 | 1.91790 | 2.05858 |
| 0.900 | 0.80200 | 1.04208 | 1.39856 | 1.51012 | 1.93718 | 2.07964 |
| 0.920 | 0.81998 | 1.05927 | 1.41122 | 1.52190 | 1.95672 | 2.10116 |
| 0.940 | 0.83802 | 1.07656 | 1.42378 | 1.53362 | 1.97648 | 2.12309 |
| 0.960 | 0.85614 | 1.09395 | 1.43624 | 1.54533 | 1.99640 | 2.14546 |
| 0.980 | 0.87433 | 1.11141 | 1.44862 | 1.55705 | 2.01644 | 2.16819 |
| 1.000 | 0.89257 | 1.12897 | 1.46091 | 1.56878 | 2.03657 | 2.19128 |

$H/R = 0.450$ $n = 4.$

| $H/L$ | $\Omega_1$ | $\Omega_2$ | $\Omega_3$ |
|---|---|---|---|
| 0. | 0.33032 | 0.57155 | 0.89501 |
| 0.001 | 0.33032 | 0.57155 | 0.89501 |
| 0.005 | 0.33034 | 0.57158 | 0.89504 |
| 0.010 | 0.33038 | 0.57166 | 0.89514 |
| 0.020 | 0.33057 | 0.57200 | 0.89553 |
| 0.030 | 0.33088 | 0.57256 | 0.89617 |
| 0.040 | 0.33131 | 0.57334 | 0.89707 |
| 0.050 | 0.33188 | 0.57434 | 0.89823 |
| 0.060 | 0.33257 | 0.57555 | 0.89964 |
| 0.070 | 0.33339 | 0.57698 | 0.90130 |
| 0.080 | 0.33435 | 0.57862 | 0.90321 |
| 0.090 | 0.33544 | 0.58047 | 0.90538 |
| 0.100 | 0.33667 | 0.58251 | 0.90778 |
| 0.120 | 0.33956 | 0.58718 | 0.91333 |
| 0.140 | 0.34304 | 0.59259 | 0.91981 |
| 0.160 | 0.34713 | 0.59871 | 0.92721 |
| 0.180 | 0.35184 | 0.60547 | 0.93549 |
| 0.200 | 0.35720 | 0.61284 | 0.94461 |
| 0.220 | 0.36320 | 0.62079 | 0.95452 |
| 0.240 | 0.36985 | 0.62926 | 0.96518 |
| 0.260 | 0.37715 | 0.63822 | 0.97653 |
| 0.280 | 0.38508 | 0.64764 | 0.98852 |
| 0.300 | 0.39364 | 0.65749 | 1.00111 |
| 0.320 | 0.40280 | 0.66774 | 1.01421 |
| 0.340 | 0.41255 | 0.67838 | 1.02779 |
| 0.360 | 0.42286 | 0.68936 | 1.04177 |
| 0.380 | 0.43370 | 0.70069 | 1.05610 |
| 0.400 | 0.44506 | 0.71234 | 1.07069 |
| 0.420 | 0.45689 | 0.72429 | 1.08551 |
| 0.440 | 0.46918 | 0.73654 | 1.10048 |
| 0.460 | 0.48190 | 0.74906 | 1.11554 |
| 0.480 | 0.49502 | 0.76185 | 1.13064 |
| 0.500 | 0.50851 | 0.77490 | 1.14573 |
| 0.520 | 0.52235 | 0.78819 | 1.16075 |
| 0.540 | 0.53657 | 0.80172 | 1.17568 |
| 0.560 | 0.55105 | 0.81548 | 1.19046 |
| 0.580 | 0.56582 | 0.82945 | 1.20507 |
| 0.600 | 0.58087 | 0.84364 | 1.21949 |
| 0.620 | 0.59617 | 0.85803 | 1.23373 |
| 0.640 | 0.61171 | 0.87262 | 1.24774 |
| 0.660 | 0.62745 | 0.88739 | 1.26155 |
| 0.680 | 0.64341 | 0.90234 | 1.27516 |
| 0.700 | 0.65957 | 0.91747 | 1.28857 |
| 0.720 | 0.67589 | 0.93276 | 1.30180 |
| 0.740 | 0.69238 | 0.94821 | 1.31486 |
| 0.760 | 0.70903 | 0.96382 | 1.32777 |
| 0.780 | 0.72583 | 0.97957 | 1.34055 |
| 0.800 | 0.74277 | 0.99546 | 1.35318 |
| 0.820 | 0.75982 | 1.01150 | 1.36571 |
| 0.840 | 0.77698 | 1.02766 | 1.37812 |
| 0.860 | 0.79428 | 1.04395 | 1.39046 |
| 0.880 | 0.81167 | 1.06036 | 1.40270 |
| 0.900 | 0.82916 | 1.07688 | 1.41486 |
| 0.920 | 0.84674 | 1.09352 | 1.42696 |
| 0.940 | 0.86441 | 1.11026 | 1.43899 |
| 0.960 | 0.88215 | 1.12711 | 1.45095 |
| 0.980 | 0.89998 | 1.14405 | 1.46287 |
| 1.000 | 0.91787 | 1.16109 | 1.47473 |

$H/R = 0.500$   $n = 4.$

| $H/L$ | $\Omega_1$ | $\Omega_2$ | $\Omega_3$ | $\Omega_4$ | $\Omega_5$ | $\Omega_6$ |
|---|---|---|---|---|---|---|
| 0. | 0.38836 | 0.63276 | 0.96798 | 1.22007 | 1.59990 | 1.78581 |
| 0.010 | 0.38842 | 0.63286 | 0.96809 | 1.22012 | 1.59995 | 1.78585 |
| 0.050 | 0.38974 | 0.63531 | 0.97081 | 1.22149 | 1.60123 | 1.78657 |
| 0.100 | 0.39397 | 0.64279 | 0.97921 | 1.22574 | 1.60519 | 1.78891 |
| 0.120 | 0.39652 | 0.64708 | 0.98408 | 1.22822 | 1.60749 | 1.79031 |
| 0.140 | 0.39959 | 0.65207 | 0.98979 | 1.23117 | 1.61019 | 1.79201 |
| 0.160 | 0.40319 | 0.65771 | 0.99629 | 1.23455 | 1.61326 | 1.79404 |
| 0.180 | 0.40735 | 0.66398 | 1.00356 | 1.23839 | 1.61671 | 1.79639 |
| 0.200 | 0.41208 | 0.67082 | 1.01157 | 1.24267 | 1.62052 | 1.79909 |
| 0.220 | 0.41739 | 0.67822 | 1.02028 | 1.24738 | 1.62468 | 1.80218 |
| 0.240 | 0.42330 | 0.68613 | 1.02964 | 1.25253 | 1.62917 | 1.80569 |
| 0.260 | 0.42979 | 0.69451 | 1.03961 | 1.25814 | 1.63397 | 1.80962 |
| 0.280 | 0.43687 | 0.70335 | 1.05013 | 1.26417 | 1.63908 | 1.81401 |
| 0.300 | 0.44454 | 0.71260 | 1.06117 | 1.27064 | 1.64448 | 1.81888 |
| 0.320 | 0.45278 | 0.72226 | 1.07267 | 1.27754 | 1.65015 | 1.82427 |
| 0.340 | 0.46159 | 0.73229 | 1.08456 | 1.28487 | 1.65611 | 1.83019 |
| 0.360 | 0.47095 | 0.74266 | 1.09680 | 1.29263 | 1.66234 | 1.83664 |
| 0.380 | 0.48083 | 0.75337 | 1.10934 | 1.30081 | 1.66885 | 1.84368 |
| 0.400 | 0.49122 | 0.76441 | 1.12212 | 1.30940 | 1.67563 | 1.85129 |
| 0.420 | 0.50210 | 0.77574 | 1.13508 | 1.31840 | 1.68270 | 1.85949 |
| 0.440 | 0.51345 | 0.78736 | 1.14818 | 1.32779 | 1.69008 | 1.86828 |
| 0.460 | 0.52524 | 0.79926 | 1.16136 | 1.33757 | 1.69778 | 1.87766 |
| 0.480 | 0.53745 | 0.81142 | 1.17458 | 1.34770 | 1.70583 | 1.88764 |
| 0.500 | 0.55006 | 0.82384 | 1.18781 | 1.35817 | 1.71424 | 1.89820 |
| 0.520 | 0.56305 | 0.83650 | 1.20101 | 1.36894 | 1.72306 | 1.90933 |
| 0.540 | 0.57639 | 0.84941 | 1.21414 | 1.38000 | 1.73230 | 1.92102 |
| 0.560 | 0.59006 | 0.86253 | 1.22719 | 1.39131 | 1.74201 | 1.93326 |
| 0.580 | 0.60405 | 0.87588 | 1.24015 | 1.40281 | 1.75222 | 1.94602 |
| 0.600 | 0.61834 | 0.88944 | 1.25299 | 1.41449 | 1.76294 | 1.95931 |
| 0.620 | 0.63290 | 0.90320 | 1.26571 | 1.42629 | 1.77421 | 1.97307 |
| 0.640 | 0.64772 | 0.91716 | 1.27831 | 1.43817 | 1.78605 | 1.98732 |
| 0.660 | 0.66279 | 0.93131 | 1.29079 | 1.45011 | 1.79849 | 2.00202 |
| 0.680 | 0.67811 | 0.94564 | 1.30316 | 1.46205 | 1.81151 | 2.01716 |
| 0.700 | 0.69360 | 0.96015 | 1.31540 | 1.47397 | 1.82514 | 2.03271 |
| 0.720 | 0.70931 | 0.97484 | 1.32755 | 1.48585 | 1.83937 | 2.04865 |
| 0.740 | 0.72521 | 0.98968 | 1.33961 | 1.49766 | 1.85419 | 2.06497 |
| 0.760 | 0.74129 | 1.00469 | 1.35158 | 1.50941 | 1.86958 | 2.08165 |
| 0.780 | 0.75753 | 1.01985 | 1.36346 | 1.52106 | 1.88552 | 2.09868 |
| 0.800 | 0.77393 | 1.03516 | 1.37527 | 1.53264 | 1.90197 | 2.11601 |
| 0.820 | 0.79047 | 1.05061 | 1.38703 | 1.54413 | 1.91892 | 2.13364 |
| 0.840 | 0.80715 | 1.06621 | 1.39873 | 1.55557 | 1.93629 | 2.15154 |
| 0.860 | 0.82396 | 1.08193 | 1.41037 | 1.56694 | 1.95408 | 2.16968 |
| 0.880 | 0.84089 | 1.09779 | 1.42197 | 1.57827 | 1.97222 | 2.18803 |
| 0.900 | 0.85793 | 1.11376 | 1.43353 | 1.58957 | 1.99066 | 2.20656 |
| 0.920 | 0.87509 | 1.12986 | 1.44506 | 1.60087 | 2.00937 | 2.22519 |
| 0.940 | 0.89233 | 1.14607 | 1.45656 | 1.61218 | 2.02828 | 2.24389 |
| 0.960 | 0.90967 | 1.16240 | 1.46803 | 1.62352 | 2.04737 | 2.26258 |
| 0.980 | 0.92709 | 1.17883 | 1.47947 | 1.63490 | 2.06655 | 2.28116 |
| 1.000 | 0.94460 | 1.19536 | 1.49090 | 1.64634 | 2.08582 | 2.29949 |

$H/R = 0.600 \qquad n = 4.$

| $H/L$ | $\Omega_1$ | $\Omega_2$ | $\Omega_3$ | $\Omega_4$ | $\Omega_5$ | $\Omega_6$ |
|---|---|---|---|---|---|---|
| 0. | 0.50814 | 0.75020 | 1.09755 | 1.32403 | 1.69343 | 1.96086 |
| 0.010 | 0.50818 | 0.75029 | 1.09763 | 1.32408 | 1.69346 | 1.96091 |
| 0.050 | 0.50920 | 0.75241 | 1.09977 | 1.32538 | 1.69412 | 1.96214 |
| 0.100 | 0.51247 | 0.75891 | 1.10638 | 1.32948 | 1.69621 | 1.96598 |
| 0.120 | 0.51444 | 0.76266 | 1.11021 | 1.33187 | 1.69744 | 1.96825 |
| 0.140 | 0.51681 | 0.76703 | 1.11470 | 1.33472 | 1.69890 | 1.97092 |
| 0.160 | 0.51960 | 0.77198 | 1.11980 | 1.33800 | 1.70061 | 1.97400 |
| 0.180 | 0.52283 | 0.77750 | 1.12551 | 1.34170 | 1.70256 | 1.97750 |
| 0.200 | 0.52651 | 0.78355 | 1.13180 | 1.34587 | 1.70476 | 1.98140 |
| 0.220 | 0.53067 | 0.79011 | 1.13862 | 1.35047 | 1.70722 | 1.98571 |
| 0.240 | 0.53530 | 0.79715 | 1.14596 | 1.35550 | 1.70996 | 1.99045 |
| 0.260 | 0.54043 | 0.80464 | 1.15377 | 1.36098 | 1.71296 | 1.99558 |
| 0.280 | 0.54605 | 0.81256 | 1.16201 | 1.36691 | 1.71626 | 2.00112 |
| 0.300 | 0.55218 | 0.82087 | 1.17064 | 1.37327 | 1.71987 | 2.00706 |
| 0.320 | 0.55881 | 0.82957 | 1.17963 | 1.38006 | 1.72379 | 2.01339 |
| 0.340 | 0.56595 | 0.83863 | 1.18893 | 1.38729 | 1.72805 | 2.02013 |
| 0.360 | 0.57358 | 0.84802 | 1.19851 | 1.39494 | 1.73267 | 2.02725 |
| 0.380 | 0.58170 | 0.85774 | 1.20831 | 1.40301 | 1.73764 | 2.03475 |
| 0.400 | 0.59031 | 0.86777 | 1.21832 | 1.41147 | 1.74302 | 2.04261 |
| 0.420 | 0.59939 | 0.87808 | 1.22849 | 1.42033 | 1.74879 | 2.05085 |
| 0.440 | 0.60893 | 0.88869 | 1.23878 | 1.42955 | 1.75502 | 2.05944 |
| 0.460 | 0.61891 | 0.89956 | 1.24916 | 1.43912 | 1.76169 | 2.06838 |
| 0.480 | 0.62933 | 0.91068 | 1.25961 | 1.44902 | 1.76883 | 2.07764 |
| 0.500 | 0.64015 | 0.92206 | 1.27011 | 1.45921 | 1.77648 | 2.08722 |
| 0.520 | 0.65138 | 0.93368 | 1.28063 | 1.46966 | 1.78464 | 2.09712 |
| 0.540 | 0.66299 | 0.94553 | 1.29117 | 1.48035 | 1.79335 | 2.10729 |
| 0.560 | 0.67497 | 0.95760 | 1.30170 | 1.49123 | 1.80262 | 2.11776 |
| 0.580 | 0.68729 | 0.96988 | 1.31223 | 1.50227 | 1.81247 | 2.12847 |
| 0.600 | 0.69995 | 0.98238 | 1.32274 | 1.51344 | 1.82290 | 2.13944 |
| 0.620 | 0.71293 | 0.99508 | 1.33321 | 1.52470 | 1.83393 | 2.15062 |
| 0.640 | 0.72621 | 1.00797 | 1.34368 | 1.53602 | 1.84556 | 2.16202 |
| 0.660 | 0.73978 | 1.02105 | 1.35411 | 1.54738 | 1.85779 | 2.17359 |
| 0.680 | 0.75361 | 1.03432 | 1.36454 | 1.55875 | 1.87062 | 2.18535 |
| 0.700 | 0.76770 | 1.04777 | 1.37493 | 1.57013 | 1.88405 | 2.19723 |
| 0.720 | 0.78204 | 1.06139 | 1.38532 | 1.58148 | 1.89803 | 2.20925 |
| 0.740 | 0.79661 | 1.07519 | 1.39568 | 1.59280 | 1.91257 | 2.22139 |
| 0.760 | 0.81140 | 1.08914 | 1.40605 | 1.60409 | 1.92764 | 2.23360 |
| 0.780 | 0.82639 | 1.10326 | 1.41642 | 1.61535 | 1.94319 | 2.24589 |
| 0.800 | 0.84157 | 1.11753 | 1.42678 | 1.62655 | 1.95921 | 2.25824 |
| 0.820 | 0.85695 | 1.13195 | 1.43716 | 1.63775 | 1.97566 | 2.27063 |
| 0.840 | 0.87249 | 1.14652 | 1.44755 | 1.64893 | 1.99248 | 2.28305 |
| 0.860 | 0.88820 | 1.16123 | 1.45795 | 1.66008 | 2.00966 | 2.29548 |
| 0.880 | 0.90406 | 1.17607 | 1.46836 | 1.67124 | 2.02713 | 2.30793 |
| 0.900 | 0.92008 | 1.19105 | 1.47880 | 1.68241 | 2.04484 | 2.32037 |
| 0.920 | 0.93622 | 1.20615 | 1.48927 | 1.69361 | 2.06278 | 2.33282 |
| 0.940 | 0.95250 | 1.22139 | 1.49977 | 1.70484 | 2.08089 | 2.34529 |
| 0.960 | 0.96891 | 1.23675 | 1.51031 | 1.71611 | 2.09912 | 2.35775 |
| 0.980 | 0.98542 | 1.25222 | 1.52088 | 1.72745 | 2.11743 | 2.37023 |
| 1.000 | 1.00205 | 1.26781 | 1.53150 | 1.73884 | 2.13579 | 2.38273 |

$H/R = 0.700 \qquad n = 4.$

| $H/L$ | $\Omega_1$ | $\Omega_2$ | $\Omega_3$ | $\Omega_4$ | $\Omega_5$ | $\Omega_6$ |
|---|---|---|---|---|---|---|
| 0. | 0.62970 | 0.85873 | 1.21216 | 1.45135 | 1.74634 | 2.20830 |
| 0.010 | 0.62973 | 0.85881 | 1.21224 | 1.45140 | 1.74637 | 2.20828 |
| 0.050 | 0.63049 | 0.86073 | 1.21398 | 1.45259 | 1.74712 | 2.20780 |
| 0.100 | 0.63293 | 0.86665 | 1.21937 | 1.45632 | 1.74946 | 2.20705 |
| 0.120 | 0.63441 | 0.87006 | 1.22249 | 1.45852 | 1.75084 | 2.20693 |
| 0.140 | 0.63620 | 0.87404 | 1.22615 | 1.46112 | 1.75249 | 2.20699 |
| 0.160 | 0.63832 | 0.87857 | 1.23031 | 1.46412 | 1.75441 | 2.20729 |
| 0.180 | 0.64078 | 0.88362 | 1.23497 | 1.46754 | 1.75659 | 2.20781 |
| 0.200 | 0.64361 | 0.88917 | 1.24010 | 1.47136 | 1.75906 | 2.20861 |
| 0.220 | 0.64681 | 0.89518 | 1.24567 | 1.47559 | 1.76181 | 2.20967 |
| 0.240 | 0.65042 | 0.90165 | 1.25165 | 1.48023 | 1.76488 | 2.21101 |
| 0.260 | 0.65442 | 0.90854 | 1.25804 | 1.48530 | 1.76823 | 2.21265 |
| 0.280 | 0.65884 | 0.91583 | 1.26477 | 1.49076 | 1.77192 | 2.21457 |
| 0.300 | 0.66370 | 0.92350 | 1.27184 | 1.49665 | 1.77592 | 2.21678 |
| 0.320 | 0.66899 | 0.93153 | 1.27920 | 1.50294 | 1.78028 | 2.21928 |
| 0.340 | 0.67471 | 0.93990 | 1.28683 | 1.50964 | 1.78498 | 2.22208 |
| 0.360 | 0.68088 | 0.94860 | 1.29469 | 1.51673 | 1.79005 | 2.22516 |
| 0.380 | 0.68750 | 0.95760 | 1.30276 | 1.52423 | 1.79549 | 2.22852 |
| 0.400 | 0.69455 | 0.96689 | 1.31101 | 1.53210 | 1.80135 | 2.23217 |
| 0.420 | 0.70204 | 0.97647 | 1.31941 | 1.54032 | 1.80762 | 2.23610 |
| 0.440 | 0.70997 | 0.98632 | 1.32795 | 1.54891 | 1.81430 | 2.24031 |
| 0.460 | 0.71832 | 0.99643 | 1.33658 | 1.55783 | 1.82145 | 2.24479 |
| 0.480 | 0.72708 | 1.00677 | 1.34530 | 1.56705 | 1.82904 | 2.24953 |
| 0.500 | 0.73626 | 1.01736 | 1.35408 | 1.57656 | 1.83712 | 2.25455 |
| 0.520 | 0.74583 | 1.02818 | 1.36293 | 1.58634 | 1.84569 | 2.25983 |
| 0.540 | 0.75579 | 1.03922 | 1.37181 | 1.59635 | 1.85476 | 2.26537 |
| 0.560 | 0.76612 | 1.05048 | 1.38072 | 1.60656 | 1.86434 | 2.27115 |
| 0.580 | 0.77681 | 1.06194 | 1.38967 | 1.61695 | 1.87445 | 2.27721 |
| 0.600 | 0.78786 | 1.07360 | 1.39863 | 1.62749 | 1.88509 | 2.28350 |
| 0.620 | 0.79924 | 1.08546 | 1.40762 | 1.63815 | 1.89626 | 2.29004 |
| 0.640 | 0.81096 | 1.09752 | 1.41662 | 1.64891 | 1.90796 | 2.29683 |
| 0.660 | 0.82297 | 1.10976 | 1.42565 | 1.65976 | 1.92020 | 2.30386 |
| 0.680 | 0.83529 | 1.12217 | 1.43471 | 1.67066 | 1.93294 | 2.31112 |
| 0.700 | 0.84790 | 1.13476 | 1.44378 | 1.68161 | 1.94620 | 2.31863 |
| 0.720 | 0.86078 | 1.14753 | 1.45288 | 1.69259 | 1.95995 | 2.32636 |
| 0.740 | 0.87393 | 1.16046 | 1.46202 | 1.70358 | 1.97417 | 2.33434 |
| 0.760 | 0.88732 | 1.17356 | 1.47118 | 1.71460 | 1.98883 | 2.34255 |
| 0.780 | 0.90095 | 1.18681 | 1.48039 | 1.72562 | 2.00392 | 2.35098 |
| 0.800 | 0.91481 | 1.20022 | 1.48965 | 1.73666 | 2.01938 | 2.35965 |
| 0.820 | 0.92889 | 1.21378 | 1.49895 | 1.74771 | 2.03521 | 2.36855 |
| 0.840 | 0.94318 | 1.22748 | 1.50831 | 1.75877 | 2.05135 | 2.37769 |
| 0.860 | 0.95766 | 1.24133 | 1.51772 | 1.76985 | 2.06779 | 2.38704 |
| 0.880 | 0.97233 | 1.25532 | 1.52720 | 1.78095 | 2.08447 | 2.39663 |
| 0.900 | 0.98718 | 1.26944 | 1.53674 | 1.79209 | 2.10136 | 2.40645 |
| 0.920 | 1.00220 | 1.28371 | 1.54637 | 1.80327 | 2.11843 | 2.41652 |
| 0.940 | 1.01738 | 1.29809 | 1.55605 | 1.81448 | 2.13562 | 2.42680 |
| 0.960 | 1.03271 | 1.31261 | 1.56582 | 1.82575 | 2.15292 | 2.43731 |
| 0.980 | 1.04819 | 1.32725 | 1.57566 | 1.83707 | 2.17029 | 2.44806 |
| 1.000 | 1.06381 | 1.34200 | 1.58560 | 1.84846 | 2.18768 | 2.45904 |

$H/R$ — 0.800 $n = 4.$

| $H/L$ | $\Omega_1$ | $\Omega_2$ | $\Omega_3$ | $\Omega_4$ | $\Omega_5$ | $\Omega_6$ |
|---|---|---|---|---|---|---|
| 0. | 0.74871 | 0.95704 | 1.32133 | 1.59949 | 1.81007 | 2.32523 |
| 0.010 | 0.74873 | 0.95713 | 1.32139 | 1.59954 | 1.81010 | 2.32524 |
| 0.050 | 0.74926 | 0.95895 | 1.32288 | 1.60057 | 1.81097 | 2.32541 |
| 0.100 | 0.75095 | 0.96458 | 1.32748 | 1.60382 | 1.81369 | 2.32601 |
| 0.120 | 0.75200 | 0.96783 | 1.33014 | 1.60574 | 1.81529 | 2.32640 |
| 0.140 | 0.75327 | 0.97162 | 1.33326 | 1.60801 | 1.81718 | 2.32688 |
| 0.160 | 0.75480 | 0.97593 | 1.33682 | 1.61064 | 1.81938 | 2.32749 |
| 0.180 | 0.75660 | 0.98073 | 1.34081 | 1.61364 | 1.82189 | 2.32822 |
| 0.200 | 0.75868 | 0.98601 | 1.34520 | 1.61699 | 1.82470 | 2.32908 |
| 0.220 | 0.76107 | 0.99173 | 1.34998 | 1.62072 | 1.82780 | 2.33010 |
| 0.240 | 0.76379 | 0.99787 | 1.35512 | 1.62481 | 1.83125 | 2.33129 |
| 0.260 | 0.76685 | 1.00441 | 1.36061 | 1.62929 | 1.83499 | 2.33264 |
| 0.280 | 0.77026 | 1.01134 | 1.36642 | 1.63414 | 1.83908 | 2.33419 |
| 0.300 | 0.77403 | 1.01862 | 1.37252 | 1.63938 | 1.84349 | 2.33592 |
| 0.320 | 0.77819 | 1.02624 | 1.37888 | 1.64500 | 1.84825 | 2.33786 |
| 0.340 | 0.78273 | 1.03419 | 1.38550 | 1.65099 | 1.85335 | 2.34001 |
| 0.360 | 0.78766 | 1.04244 | 1.39233 | 1.65735 | 1.85880 | 2.34236 |
| 0.380 | 0.79299 | 1.05098 | 1.39936 | 1.66409 | 1.86462 | 2.34496 |
| 0.400 | 0.79871 | 1.05979 | 1.40656 | 1.67120 | 1.87081 | 2.34777 |
| 0.420 | 0.80485 | 1.06887 | 1.41392 | 1.67867 | 1.87739 | 2.35081 |
| 0.440 | 0.81138 | 1.07821 | 1.42141 | 1.68649 | 1.88437 | 2.35408 |
| 0.460 | 0.81830 | 1.08778 | 1.42901 | 1.69464 | 1.89175 | 2.35759 |
| 0.480 | 0.82563 | 1.09759 | 1.43670 | 1.70310 | 1.89953 | 2.36135 |
| 0.500 | 0.83334 | 1.10762 | 1.44448 | 1.71186 | 1.90775 | 2.36535 |
| 0.520 | 0.84144 | 1.11786 | 1.45233 | 1.72090 | 1.91640 | 2.36959 |
| 0.540 | 0.84991 | 1.12831 | 1.46025 | 1.73021 | 1.92549 | 2.37407 |
| 0.560 | 0.85876 | 1.13896 | 1.46820 | 1.73975 | 1.93503 | 2.37880 |
| 0.580 | 0.86797 | 1.14981 | 1.47621 | 1.74950 | 1.94502 | 2.38378 |
| 0.600 | 0.87752 | 1.16085 | 1.48425 | 1.75943 | 1.95549 | 2.38902 |
| 0.620 | 0.88743 | 1.17207 | 1.49234 | 1.76954 | 1.96640 | 2.39449 |
| 0.640 | 0.89766 | 1.18348 | 1.50047 | 1.77979 | 1.97779 | 2.40023 |
| 0.660 | 0.90822 | 1.19506 | 1.50864 | 1.79016 | 1.98963 | 2.40621 |
| 0.680 | 0.91909 | 1.20681 | 1.51684 | 1.80063 | 2.00191 | 2.41245 |
| 0.700 | 0.93027 | 1.21873 | 1.52509 | 1.81118 | 2.01465 | 2.41894 |
| 0.720 | 0.94174 | 1.23081 | 1.53339 | 1.82181 | 2.02782 | 2.42568 |
| 0.740 | 0.95349 | 1.24306 | 1.54175 | 1.83251 | 2.04139 | 2.43270 |
| 0.760 | 0.96551 | 1.25546 | 1.55015 | 1.84325 | 2.05535 | 2.43995 |
| 0.780 | 0.97779 | 1.26801 | 1.55862 | 1.85403 | 2.06968 | 2.44748 |
| 0.800 | 0.99032 | 1.28072 | 1.56715 | 1.86485 | 2.08436 | 2.45525 |
| 0.820 | 1.00309 | 1.29356 | 1.57574 | 1.87572 | 2.09934 | 2.46330 |
| 0.840 | 1.01610 | 1.30656 | 1.58443 | 1.88662 | 2.11463 | 2.47161 |
| 0.860 | 1.02933 | 1.31969 | 1.59318 | 1.89754 | 2.13017 | 2.48016 |
| 0.880 | 1.04277 | 1.33296 | 1.60202 | 1.90852 | 2.14593 | 2.48901 |
| 0.900 | 1.05642 | 1.34637 | 1.61094 | 1.91953 | 2.16189 | 2.49811 |
| 0.920 | 1.07026 | 1.35990 | 1.61996 | 1.93057 | 2.17801 | 2.50747 |
| 0.940 | 1.08430 | 1.37358 | 1.62909 | 1.94168 | 2.19425 | 2.51710 |
| 0.960 | 1.09850 | 1.38737 | 1.63830 | 1.95283 | 2.21061 | 2.52700 |
| 0.980 | 1.11288 | 1.40128 | 1.64761 | 1.96403 | 2.22701 | 2.53718 |
| 1.000 | 1.12743 | 1.41532 | 1.65704 | 1.97528 | 2.24345 | 2.54762 |

$H/R = 0.900 \qquad n = 4.$

| $H/L$ | $\Omega_1$ | $\Omega_2$ | $\Omega_3$ | $\Omega_4$ | $\Omega_5$ | $\Omega_6$ |
|---|---|---|---|---|---|---|
| 0. | 0.86001 | 1.04573 | 1.43421 | 1.76067 | 1.89377 | 2.44077 |
| 0.010 | 0.86002 | 1.04580 | 1.43426 | 1.76070 | 1.89381 | 2.44080 |
| 0.050 | 0.86031 | 1.04762 | 1.43558 | 1.76155 | 1.89475 | 2.44113 |
| 0.100 | 0.86132 | 1.05319 | 1.43970 | 1.76423 | 1.89774 | 2.44221 |
| 0.120 | 0.86197 | 1.05640 | 1.44209 | 1.76582 | 1.89950 | 2.44285 |
| 0.140 | 0.86278 | 1.06013 | 1.44490 | 1.76770 | 1.90156 | 2.44359 |
| 0.160 | 0.86378 | 1.06437 | 1.44811 | 1.76989 | 1.90393 | 2.44449 |
| 0.180 | 0.86498 | 1.06909 | 1.45170 | 1.77239 | 1.90663 | 2.44551 |
| 0.200 | 0.86642 | 1.07426 | 1.45566 | 1.77521 | 1.90964 | 2.44666 |
| 0.220 | 0.86811 | 1.07987 | 1.45999 | 1.77836 | 1.91295 | 2.44797 |
| 0.240 | 0.87006 | 1.08588 | 1.46465 | 1.78184 | 1.91658 | 2.44943 |
| 0.260 | 0.87231 | 1.09228 | 1.46963 | 1.78567 | 1.92052 | 2.45103 |
| 0.280 | 0.87485 | 1.09904 | 1.47492 | 1.78984 | 1.92478 | 2.45280 |
| 0.300 | 0.87772 | 1.10613 | 1.48048 | 1.79436 | 1.92935 | 2.45472 |
| 0.320 | 0.88092 | 1.11355 | 1.48630 | 1.79923 | 1.93423 | 2.45683 |
| 0.340 | 0.88446 | 1.12127 | 1.49238 | 1.80447 | 1.93943 | 2.45909 |
| 0.360 | 0.88835 | 1.12927 | 1.49866 | 1.81008 | 1.94494 | 2.46155 |
| 0.380 | 0.89261 | 1.13755 | 1.50515 | 1.81607 | 1.95078 | 2.46418 |
| 0.400 | 0.89723 | 1.14607 | 1.51181 | 1.82241 | 1.95695 | 2.46700 |
| 0.420 | 0.90222 | 1.15485 | 1.51864 | 1.82911 | 1.96344 | 2.47002 |
| 0.440 | 0.90757 | 1.16386 | 1.52562 | 1.83618 | 1.97026 | 2.47325 |
| 0.460 | 0.91331 | 1.17309 | 1.53270 | 1.84359 | 1.97742 | 2.47665 |
| 0.480 | 0.91941 | 1.18253 | 1.53991 | 1.85135 | 1.98493 | 2.48029 |
| 0.500 | 0.92589 | 1.19219 | 1.54721 | 1.85943 | 1.99280 | 2.48411 |
| 0.520 | 0.93273 | 1.20203 | 1.55459 | 1.86782 | 2.00104 | 2.48816 |
| 0.540 | 0.93994 | 1.21208 | 1.56204 | 1.87652 | 2.00964 | 2.49242 |
| 0.560 | 0.94750 | 1.22230 | 1.56955 | 1.88548 | 2.01861 | 2.49690 |
| 0.580 | 0.95542 | 1.23270 | 1.57713 | 1.89471 | 2.02798 | 2.50160 |
| 0.600 | 0.96369 | 1.24329 | 1.58476 | 1.90416 | 2.03772 | 2.50652 |
| 0.620 | 0.97229 | 1.25404 | 1.59244 | 1.91381 | 2.04787 | 2.51168 |
| 0.640 | 0.98123 | 1.26496 | 1.60016 | 1.92366 | 2.05843 | 2.51708 |
| 0.660 | 0.99050 | 1.27603 | 1.60793 | 1.93367 | 2.06939 | 2.52270 |
| 0.680 | 1.00008 | 1.28728 | 1.61575 | 1.94381 | 2.08074 | 2.52856 |
| 0.700 | 1.00997 | 1.29867 | 1.62361 | 1.95407 | 2.09250 | 2.53465 |
| 0.720 | 1.02016 | 1.31021 | 1.63153 | 1.96443 | 2.10465 | 2.54100 |
| 0.740 | 1.03065 | 1.32192 | 1.63950 | 1.97487 | 2.11719 | 2.54758 |
| 0.760 | 1.04141 | 1.33376 | 1.64753 | 1.98538 | 2.13011 | 2.55441 |
| 0.780 | 1.05245 | 1.34574 | 1.65563 | 1.99594 | 2.14337 | 2.56150 |
| 0.800 | 1.06376 | 1.35787 | 1.66380 | 2.00655 | 2.15697 | 2.56885 |
| 0.820 | 1.07532 | 1.37014 | 1.67203 | 2.01720 | 2.17089 | 2.57645 |
| 0.840 | 1.08713 | 1.38253 | 1.68035 | 2.02790 | 2.18510 | 2.58429 |
| 0.860 | 1.09918 | 1.39507 | 1.68874 | 2.03863 | 2.19959 | 2.59241 |
| 0.880 | 1.11146 | 1.40774 | 1.69722 | 2.04938 | 2.21432 | 2.60078 |
| 0.900 | 1.12396 | 1.42054 | 1.70579 | 2.06019 | 2.22925 | 2.60944 |
| 0.920 | 1.13668 | 1.43346 | 1.71446 | 2.07102 | 2.24437 | 2.61833 |
| 0.940 | 1.14961 | 1.44651 | 1.72322 | 2.08189 | 2.25964 | 2.62753 |
| 0.960 | 1.16273 | 1.45968 | 1.73210 | 2.09280 | 2.27503 | 2.63696 |
| 0.980 | 1.17604 | 1.47297 | 1.74108 | 2.10376 | 2.29053 | 2.64670 |
| 1.000 | 1.18954 | 1.48638 | 1.75015 | 2.11478 | 2.30609 | 2.65667 |

$H/R = 1.000 \qquad n = 4.$

| $H/L$ | $\Omega_1$ | $\Omega_2$ | $\Omega_3$ | $\Omega_4$ | $\Omega_5$ | $\Omega_6$ |
|---|---|---|---|---|---|---|
| 0. | 0.95897 | 1.12632 | 1.55537 | 1.92242 | 2.00330 | 2.58987 |
| 0.010 | 0.95897 | 1.12641 | 1.55542 | 1.92244 | 2.00334 | 2.58988 |
| 0.050 | 0.95908 | 1.12823 | 1.55669 | 1.92306 | 2.00437 | 2.59020 |
| 0.100 | 0.95948 | 1.13385 | 1.56058 | 1.92503 | 2.00757 | 2.59128 |
| 0.120 | 0.95979 | 1.13706 | 1.56286 | 1.92620 | 2.00943 | 2.59192 |
| 0.140 | 0.96021 | 1.14081 | 1.56552 | 1.92762 | 2.01162 | 2.59266 |
| 0.160 | 0.96076 | 1.14506 | 1.56857 | 1.92928 | 2.01411 | 2.59354 |
| 0.180 | 0.96147 | 1.14977 | 1.57200 | 1.93121 | 2.01693 | 2.59453 |
| 0.200 | 0.96238 | 1.15493 | 1.57579 | 1.93339 | 2.02004 | 2.59566 |
| 0.220 | 0.96348 | 1.16051 | 1.57993 | 1.93588 | 2.02345 | 2.59693 |
| 0.240 | 0.96481 | 1.16648 | 1.58440 | 1.93864 | 2.02715 | 2.59833 |
| 0.260 | 0.96640 | 1.17281 | 1.58920 | 1.94172 | 2.03114 | 2.59988 |
| 0.280 | 0.96825 | 1.17949 | 1.59429 | 1.94511 | 2.03541 | 2.60159 |
| 0.300 | 0.97038 | 1.18650 | 1.59968 | 1.94884 | 2.03995 | 2.60343 |
| 0.320 | 0.97281 | 1.19380 | 1.60534 | 1.95291 | 2.04475 | 2.60541 |
| 0.340 | 0.97556 | 1.20139 | 1.61125 | 1.95734 | 2.04981 | 2.60757 |
| 0.360 | 0.97862 | 1.20925 | 1.61738 | 1.96212 | 2.05514 | 2.60989 |
| 0.380 | 0.98201 | 1.21736 | 1.62374 | 1.96728 | 2.06074 | 2.61237 |
| 0.400 | 0.98574 | 1.22571 | 1.63028 | 1.97281 | 2.06658 | 2.61502 |
| 0.420 | 0.98982 | 1.23429 | 1.63700 | 1.97872 | 2.07268 | 2.61783 |
| 0.440 | 0.99424 | 1.24309 | 1.64390 | 1.98500 | 2.07904 | 2.62085 |
| 0.460 | 0.99901 | 1.25209 | 1.65092 | 1.99168 | 2.08564 | 2.62404 |
| 0.480 | 1.00413 | 1.26129 | 1.65808 | 1.99874 | 2.09252 | 2.62740 |
| 0.500 | 1.00961 | 1.27067 | 1.66535 | 2.00616 | 2.09966 | 2.63097 |
| 0.520 | 1.01544 | 1.28025 | 1.67272 | 2.01397 | 2.10706 | 2.63473 |
| 0.540 | 1.02161 | 1.28999 | 1.68016 | 2.02211 | 2.11472 | 2.63869 |
| 0.560 | 1.02813 | 1.29990 | 1.68769 | 2.03060 | 2.12268 | 2.64284 |
| 0.580 | 1.03500 | 1.30998 | 1.69528 | 2.03941 | 2.13093 | 2.64720 |
| 0.600 | 1.04220 | 1.32021 | 1.70292 | 2.04853 | 2.13948 | 2.65177 |
| 0.620 | 1.04974 | 1.33060 | 1.71063 | 2.05792 | 2.14832 | 2.65656 |
| 0.640 | 1.05760 | 1.34115 | 1.71838 | 2.06757 | 2.15750 | 2.66157 |
| 0.660 | 1.06579 | 1.35184 | 1.72617 | 2.07744 | 2.16700 | 2.66679 |
| 0.680 | 1.07429 | 1.36269 | 1.73401 | 2.08752 | 2.17684 | 2.67222 |
| 0.700 | 1.08310 | 1.37367 | 1.74190 | 2.09776 | 2.18704 | 2.67789 |
| 0.720 | 1.09221 | 1.38478 | 1.74982 | 2.10814 | 2.19760 | 2.68380 |
| 0.740 | 1.10162 | 1.39605 | 1.75778 | 2.11862 | 2.20853 | 2.68992 |
| 0.760 | 1.11132 | 1.40744 | 1.76580 | 2.12918 | 2.21982 | 2.69630 |
| 0.780 | 1.12130 | 1.41897 | 1.77386 | 2.13982 | 2.23150 | 2.70290 |
| 0.800 | 1.13155 | 1.43062 | 1.78199 | 2.15049 | 2.24353 | 2.70975 |
| 0.820 | 1.14205 | 1.44241 | 1.79017 | 2.16118 | 2.25593 | 2.71686 |
| 0.840 | 1.15283 | 1.45433 | 1.79841 | 2.17188 | 2.26867 | 2.72420 |
| 0.860 | 1.16385 | 1.46637 | 1.80672 | 2.18259 | 2.28173 | 2.73182 |
| 0.880 | 1.17511 | 1.47853 | 1.81510 | 2.19329 | 2.29510 | 2.73967 |
| 0.900 | 1.18661 | 1.49081 | 1.82356 | 2.20401 | 2.30876 | 2.74779 |
| 0.920 | 1.19833 | 1.50321 | 1.83210 | 2.21472 | 2.32267 | 2.75618 |
| 0.940 | 1.21027 | 1.51574 | 1.84072 | 2.22543 | 2.33681 | 2.76480 |
| 0.960 | 1.22242 | 1.52838 | 1.84944 | 2.23614 | 2.35113 | 2.77373 |
| 0.980 | 1.23479 | 1.54114 | 1.85826 | 2.24687 | 2.36563 | 2.78289 |
| 1.000 | 1.24734 | 1.55401 | 1.86716 | 2.25763 | 2.38029 | 2.79234 |

$H/R = 1.100 \qquad n = 4.$

| $H/L$ | $\Omega_1$ | $\Omega_2$ | $\Omega_3$ | $\Omega_4$ | $\Omega_5$ | $\Omega_6$ |
|---|---|---|---|---|---|---|
| 0. | 1.04384 | 1.20041 | 1.68035 | 2.07256 | 2.14718 | 2.77197 |
| 0.010 | 1.04384 | 1.20049 | 1.68040 | 2.07258 | 2.14723 | 2.77197 |
| 0.050 | 1.04379 | 1.20233 | 1.68165 | 2.07305 | 2.14822 | 2.77223 |
| 0.100 | 1.04376 | 1.20795 | 1.68551 | 2.07454 | 2.15129 | 2.77308 |
| 0.120 | 1.04382 | 1.21118 | 1.68778 | 2.07545 | 2.15307 | 2.77359 |
| 0.140 | 1.04395 | 1.21492 | 1.69042 | 2.07655 | 2.15516 | 2.77417 |
| 0.160 | 1.04418 | 1.21915 | 1.69347 | 2.07785 | 2.15755 | 2.77487 |
| 0.180 | 1.04454 | 1.22385 | 1.69689 | 2.07935 | 2.16021 | 2.77566 |
| 0.200 | 1.04505 | 1.22898 | 1.70068 | 2.08110 | 2.16317 | 2.77657 |
| 0.220 | 1.04573 | 1.23451 | 1.70483 | 2.08308 | 2.16639 | 2.77759 |
| 0.240 | 1.04661 | 1.24043 | 1.70933 | 2.08530 | 2.16988 | 2.77873 |
| 0.260 | 1.04771 | 1.24671 | 1.71416 | 2.08779 | 2.17363 | 2.77999 |
| 0.280 | 1.04904 | 1.25330 | 1.71931 | 2.09056 | 2.17763 | 2.78135 |
| 0.300 | 1.05063 | 1.26022 | 1.72478 | 2.09362 | 2.18187 | 2.78287 |
| 0.320 | 1.05250 | 1.26742 | 1.73053 | 2.09699 | 2.18633 | 2.78451 |
| 0.340 | 1.05464 | 1.27490 | 1.73656 | 2.10067 | 2.19103 | 2.78628 |
| 0.360 | 1.05709 | 1.28263 | 1.74285 | 2.10468 | 2.19595 | 2.78823 |
| 0.380 | 1.05984 | 1.29059 | 1.74938 | 2.10903 | 2.20108 | 2.79028 |
| 0.400 | 1.06291 | 1.29880 | 1.75614 | 2.11373 | 2.20643 | 2.79251 |
| 0.420 | 1.06629 | 1.30721 | 1.76311 | 2.11879 | 2.21199 | 2.79489 |
| 0.440 | 1.07000 | 1.31583 | 1.77026 | 2.12421 | 2.21775 | 2.79745 |
| 0.460 | 1.07404 | 1.32463 | 1.77759 | 2.13001 | 2.22372 | 2.80015 |
| 0.480 | 1.07841 | 1.33363 | 1.78507 | 2.13618 | 2.22989 | 2.80305 |
| 0.500 | 1.08312 | 1.34280 | 1.79268 | 2.14275 | 2.23625 | 2.80609 |
| 0.520 | 1.08816 | 1.35214 | 1.80043 | 2.14969 | 2.24283 | 2.80935 |
| 0.540 | 1.09354 | 1.36164 | 1.80827 | 2.15702 | 2.24959 | 2.81276 |
| 0.560 | 1.09924 | 1.37130 | 1.81623 | 2.16473 | 2.25656 | 2.81637 |
| 0.580 | 1.10529 | 1.38112 | 1.82425 | 2.17281 | 2.26373 | 2.82017 |
| 0.600 | 1.11165 | 1.39108 | 1.83234 | 2.18126 | 2.27111 | 2.82418 |
| 0.620 | 1.11834 | 1.40119 | 1.84049 | 2.19008 | 2.27869 | 2.82836 |
| 0.640 | 1.12535 | 1.41145 | 1.84869 | 2.19923 | 2.28649 | 2.83277 |
| 0.660 | 1.13268 | 1.42183 | 1.85694 | 2.20871 | 2.29452 | 2.83738 |
| 0.680 | 1.14031 | 1.43236 | 1.86522 | 2.21851 | 2.30276 | 2.84221 |
| 0.700 | 1.14826 | 1.44301 | 1.87354 | 2.22859 | 2.31125 | 2.84725 |
| 0.720 | 1.15649 | 1.45379 | 1.88188 | 2.23892 | 2.32000 | 2.85251 |
| 0.740 | 1.16503 | 1.46469 | 1.89026 | 2.24950 | 2.32901 | 2.85800 |
| 0.760 | 1.17385 | 1.47573 | 1.89866 | 2.26027 | 2.33828 | 2.86371 |
| 0.780 | 1.18296 | 1.48689 | 1.90708 | 2.27122 | 2.34785 | 2.86967 |
| 0.800 | 1.19234 | 1.49817 | 1.91555 | 2.28230 | 2.35773 | 2.87584 |
| 0.820 | 1.20198 | 1.50958 | 1.92403 | 2.29347 | 2.36792 | 2.88227 |
| 0.840 | 1.21189 | 1.52111 | 1.93256 | 2.30472 | 2.37845 | 2.88894 |
| 0.860 | 1.22206 | 1.53274 | 1.94112 | 2.31599 | 2.38932 | 2.89585 |
| 0.880 | 1.23247 | 1.54450 | 1.94972 | 2.32727 | 2.40053 | 2.90304 |
| 0.900 | 1.24312 | 1.55638 | 1.95838 | 2.33852 | 2.41210 | 2.91045 |
| 0.920 | 1.25400 | 1.56836 | 1.96708 | 2.34974 | 2.42402 | 2.91815 |
| 0.940 | 1.26511 | 1.58046 | 1.97585 | 2.36091 | 2.43626 | 2.92610 |
| 0.960 | 1.27645 | 1.59267 | 1.98466 | 2.37200 | 2.44885 | 2.93431 |
| 0.980 | 1.28800 | 1.60498 | 1.99355 | 2.38305 | 2.46170 | 2.94280 |
| 1.000 | 1.29975 | 1.61741 | 2.00250 | 2.39403 | 2.47485 | 2.95158 |

$H/R = 1.200 \qquad n = 4.$

| $H/L$ | $\Omega_1$ | $\Omega_2$ | $\Omega_3$ | $\Omega_4$ | $\Omega_5$ | $\Omega_6$ |
|---|---|---|---|---|---|---|
| 0. | 1.11617 | 1.26920 | 1.79862 | 2.20592 | 2.32973 | 2.96929 |
| 0.010 | 1.11616 | 1.26928 | 1.79866 | 2.20594 | 2.32977 | 2.96929 |
| 0.050 | 1.11603 | 1.27109 | 1.79991 | 2.20645 | 2.33056 | 2.96937 |
| 0.100 | 1.11575 | 1.27664 | 1.80379 | 2.20805 | 2.33300 | 2.96963 |
| 0.120 | 1.11566 | 1.27982 | 1.80606 | 2.20900 | 2.33443 | 2.96982 |
| 0.140 | 1.11562 | 1.28350 | 1.80873 | 2.21013 | 2.33611 | 2.97004 |
| 0.160 | 1.11565 | 1.28767 | 1.81178 | 2.21147 | 2.33805 | 2.97031 |
| 0.180 | 1.11578 | 1.29231 | 1.81523 | 2.21299 | 2.34023 | 2.97066 |
| 0.200 | 1.11603 | 1.29735 | 1.81905 | 2.21471 | 2.34264 | 2.97107 |
| 0.220 | 1.11644 | 1.30279 | 1.82325 | 2.21665 | 2.34530 | 2.97156 |
| 0.240 | 1.11702 | 1.30861 | 1.82781 | 2.21881 | 2.34819 | 2.97213 |
| 0.260 | 1.11779 | 1.31478 | 1.83274 | 2.22118 | 2.35132 | 2.97277 |
| 0.280 | 1.11878 | 1.32126 | 1.83799 | 2.22379 | 2.35467 | 2.97355 |
| 0.300 | 1.12000 | 1.32805 | 1.84358 | 2.22665 | 2.35825 | 2.97442 |
| 0.320 | 1.12147 | 1.33512 | 1.84948 | 2.22977 | 2.36203 | 2.97538 |
| 0.340 | 1.12320 | 1.34246 | 1.85569 | 2.23313 | 2.36604 | 2.97648 |
| 0.360 | 1.12521 | 1.35005 | 1.86218 | 2.23677 | 2.37026 | 2.97772 |
| 0.380 | 1.12750 | 1.35787 | 1.86896 | 2.24070 | 2.37469 | 2.97907 |
| 0.400 | 1.13007 | 1.36590 | 1.87598 | 2.24491 | 2.37931 | 2.98057 |
| 0.420 | 1.13296 | 1.37415 | 1.88326 | 2.24943 | 2.38414 | 2.98223 |
| 0.440 | 1.13614 | 1.38260 | 1.89075 | 2.25424 | 2.38917 | 2.98401 |
| 0.460 | 1.13964 | 1.39122 | 1.89847 | 2.25939 | 2.39438 | 2.98597 |
| 0.480 | 1.14345 | 1.40003 | 1.90637 | 2.26484 | 2.39978 | 2.98808 |
| 0.500 | 1.14758 | 1.40901 | 1.91445 | 2.27062 | 2.40539 | 2.99036 |
| 0.520 | 1.15203 | 1.41815 | 1.92270 | 2.27675 | 2.41117 | 2.99282 |
| 0.540 | 1.15679 | 1.42745 | 1.93109 | 2.28322 | 2.41714 | 2.99545 |
| 0.560 | 1.16188 | 1.43689 | 1.93960 | 2.29002 | 2.42329 | 2.99828 |
| 0.580 | 1.16728 | 1.44649 | 1.94823 | 2.29718 | 2.42963 | 3.00128 |
| 0.600 | 1.17299 | 1.45622 | 1.95696 | 2.30468 | 2.43615 | 3.00445 |
| 0.620 | 1.17903 | 1.46610 | 1.96578 | 2.31253 | 2.44284 | 3.00785 |
| 0.640 | 1.18537 | 1.47611 | 1.97466 | 2.32074 | 2.44973 | 3.01142 |
| 0.660 | 1.19202 | 1.48624 | 1.98360 | 2.32928 | 2.45681 | 3.01522 |
| 0.680 | 1.19897 | 1.49651 | 1.99259 | 2.33817 | 2.46404 | 3.01921 |
| 0.700 | 1.20622 | 1.50690 | 2.00162 | 2.34738 | 2.47150 | 3.02342 |
| 0.720 | 1.21376 | 1.51742 | 2.01068 | 2.35691 | 2.47912 | 3.02784 |
| 0.740 | 1.22161 | 1.52804 | 2.01976 | 2.36676 | 2.48694 | 3.03246 |
| 0.760 | 1.22972 | 1.53880 | 2.02886 | 2.37690 | 2.49493 | 3.03733 |
| 0.780 | 1.23812 | 1.54967 | 2.03798 | 2.38731 | 2.50315 | 3.04242 |
| 0.800 | 1.24679 | 1.56066 | 2.04711 | 2.39798 | 2.51157 | 3.04776 |
| 0.820 | 1.25574 | 1.57176 | 2.05625 | 2.40889 | 2.52019 | 3.05331 |
| 0.840 | 1.26493 | 1.58297 | 2.06540 | 2.42003 | 2.52903 | 3.05912 |
| 0.860 | 1.27439 | 1.59430 | 2.07457 | 2.43135 | 2.53808 | 3.06517 |
| 0.880 | 1.28411 | 1.60573 | 2.08373 | 2.44285 | 2.54738 | 3.07147 |
| 0.900 | 1.29406 | 1.61728 | 2.09293 | 2.45449 | 2.55690 | 3.07803 |
| 0.920 | 1.30425 | 1.62893 | 2.10213 | 2.46625 | 2.56668 | 3.08484 |
| 0.940 | 1.31467 | 1.64070 | 2.11134 | 2.47808 | 2.57668 | 3.09193 |
| 0.960 | 1.32532 | 1.65257 | 2.12060 | 2.49000 | 2.58696 | 3.09928 |
| 0.980 | 1.33618 | 1.66454 | 2.12986 | 2.50192 | 2.59749 | 3.10688 |
| 1.000 | 1.34727 | 1.67662 | 2.13917 | 2.51389 | 2.60829 | 3.11480 |

$H/R = 1.300 \qquad n = 4.$

| $H/L$ | $\Omega_1$ | $\Omega_2$ | $\Omega_3$ | $\Omega_4$ | $\Omega_5$ | $\Omega_6$ |
|---|---|---|---|---|---|---|
| 0. | 1.17903 | 1.33351 | 1.90391 | 2.32450 | 2.53464 | 3.15474 |
| 0.010 | 1.17903 | 1.33358 | 1.90397 | 2.32452 | 2.53467 | 3.15473 |
| 0.050 | 1.17886 | 1.33535 | 1.90520 | 2.32506 | 2.53531 | 3.15461 |
| 0.100 | 1.17845 | 1.34075 | 1.90903 | 2.32674 | 2.53733 | 3.15432 |
| 0.120 | 1.17829 | 1.34386 | 1.91127 | 2.32775 | 2.53850 | 3.15418 |
| 0.140 | 1.17816 | 1.34744 | 1.91390 | 2.32893 | 2.53991 | 3.15404 |
| 0.160 | 1.17808 | 1.35152 | 1.91692 | 2.33030 | 2.54150 | 3.15395 |
| 0.180 | 1.17808 | 1.35602 | 1.92034 | 2.33187 | 2.54331 | 3.15385 |
| 0.200 | 1.17819 | 1.36095 | 1.92414 | 2.33363 | 2.54534 | 3.15381 |
| 0.220 | 1.17843 | 1.36627 | 1.92831 | 2.33559 | 2.54754 | 3.15381 |
| 0.240 | 1.17882 | 1.37195 | 1.93285 | 2.33776 | 2.54998 | 3.15391 |
| 0.260 | 1.17939 | 1.37797 | 1.93775 | 2.34014 | 2.55263 | 3.15406 |
| 0.280 | 1.18015 | 1.38432 | 1.94300 | 2.34273 | 2.55547 | 3.15428 |
| 0.300 | 1.18112 | 1.39096 | 1.94859 | 2.34555 | 2.55851 | 3.15463 |
| 0.320 | 1.18232 | 1.39788 | 1.95450 | 2.34860 | 2.56175 | 3.15507 |
| 0.340 | 1.18376 | 1.40506 | 1.96074 | 2.35187 | 2.56516 | 3.15561 |
| 0.360 | 1.18546 | 1.41248 | 1.96730 | 2.35537 | 2.56881 | 3.15630 |
| 0.380 | 1.18741 | 1.42014 | 1.97413 | 2.35913 | 2.57262 | 3.15711 |
| 0.400 | 1.18964 | 1.42801 | 1.98126 | 2.36313 | 2.57665 | 3.15803 |
| 0.420 | 1.19215 | 1.43609 | 1.98864 | 2.36740 | 2.58084 | 3.15911 |
| 0.440 | 1.19495 | 1.44436 | 1.99629 | 2.37192 | 2.58524 | 3.16033 |
| 0.460 | 1.19803 | 1.45282 | 2.00418 | 2.37672 | 2.58984 | 3.16171 |
| 0.480 | 1.20142 | 1.46145 | 2.01229 | 2.38179 | 2.59460 | 3.16324 |
| 0.500 | 1.20511 | 1.47025 | 2.02061 | 2.38715 | 2.59954 | 3.16494 |
| 0.520 | 1.20910 | 1.47921 | 2.02912 | 2.39279 | 2.60469 | 3.16679 |
| 0.540 | 1.21340 | 1.48831 | 2.03782 | 2.39873 | 2.61000 | 3.16880 |
| 0.560 | 1.21799 | 1.49757 | 2.04667 | 2.40496 | 2.61550 | 3.17101 |
| 0.580 | 1.22290 | 1.50698 | 2.05568 | 2.41151 | 2.62119 | 3.17338 |
| 0.600 | 1.22810 | 1.51652 | 2.06481 | 2.41836 | 2.62704 | 3.17594 |
| 0.620 | 1.23361 | 1.52619 | 2.07406 | 2.42551 | 2.63309 | 3.17869 |
| 0.640 | 1.23942 | 1.53599 | 2.08343 | 2.43299 | 2.63930 | 3.18160 |
| 0.660 | 1.24552 | 1.54593 | 2.09288 | 2.44078 | 2.64570 | 3.18473 |
| 0.680 | 1.25193 | 1.55597 | 2.10240 | 2.44887 | 2.65228 | 3.18804 |
| 0.700 | 1.25862 | 1.56615 | 2.11198 | 2.45731 | 2.65903 | 3.19154 |
| 0.720 | 1.26560 | 1.57644 | 2.12162 | 2.46602 | 2.66596 | 3.19525 |
| 0.740 | 1.27286 | 1.58686 | 2.13132 | 2.47506 | 2.67307 | 3.19916 |
| 0.760 | 1.28041 | 1.59738 | 2.14102 | 2.48439 | 2.68036 | 3.20330 |
| 0.780 | 1.28823 | 1.60801 | 2.15076 | 2.49404 | 2.68783 | 3.20762 |
| 0.800 | 1.29632 | 1.61877 | 2.16052 | 2.50396 | 2.69548 | 3.21219 |
| 0.820 | 1.30468 | 1.62962 | 2.17028 | 2.51414 | 2.70332 | 3.21696 |
| 0.840 | 1.31329 | 1.64059 | 2.18006 | 2.52463 | 2.71134 | 3.22195 |
| 0.860 | 1.32217 | 1.65166 | 2.18984 | 2.53533 | 2.71953 | 3.22718 |
| 0.880 | 1.33129 | 1.66284 | 2.19961 | 2.54628 | 2.72791 | 3.23264 |
| 0.900 | 1.34065 | 1.67413 | 2.20938 | 2.55747 | 2.73649 | 3.23833 |
| 0.920 | 1.35027 | 1.68551 | 2.21916 | 2.56888 | 2.74526 | 3.24427 |
| 0.940 | 1.36011 | 1.69701 | 2.22894 | 2.58046 | 2.75421 | 3.25044 |
| 0.960 | 1.37018 | 1.70861 | 2.23870 | 2.59221 | 2.76336 | 3.25686 |
| 0.980 | 1.38047 | 1.72031 | 2.24848 | 2.60413 | 2.77270 | 3.26356 |
| 1.000 | 1.39098 | 1.73211 | 2.25826 | 2.61618 | 2.78221 | 3.27049 |

$H/R = 1.400 \qquad n = 4.$

| $H/L$ | $\Omega_1$ | $\Omega_2$ | $\Omega_3$ | $\Omega_4$ | $\Omega_5$ | $\Omega_6$ |
|---|---|---|---|---|---|---|
| 0. | 1.23525 | 1.39392 | 1.99717 | 2.43233 | 2.72404 | 3.31613 |
| 0.010 | 1.23525 | 1.39398 | 1.99722 | 2.43234 | 2.72406 | 3.31614 |
| 0.050 | 1.23507 | 1.39569 | 1.99842 | 2.43288 | 2.72464 | 3.31614 |
| 0.100 | 1.23462 | 1.40092 | 2.00214 | 2.43457 | 2.72645 | 3.31616 |
| 0.120 | 1.23442 | 1.40392 | 2.00431 | 2.43555 | 2.72751 | 3.31620 |
| 0.140 | 1.23424 | 1.40741 | 2.00689 | 2.43673 | 2.72876 | 3.31625 |
| 0.160 | 1.23411 | 1.41135 | 2.00983 | 2.43809 | 2.73020 | 3.31634 |
| 0.180 | 1.23405 | 1.41574 | 2.01316 | 2.43964 | 2.73185 | 3.31645 |
| 0.200 | 1.23407 | 1.42052 | 2.01686 | 2.44139 | 2.73367 | 3.31663 |
| 0.220 | 1.23420 | 1.42570 | 2.02094 | 2.44332 | 2.73567 | 3.31683 |
| 0.240 | 1.23448 | 1.43123 | 2.02538 | 2.44544 | 2.73786 | 3.31710 |
| 0.260 | 1.23491 | 1.43710 | 2.03017 | 2.44779 | 2.74026 | 3.31744 |
| 0.280 | 1.23552 | 1.44329 | 2.03532 | 2.45032 | 2.74284 | 3.31786 |
| 0.300 | 1.23632 | 1.44977 | 2.04079 | 2.45306 | 2.74561 | 3.31836 |
| 0.320 | 1.23733 | 1.45653 | 2.04661 | 2.45600 | 2.74854 | 3.31894 |
| 0.340 | 1.23856 | 1.46355 | 2.05274 | 2.45917 | 2.75167 | 3.31963 |
| 0.360 | 1.24003 | 1.47082 | 2.05920 | 2.46258 | 2.75500 | 3.32042 |
| 0.380 | 1.24174 | 1.47830 | 2.06593 | 2.46619 | 2.75849 | 3.32130 |
| 0.400 | 1.24370 | 1.48602 | 2.07298 | 2.47002 | 2.76217 | 3.32231 |
| 0.420 | 1.24593 | 1.49392 | 2.08029 | 2.47410 | 2.76604 | 3.32345 |
| 0.440 | 1.24843 | 1.50203 | 2.08788 | 2.47842 | 2.77009 | 3.32469 |
| 0.460 | 1.25120 | 1.51032 | 2.09572 | 2.48299 | 2.77431 | 3.32606 |
| 0.480 | 1.25426 | 1.51879 | 2.10379 | 2.48781 | 2.77874 | 3.32759 |
| 0.500 | 1.25760 | 1.52741 | 2.11210 | 2.49288 | 2.78332 | 3.32922 |
| 0.520 | 1.26122 | 1.53621 | 2.12062 | 2.49820 | 2.78808 | 3.33102 |
| 0.540 | 1.26514 | 1.54514 | 2.12934 | 2.50379 | 2.79304 | 3.33297 |
| 0.560 | 1.26935 | 1.55424 | 2.13824 | 2.50966 | 2.79815 | 3.33506 |
| 0.580 | 1.27384 | 1.56347 | 2.14732 | 2.51582 | 2.80348 | 3.33730 |
| 0.600 | 1.27863 | 1.57283 | 2.15655 | 2.52224 | 2.80895 | 3.33972 |
| 0.620 | 1.28372 | 1.58233 | 2.16592 | 2.52893 | 2.81461 | 3.34228 |
| 0.640 | 1.28909 | 1.59196 | 2.17542 | 2.53591 | 2.82044 | 3.34499 |
| 0.660 | 1.29474 | 1.60170 | 2.18504 | 2.54321 | 2.82645 | 3.34790 |
| 0.680 | 1.30070 | 1.61157 | 2.19476 | 2.55078 | 2.83265 | 3.35095 |
| 0.700 | 1.30692 | 1.62156 | 2.20456 | 2.55863 | 2.83901 | 3.35420 |
| 0.720 | 1.31343 | 1.63167 | 2.21445 | 2.56678 | 2.84555 | 3.35760 |
| 0.740 | 1.32021 | 1.64188 | 2.22438 | 2.57523 | 2.85227 | 3.36122 |
| 0.760 | 1.32728 | 1.65222 | 2.23438 | 2.58398 | 2.85919 | 3.36499 |
| 0.780 | 1.33462 | 1.66266 | 2.24441 | 2.59299 | 2.86627 | 3.36894 |
| 0.800 | 1.34221 | 1.67321 | 2.25449 | 2.60230 | 2.87351 | 3.37310 |
| 0.820 | 1.35007 | 1.68386 | 2.26458 | 2.61191 | 2.88097 | 3.37744 |
| 0.840 | 1.35820 | 1.69463 | 2.27470 | 2.62176 | 2.88860 | 3.38197 |
| 0.860 | 1.36658 | 1.70550 | 2.28482 | 2.63190 | 2.89640 | 3.38672 |
| 0.880 | 1.37520 | 1.71647 | 2.29495 | 2.64228 | 2.90438 | 3.39165 |
| 0.900 | 1.38406 | 1.72754 | 2.30507 | 2.65291 | 2.91255 | 3.39679 |
| 0.920 | 1.39318 | 1.73871 | 2.31520 | 2.66380 | 2.92090 | 3.40214 |
| 0.940 | 1.40251 | 1.74999 | 2.32532 | 2.67490 | 2.92943 | 3.40769 |
| 0.960 | 1.41208 | 1.76137 | 2.33544 | 2.68621 | 2.93813 | 3.41349 |
| 0.980 | 1.42187 | 1.77284 | 2.34555 | 2.69771 | 2.94704 | 3.41947 |
| 1.000 | 1.43189 | 1.78440 | 2.35565 | 2.70942 | 2.95614 | 3.42569 |

$H/R = 1.500 \qquad n = 4.$

| $H/L$ | $\Omega_1$ | $\Omega_2$ | $\Omega_3$ | $\Omega_4$ | $\Omega_5$ | $\Omega_6$ |
|---|---|---|---|---|---|---|
| 0. | 1.28678 | 1.45083 | 2.08157 | 2.53239 | 2.87502 | 3.45816 |
| 0.010 | 1.28678 | 1.45089 | 2.08162 | 2.53242 | 2.87503 | 3.45815 |
| 0.050 | 1.28660 | 1.45254 | 2.08278 | 2.53294 | 2.87558 | 3.45824 |
| 0.100 | 1.28613 | 1.45760 | 2.08638 | 2.53457 | 2.87728 | 3.45854 |
| 0.120 | 1.28593 | 1.46049 | 2.08849 | 2.53555 | 2.87827 | 3.45873 |
| 0.140 | 1.28573 | 1.46387 | 2.09097 | 2.53669 | 2.87945 | 3.45896 |
| 0.160 | 1.28556 | 1.46769 | 2.09383 | 2.53801 | 2.88080 | 3.45923 |
| 0.180 | 1.28546 | 1.47194 | 2.09705 | 2.53952 | 2.88233 | 3.45955 |
| 0.200 | 1.28543 | 1.47660 | 2.10065 | 2.54121 | 2.88405 | 3.45991 |
| 0.220 | 1.28551 | 1.48162 | 2.10459 | 2.54309 | 2.88595 | 3.46035 |
| 0.240 | 1.28571 | 1.48701 | 2.10889 | 2.54516 | 2.88800 | 3.46084 |
| 0.260 | 1.28605 | 1.49273 | 2.11354 | 2.54739 | 2.89026 | 3.46141 |
| 0.280 | 1.28655 | 1.49876 | 2.11853 | 2.54985 | 2.89269 | 3.46206 |
| 0.300 | 1.28722 | 1.50508 | 2.12387 | 2.55250 | 2.89529 | 3.46277 |
| 0.320 | 1.28809 | 1.51168 | 2.12952 | 2.55537 | 2.89808 | 3.46357 |
| 0.340 | 1.28916 | 1.51854 | 2.13549 | 2.55841 | 2.90102 | 3.46446 |
| 0.360 | 1.29044 | 1.52565 | 2.14178 | 2.56166 | 2.90416 | 3.46545 |
| 0.380 | 1.29197 | 1.53299 | 2.14836 | 2.56514 | 2.90747 | 3.46654 |
| 0.400 | 1.29372 | 1.54054 | 2.15523 | 2.56884 | 2.91097 | 3.46772 |
| 0.420 | 1.29573 | 1.54829 | 2.16238 | 2.57275 | 2.91463 | 3.46902 |
| 0.440 | 1.29799 | 1.55624 | 2.16980 | 2.57690 | 2.91847 | 3.47043 |
| 0.460 | 1.30050 | 1.56438 | 2.17749 | 2.58124 | 2.92247 | 3.47195 |
| 0.480 | 1.30328 | 1.57268 | 2.18542 | 2.58585 | 2.92666 | 3.47360 |
| 0.500 | 1.30634 | 1.58116 | 2.19359 | 2.59070 | 2.93102 | 3.47537 |
| 0.520 | 1.30967 | 1.58980 | 2.20197 | 2.59576 | 2.93557 | 3.47726 |
| 0.540 | 1.31327 | 1.59859 | 2.21056 | 2.60108 | 2.94026 | 3.47928 |
| 0.560 | 1.31715 | 1.60752 | 2.21935 | 2.60667 | 2.94514 | 3.48143 |
| 0.580 | 1.32131 | 1.61659 | 2.22834 | 2.61251 | 2.95019 | 3.48375 |
| 0.600 | 1.32576 | 1.62580 | 2.23748 | 2.61860 | 2.95543 | 3.48618 |
| 0.620 | 1.33048 | 1.63513 | 2.24679 | 2.62496 | 2.96084 | 3.48876 |
| 0.640 | 1.33547 | 1.64461 | 2.25624 | 2.63157 | 2.96643 | 3.49149 |
| 0.660 | 1.34076 | 1.65420 | 2.26581 | 2.63848 | 2.97217 | 3.49435 |
| 0.680 | 1.34633 | 1.66391 | 2.27552 | 2.64565 | 2.97809 | 3.49739 |
| 0.700 | 1.35215 | 1.67374 | 2.28532 | 2.65309 | 2.98420 | 3.50056 |
| 0.720 | 1.35826 | 1.68368 | 2.29522 | 2.66081 | 2.99046 | 3.50391 |
| 0.740 | 1.36465 | 1.69374 | 2.30519 | 2.66881 | 2.99692 | 3.50741 |
| 0.760 | 1.37129 | 1.70390 | 2.31524 | 2.67709 | 3.00355 | 3.51106 |
| 0.780 | 1.37821 | 1.71417 | 2.32535 | 2.68564 | 3.01036 | 3.51489 |
| 0.800 | 1.38538 | 1.72455 | 2.33549 | 2.69447 | 3.01733 | 3.51887 |
| 0.820 | 1.39281 | 1.73504 | 2.34568 | 2.70358 | 3.02450 | 3.52305 |
| 0.840 | 1.40050 | 1.74562 | 2.35590 | 2.71295 | 3.03183 | 3.52738 |
| 0.860 | 1.40845 | 1.75631 | 2.36615 | 2.72259 | 3.03935 | 3.53190 |
| 0.880 | 1.41663 | 1.76710 | 2.37641 | 2.73250 | 3.04707 | 3.53658 |
| 0.900 | 1.42506 | 1.77800 | 2.38668 | 2.74265 | 3.05494 | 3.54143 |
| 0.920 | 1.43373 | 1.78898 | 2.39695 | 2.75305 | 3.06300 | 3.54648 |
| 0.940 | 1.44263 | 1.80006 | 2.40722 | 2.76370 | 3.07124 | 3.55172 |
| 0.960 | 1.45176 | 1.81124 | 2.41749 | 2.77456 | 3.07967 | 3.55715 |
| 0.980 | 1.46111 | 1.82252 | 2.42775 | 2.78565 | 3.08828 | 3.56278 |
| 1.000 | 1.47069 | 1.83390 | 2.43801 | 2.79696 | 3.09707 | 3.56859 |

$H/R = 1.600 \qquad n = 4.$

| $H/L$ | $\Omega_1$ | $\Omega_2$ | $\Omega_3$ | $\Omega_4$ | $\Omega_5$ | $\Omega_6$ |
|---|---|---|---|---|---|---|
| 0. | 1.33480 | 1.50457 | 2.15965 | 2.62636 | 2.99606 | 3.58794 |
| 0.010 | 1.33480 | 1.50463 | 2.15968 | 2.62637 | 2.99609 | 3.58795 |
| 0.050 | 1.33462 | 1.50623 | 2.16081 | 2.62689 | 2.99660 | 3.58807 |
| 0.100 | 1.33417 | 1.51110 | 2.16429 | 2.62849 | 2.99823 | 3.58843 |
| 0.120 | 1.33395 | 1.51391 | 2.16633 | 2.62943 | 2.99917 | 3.58866 |
| 0.140 | 1.33374 | 1.51718 | 2.16873 | 2.63051 | 3.00030 | 3.58894 |
| 0.160 | 1.33356 | 1.52089 | 2.17149 | 2.63181 | 3.00160 | 3.58926 |
| 0.180 | 1.33343 | 1.52501 | 2.17460 | 2.63326 | 3.00305 | 3.58964 |
| 0.200 | 1.33338 | 1.52954 | 2.17808 | 2.63489 | 3.00470 | 3.59006 |
| 0.220 | 1.33340 | 1.53444 | 2.18190 | 2.63670 | 3.00652 | 3.59056 |
| 0.240 | 1.33354 | 1.53968 | 2.18606 | 2.63872 | 3.00851 | 3.59111 |
| 0.260 | 1.33381 | 1.54525 | 2.19058 | 2.64089 | 3.01065 | 3.59175 |
| 0.280 | 1.33423 | 1.55114 | 2.19541 | 2.64326 | 3.01297 | 3.59243 |
| 0.300 | 1.33480 | 1.55733 | 2.20058 | 2.64581 | 3.01548 | 3.59321 |
| 0.320 | 1.33556 | 1.56377 | 2.20607 | 2.64854 | 3.01814 | 3.59408 |
| 0.340 | 1.33650 | 1.57049 | 2.21187 | 2.65148 | 3.02098 | 3.59502 |
| 0.360 | 1.33765 | 1.57744 | 2.21797 | 2.65463 | 3.02398 | 3.59604 |
| 0.380 | 1.33901 | 1.58463 | 2.22438 | 2.65799 | 3.02715 | 3.59718 |
| 0.400 | 1.34060 | 1.59204 | 2.23107 | 2.66153 | 3.03049 | 3.59839 |
| 0.420 | 1.34242 | 1.59965 | 2.23803 | 2.66529 | 3.03401 | 3.59972 |
| 0.440 | 1.34447 | 1.60745 | 2.24527 | 2.66926 | 3.03770 | 3.60115 |
| 0.460 | 1.34678 | 1.61544 | 2.25276 | 2.67343 | 3.04154 | 3.60269 |
| 0.480 | 1.34933 | 1.62360 | 2.26051 | 2.67786 | 3.04555 | 3.60433 |
| 0.500 | 1.35215 | 1.63194 | 2.26850 | 2.68248 | 3.04975 | 3.60610 |
| 0.520 | 1.35522 | 1.64042 | 2.27670 | 2.68734 | 3.05410 | 3.60800 |
| 0.540 | 1.35856 | 1.64907 | 2.28512 | 2.69244 | 3.05864 | 3.61000 |
| 0.560 | 1.36217 | 1.65786 | 2.29375 | 2.69777 | 3.06331 | 3.61213 |
| 0.580 | 1.36604 | 1.66679 | 2.30257 | 2.70336 | 3.06820 | 3.61438 |
| 0.600 | 1.37018 | 1.67586 | 2.31157 | 2.70916 | 3.07322 | 3.61677 |
| 0.620 | 1.37460 | 1.68505 | 2.32073 | 2.71525 | 3.07840 | 3.61929 |
| 0.640 | 1.37928 | 1.69437 | 2.33005 | 2.72156 | 3.08379 | 3.62196 |
| 0.660 | 1.38424 | 1.70382 | 2.33952 | 2.72813 | 3.08933 | 3.62474 |
| 0.680 | 1.38947 | 1.71339 | 2.34910 | 2.73498 | 3.09503 | 3.62769 |
| 0.700 | 1.39495 | 1.72308 | 2.35881 | 2.74207 | 3.10090 | 3.63076 |
| 0.720 | 1.40071 | 1.73288 | 2.36863 | 2.74943 | 3.10694 | 3.63401 |
| 0.740 | 1.40673 | 1.74278 | 2.37853 | 2.75706 | 3.11316 | 3.63737 |
| 0.760 | 1.41301 | 1.75280 | 2.38853 | 2.76496 | 3.11954 | 3.64089 |
| 0.780 | 1.41956 | 1.76292 | 2.39859 | 2.77314 | 3.12611 | 3.64458 |
| 0.800 | 1.42637 | 1.77315 | 2.40871 | 2.78158 | 3.13284 | 3.64841 |
| 0.820 | 1.43342 | 1.78349 | 2.41889 | 2.79027 | 3.13974 | 3.65240 |
| 0.840 | 1.44072 | 1.79391 | 2.42911 | 2.79921 | 3.14681 | 3.65654 |
| 0.860 | 1.44828 | 1.80445 | 2.43936 | 2.80844 | 3.15406 | 3.66086 |
| 0.880 | 1.45608 | 1.81508 | 2.44964 | 2.81794 | 3.16148 | 3.66533 |
| 0.900 | 1.46411 | 1.82581 | 2.45993 | 2.82767 | 3.16909 | 3.66996 |
| 0.920 | 1.47239 | 1.83664 | 2.47024 | 2.83764 | 3.17686 | 3.67478 |
| 0.940 | 1.48090 | 1.84756 | 2.48056 | 2.84788 | 3.18484 | 3.67975 |
| 0.960 | 1.48964 | 1.85857 | 2.49089 | 2.85836 | 3.19297 | 3.68490 |
| 0.980 | 1.49860 | 1.86968 | 2.50121 | 2.86905 | 3.20127 | 3.69025 |
| 1.000 | 1.50778 | 1.88089 | 2.51152 | 2.87994 | 3.2097[illegible] | 3.69574 |

$H/R = 1.700 \qquad n = 4.$

| $H/L$ | $\Omega_1$ | $\Omega_2$ | $\Omega_3$ | $\Omega_4$ | $\Omega_5$ | $\Omega_6$ |
|---|---|---|---|---|---|---|
| 0. | 1.37999 | 1.55540 | 2.23285 | 2.71511 | 3.10000 | 3.70945 |
| 0.010 | 1.37998 | 1.55546 | 2.23290 | 2.71514 | 3.10119 | 3.70946 |
| 0.050 | 1.37981 | 1.55700 | 2.23398 | 2.71563 | 3.10171 | 3.70958 |
| 0.100 | 1.37936 | 1.56173 | 2.23736 | 2.71716 | 3.10326 | 3.70996 |
| 0.120 | 1.37915 | 1.56446 | 2.23932 | 2.71807 | 3.10417 | 3.71019 |
| 0.140 | 1.37893 | 1.56763 | 2.24165 | 2.71914 | 3.10526 | 3.71047 |
| 0.160 | 1.37875 | 1.57123 | 2.24433 | 2.72039 | 3.10652 | 3.71080 |
| 0.180 | 1.37860 | 1.57525 | 2.24734 | 2.72180 | 3.10794 | 3.71118 |
| 0.200 | 1.37851 | 1.57965 | 2.25071 | 2.72338 | 3.10953 | 3.71159 |
| 0.220 | 1.37852 | 1.58441 | 2.25442 | 2.72514 | 3.11126 | 3.71211 |
| 0.240 | 1.37861 | 1.58953 | 2.25846 | 2.72706 | 3.11318 | 3.71266 |
| 0.260 | 1.37882 | 1.59499 | 2.26283 | 2.72916 | 3.11526 | 3.71327 |
| 0.280 | 1.37916 | 1.60073 | 2.26752 | 2.73145 | 3.11750 | 3.71398 |
| 0.300 | 1.37966 | 1.60678 | 2.27254 | 2.73393 | 3.11991 | 3.71472 |
| 0.320 | 1.38032 | 1.61310 | 2.27787 | 2.73658 | 3.12248 | 3.71559 |
| 0.340 | 1.38116 | 1.61967 | 2.28350 | 2.73942 | 3.12523 | 3.71650 |
| 0.360 | 1.38219 | 1.62649 | 2.28945 | 2.74244 | 3.12812 | 3.71752 |
| 0.380 | 1.38342 | 1.63355 | 2.29567 | 2.74566 | 3.13120 | 3.71862 |
| 0.400 | 1.38487 | 1.64082 | 2.30218 | 2.74907 | 3.13444 | 3.71979 |
| 0.420 | 1.38652 | 1.64828 | 2.30896 | 2.75270 | 3.13783 | 3.72109 |
| 0.440 | 1.38842 | 1.65595 | 2.31602 | 2.75651 | 3.14139 | 3.72246 |
| 0.460 | 1.39053 | 1.66381 | 2.32333 | 2.76055 | 3.14511 | 3.72396 |
| 0.480 | 1.39290 | 1.67184 | 2.33089 | 2.76479 | 3.14898 | 3.72556 |
| 0.500 | 1.39550 | 1.68003 | 2.33869 | 2.76925 | 3.15303 | 3.72724 |
| 0.520 | 1.39835 | 1.68839 | 2.34672 | 2.77393 | 3.15725 | 3.72906 |
| 0.540 | 1.40146 | 1.69690 | 2.35496 | 2.77883 | 3.16163 | 3.73101 |
| 0.560 | 1.40483 | 1.70557 | 2.36341 | 2.78393 | 3.16615 | 3.73303 |
| 0.580 | 1.40846 | 1.71437 | 2.37206 | 2.78929 | 3.17085 | 3.73522 |
| 0.600 | 1.41233 | 1.72330 | 2.38089 | 2.79487 | 3.17572 | 3.73749 |
| 0.620 | 1.41648 | 1.73236 | 2.38990 | 2.80070 | 3.18075 | 3.73992 |
| 0.640 | 1.42089 | 1.74157 | 2.39906 | 2.80677 | 3.18593 | 3.74247 |
| 0.660 | 1.42555 | 1.75088 | 2.40838 | 2.81306 | 3.19129 | 3.74514 |
| 0.680 | 1.43048 | 1.76032 | 2.41784 | 2.81961 | 3.19680 | 3.74796 |
| 0.700 | 1.43567 | 1.76987 | 2.42742 | 2.82641 | 3.20248 | 3.75089 |
| 0.720 | 1.44111 | 1.77953 | 2.43713 | 2.83348 | 3.20832 | 3.75399 |
| 0.740 | 1.44683 | 1.78931 | 2.44694 | 2.84078 | 3.21433 | 3.75721 |
| 0.760 | 1.45278 | 1.79918 | 2.45684 | 2.84836 | 3.22052 | 3.76057 |
| 0.780 | 1.45900 | 1.80918 | 2.46681 | 2.85618 | 3.22685 | 3.76407 |
| 0.800 | 1.46547 | 1.81927 | 2.47687 | 2.86427 | 3.23335 | 3.76773 |
| 0.820 | 1.47219 | 1.82947 | 2.48698 | 2.87261 | 3.24004 | 3.77154 |
| 0.840 | 1.47915 | 1.83976 | 2.49715 | 2.88122 | 3.24686 | 3.77549 |
| 0.860 | 1.48636 | 1.85015 | 2.50737 | 2.89007 | 3.25388 | 3.77960 |
| 0.880 | 1.49381 | 1.86065 | 2.51762 | 2.89918 | 3.26106 | 3.78387 |
| 0.900 | 1.50150 | 1.87123 | 2.52789 | 2.90854 | 3.26840 | 3.78828 |
| 0.920 | 1.50942 | 1.88191 | 2.53820 | 2.91815 | 3.27592 | 3.79286 |
| 0.940 | 1.51758 | 1.89269 | 2.54852 | 2.92800 | 3.28360 | 3.79761 |
| 0.960 | 1.52595 | 1.90355 | 2.55884 | 2.93811 | 3.29146 | 3.80249 |
| 0.980 | 1.53456 | 1.91452 | 2.56917 | 2.94843 | 3.29950 | 3.80757 |
| 1.000 | 1.54337 | 1.92557 | 2.57949 | 2.95897 | 3.30771 | 3.81281 |

$H/R = 1.800 \qquad n = 4.$

| $H/L$ | $\Omega_1$ | $\Omega_2$ | $\Omega_3$ | $\Omega_4$ | $\Omega_5$ | $\Omega_6$ |
|---|---|---|---|---|---|---|
| 0. | 1.42273 | 1.60355 | 2.30203 | 2.79917 | 3.19789 | 3.82432 |
| 0.010 | 1.42272 | 1.60362 | 2.30207 | 2.79920 | 3.19790 | 3.82433 |
| 0.050 | 1.42256 | 1.60510 | 2.30312 | 2.79969 | 3.19839 | 3.82445 |
| 0.100 | 1.42212 | 1.60969 | 2.30639 | 2.80118 | 3.19990 | 3.82482 |
| 0.120 | 1.42190 | 1.61234 | 2.30831 | 2.80205 | 3.20079 | 3.82505 |
| 0.140 | 1.42169 | 1.61543 | 2.31056 | 2.80308 | 3.20185 | 3.82530 |
| 0.160 | 1.42150 | 1.61895 | 2.31316 | 2.80429 | 3.20306 | 3.82563 |
| 0.180 | 1.42134 | 1.62285 | 2.31609 | 2.80566 | 3.20443 | 3.82600 |
| 0.200 | 1.42123 | 1.62715 | 2.31937 | 2.80720 | 3.20597 | 3.82642 |
| 0.220 | 1.42119 | 1.63180 | 2.32297 | 2.80889 | 3.20766 | 3.82689 |
| 0.240 | 1.42125 | 1.63681 | 2.32688 | 2.81076 | 3.20953 | 3.82744 |
| 0.260 | 1.42142 | 1.64213 | 2.33113 | 2.81279 | 3.21153 | 3.82805 |
| 0.280 | 1.42172 | 1.64776 | 2.33570 | 2.81501 | 3.21371 | 3.82871 |
| 0.300 | 1.42214 | 1.65368 | 2.34058 | 2.81740 | 3.21604 | 3.82944 |
| 0.320 | 1.42273 | 1.65988 | 2.34576 | 2.81997 | 3.21854 | 3.83025 |
| 0.340 | 1.42347 | 1.66632 | 2.35125 | 2.82270 | 3.22120 | 3.83115 |
| 0.360 | 1.42440 | 1.67302 | 2.35702 | 2.82564 | 3.22402 | 3.83212 |
| 0.380 | 1.42551 | 1.67995 | 2.36309 | 2.82875 | 3.22699 | 3.83318 |
| 0.400 | 1.42684 | 1.68708 | 2.36944 | 2.83206 | 3.23014 | 3.83430 |
| 0.420 | 1.42836 | 1.69444 | 2.37605 | 2.83557 | 3.23343 | 3.83555 |
| 0.440 | 1.43009 | 1.70198 | 2.38294 | 2.83925 | 3.23687 | 3.83686 |
| 0.460 | 1.43206 | 1.70971 | 2.39008 | 2.84315 | 3.24048 | 3.83829 |
| 0.480 | 1.43425 | 1.71762 | 2.39747 | 2.84724 | 3.24425 | 3.83982 |
| 0.500 | 1.43668 | 1.72569 | 2.40509 | 2.85151 | 3.24818 | 3.84143 |
| 0.520 | 1.43934 | 1.73393 | 2.41294 | 2.85602 | 3.25226 | 3.84316 |
| 0.540 | 1.44225 | 1.74231 | 2.42101 | 2.86075 | 3.25652 | 3.84501 |
| 0.560 | 1.44540 | 1.75086 | 2.42930 | 2.86569 | 3.26092 | 3.84697 |
| 0.580 | 1.44880 | 1.75954 | 2.43779 | 2.87083 | 3.26548 | 3.84904 |
| 0.600 | 1.45246 | 1.76835 | 2.44647 | 2.87622 | 3.27018 | 3.85122 |
| 0.620 | 1.45636 | 1.77730 | 2.45531 | 2.88182 | 3.27507 | 3.85353 |
| 0.640 | 1.46052 | 1.78637 | 2.46431 | 2.88767 | 3.28009 | 3.85598 |
| 0.660 | 1.46493 | 1.79557 | 2.47349 | 2.89372 | 3.28529 | 3.85852 |
| 0.680 | 1.46959 | 1.80489 | 2.48280 | 2.90003 | 3.29065 | 3.86119 |
| 0.700 | 1.47451 | 1.81431 | 2.49228 | 2.90658 | 3.29614 | 3.86401 |
| 0.720 | 1.47969 | 1.82385 | 2.50185 | 2.91336 | 3.30180 | 3.86696 |
| 0.740 | 1.48511 | 1.83351 | 2.51153 | 2.92038 | 3.30763 | 3.87004 |
| 0.760 | 1.49078 | 1.84328 | 2.52134 | 2.92768 | 3.31363 | 3.87324 |
| 0.780 | 1.49670 | 1.85314 | 2.53121 | 2.93520 | 3.31978 | 3.87659 |
| 0.800 | 1.50286 | 1.86311 | 2.54118 | 2.94297 | 3.32609 | 3.88008 |
| 0.820 | 1.50929 | 1.87318 | 2.55123 | 2.95100 | 3.33255 | 3.88372 |
| 0.840 | 1.51594 | 1.88335 | 2.56132 | 2.95928 | 3.33918 | 3.88749 |
| 0.860 | 1.52284 | 1.89362 | 2.57149 | 2.96782 | 3.34598 | 3.89141 |
| 0.880 | 1.52997 | 1.90398 | 2.58168 | 2.97659 | 3.35293 | 3.89547 |
| 0.900 | 1.53734 | 1.91443 | 2.59194 | 2.98561 | 3.36006 | 3.89967 |
| 0.920 | 1.54495 | 1.92499 | 2.60220 | 2.99489 | 3.36734 | 3.90403 |
| 0.940 | 1.55277 | 1.93563 | 2.61250 | 3.00439 | 3.37479 | 3.90854 |
| 0.960 | 1.56083 | 1.94638 | 2.62281 | 3.01415 | 3.38241 | 3.91323 |
| 0.980 | 1.56910 | 1.95719 | 2.63313 | 3.02414 | 3.39019 | 3.91806 |
| 1.000 | 1.57759 | 1.96812 | 2.64345 | 3.03434 | 3.39814 | 3.92303 |

$H/R = 1.900 \qquad n = 4.$

| $H/L$ | $\Omega_1$ | $\Omega_2$ | $\Omega_3$ | $\Omega_4$ | $\Omega_5$ | $\Omega_6$ |
|---|---|---|---|---|---|---|
| 0. | 1.46327 | 1.64924 | 2.36762 | 2.87892 | 3.28904 | 3.93327 |
| 0.010 | 1.46326 | 1.64929 | 2.36766 | 2.87895 | 3.28907 | 3.93328 |
| 0.050 | 1.46310 | 1.65075 | 2.36868 | 2.87941 | 3.28954 | 3.93340 |
| 0.100 | 1.46266 | 1.65521 | 2.37186 | 2.88085 | 3.29102 | 3.93377 |
| 0.120 | 1.46245 | 1.65780 | 2.37372 | 2.88172 | 3.29189 | 3.93399 |
| 0.140 | 1.46223 | 1.66081 | 2.37592 | 2.88272 | 3.29289 | 3.93423 |
| 0.160 | 1.46204 | 1.66424 | 2.37845 | 2.88391 | 3.29408 | 3.93454 |
| 0.180 | 1.46187 | 1.66806 | 2.38129 | 2.88521 | 3.29541 | 3.93490 |
| 0.200 | 1.46175 | 1.67225 | 2.38448 | 2.88672 | 3.29691 | 3.93530 |
| 0.220 | 1.46168 | 1.67680 | 2.38798 | 2.88836 | 3.29856 | 3.93577 |
| 0.240 | 1.46172 | 1.68169 | 2.39180 | 2.89018 | 3.30038 | 3.93630 |
| 0.260 | 1.46184 | 1.68690 | 2.39594 | 2.89217 | 3.30232 | 3.93687 |
| 0.280 | 1.46208 | 1.69242 | 2.40038 | 2.89432 | 3.30444 | 3.93752 |
| 0.300 | 1.46246 | 1.69823 | 2.40514 | 2.89664 | 3.30671 | 3.93823 |
| 0.320 | 1.46298 | 1.70431 | 2.41020 | 2.89910 | 3.30915 | 3.93901 |
| 0.340 | 1.46365 | 1.71065 | 2.41554 | 2.90179 | 3.31174 | 3.93985 |
| 0.360 | 1.46448 | 1.71721 | 2.42118 | 2.90462 | 3.31447 | 3.94081 |
| 0.380 | 1.46550 | 1.72403 | 2.42710 | 2.90764 | 3.31737 | 3.94180 |
| 0.400 | 1.46671 | 1.73106 | 2.43330 | 2.91085 | 3.32041 | 3.94290 |
| 0.420 | 1.46812 | 1.73828 | 2.43976 | 2.91422 | 3.32361 | 3.94408 |
| 0.440 | 1.46973 | 1.74572 | 2.44649 | 2.91779 | 3.32696 | 3.94533 |
| 0.460 | 1.47155 | 1.75333 | 2.45347 | 2.92158 | 3.33048 | 3.94670 |
| 0.480 | 1.47358 | 1.76112 | 2.46070 | 2.92553 | 3.33414 | 3.94818 |
| 0.500 | 1.47586 | 1.76909 | 2.46817 | 2.92969 | 3.33795 | 3.94973 |
| 0.520 | 1.47834 | 1.77721 | 2.47586 | 2.93405 | 3.34193 | 3.95139 |
| 0.540 | 1.48108 | 1.78549 | 2.48377 | 2.93859 | 3.34605 | 3.95315 |
| 0.560 | 1.48405 | 1.79391 | 2.49190 | 2.94338 | 3.35035 | 3.95500 |
| 0.580 | 1.48724 | 1.80248 | 2.50022 | 2.94836 | 3.35477 | 3.95698 |
| 0.600 | 1.49069 | 1.81119 | 2.50874 | 2.95355 | 3.35938 | 3.95908 |
| 0.620 | 1.49438 | 1.82002 | 2.51745 | 2.95897 | 3.36411 | 3.96129 |
| 0.640 | 1.49833 | 1.82898 | 2.52631 | 2.96458 | 3.36902 | 3.96361 |
| 0.660 | 1.50250 | 1.83808 | 2.53533 | 2.97045 | 3.37406 | 3.96606 |
| 0.680 | 1.50693 | 1.84728 | 2.54454 | 2.97653 | 3.37926 | 3.96863 |
| 0.700 | 1.51162 | 1.85660 | 2.55386 | 2.98285 | 3.38460 | 3.97131 |
| 0.720 | 1.51654 | 1.86603 | 2.56331 | 2.98940 | 3.39012 | 3.97413 |
| 0.740 | 1.52170 | 1.87558 | 2.57288 | 2.99619 | 3.39579 | 3.97707 |
| 0.760 | 1.52711 | 1.88523 | 2.58256 | 3.00321 | 3.40161 | 3.98012 |
| 0.780 | 1.53278 | 1.89499 | 2.59236 | 3.01047 | 3.40758 | 3.98334 |
| 0.800 | 1.53867 | 1.90484 | 2.60224 | 3.01798 | 3.41372 | 3.98667 |
| 0.820 | 1.54481 | 1.91480 | 2.61220 | 3.02572 | 3.42002 | 3.99016 |
| 0.840 | 1.55118 | 1.92484 | 2.62222 | 3.03371 | 3.42645 | 3.99376 |
| 0.860 | 1.55780 | 1.93500 | 2.63233 | 3.04194 | 3.43307 | 3.99750 |
| 0.880 | 1.56465 | 1.94525 | 2.64246 | 3.05041 | 3.43983 | 4.00139 |
| 0.900 | 1.57173 | 1.95558 | 2.65267 | 3.05915 | 3.44674 | 4.00542 |
| 0.920 | 1.57904 | 1.96602 | 2.66288 | 3.06810 | 3.45381 | 4.00959 |
| 0.940 | 1.58657 | 1.97655 | 2.67315 | 3.07731 | 3.46104 | 4.01392 |
| 0.960 | 1.59434 | 1.98716 | 2.68344 | 3.08675 | 3.46845 | 4.01838 |
| 0.980 | 1.60231 | 1.99787 | 2.69375 | 3.09641 | 3.47601 | 4.02300 |
| 1.000 | 1.61050 | 2.00866 | 2.70406 | 3.10630 | 3.48373 | 4.02776 |

$H/R = 2.000 \qquad n = 4.$

| $H/L$ | $\Omega_1$ | $\Omega_2$ | $\Omega_3$ | $\Omega_4$ | $\Omega_5$ | $\Omega_6$ |
|---|---|---|---|---|---|---|
| 0. | 1.50178 | 1.69263 | 2.42993 | 2.95468 | 3.37560 | 4.03678 |
| 0.010 | 1.50177 | 1.69270 | 2.42997 | 2.95471 | 3.37563 | 4.03679 |
| 0.050 | 1.50161 | 1.69411 | 2.43096 | 2.95517 | 3.37608 | 4.03691 |
| 0.100 | 1.50118 | 1.69847 | 2.43406 | 2.95658 | 3.37752 | 4.03726 |
| 0.120 | 1.50097 | 1.70099 | 2.43588 | 2.95740 | 3.37836 | 4.03748 |
| 0.140 | 1.50075 | 1.70393 | 2.43801 | 2.95839 | 3.37935 | 4.03772 |
| 0.160 | 1.50054 | 1.70727 | 2.44048 | 2.95954 | 3.38050 | 4.03801 |
| 0.180 | 1.50037 | 1.71101 | 2.44326 | 2.96083 | 3.38182 | 4.03838 |
| 0.200 | 1.50024 | 1.71511 | 2.44637 | 2.96228 | 3.38327 | 4.03877 |
| 0.220 | 1.50017 | 1.71957 | 2.44978 | 2.96388 | 3.38487 | 4.03921 |
| 0.240 | 1.50017 | 1.72436 | 2.45350 | 2.96565 | 3.38664 | 4.03972 |
| 0.260 | 1.50025 | 1.72948 | 2.45754 | 2.96757 | 3.38853 | 4.04028 |
| 0.280 | 1.50046 | 1.73490 | 2.46189 | 2.96967 | 3.39061 | 4.04089 |
| 0.300 | 1.50078 | 1.74060 | 2.46652 | 2.97193 | 3.39281 | 4.04158 |
| 0.320 | 1.50124 | 1.74657 | 2.47146 | 2.97433 | 3.39519 | 4.04233 |
| 0.340 | 1.50184 | 1.75279 | 2.47668 | 2.97692 | 3.39771 | 4.04314 |
| 0.360 | 1.50261 | 1.75926 | 2.48220 | 2.97968 | 3.40037 | 4.04404 |
| 0.380 | 1.50354 | 1.76596 | 2.48798 | 2.98262 | 3.40319 | 4.04502 |
| 0.400 | 1.50465 | 1.77289 | 2.49404 | 2.98573 | 3.40616 | 4.04607 |
| 0.420 | 1.50594 | 1.78000 | 2.50035 | 2.98902 | 3.40929 | 4.04719 |
| 0.440 | 1.50744 | 1.78733 | 2.50695 | 2.99248 | 3.41257 | 4.04841 |
| 0.460 | 1.50914 | 1.79483 | 2.51379 | 2.99616 | 3.41597 | 4.04973 |
| 0.480 | 1.51105 | 1.80252 | 2.52087 | 2.99999 | 3.41956 | 4.05112 |
| 0.500 | 1.51318 | 1.81039 | 2.52817 | 3.00402 | 3.42327 | 4.05264 |
| 0.520 | 1.51552 | 1.81840 | 2.53573 | 3.00825 | 3.42714 | 4.05422 |
| 0.540 | 1.51808 | 1.82657 | 2.54348 | 3.01266 | 3.43118 | 4.05591 |
| 0.560 | 1.52089 | 1.83489 | 2.55146 | 3.01730 | 3.43535 | 4.05771 |
| 0.580 | 1.52392 | 1.84337 | 2.55964 | 3.02211 | 3.43967 | 4.05959 |
| 0.600 | 1.52718 | 1.85196 | 2.56801 | 3.02716 | 3.44414 | 4.06162 |
| 0.620 | 1.53068 | 1.86070 | 2.57658 | 3.03238 | 3.44876 | 4.06372 |
| 0.640 | 1.53442 | 1.86956 | 2.58530 | 3.03785 | 3.45352 | 4.06597 |
| 0.660 | 1.53840 | 1.87855 | 2.59421 | 3.04352 | 3.45845 | 4.06831 |
| 0.680 | 1.54262 | 1.88764 | 2.60326 | 3.04941 | 3.46352 | 4.07075 |
| 0.700 | 1.54708 | 1.89687 | 2.61246 | 3.05552 | 3.46874 | 4.07334 |
| 0.720 | 1.55177 | 1.90620 | 2.62179 | 3.06186 | 3.47411 | 4.07603 |
| 0.740 | 1.55671 | 1.91564 | 2.63126 | 3.06841 | 3.47963 | 4.07886 |
| 0.760 | 1.56188 | 1.92519 | 2.64085 | 3.07522 | 3.48529 | 4.08181 |
| 0.780 | 1.56730 | 1.93483 | 2.65052 | 3.08225 | 3.49113 | 4.08486 |
| 0.800 | 1.57295 | 1.94459 | 2.66032 | 3.08949 | 3.49710 | 4.08809 |
| 0.820 | 1.57885 | 1.95444 | 2.67020 | 3.09700 | 3.50324 | 4.09141 |
| 0.840 | 1.58497 | 1.96440 | 2.68014 | 3.10474 | 3.50951 | 4.09487 |
| 0.860 | 1.59133 | 1.97445 | 2.69016 | 3.11268 | 3.51594 | 4.09846 |
| 0.880 | 1.59792 | 1.98458 | 2.70027 | 3.12089 | 3.52251 | 4.10220 |
| 0.900 | 1.60473 | 1.99482 | 2.71039 | 3.12935 | 3.52926 | 4.10605 |
| 0.920 | 1.61177 | 2.00513 | 2.72058 | 3.13802 | 3.53616 | 4.11006 |
| 0.940 | 1.61904 | 2.01556 | 2.73081 | 3.14694 | 3.54319 | 4.11421 |
| 0.960 | 1.62652 | 2.02606 | 2.74106 | 3.15609 | 3.55039 | 4.11851 |
| 0.980 | 1.63422 | 2.03665 | 2.75134 | 3.16547 | 3.55777 | 4.12292 |
| 1.000 | 1.64214 | 2.04733 | 2.76165 | 3.17507 | 3.56527 | 4.12751 |

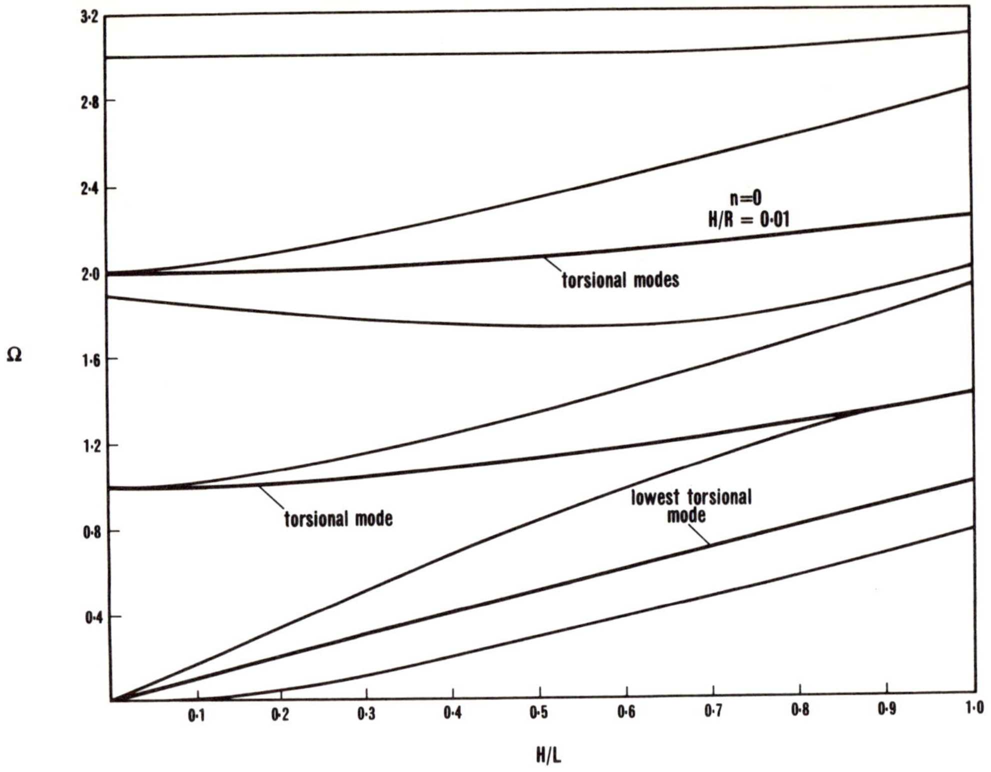
3·2
2·8
2·4
2·0
1·6
1·2
0·8
0·4
Ω
0·1
0·2
0·3
0·4
0·5
0·6
0·7
0·8
0·9
1·0
H/L
n=0
H/R = 0·01
torsional modes
torsional mode
lowest torsional mode

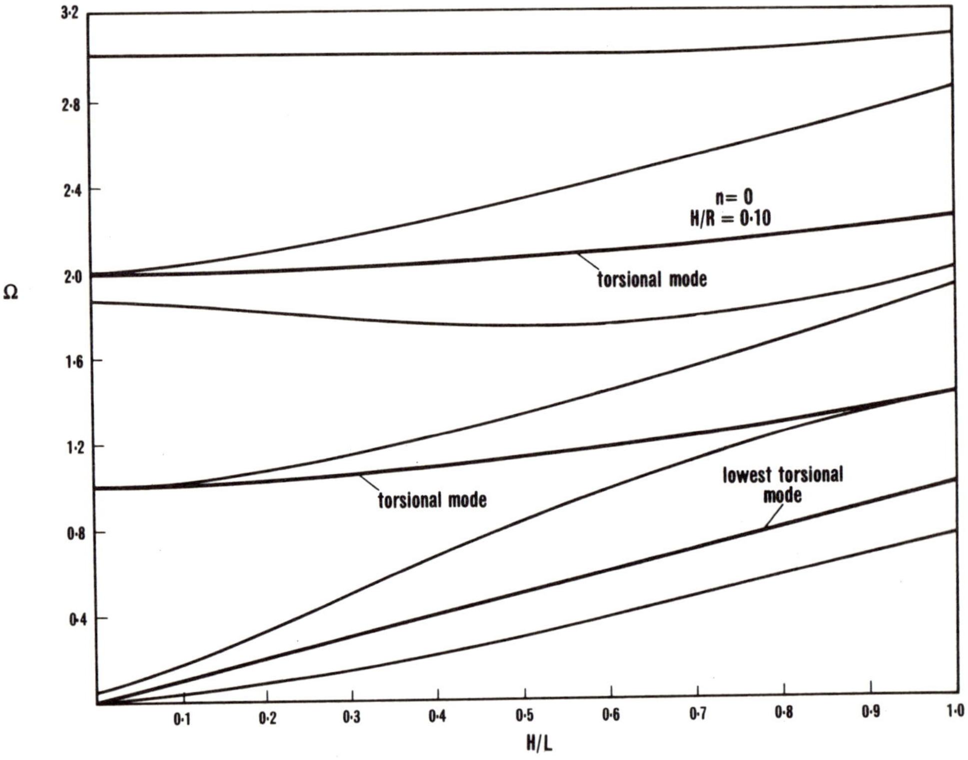
3·2
2·8
2·4
2·0
1·6
1·2
0·8
0·4
Ω
0·1
0·2
0·3
0·4
0·5
0·6
0·7
0·8
0·9
1·0
H/L
n= 0
H/R = 0·10
torsional mode
torsional mode
lowest torsional mode

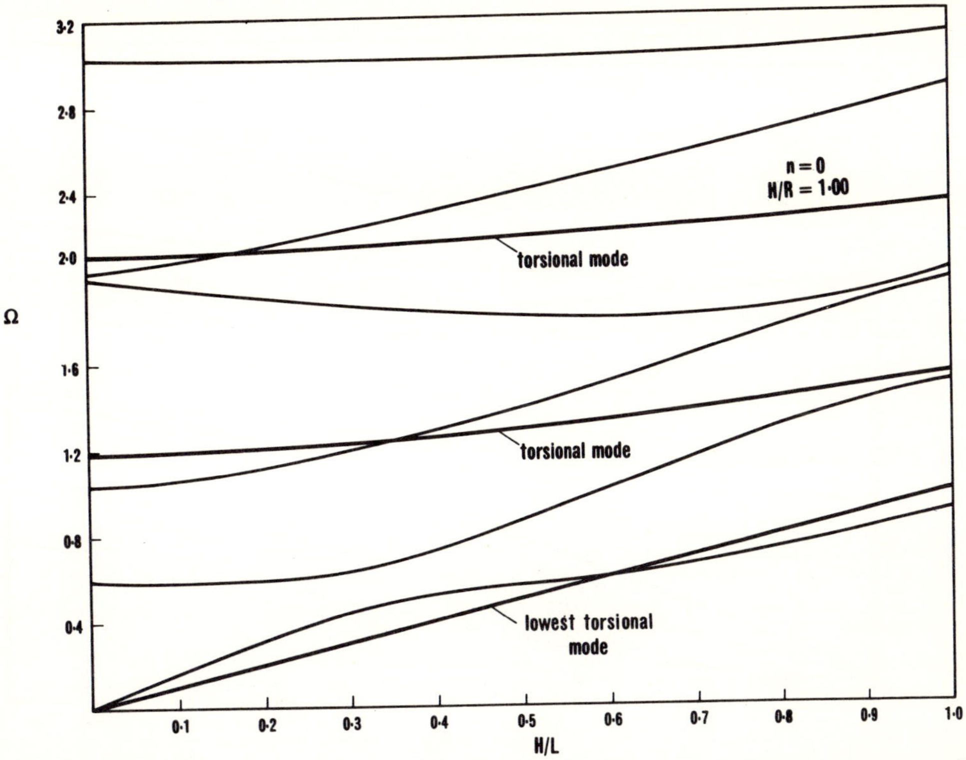
3·2
2·8
2·4
2·0
Ω
1·6
1·2
0·8
0·4
n = 0
H/R = 1·00
torsional mode
torsional mode
lowest torsional mode
0·1
0·2
0·3
0·4
0·5
0·6
0·7
0·8
0·9
1·0
H/L

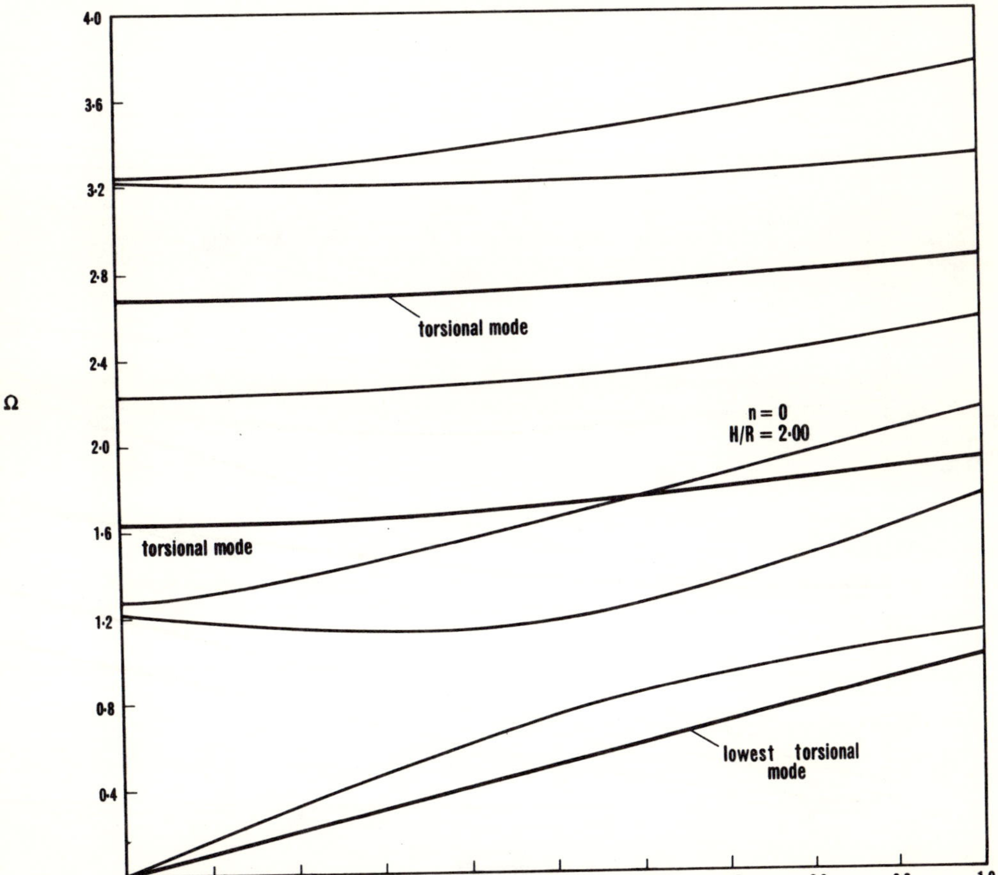
4·0
3·6
3·2
2·8
2·4
Ω
2·0
1·6
1·2
0·8
0·4
torsional mode
n = 0
H/R = 2·00
torsional mode
lowest torsional mode
0·1
0·2
0·3
0·4
0·5
0·6
0·7
0·8
0·9
1·0
H/L

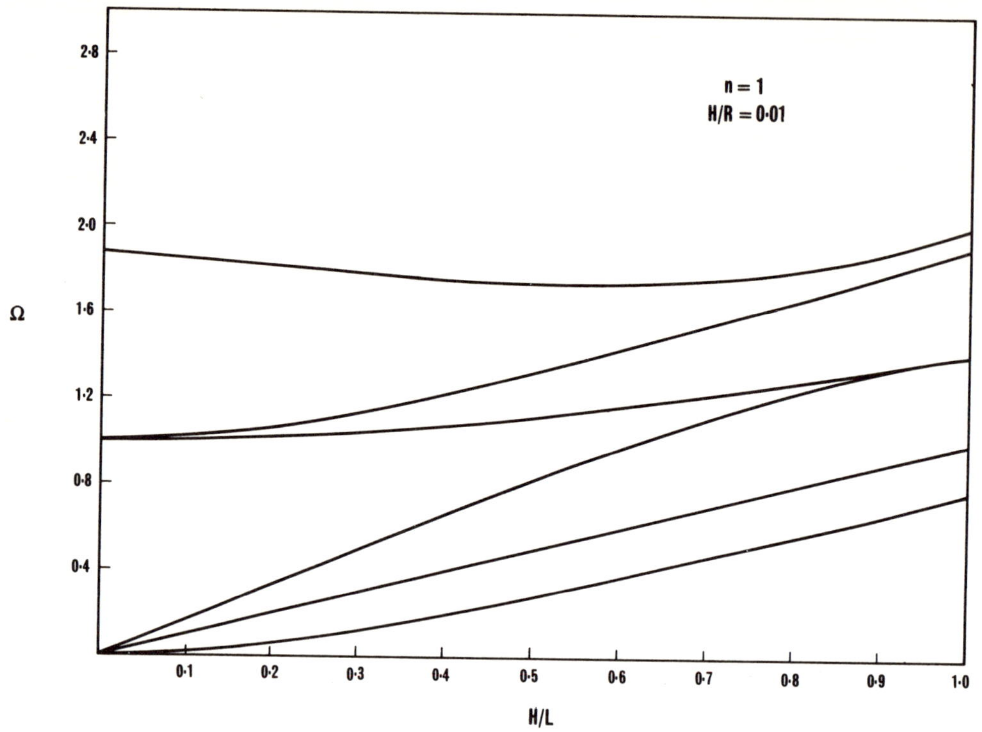
n = 1
H/R = 0·01
2·8
2·4
2·0
1·6
1·2
0·8
0·4
Ω
0·1
0·2
0·3
0·4
0·5
0·6
0·7
0·8
0·9
1·0
H/L

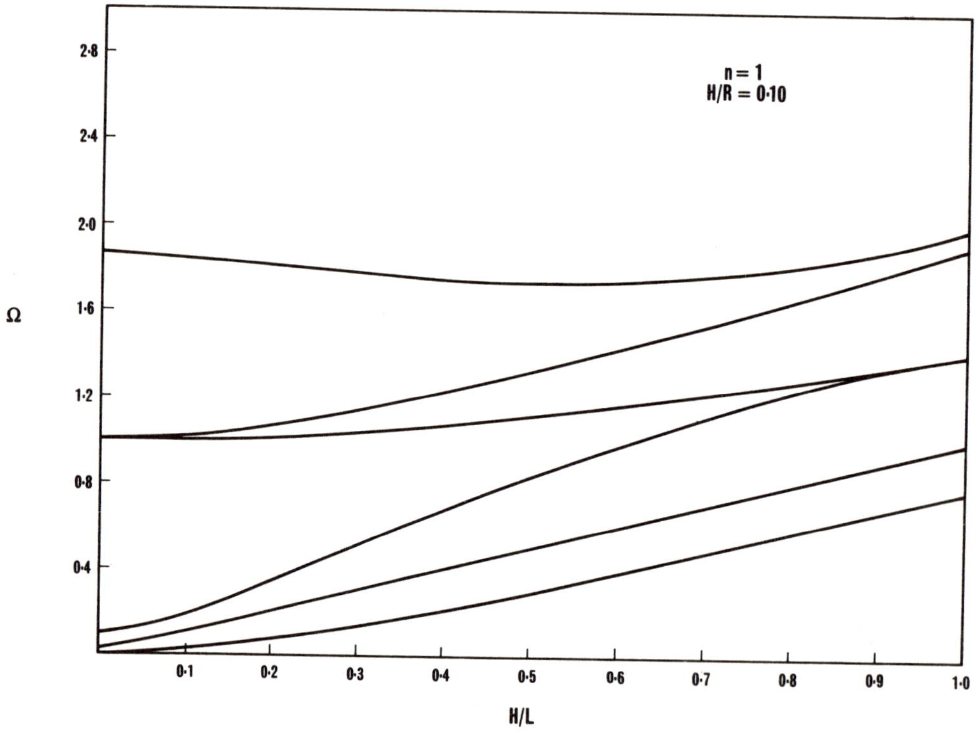
n = 1
H/R = 0·10
2·8
2·4
2·0
1·6
1·2
0·8
0·4
Ω
0·1
0·2
0·3
0·4
0·5
0·6
0·7
0·8
0·9
1·0
H/L

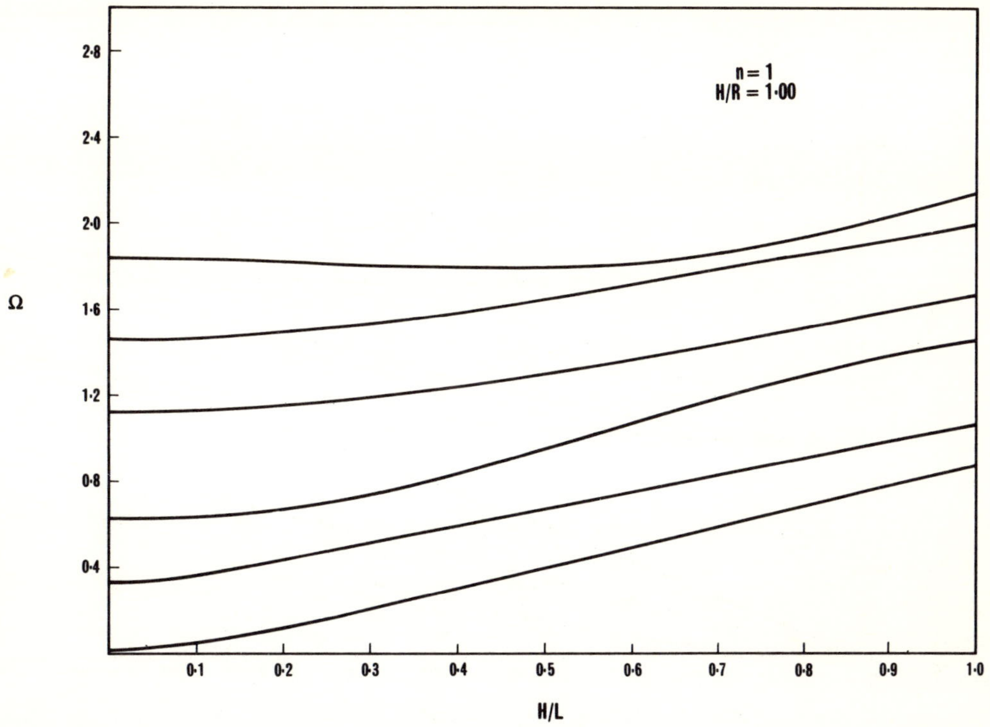
n = 1
H/R = 1·00
Ω
2·8
2·4
2·0
1·6
1·2
0·8
0·4
0·1
0·2
0·3
0·4
0·5
0·6
0·7
0·8
0·9
1·0
H/L

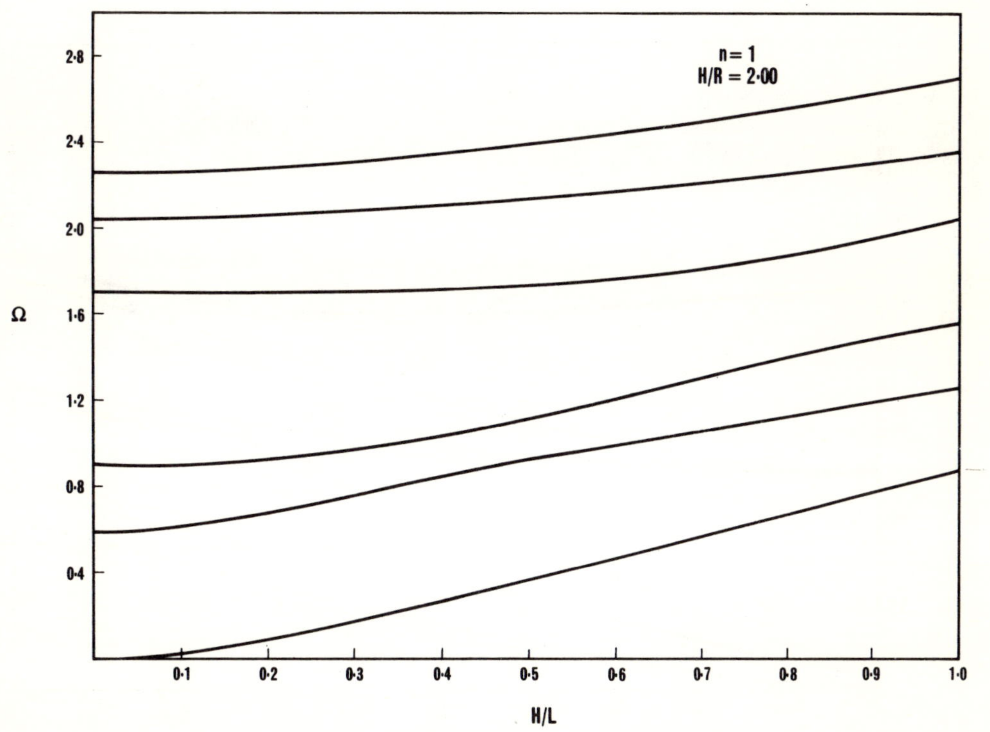
n = 1
H/R = 2·00
Ω
2·8
2·4
2·0
1·6
1·2
0·8
0·4
0·1
0·2
0·3
0·4
0·5
0·6
0·7
0·8
0·9
1·0
H/L

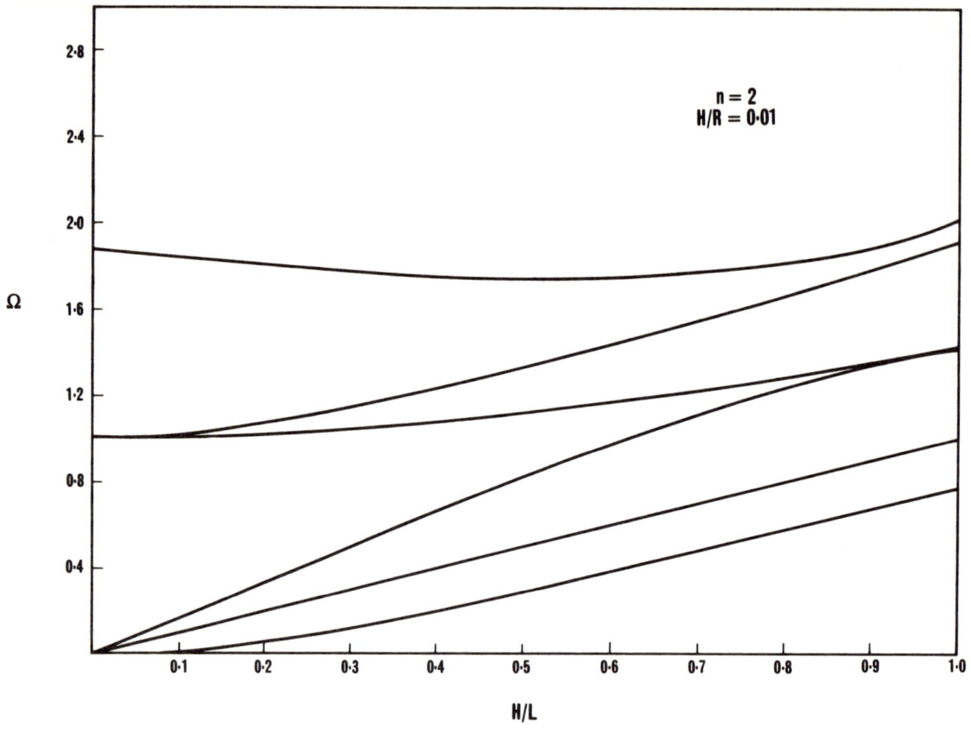
n = 2
H/R = 0·01
2·8
2·4
2·0
1·6
1·2
0·8
0·4
Ω
0·1
0·2
0·3
0·4
0·5
0·6
0·7
0·8
0·9
1·0
H/L

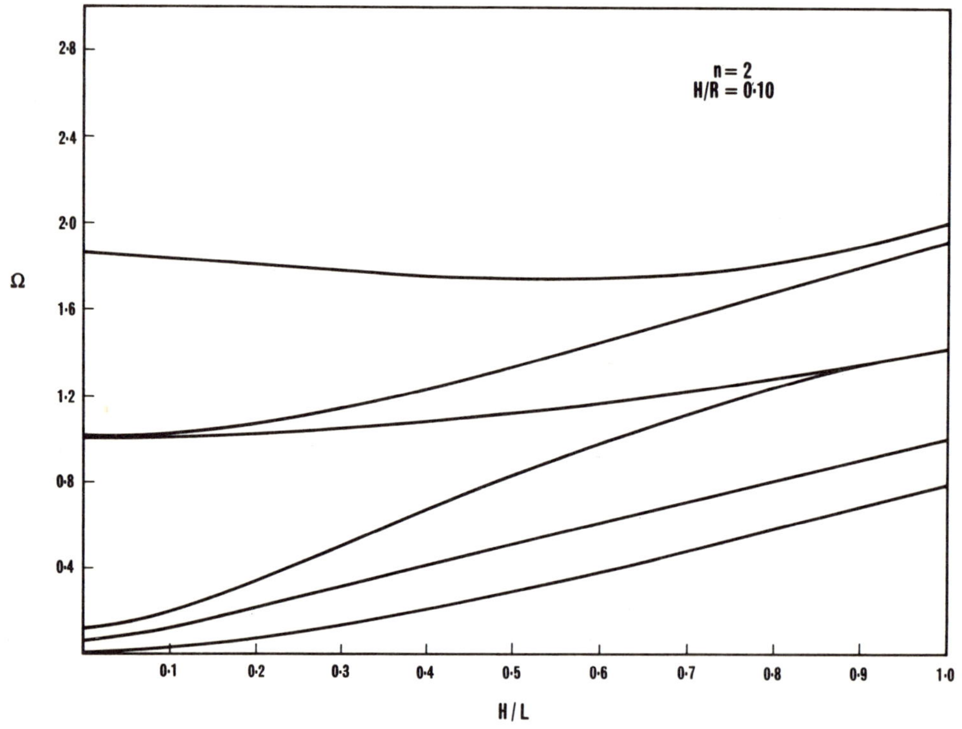
n = 2
H/R = 0·10
2·8
2·4
2·0
1·6
1·2
0·8
0·4
Ω
0·1
0·2
0·3
0·4
0·5
0·6
0·7
0·8
0·9
1·0
H/L

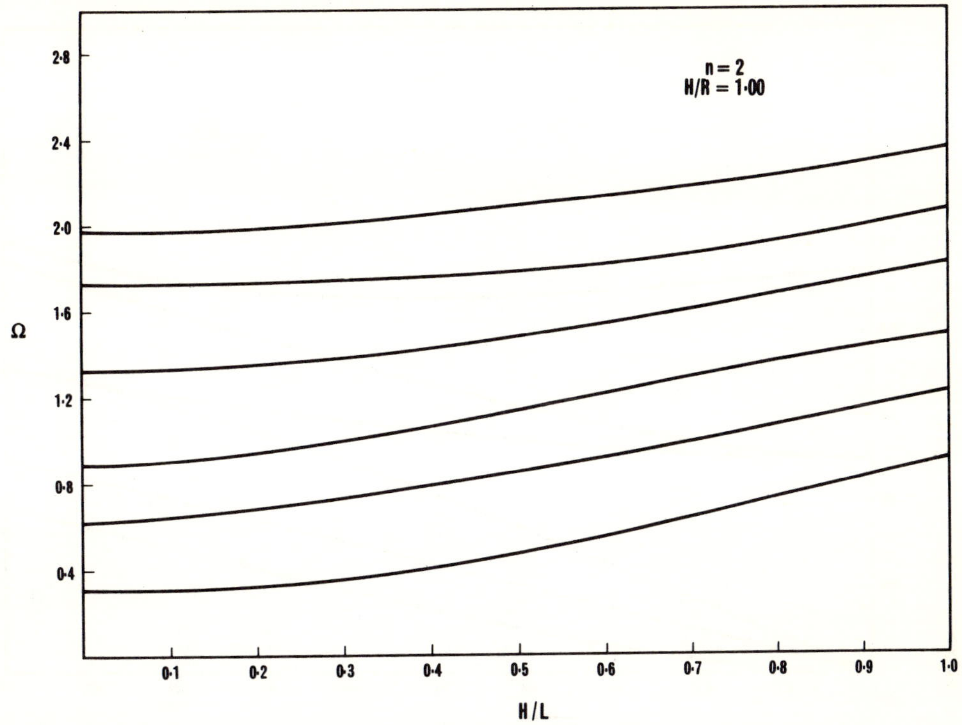
n = 2
H/R = 1·00
Ω
2·8
2·4
2·0
1·6
1·2
0·8
0·4
0·1
0·2
0·3
0·4
0·5
0·6
0·7
0·8
0·9
1·0
H/L

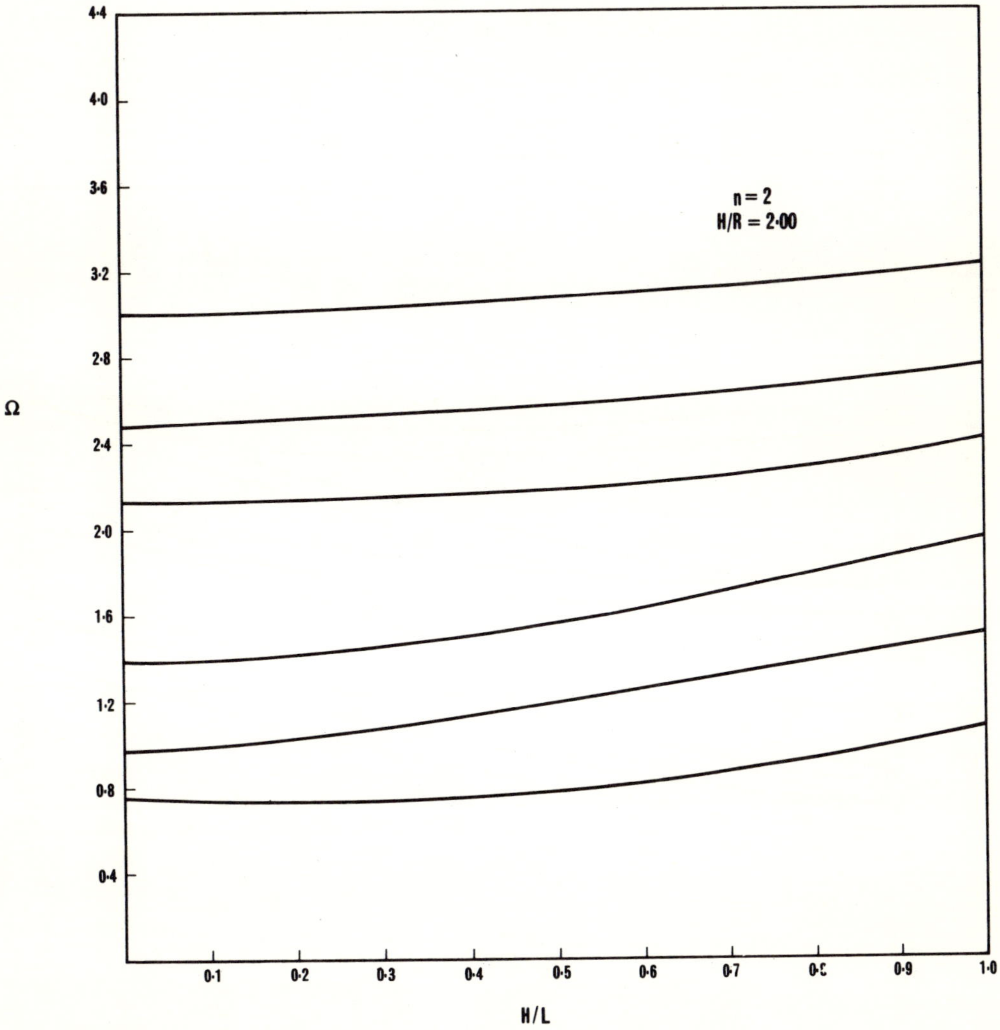
n = 2
H/R = 2·00
Ω
4·4
4·0
3·6
3·2
2·8
2·4
2·0
1·6
1·2
0·8
0·4
0·1
0·2
0·3
0·4
0·5
0·6
0·7
0·9
1·0
H/L

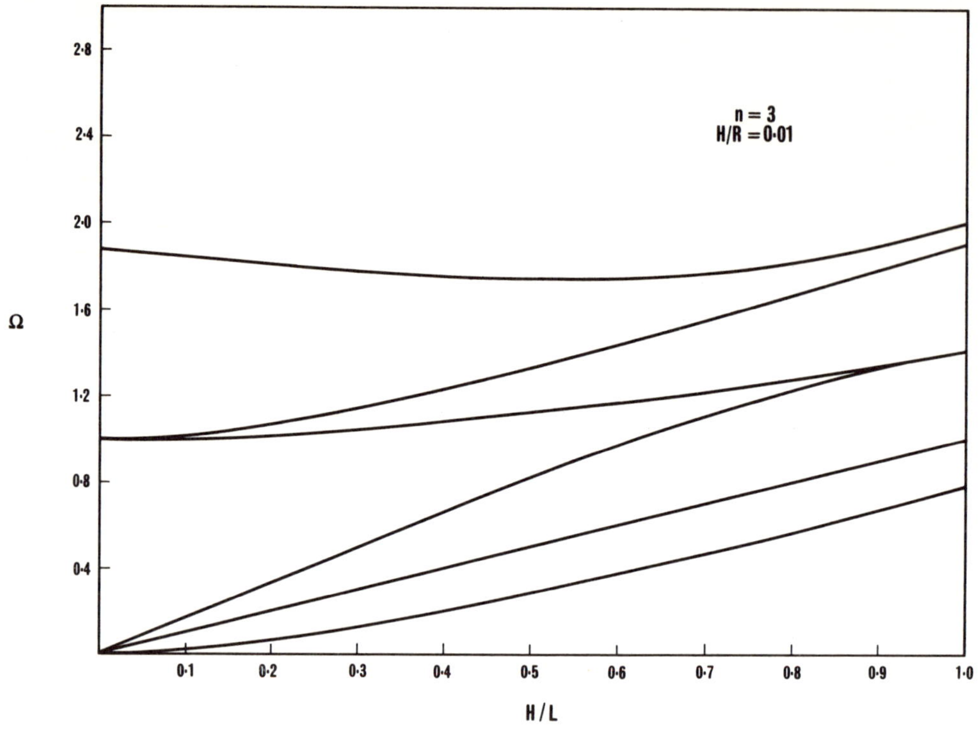
n = 3
H/R = 0·01
2·8
2·4
2·0
1·6
1·2
0·8
0·4
Ω
0·1
0·2
0·3
0·4
0·5
0·6
0·7
0·8
0·9
1·0
H / L

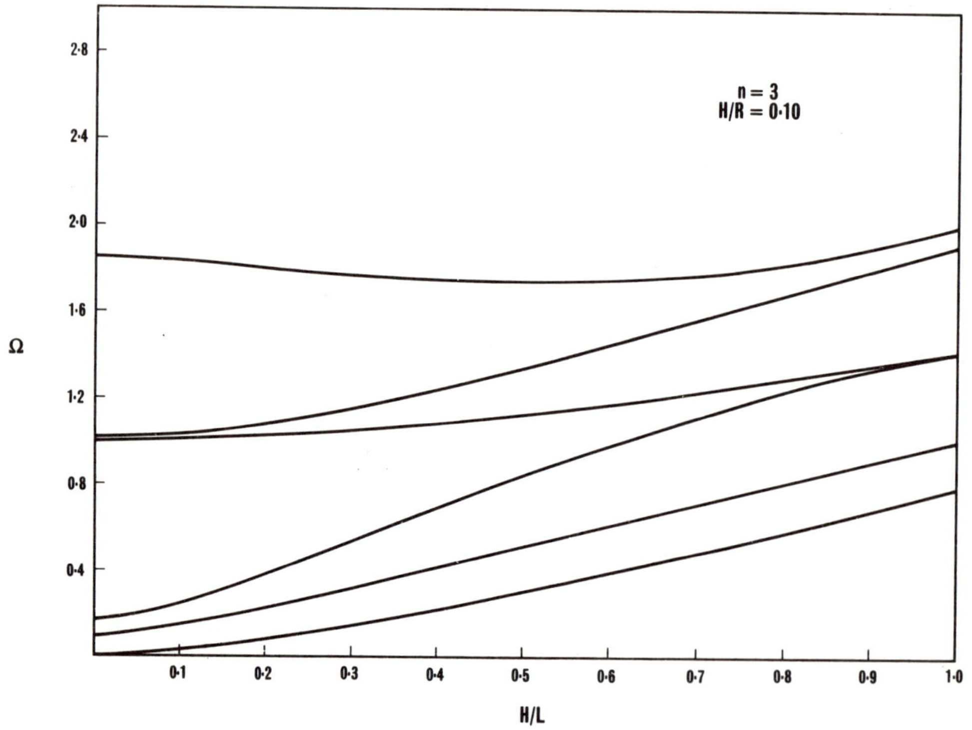
n = 3
H/R = 0·10
2·8
2·4
2·0
1·6
1·2
0·8
0·4
Ω
0·1
0·2
0·3
0·4
0·5
0·6
0·7
0·8
0·9
1·0
H/L

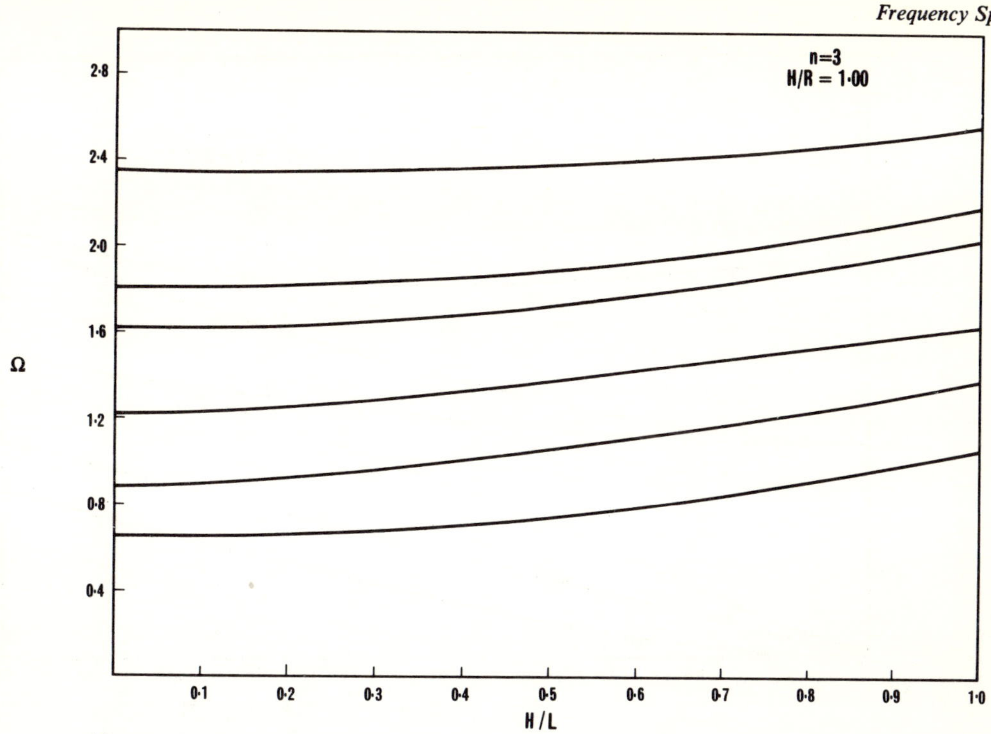
n=3
H/R = 1·00
2·8
2·4
2·0
1·6
1·2
0·8
0·4
Ω
0·1
0·2
0·3
0·4
0·5
0·6
0·7
0·8
0·9
1·0
H / L

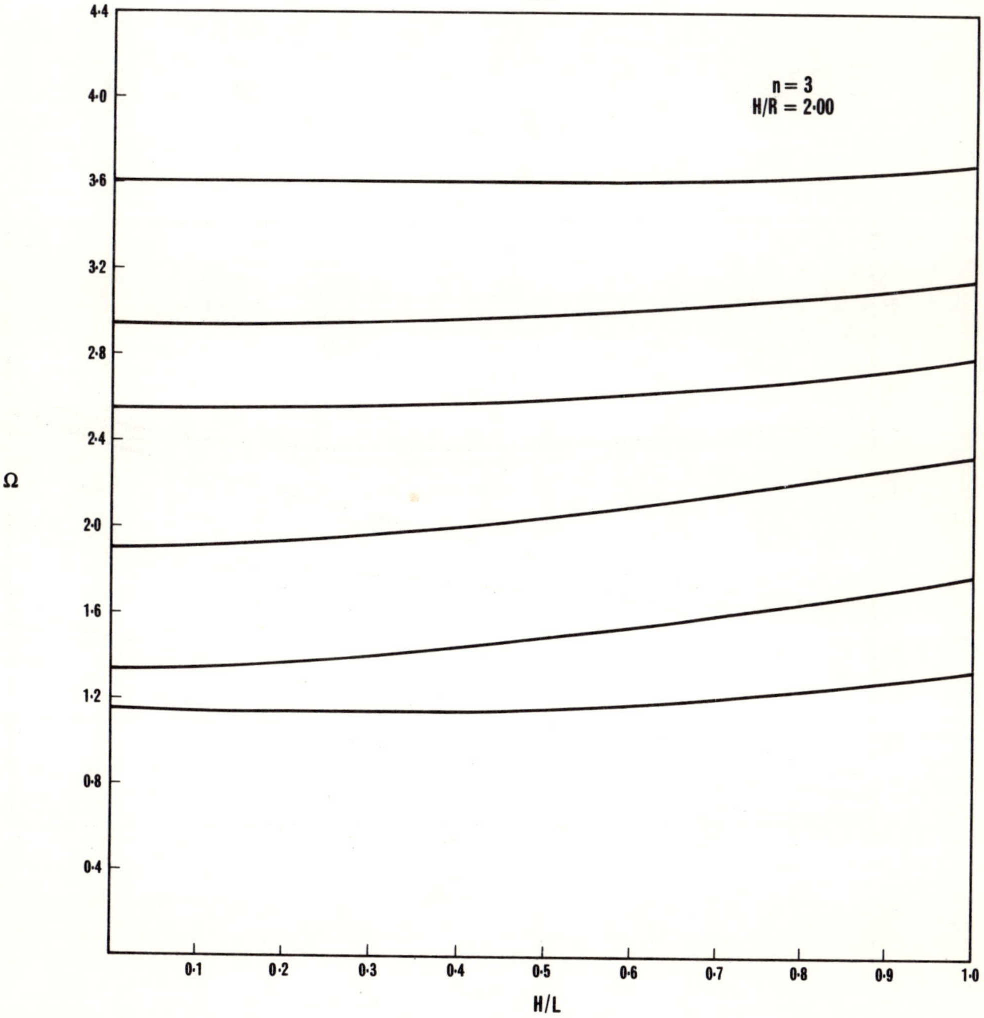
n = 3
H/R = 2·00
4·4
4·0
3·6
3·2
2·8
2·4
2·0
1·6
1·2
0·8
0·4
Ω
0·1
0·2
0·3
0·4
0·5
0·6
0·7
0·8
0·9
1·0
H/L

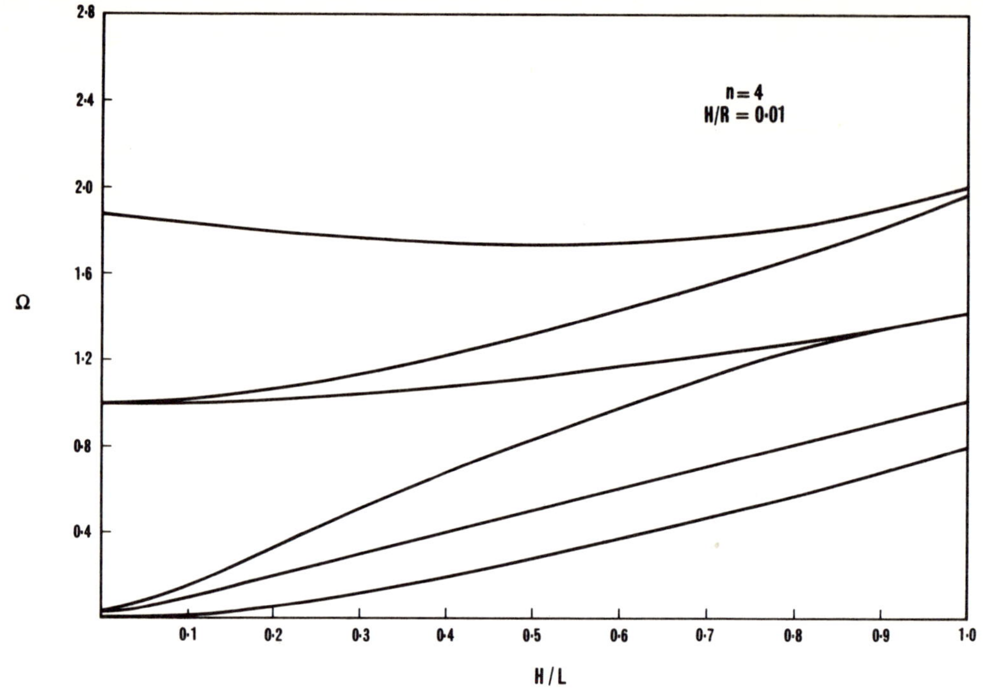
n = 4
H/R = 0·01
2·8
2·4
2·0
1·6
1·2
0·8
0·4
Ω
0·1
0·2
0·3
0·4
0·5
0·6
0·7
0·8
0·9
1·0
H/L

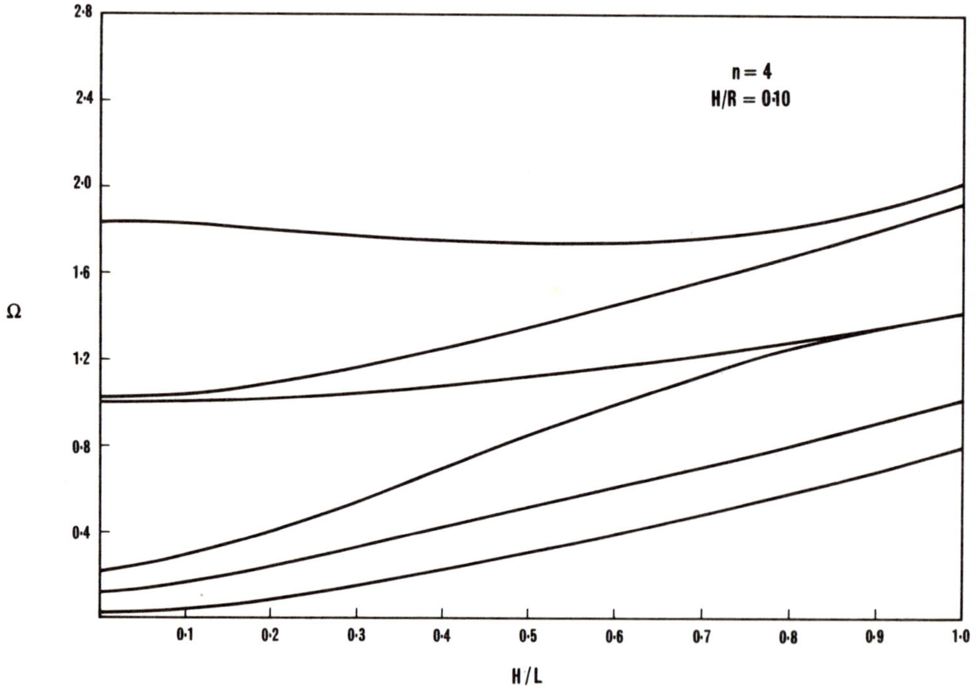
n = 4
H/R = 0·10
2·8
2·4
2·0
1·6
1·2
0·8
0·4
Ω
0·1
0·2
0·3
0·4
0·5
0·6
0·7
0·8
0·9
1·0
H/L

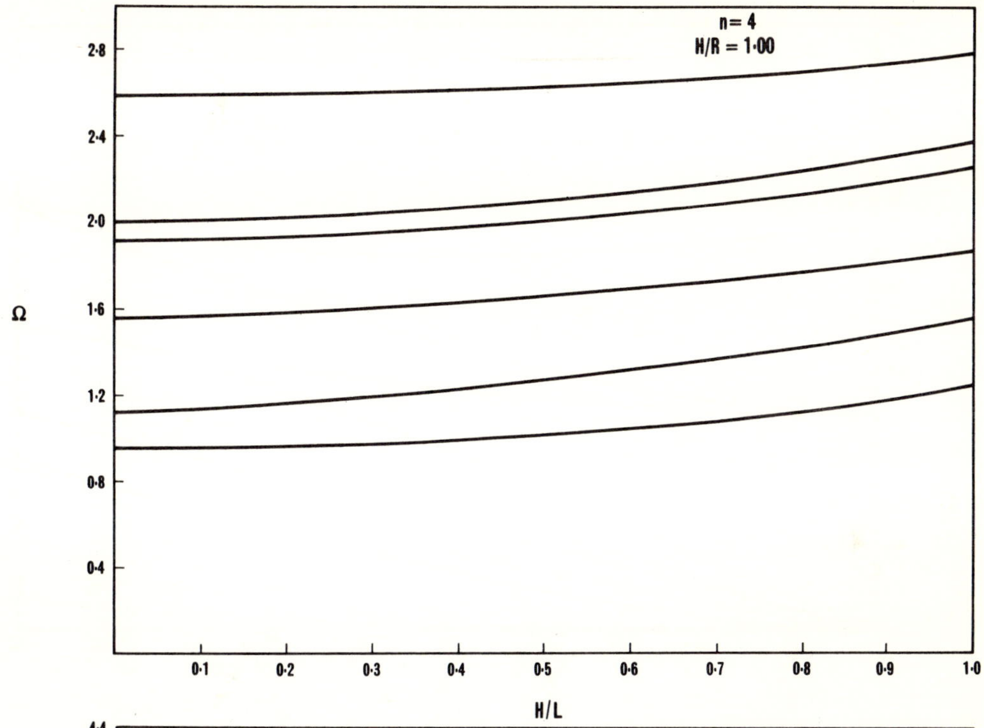
n= 4
H/R = 1·00
2·8
2·4
2·0
1·6
1·2
0·8
0·4
Ω
0·1
0·2
0·3
0·4
0·5
0·6
0·7
0·8
0·9
1·0
H/L

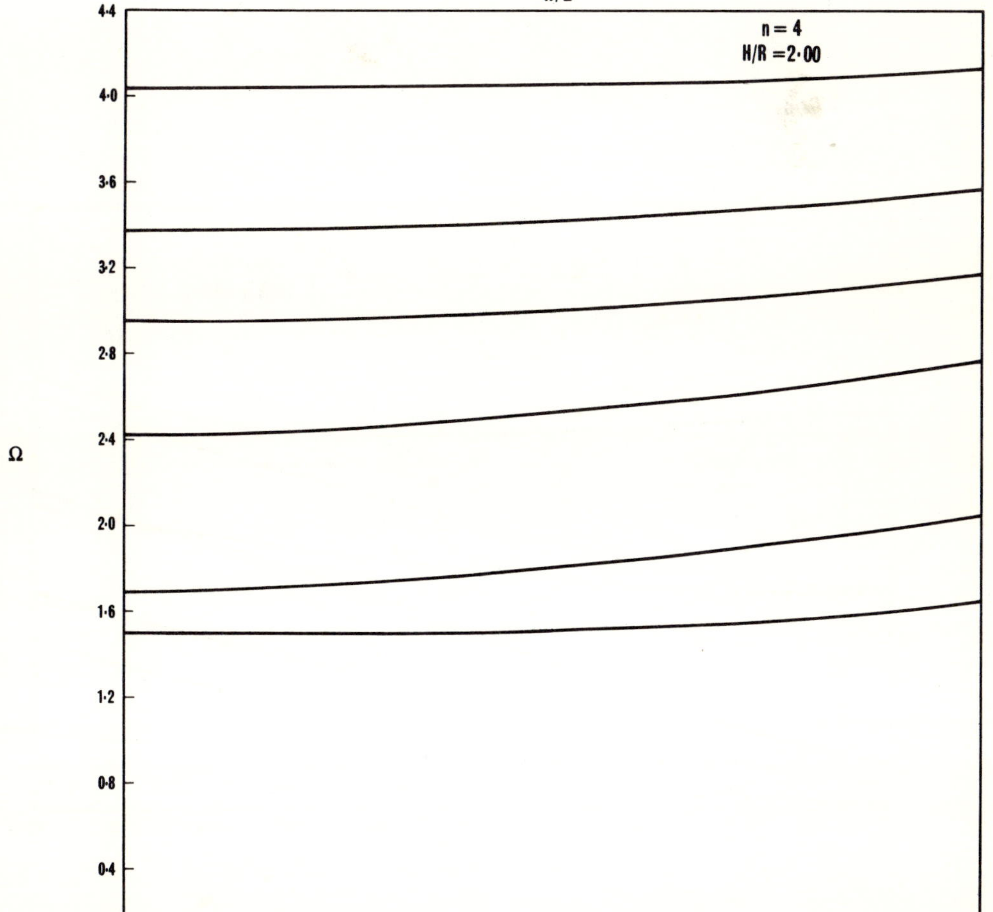
n = 4
H/R =2·00
4·4
4·0
3·6
3·2
2·8
2·4
2·0
1·6
1·2
0·8
0·4
Ω
0·1
0·2
0·3
0·4
0·5
0·6
0·7
0·8
0·9
1·0
H/L

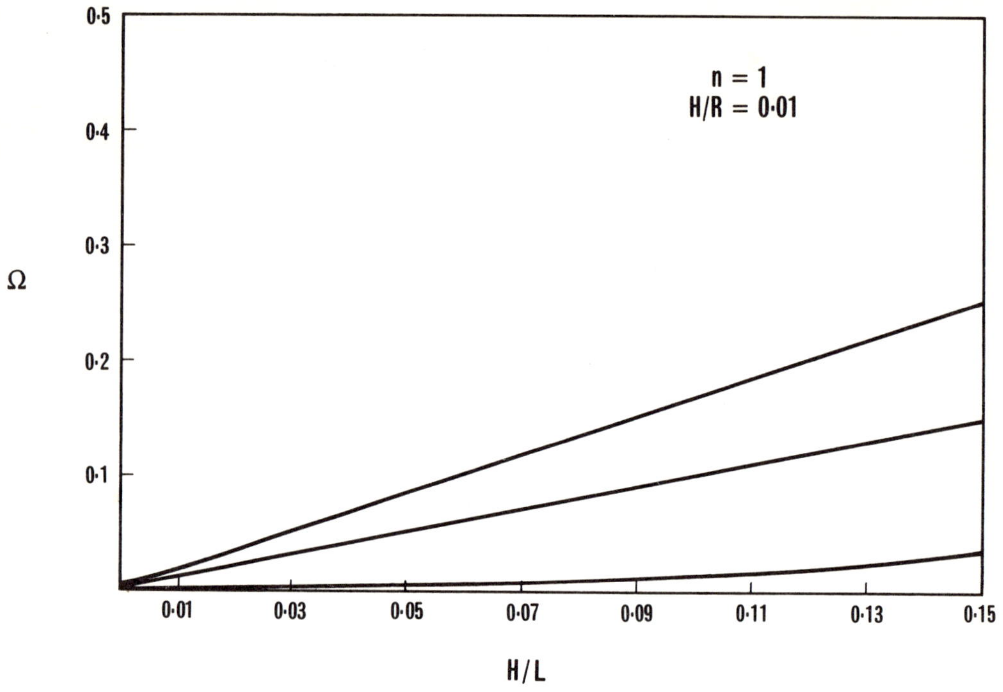
n = 1
H/R = 0·01
Ω
0·5
0·4
0·3
0·2
0·1
0·01
0·03
0·05
0·07
0·09
0·11
0·13
0·15
H/L

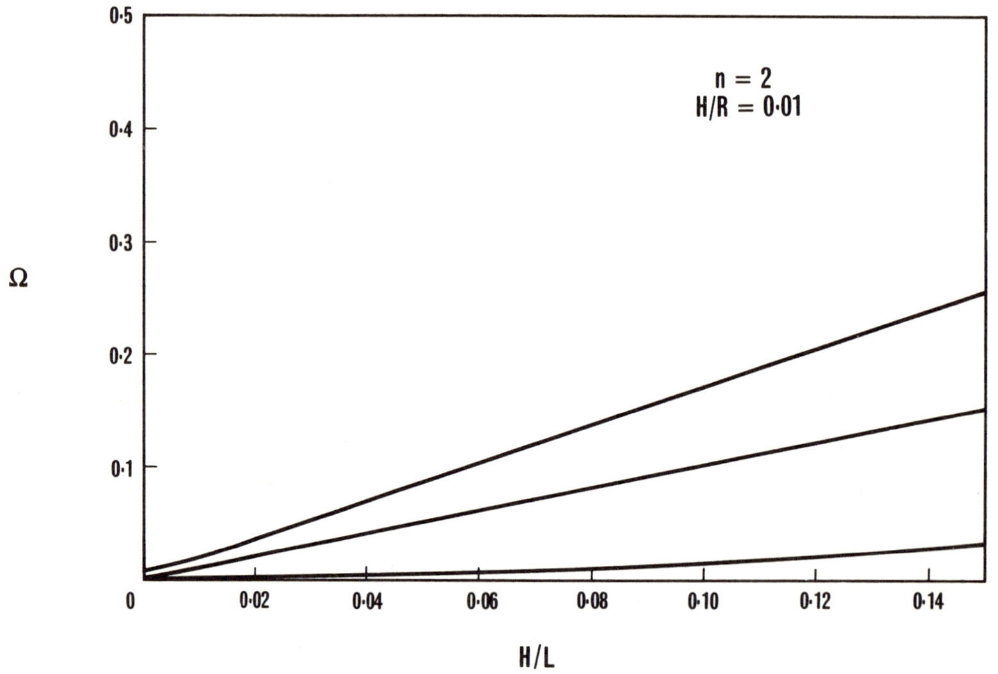
n = 2
H/R = 0·01
Ω
0·5
0·4
0·3
0·2
0·1
0
0·02
0·04
0·06
0·08
0·10
0·12
0·14
H/L

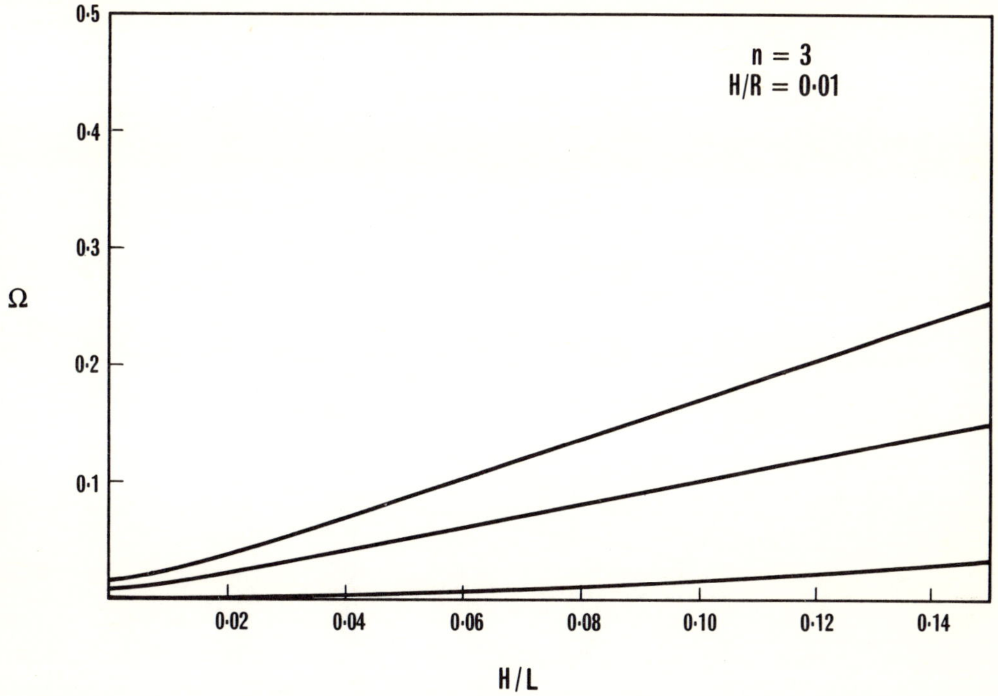
n = 3
H/R = 0·01
Ω
0·5
0·4
0·3
0·2
0·1
0·02
0·04
0·06
0·08
0·10
0·12
0·14
H/L

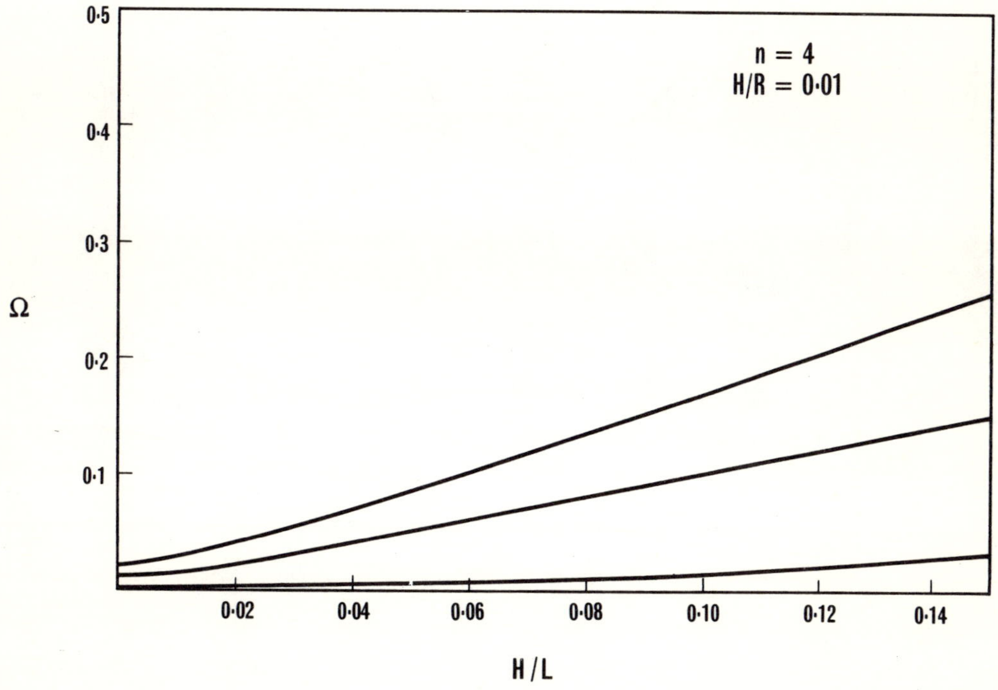
n = 4
H/R = 0·01
Ω
0·5
0·4
0·3
0·2
0·1
0·02
0·04
0·06
0·08
0·10
0·12
0·14
H/L

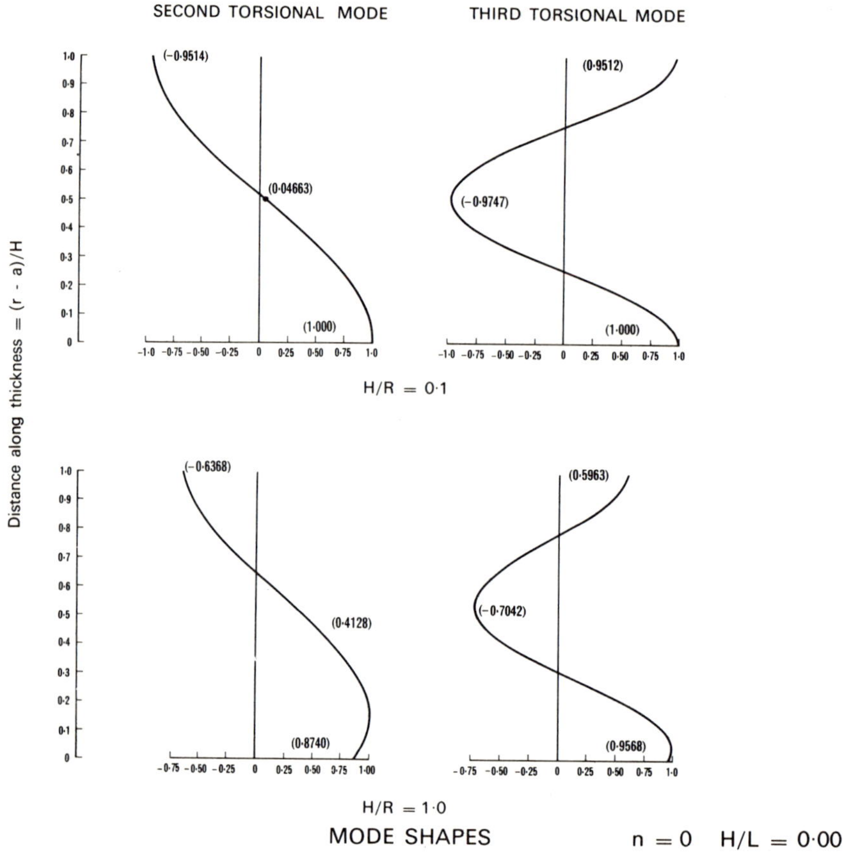


MODE SHAPES n = 0 H/L = 0·00

Variation of Normalized tangential displacement along the thickness

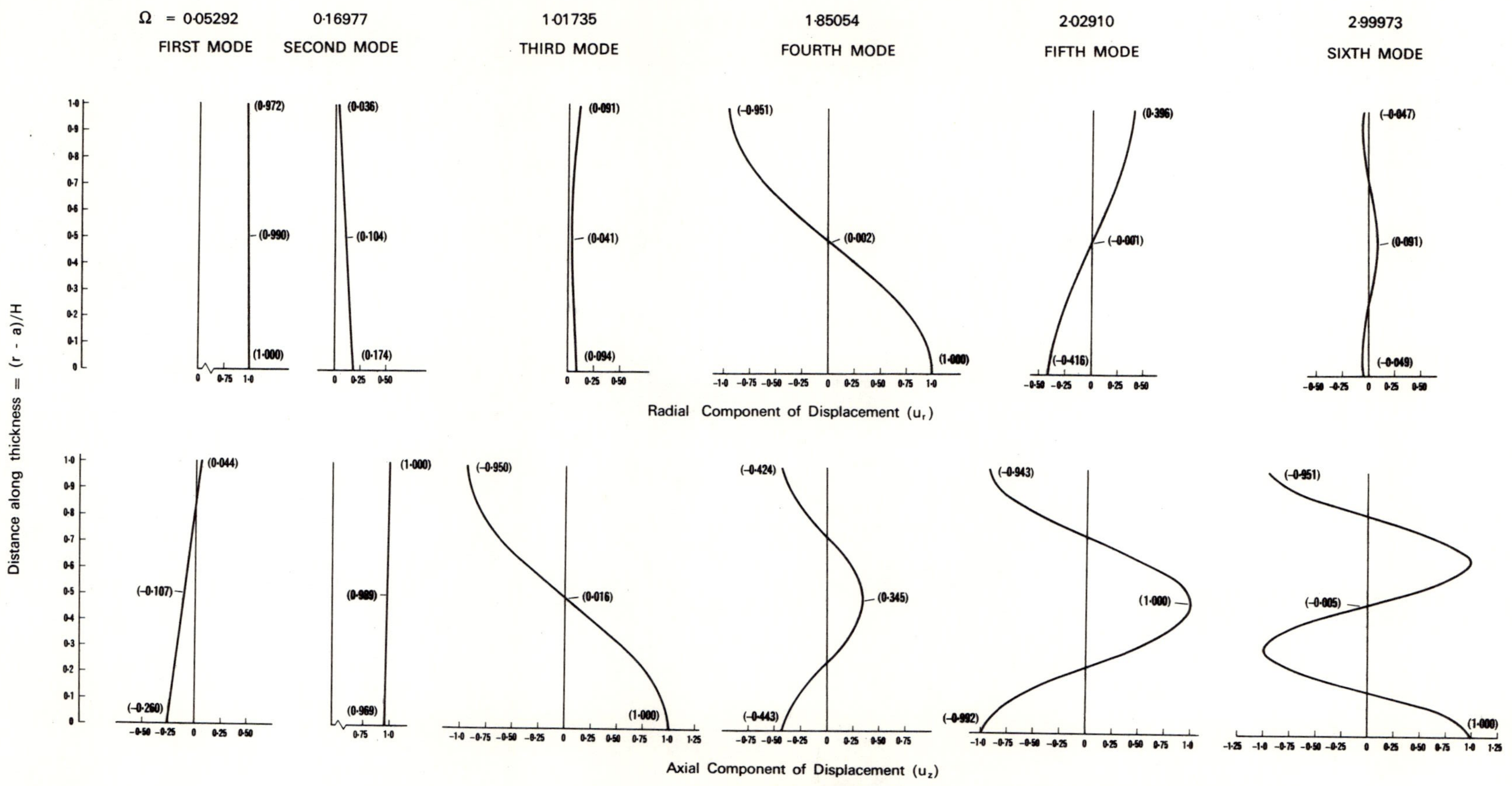


MODE SHAPES

$n = 0 \quad H/R = 0{\cdot}10 \quad H/L = 0{\cdot}10$

(Variation of Normalized displacements along the thickness)

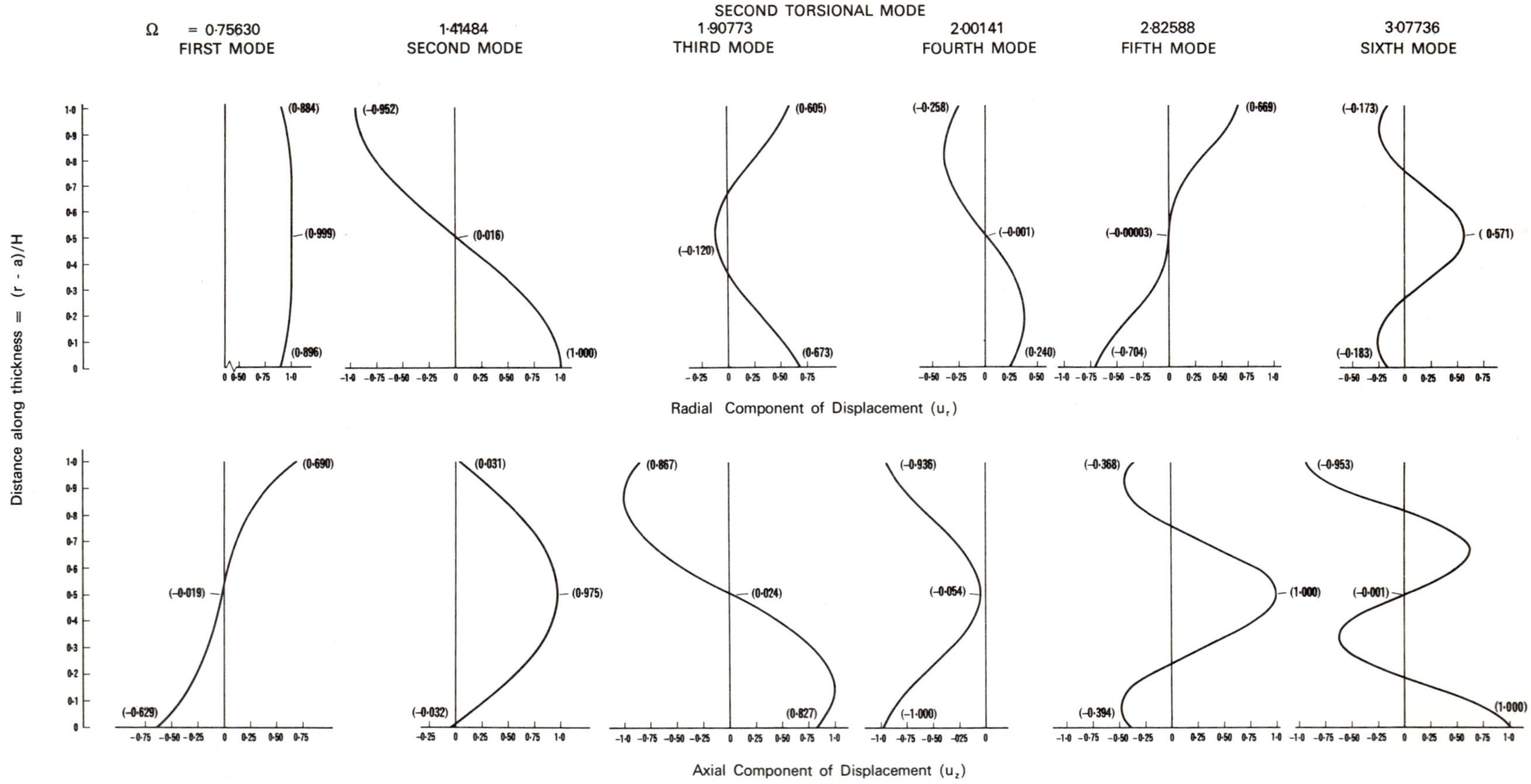


MODE SHAPES $n = 0 \quad H/R = 0{\cdot}10 \quad H/L = 1{\cdot}00$

(Variation of Normalized displacements along the thickness)

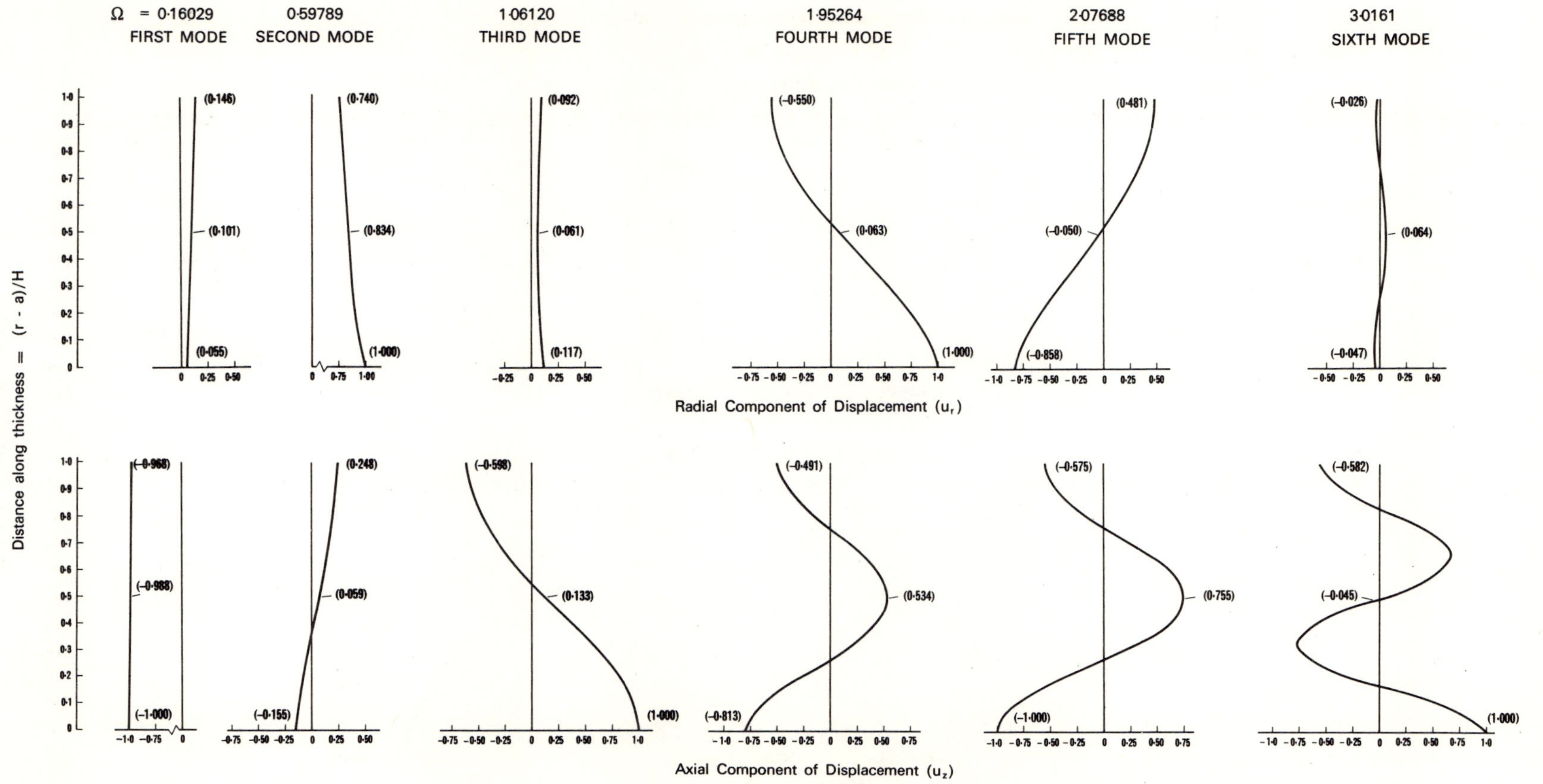


MODE SHAPES $n = 0$ $H/R = 1{\cdot}00$ $H/L = 0{\cdot}10$

(Variation of Normalized displacements along the thickness)

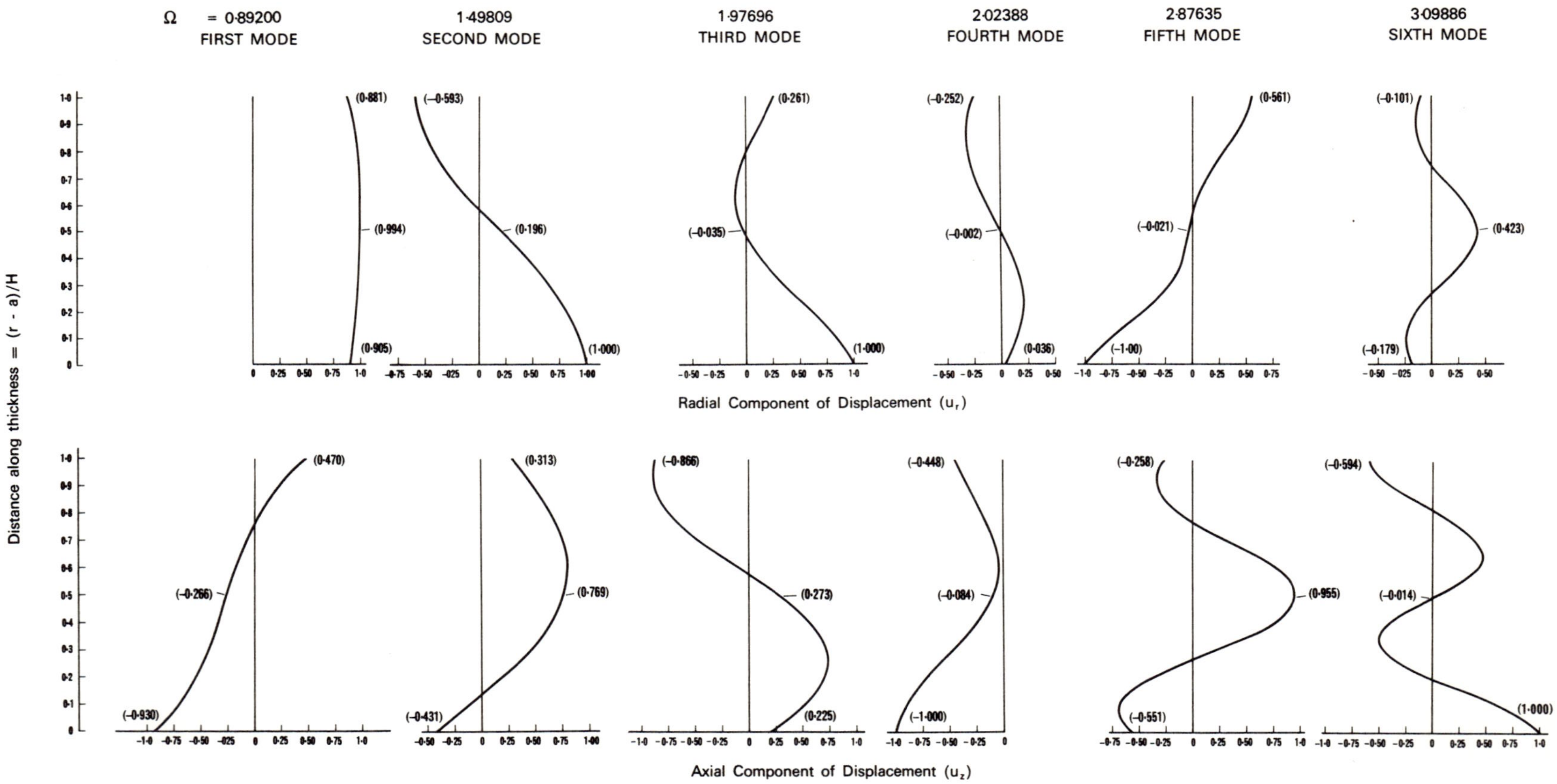


MODE SHAPES $n = 0$ $H/R = 1{\cdot}00$ $H/L = 1{\cdot}00$

(Variation of Normalized displacements along the thickness)

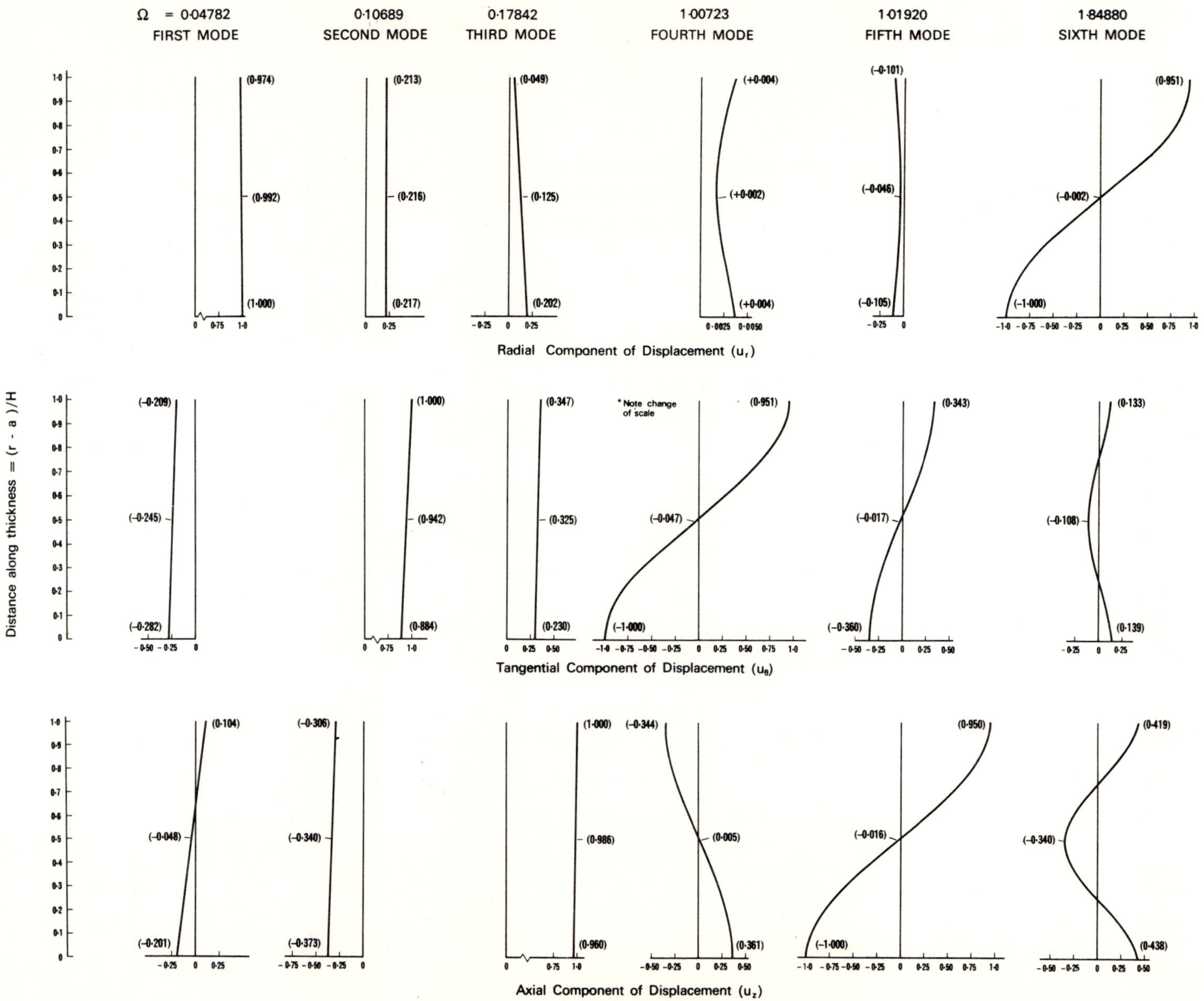


MODE SHAPES n = 1 H/R = 0·10 H/L = 0·10

(Variation of Normalized displacements along the thickness)

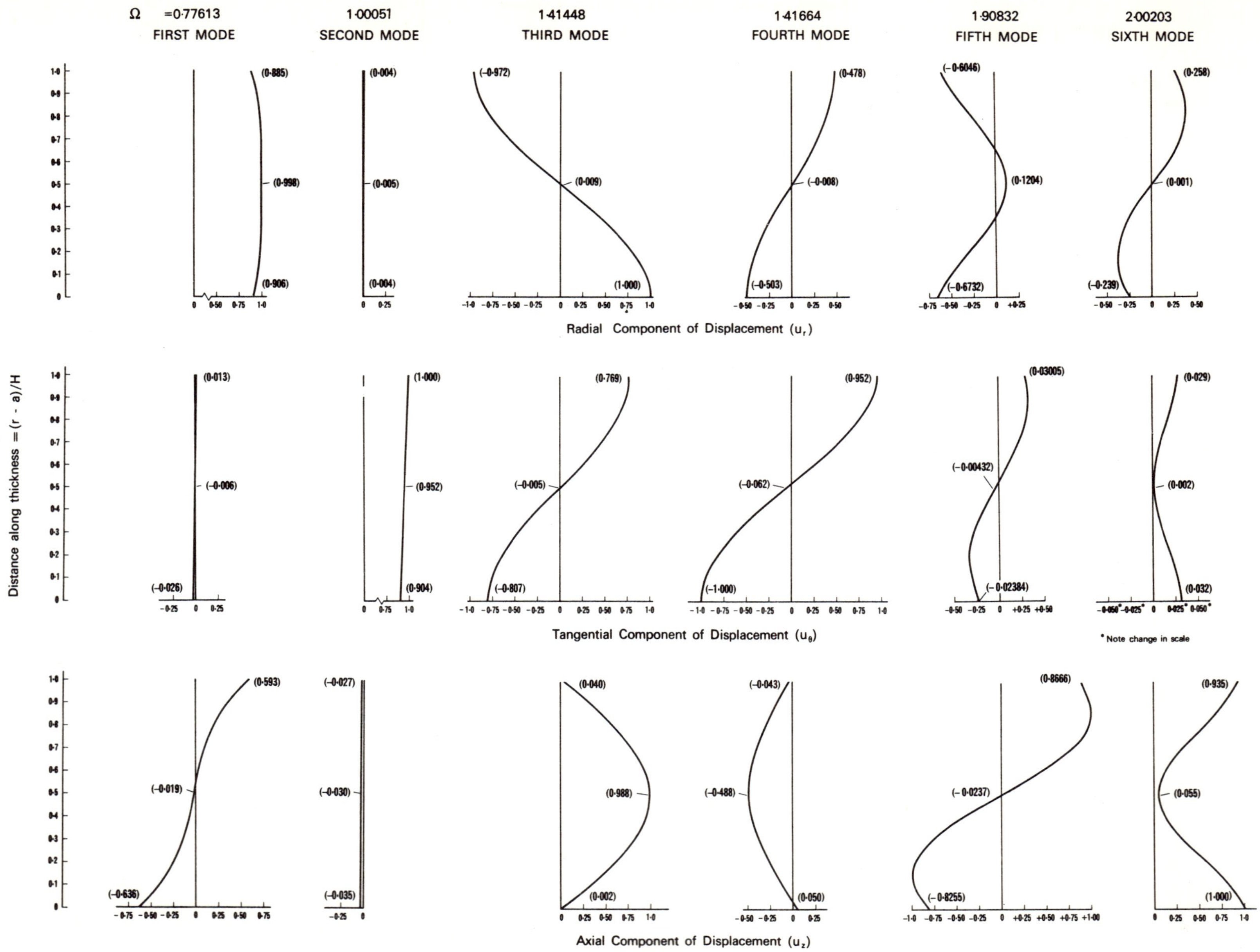


MODE SHAPES $n = 1$ $H/R = 0{\cdot}10$ $H/L = 1{\cdot}00$

(Variation of Normalized displacements along the thickness)

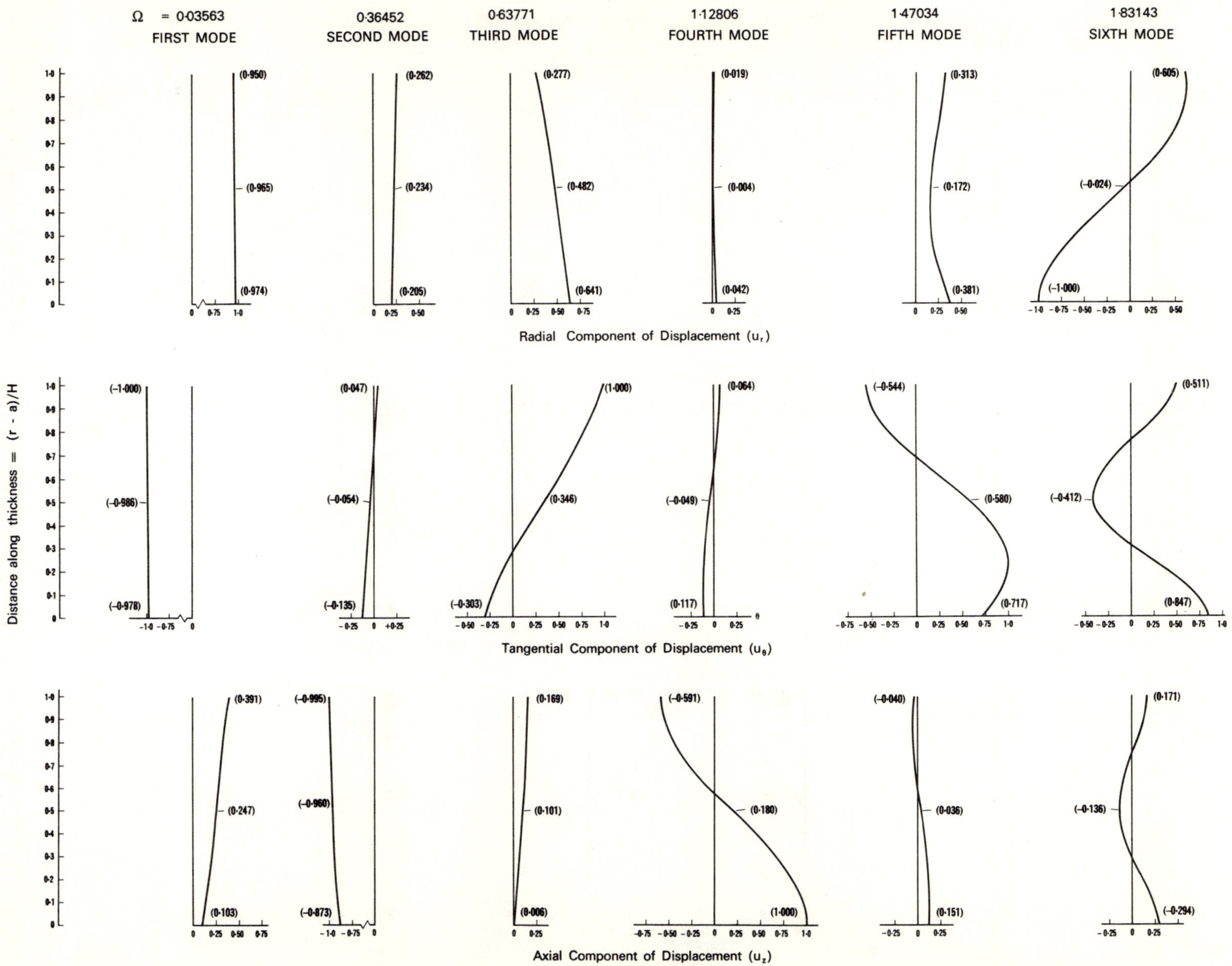


MODE SHAPES $n = 1$ $H/R = 1{\cdot}00$ $H/L = 0{\cdot}10$

(Variation of Normalized displacements along the thickness)

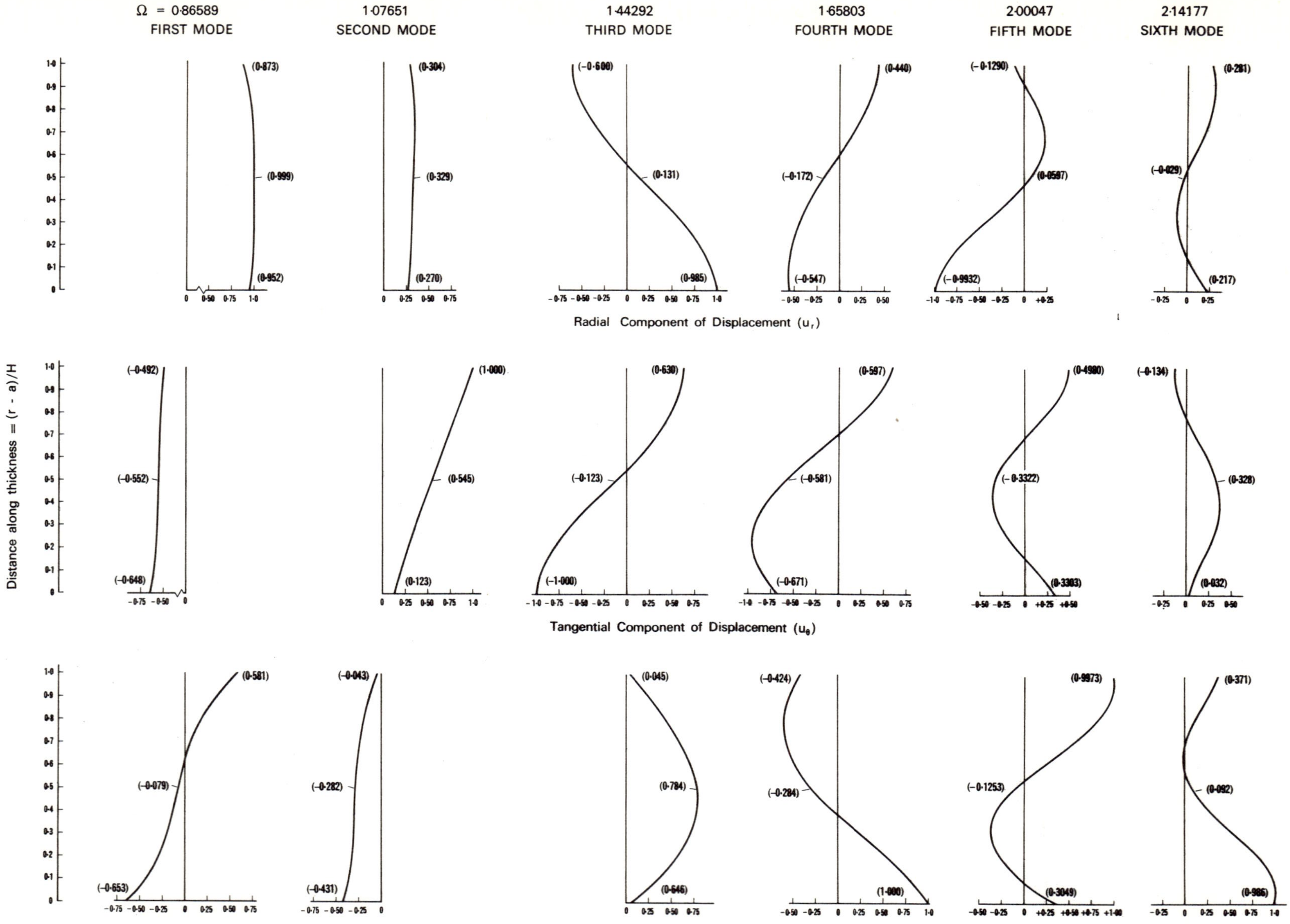


MODE SHAPES n = 1 H/R = 1·00 H/L = 1·00

(Variation of Normalized displacements along the thickness)

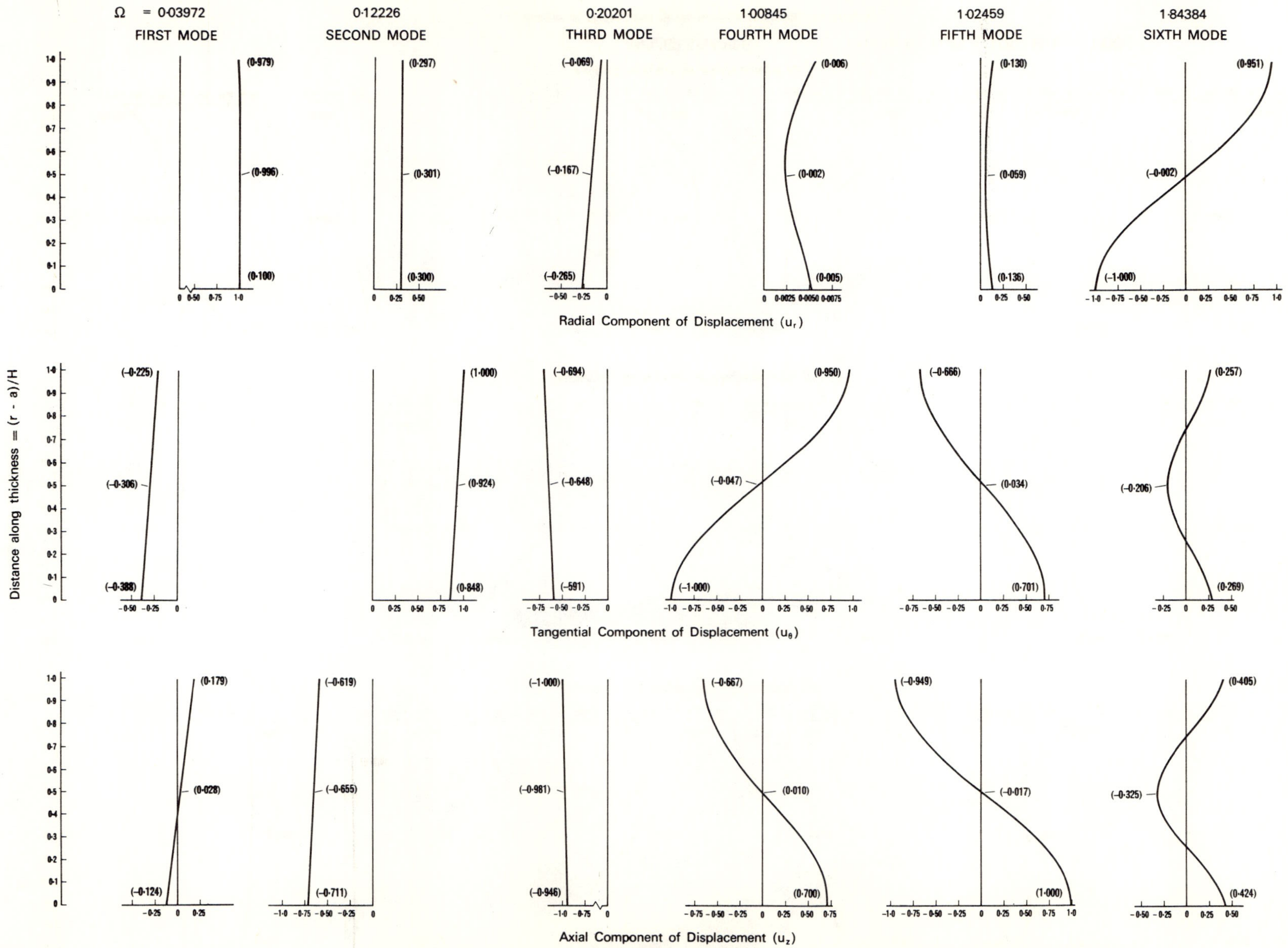


MODE SHAPES $n = 2$ $H/R = 0{\cdot}10$ $H/L = 0{\cdot}10$

(Variation of Normalized displacements along the thickness)

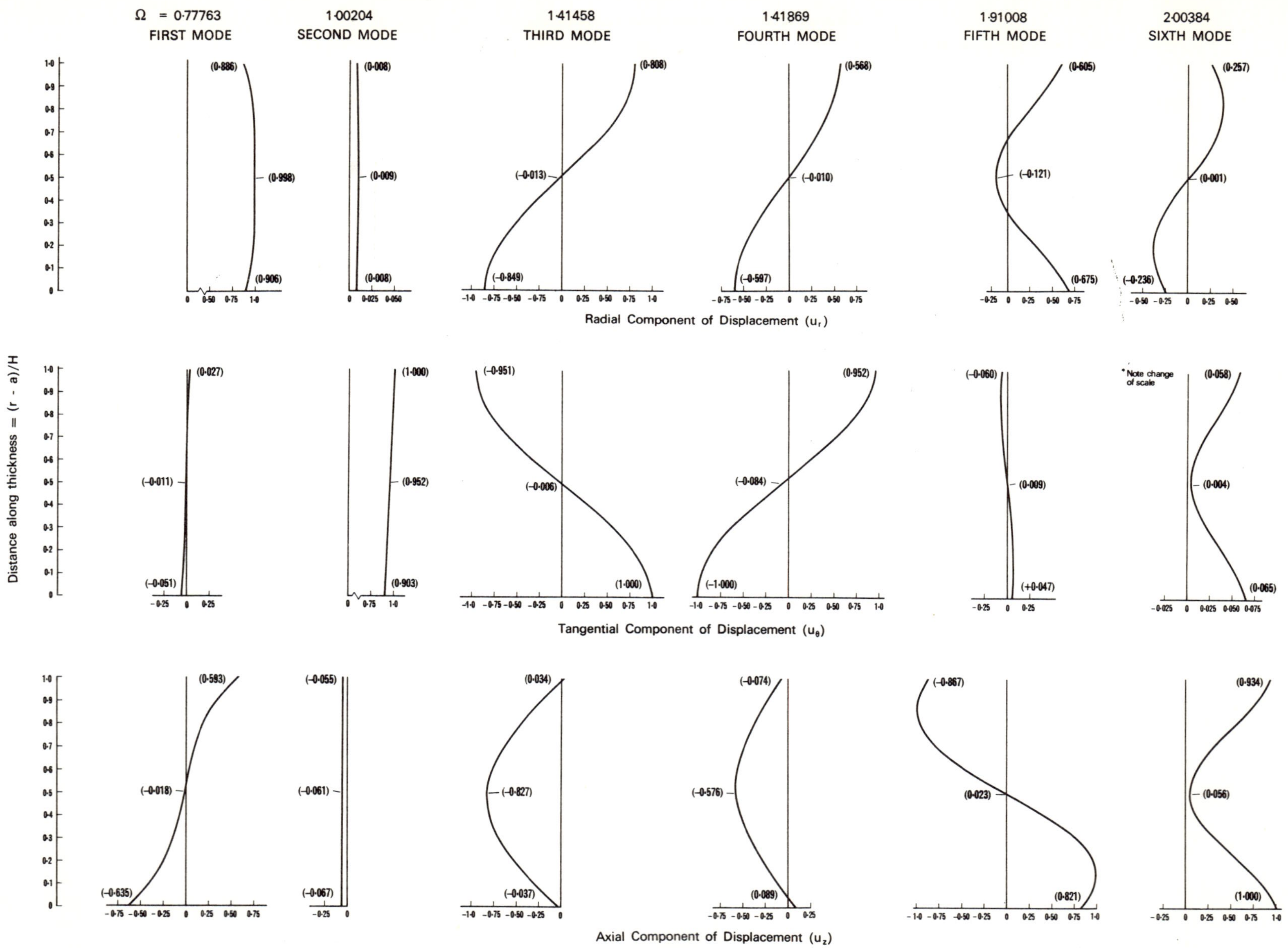


MODE SHAPES n = 2 H/R = 0·10 H/L = 1·00

(Variation of Normalized displacements along the thickness)

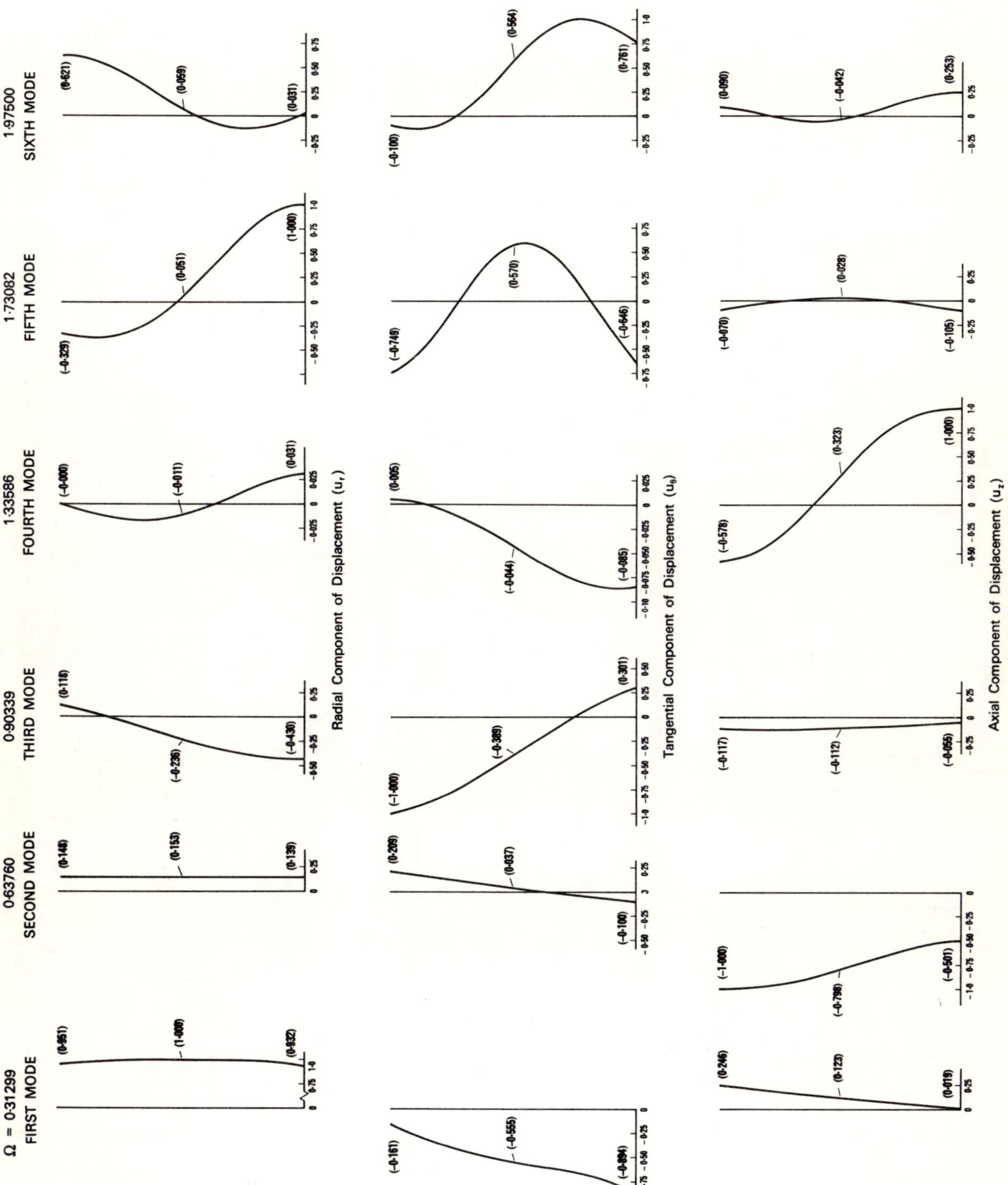


MODE SHAPES n = 2 H/R = 1·00 H/L = 0·10

(Variation of Normalized displacements along the thickness)

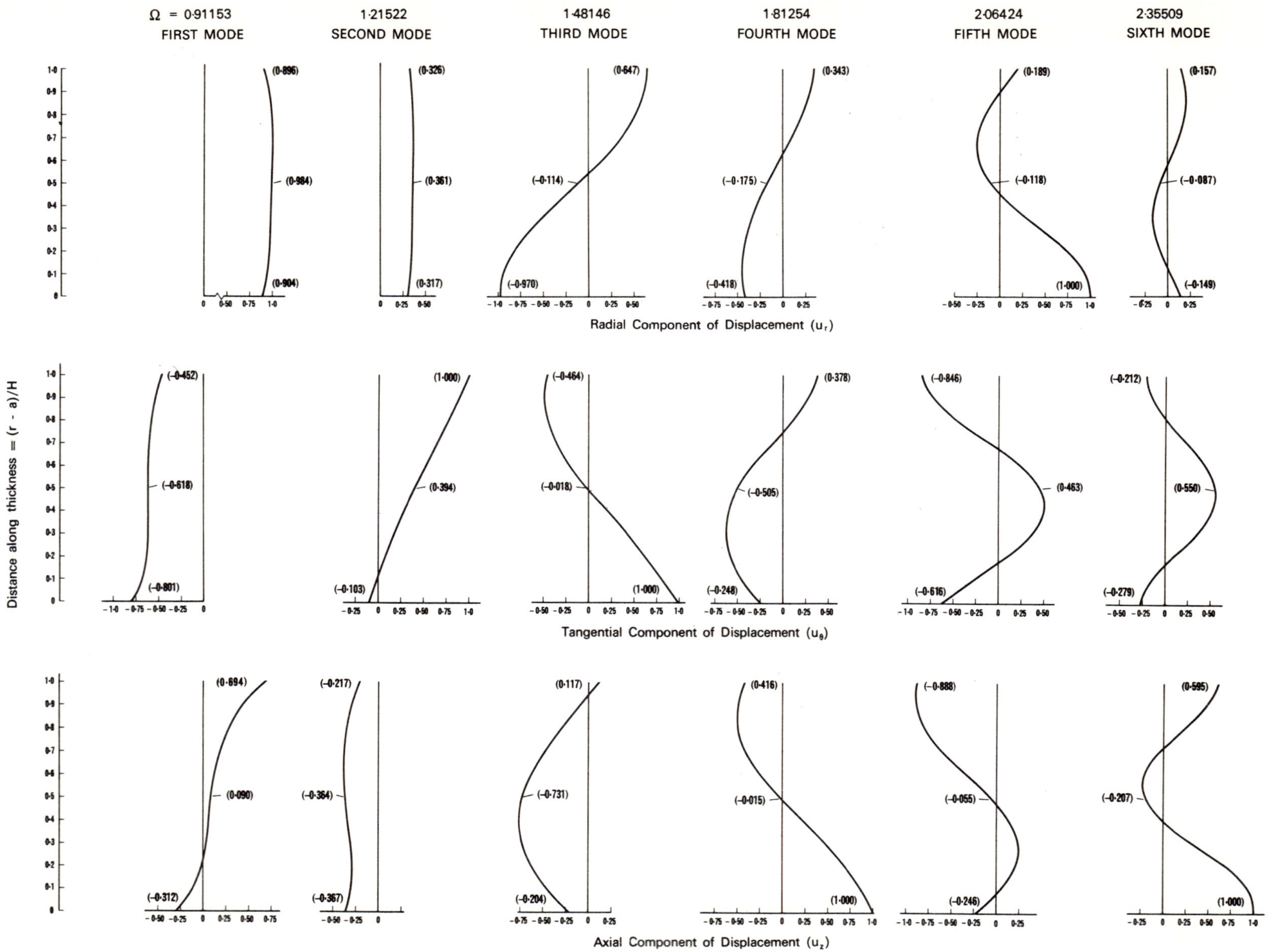


MODE SHAPES $n = 2$ $H/R = 1{\cdot}00$ $H/L = 1{\cdot}00$

(Variation of Normalized displacements along the thickness)

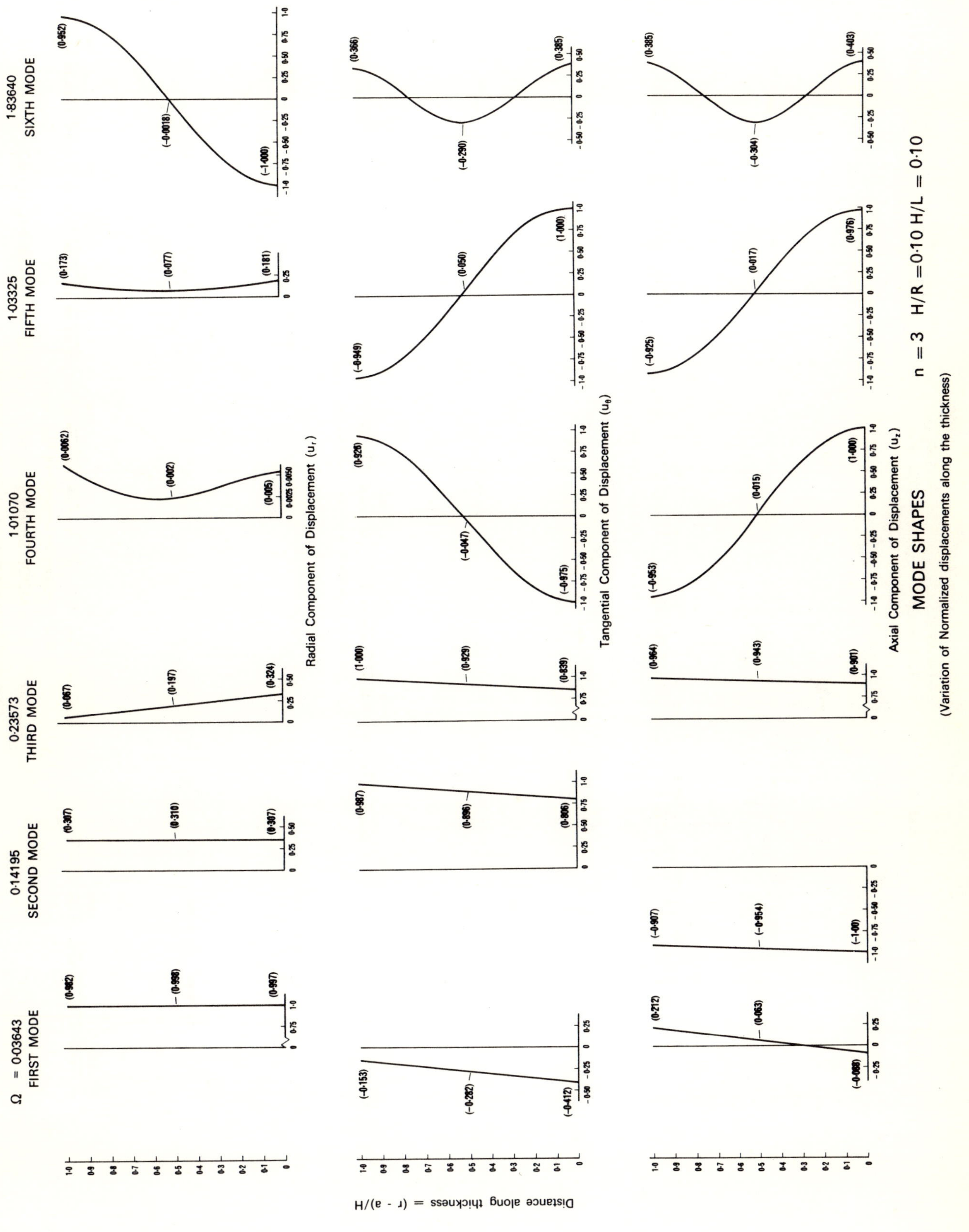

Ω = 0·03643 FIRST MODE
0·14195 SECOND MODE
0·23573 THIRD MODE
1·01070 FOURTH MODE
1·03325 FIFTH MODE
1·83640 SIXTH MODE
Radial Component of Displacement ($u_r$)
Tangential Component of Displacement ($u_\theta$)
Axial Component of Displacement ($u_z$)
Distance along thickness = (r - a)/H
MODE SHAPES
(Variation of Normalized displacements along the thickness)
n = 3 H/R = 0·10 H/L = 0·10

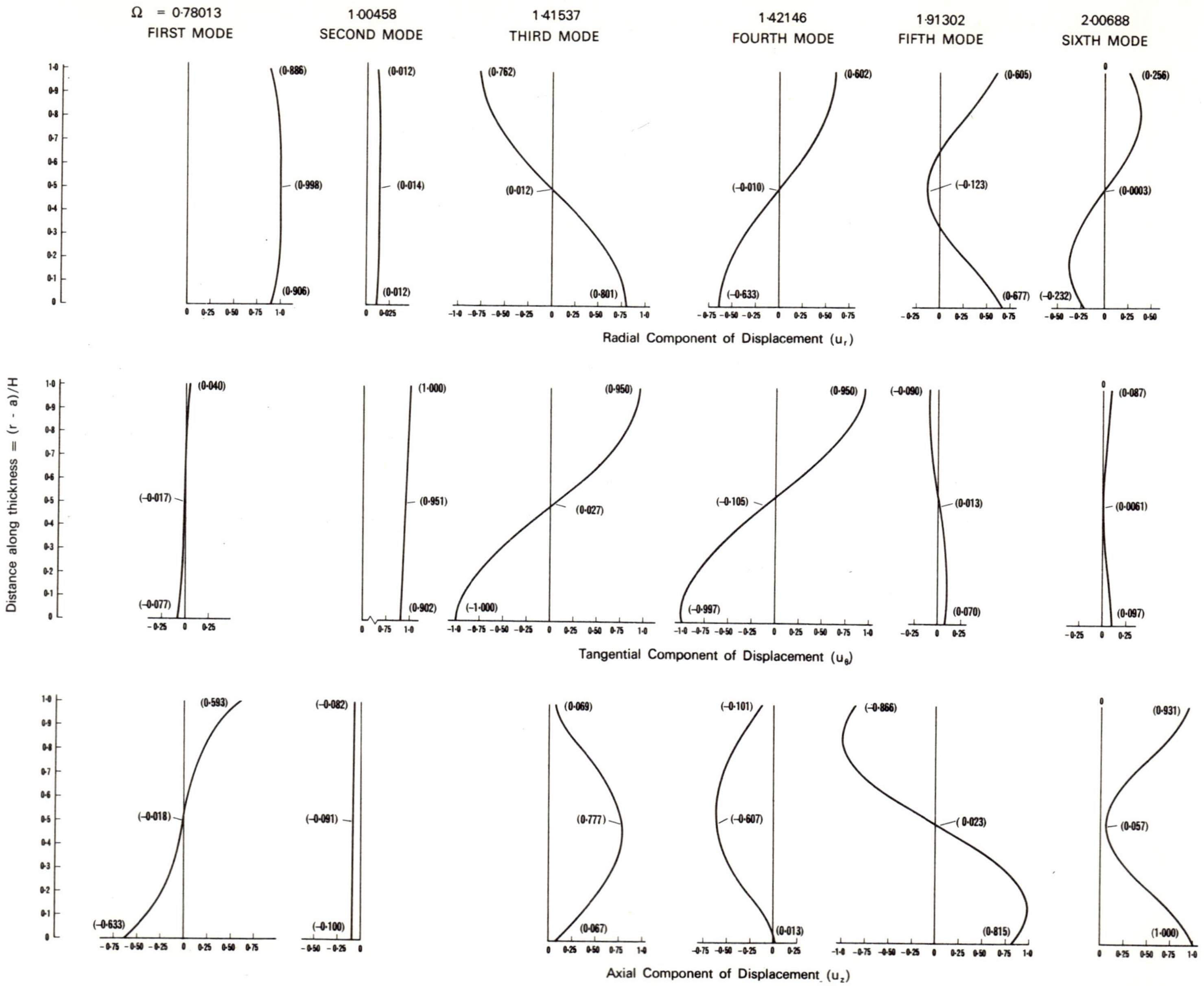


MODE SHAPES $n = 3$ $H/R = 0{\cdot}10$ $H/L = 1{\cdot}00$

(Variation of Normalized displacements along the thickness)

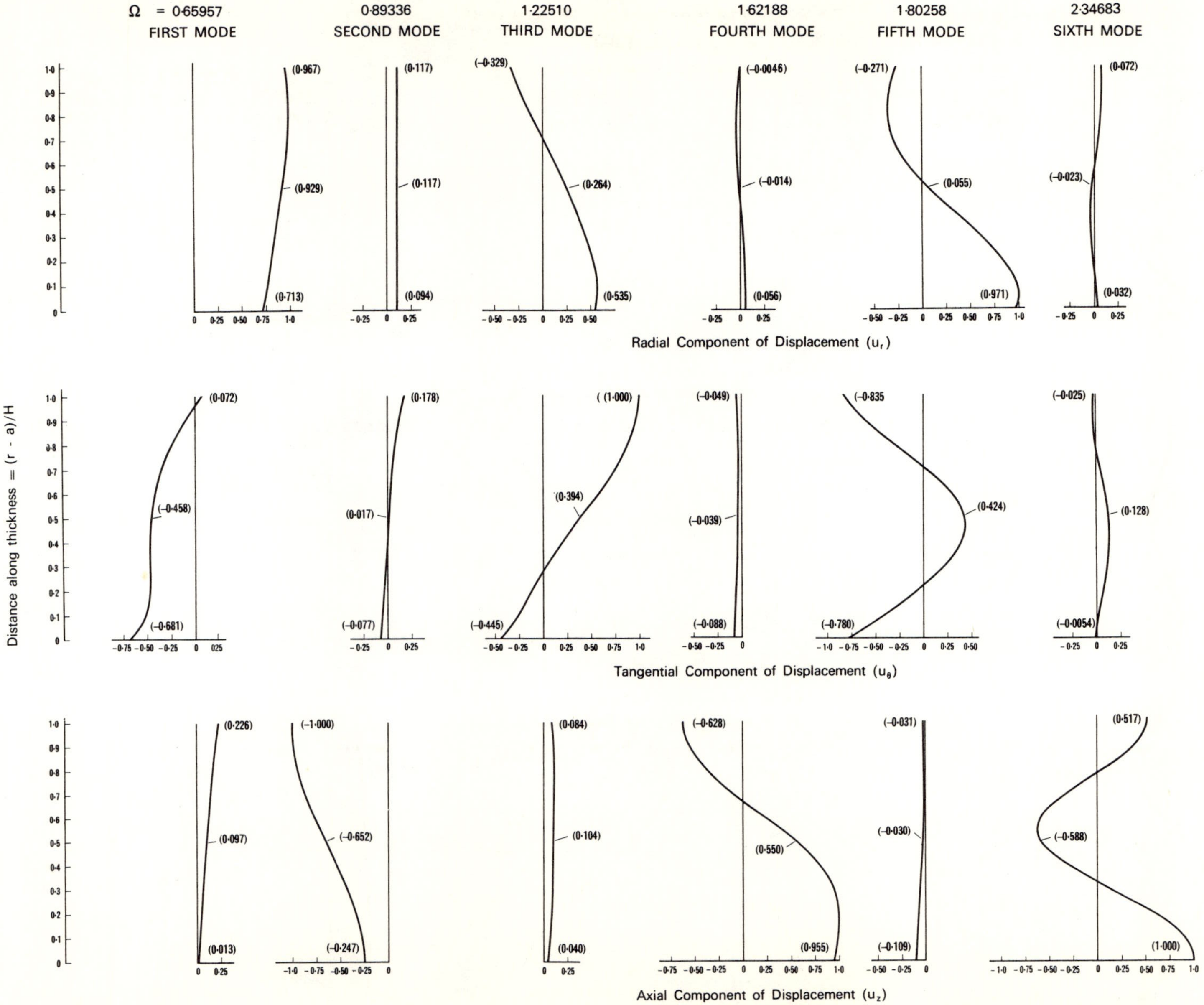


MODE SHAPES n = 3 H/R = 1·00 H/L = 0·10

(Variation of Normalized displacements along the thickness)

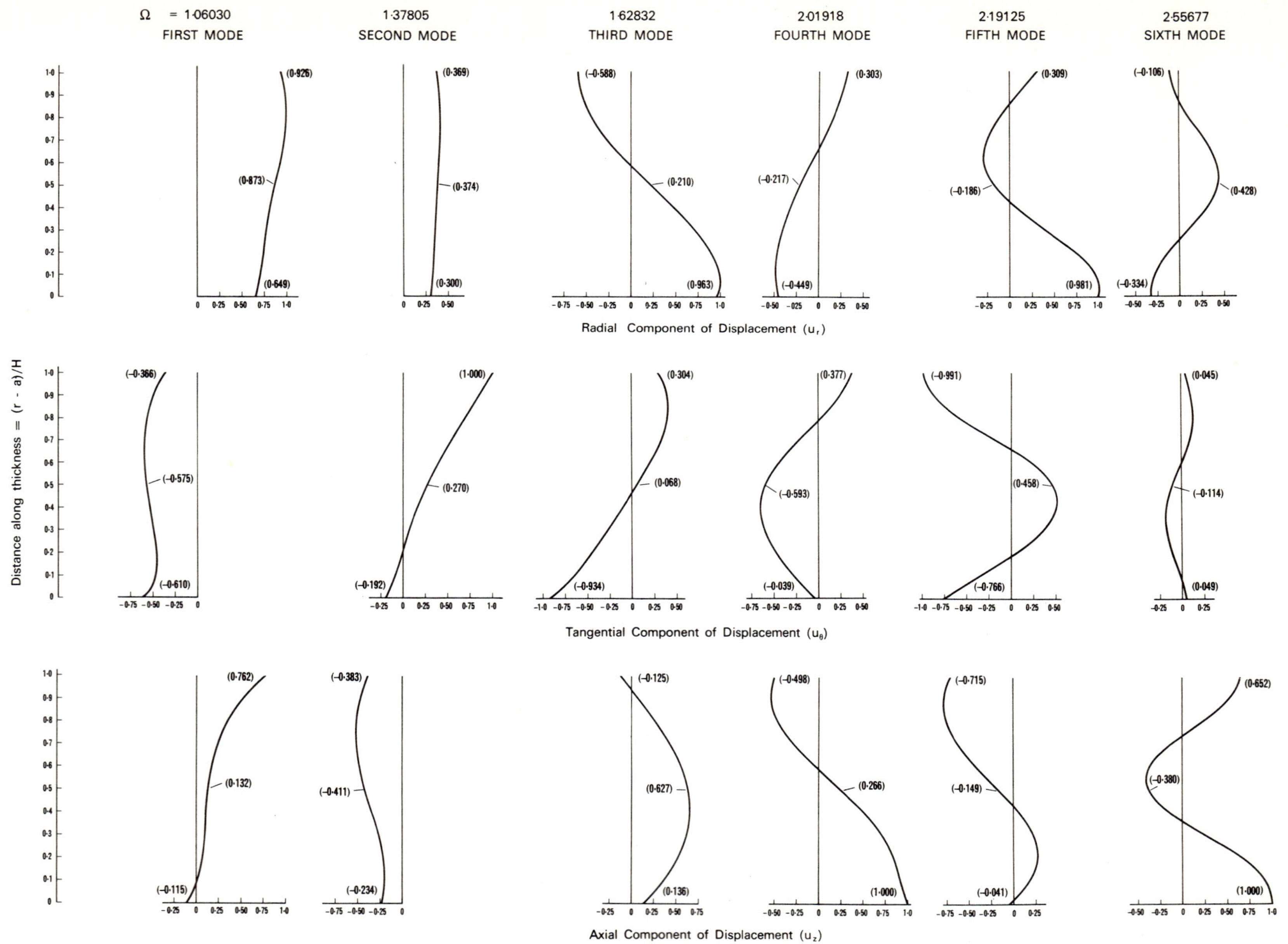


MODE SHAPES $n = 3$ $H/R = 1{\cdot}00$ $H/L = 1{\cdot}00$

(Variation of Normalized displacements along the thickness)

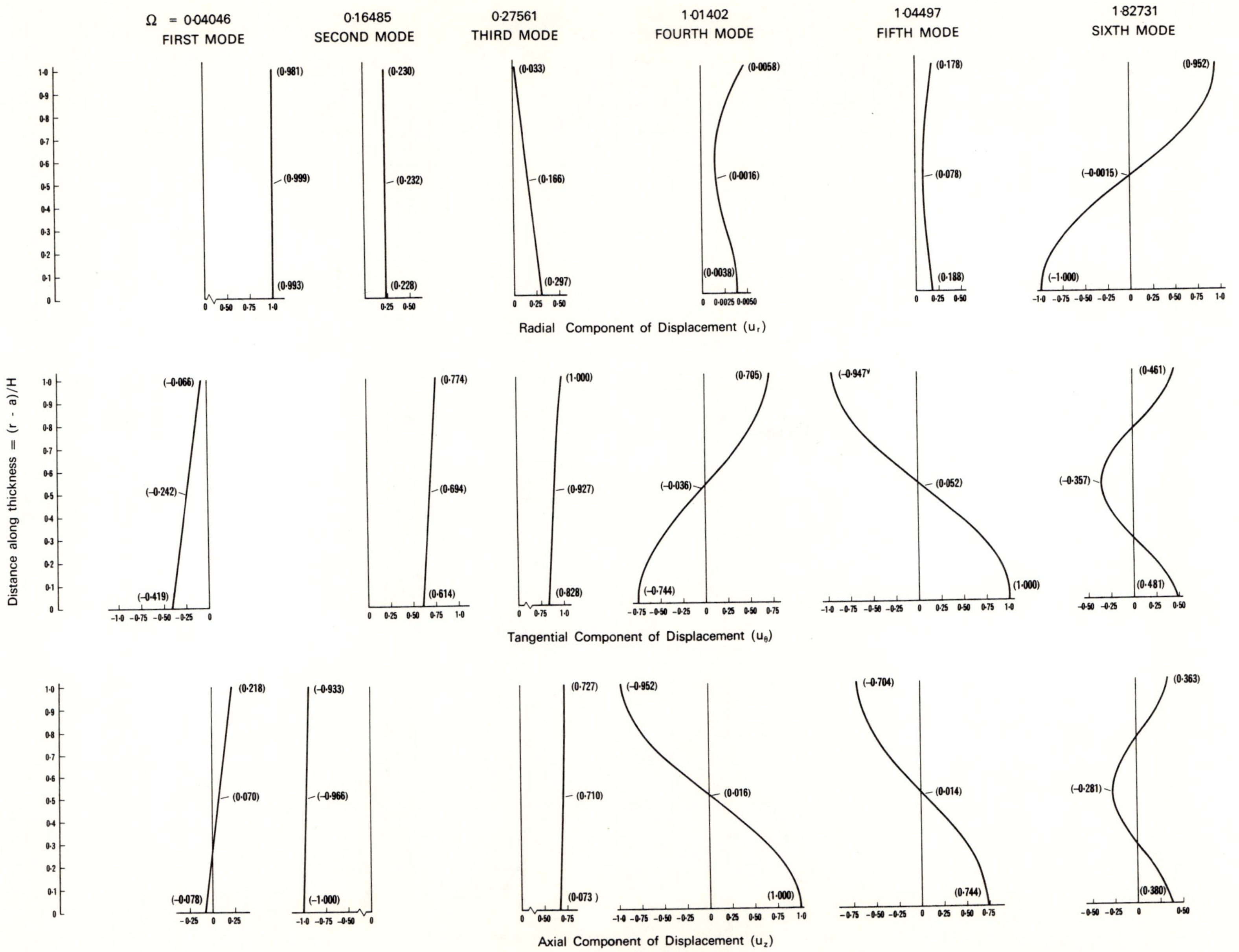


MODE SHAPES n = 4 H/R = 0·10 H/L = 0·10

(Variation of Normalized displacements along the thickness)

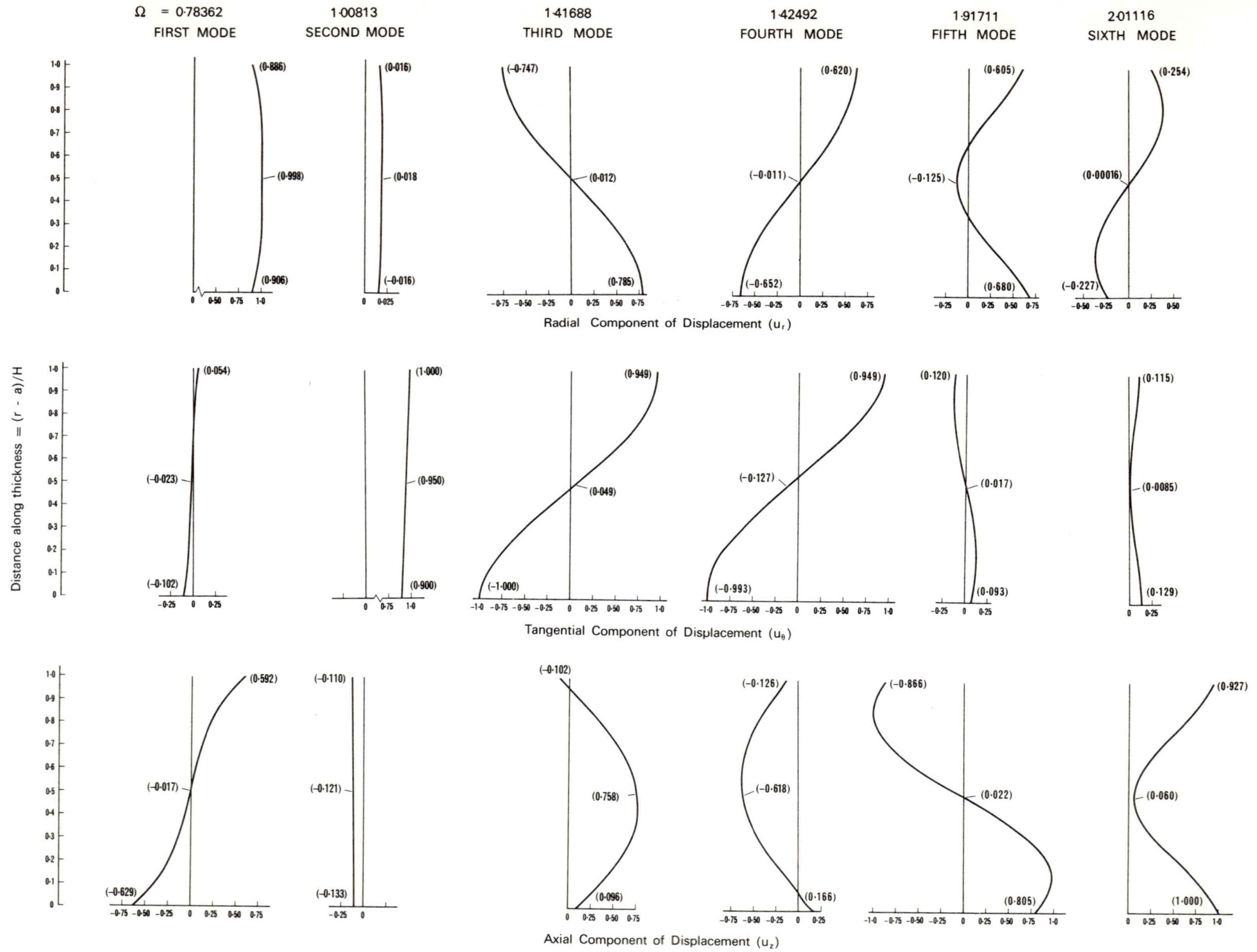


MODE SHAPES $n = 4 \quad H/R = 0{\cdot}10 \quad H/L = 1{\cdot}00$

(Variation of Normalized displacements along the thickness)

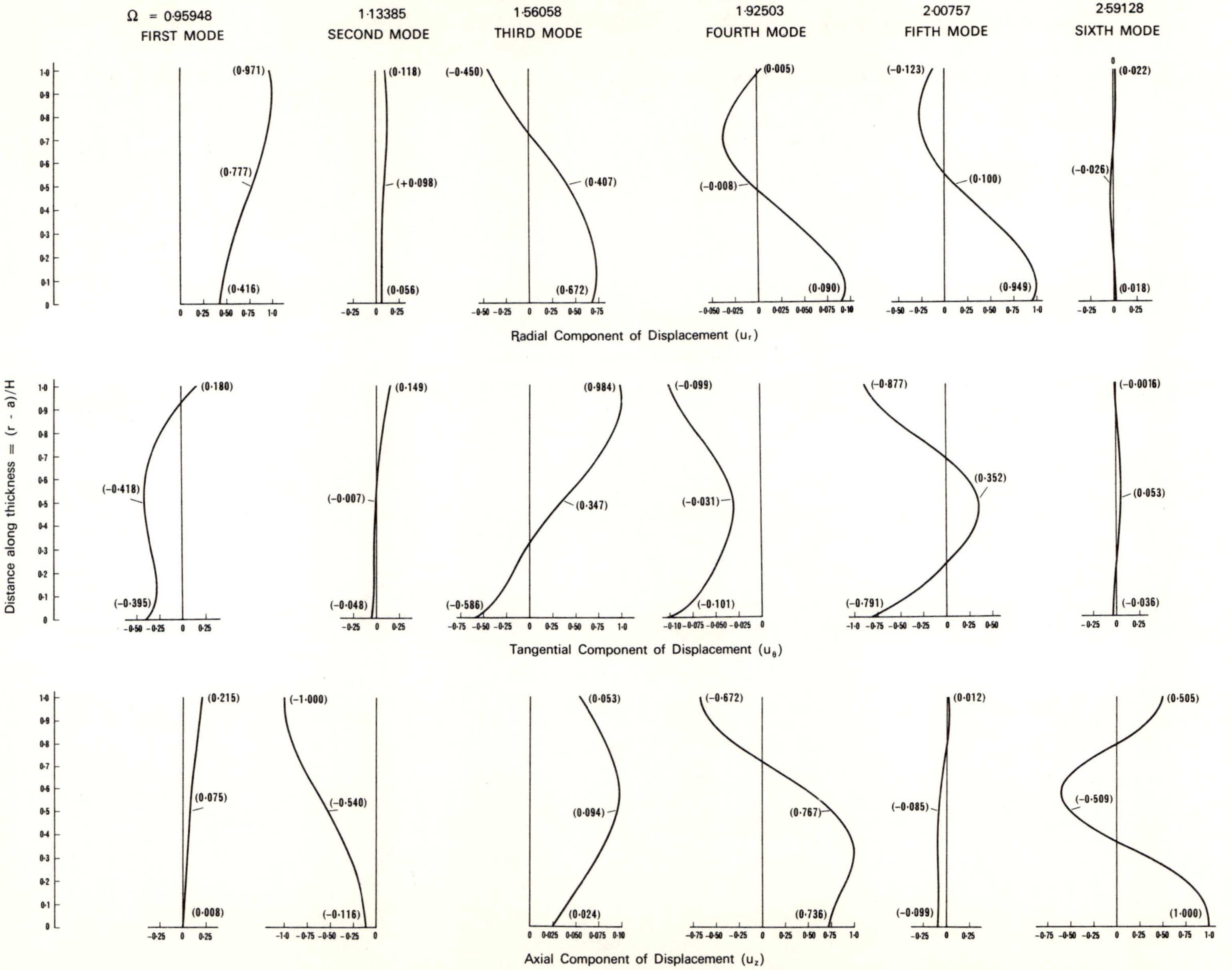


MODE SHAPES n = 4 H/R = 1·00 H/L = 0·10

(Variation of Normalized displacements along the thickness)

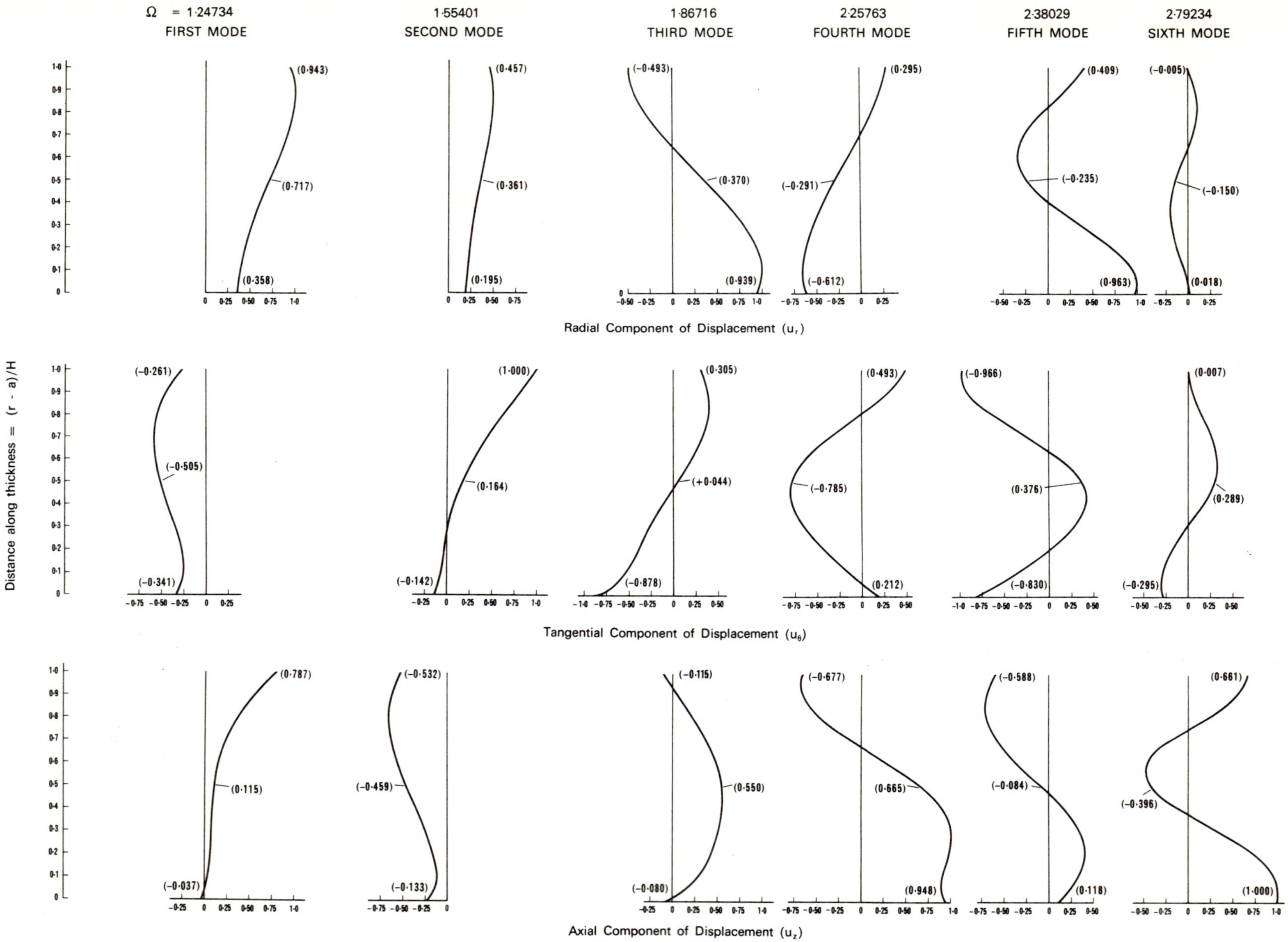


MODE SHAPES n = 4 H/R = 1·00 H/L = 1·00

(Variation of Normalized displacements along the thickness)

APPENDIX

# THE COMPUTER PROGRAM

THE FORTRAN program appended here was used in all the computations except a few in the range of very small values of $H/L$. In the latter case the program was adapted for double precision computations in order to avoid round-off errors.

As indicated in the introductory text, when the other geometrical and physical parameters are given, the frequency equation may be considered as an implicit function of $X$ and $Y$, namely

$$f(X, Y) = 0 \tag{A1}$$

where $X$ can be taken as the ratio of thickness to wavelength. The quantity $Y$ can be taken as the normalized frequency, $\omega/\omega_s$, or normalized phase-velocity, $c/c_s$. The roots of equation (A1) can be computed by an interval-halving subroutine, as explained in the text, by fixing the value of one of the quantities $X$, $Y$ and scanning a given interval of the other.

The program computes all roots when $n$ is different from zero. When $n = 0$, the program computes only the longitudinal modes, for all values of $H/L$, and the torsional cut-off frequencies only, for $H/L = 0$.

*Explanation of the Input Parameters*

ANU = Poisson's ratio, $\nu$.

N = circumferential order, i.e. number of waves around the circumference of the cylinder.

NR = number of values of $H/R$, i.e. ratio of thickness to median radius, used in computations.

MI = { 1. Roots computed along $Y$ for constant values of $X$.
2. Roots computed along $X$ for constant values of $Y$.

LM = number of roots computed in the appropriate interval of $X$ or $Y$ (LM $\leqq$ 10).

MCT = maximum number of iterations allowed in the interval-halving subroutine.

R(I) = up to ten values of $H/R$, the exact number given by NR above.

XL, DX, XM = left bound, increment, and right bound of abscissas, $H/L$, i.e. ratio of thickness to half-wavelength.

YL, DY, YM = lower bound, increment, and upper bound of the ordinate.

The values DELX, EX1, and EX2 are used if roots are computed along $X$ for constant $Y$. The values of DELY, EY1, and EY2 are used if roots are computed along $Y$ for constant $X$. The meaning of these values is as follows:

DELX or DELY = small increment added to the value of a root in order to obtain a new starting point for the computation of the next root.

EX1, EX2 or EY1, EY2 = Measures of accuracy to be used in convergence tests of the interval-halving subroutine. If EY1 and EY2 are left unspecified, they are taken as $10^{-5}$.

NC1: This is operative only if NC4 = 3. Then,

(a) if NC1 = 1, a new run starts by reading instruction 120 and proceeding from that point on.

(b) if NC1 = 2, a new run starts by reading instruction 130 and proceeding from that point on.

NC2 = { 1. The terms of the frequency determinant are not printed. 2. The terms of the frequency determinant are printed every time a root is obtained. }

NC4 = { 1. Start of new run by reading statement 100. 2. Start of new run by reading statement 110. 3. Start of new run according to NC1.

NC5 = { 1. The ordinate $Y$ is the normalized phase-velocity, $c/c_s$. 2. The ordinate $Y$ is the normalized frequency, $\omega/\omega_s$.

MASK = A fixed point number denoting the number of end bits masked out in a convergence test of the Bessel function subroutine, e.g. 3.

DC = Scaling factor multiplying the determinant.

*Subroutines needed*

Three subroutines are needed, namely, BESSF, XITR2, and XDETRM. The function of these subroutines is as follows:

BESSF: computes the Bessel function of any order, kind, and argument. Instead of the modified functions $I$ and $K$, the values $e^{-x}I$ and $e^{x}K$ are computed, in order to increase the range of computability of the Bessel functions without overflow.

XITR2: Interval-halving subroutine.

XDETRM: determinant evaluation subroutine.

The subroutines XITR2 and XDETRM are available through SHARE. The subroutine BESSF is a modification of an old SHARE subroutine. The user is advised to use one of the many others available today. The meaning of the arguments of BESSF is as follows:

Calling sequence:

N = XBESSF (ARG1, ARG2, ARG3, XLOCF(ARG4), ARG5).

ARG1 = Argument of Bessel function, a floating point number.

ARG2 = Fixed point number determining the kind of Bessel function to be computed. Its value is 1 to 6, and the corresponding functions are $J$, $I$, $K$, $Y$, $e^{-x}I$ and $e^{x}K$.

ARG3 = Order of the Bessel function, a fixed point number.

ARG4 = Storage location of the computed Bessel function, a floating point number.

ARG5 = Number of end bits masked out before convergence test, a fixed point number.

N = Fixed point number equal to:

1. For normal return.
2. For error return.

```
      DIMENSION CN(3,3,2),CN1(3,3,2),CM(3,3,2),CM1(3,3,2),TERM(6,6)
      DIMENSION BFAN(2,2),BFAN1(2,2),BFBN(2,2),BFBN1(2,2),DET(6,6)
      DIMENSION AR(2),ARSQ(2),BR(2),BRSQ(2),ZR(2),ZRSQ(2),R(50),ROOT(10)
      DIMENSION KA(2),KB(2),REDRT(10),EX(10),CC2(10),OOS(10)
      DIMENSION RN(50)
      DIMENSION RTSV(50)
C     POSITION OUTPUT TAPE
      CALL POSIT(15)
      REWIND 5
      PI=3.14159265
  100 READ INPUT TAPE 2,10,ANU,EN,N,NR,MI,LM,MCT
   10 FORMAT(2F10.5,5I5)
      WRITE OUTPUT TAPE 5,6001,ANU,EN
 6001 FORMAT(6H1ANU =F10.5,5X3HN =F5.1)
  110 READ INPUT TAPE 2,11,(RN(I),I=1,NR)
   11 FORMAT(5F14.7)
      DO 111 I=1,NR
  111 R(I)=1./RN(I)-.5
  120 READ INPUT TAPE 2,12,XLN,DXN,XMN,DELXN,EX1,EX2
      IF(EX1) 8001,8000,8001
 8000   EX1 = 1.E-5
 8001     IF(EX2) 8003,8002,8003
 8002  EX2 = 1.E-5
 8003     CONTINUE
   12 FORMAT(4F10.5,2E10.1)
      XL=PI*XLN
      DX=PI*DXN
      XM=PI*XMN
      DELX=PI*DELXN
  130 READ INPUT TAPE 2,13,YL,DY,YM,DELY,EY1,EY2
        IF(EY1) 8005,8004,8005
 8004   EY1 = 1.E-5
 8005      IF(EY2)  8007,8006,8007
 8006    EY2 = 1.E-5
 8007     CONTINUE
   13 FORMAT(4F10.5,2E10.1)
  140 READ INPUT TAPE 2,14,NC1,NC2,NC4,NC5,MASK,DC
   14 FORMAT(5I5,E10.3)
  101 WRITE OUTPUT TAPE 3,1101,ANU,EN,N,NR,MI,LM,MCT
 1101 FORMAT(21H1LIST OF INPUT VALUES/22H0ANU,EN,N,NR,MI,LM,MCT/2F15.5,5
     1I5)
      WRITE OUTPUT TAPE 3,1111,(RN(I),I=1,NR)
 1111 FORMAT(5H0R(I)/(5E20.8))
  121 WRITE OUTPUT TAPE 3,1121,XLN,DXN,XMN,DELXN,EX1,EX2
 1121 FORMAT(22H0XL,DX,XM,DELX,EX1,EX2/(6E20.8))
  131 WRITE OUTPUT TAPE 3,1131,YL,DY,YM,DELY,EY1,EY2
 1131 FORMAT(22H0YL,DY,YM,DELY,EY1,EY2/(6E20.8))
  141 WRITE OUTPUT TAPE 3,1141,NC1,NC2,NC4,NC5,MASK,DC
 1141 FORMAT(24H0NC1,NC2,NC4,NC5,MASK,DC/(5I5,E20.8))
      DO 390 IR=1,NR
      WRITE OUTPUT TAPE 5,6002,RN(IR)
 6002 FORMAT(6H0H/R =F10.6/1H06X3HH/L6X4HW/WS8X1H114X1H214X1H314X1H4
     114X1H514X1H6)
      IF(R(IR))145,145,146
```

```
145 NC3=2
    GO TO 150
146 NC3=1
150 WRITE OUTPUT TAPE  3,15,ANU,N,LM,RN(IR)
 15 FORMAT(1H1,20H  POISSON RATIO  NU=F8.5,10H  ORDER N=I3,9H.  FIRST
   XI3,17H  ROOTS FOR  H/R=F10.5)
    GO TO(160,170),MI
160 WRITE OUTPUT TAPE  3,16,YL,YM,EY1,EY2
 16 FORMAT(1H0,68H  ROOTS COMPUTED FOR  H/L   CONSTANT AND INCREASING
   1ORDINATE BETWEENF10.5,6H  AND F10.5//30H   ITERATION CARRIED TO
   2(1)=E10.1,12H  AND  E(2)=E10.1)
    GO TO 180
170 WRITE OUTPUT TAPE  3,17,XLN,XMN,EX1,EX2
 17 FORMAT(1H0,68H  ROOTS COMPUTED FOR ORDINATE CONSTANT AND INCREASIN
   1G  H/L   BETWEENF10.5,6H  AND F10.5//30H   ITERATION CARRIED TO
   2(1)=E10.1,12H  AND  E(2)=E10.1)
180 GO TO(184,186),NC5
184 WRITE OUTPUT TAPE  3,1184
1184 FORMAT(1H0,17H  ORDINATE = C/C2)
    GO TO 190
186 WRITE OUTPUT TAPE  3,1186
1186 FORMAT(1H0,27H  ORDINATE = OMEGA/OMEGA(S))
190 N1=N+1
    CONS1=2.*EN*(EN-1.)
    ESQ=2.*(1.-ANU)/(1.-2.*ANU)
    EEE=SQRTF(ESQ)
    GO TO (200,300),MI
200 X=XL
210 LRT=0
    IF(SENSE SWITCH 5) 6000,6010
6010  CONTINUE
    NZM=1
215 YLFT=YL
    IF(X)216,216,217
216 NX=1
    GO TO 220
217 NX=2
220 A7=ITR2GF(YLFT,DY,YM,EY1,EY2,MCT,XLOCF(Y))
230 GO TO 400
235 N7=XITR2F(RESLT)
    GO TO(230,240,250,260),N7
240 IF(LRT)245,245,270
245 XN=X/PI
    WRITE OUTPUT TAPE  3,24,XN,YL,YM
 24 FORMAT (///18H  NO ROOTS FOR X= F10.5,11H BETWEEN Y=F10.5,10H   AN
   1D Y= F10.5)
    GO TO 280
250 XN=X/PI
    WRITE OUTPUT TAPE 3,24,XN,Y,((TERM(J,K),K=1,6),J=1,6)
 25 FORMAT (///31H  ITERATION COUNT EXCEEDED   X=F12.6,7H     Y=F12.6/
   1/(6E20.8))
    GO TO 266
260 GO TO(266,265),NC2
265 XN=X/PI
    WRITE OUTPUT TAPE 3,26,XN,Y,((TERM(J,K),K=1,6),J=1,6)
 26 FORMAT(1H0,22H  MATRIX TERMS FOR  X=F12.6,10H   AND Y= F12.6//(6E2
```

```
     X0.8))
 266 LRT=LRT+1
     IF(LRT-LM)268,269,267
 267 LRT=LRT-1
     GO TO 270
 268 ROOT(LRT)=Y
1268 YLFT=Y+DELY
     IF(YLFT-YM)220,220,240
 269 ROOT(LRT)=Y
 270 GO TO(1271,271),NX
 271 FACTOR=PI/X
     GO TO(272,276),NC5
 272 DO 274 K=1,LRT
     REDRT(K)=ROOT(K)/FACTOR
 274 CONTINUE
     GO TO 279
 276 DO 278 K=1,LRT
     REDRT(K)=ROOT(K)
     ROOT(K)=ROOT(K)*FACTOR
 278 CONTINUE
 279 XN=X/PI
     WRITE OUTPUT TAPE 3,27,XN,LRT,((ROOT(K),REDRT(K)),K=1,LRT)
  27 FORMAT (///11H  FOR H/L =F10.5,I10,14H    ROOTS FOUND//46X,4HC/C217
    1X,14HOMEGA/OMEGA(S)//(E55.7,E26.7))
C    WRITE VALUES ON DATA TAPE
     WRITE OUTPUT TAPE 5,6003,XN,(REDRT(K),K=1,LRT)
6003 FORMAT(5XF10.5,5X,6E15.6)
     CALL RITE(15,EN,RN(IR),XN,LRT,REDRT)
 280 X=X+DX
     IF(X-XM)210,210,290
 290 GO TO 390
 300 Y=YL
 310 LRT=0
     IF(SENSE SWITCH 5) 6000,6011
6011  CONTINUE
     XLFT=XL
 320 A7=ITR2GF(XLFT,DX,XM,EX1,EX2,MCT,XLOCF(X))
 330 GO TO 400
 335 N7=XITR2F(RESLT)
     GO TO(330,340,350,360),N7
 340 IF(LRT)345,345,370
 345 WRITE OUTPUT TAPE  3,34,Y,XLN,XMN
  34 FORMAT (///18H  NO ROOTS FOR Y= F10.5,11H BETWEEN X=F10.5,10H   AN
    1D X= F10.5)
     GO TO 380
 350 XN=X/PI
     WRITE OUTPUT TAPE 3,35,XN,Y,((TERM(J,K),K=1,6),J=1,6)
  35 FORMAT (///31H  ITERATION COUNT EXCEEDED   X=F12.6,7H     Y=F12.6/
    1/(6E20.8))
     GO TO 366
 360 GO TO(366,365),NC2
 365 XN=X/PI
     WRITE OUTPUT TAPE 3,36,XN,Y,((TERM(J,K),K=1,6),J=1,6)
  36 FORMAT(1H0,22H  MATRIX TERMS FOR  X=F12.6,10H   AND Y= F12.6//(6E2
    X0.8))
 366 LRT=LRT+1
```

```
      IF(LRT-LM)368,369,367
  367 LRT=LRT-1
      GO TO 370
  368 ROOT(LRT)=X
 1368 XLFT=X+DELX
      IF(XLFT-XM)320,320,340
  369 ROOT(LRT)=X
  370 GO TO(372,376),NC5
  372 DO 374 K=1,LRT
      EX(K)=ROOT(K)
      CC2(K)=Y
      OOS(K)=Y/PI*ROOT(K)
  374 CONTINUE
      GO TO 379
  376 DO 378 K=1,LRT
      EX(K)=ROOT(K)
      OOS(K)=Y
      CC2(K)=PI*Y/ROOT(K)
  378 CONTINUE
  379 WRITE OUTPUT TAPE  3,37,((EX(K),CC2(K),OOS(K)),K=1,LRT)
   37 FORMAT (///34X,4HZI*H21X,4HC/C216X,14HOMEGA/OMEGA(S)//(E42.7,2E25.
     17))
  380 Y=Y+DY
      IF( Y-YM)310,310,390
  390 CONTINUE
      GO TO (100,110,395,6000,7000),NC4
  395 GO TO(120,130),NC1
C     COMPUTATION OF THE DETERMINANT
  400 GO TO(1400,2400),NX
 1400 GO TO(1410,1420),NC5
 1410 GO TO(280,1368),MI
 1420 BR(1)=PI*Y*R(IR)
      BR(2)=PI*Y*(R(IR)+1.)
      DO 1430 K=1,2
      AR(K)=BR(K)/EEE
      ARSQ(K)=AR(K)**2
      BRSQ(K)=BR(K)**2
      ZR(K)=0.
      ZRSQ(K)=0.
 1430 CONTINUE
      MB=1
      GO TO 435
 2400 GO TO(401,402),NC5
  401 YE=Y
      GO TO 403
  402 YE=PI*Y/X
  403 ZR(1)=X*R(IR)
      ZR(2)=ZR(1)+X
      GAZ1=YE**2-1.
      GAZ2=YE**2/ESQ-1.
      DO 410 K=1,2
      ZRSQ(K)=ZR(K)**2
      ARSQ(K)=              GAZ2*ZRSQ(K)
      BRSQ(K)=        GAZ1*ZRSQ(K)
       AR(K) = SQRTF(ABSF(ARSQ(K)))
       BR(K) = SQRTF(ABSF(BRSQ(K)))
```

```
  410 CONTINUE
      IF(     GAZ2      )420,415,430
  415 Y=Y+EY1
      GO  TO 400
  420 IF(   GAZ1   )425,415,426
  425 MB=3
      GO  TO 435
  426 MB=2
      GO  TO 435
  430 MB=1
  435 MBM=MB-1
      DO  560  I=NC3,2
      GO  TO(440,450,450),MB
  440 EL1=1.
      KA(1)=1
      KA(2)=4
      GO  TO 460
  450 EL1=-1.
      KA(1)=5
      KA(2)=6
      GO  TO(460,470),MBM
  460 EL2=1.
      KB(1)=1
      KB(2)=4
      GO  TO 480
  470 EL2=-1.
      KB(1)=5
      KB(2)=6
  480 NCRT=3-NC3
      DO  490  L=1,NCRT
      K7=XBESSF(AR(I),KA(L),N,XLOCF(BFCT),MASK)
      BFAN(L,I)=BFCT
      GO  TO(484,494),K7
  484 K7=XBESSF(BR(I),KB(L),N,XLOCF(BFCT),MASK)
      BFBN(L,I)=BFCT
      GO  TO(485,495),K7
  485 K7=XBESSF(AR(I),KA(L),N1,XLOCF(BFCT),MASK)
      BFAN1(L,I)=BFCT
      GO  TO(486,496),K7
  486 K7=XBESSF(BR(I),KB(L),N1,XLOCF(BFCT),MASK)
      BFBN1(L,I)=BFCT
      GO  TO(490,497),K7
  490 CONTINUE
C     PARENTESIS AFTER E47A1430,   FROM E47CP001 TO E47CP015
C     WEIGHTING OF   I   AND   K   FUNCTIONS OF ARGUMENTS (ALPHA*B),(BETA*E
      GO  TO(1498,1491),I
 1491 GO  TO(1492,1498),NC3
 1492 GO  TO(1498,1496,1494),MB
 1494 ROFB=EXPF(SQRTF(1.-YE**2)*X)
      BFBN(1,2)=BFBN(1,2)*ROFB
      BFBN(2,2)=BFBN(2,2)/ROFB
      BFBN1(1,2)=BFBN1(1,2)*ROFB
      BFBN1(2,2)=BFBN1(2,2)/ROFB
 1496 ROFA=EXPF(SQRTF(1.-YE**2/ESQ)*X)
      BFAN(1,2)=BFAN(1,2)*ROFA
      BFAN(2,2)=BFAN(2,2)/ROFA
```

```
      BFAN1(1,2)=BFAN1(1,2)*ROFA
      BFAN1(2,2)=BFAN1(2,2)/ROFA
 1498 GO TO 500
  494 INDEX=1
      GO TO 499
  495 INDEX=2
      GO TO 499
  496 INDEX=3
      GO TO 499
  497 INDEX=4
  499 WRITE OUTPUT TAPE  3,49,INDEX,I,L,KA(L),KB(L),AR(I),BR(I),N,N1
   49 FORMAT(1H0,20H BESSEL ERROR RETURN//105H   INDEX      I          L
     1     P(ALPHA)  P(BETA)      ARG(ALPHA*R)          ARG(BETA*R)
     2 N          N+1/I6,I8,3I10,E21.8,E20.8,I9,I10)
      GO TO(1268,1368),MI
  500 CN(1,1,I)=2.*CONS1-(BRSQ(I)-ZRSQ(I))
      CN(1,2,I)=BR(I)*ZR(I)
      CN(1,3,I)=-2.*CONS1+BRSQ(I)
      CN(2,1,I)=-BRSQ(I)+ZRSQ(I)
      CN(2,2,I)=3.*BR(I)*ZR(I)
      CN(2,3,I)=-BRSQ(I)
      CN(3,1,I)=2.*ZR(I)*EN
      CN(3,2,I)=BR(I)*EN
      CN(3,3,I)=-ZR(I)*EN
      CN1(1,1,I)=-2.*AR(I)*(EN-1.)*EL1
      CN1(1,2,I)=0.
      CN1(1,3,I)=2.*(EN-1.)*BR(I)*EL2
      CN1(2,1,I)=2.*(EN+1.)*AR(I)*EL1
      CN1(2,2,I)=-4.*(EN+1.)*ZR(I)
      CN1(2,3,I)=2.*(EN+1.)*BR(I)*EL2
      CN1(3,1,I)=-2.*ZR(I)*AR(I)*EL1
      CN1(3,2,I)=-BRSQ(I)+ZRSQ(I)
      CN1(3,3,I)=0.
      DO 505 J=1,3
      DO 504 K=1,3
      CM(J,K,I)=CN(J,K,I)
      CM1(J,K,I)=CN1(J,K,I)
  504 CONTINUE
  505 CONTINUE
      DO 506 J=1,3
      CM1(J,1,I)=EL1*CM1(J,1,I)
      CM(J,2,I)=EL2*CM(J,2,I)
      CM1(J,3,I)=EL2*CM1(J,3,I)
  506 CONTINUE
      DO 550 JN=1,3
      GO TO(510,520),I
  510 J=JN
      GO TO 530
  520 J=JN+3
  530 DO 540 K=1,3
      GO TO(532,533,533),K
  532 TERM(J,K)=CN(JN,K,I)*BFAN(1,I)+CN1(JN,K,I)*BFAN1(1,I)
      GO TO 534
  533 TERM(J,K)=CN(JN,K,I)*BFBN(1,I)+CN1(JN,K,I)*BFBN1(1,I)
  534 GO TO(535,540),NC3
  535 L=K+3
```

```
      GO TO(536,537,537),K
  536 TERM(J,L)=CM(JN,K,I)*BFAN(2,I)+CM1(JN,K,I)*BFAN1(2,I)
      GO TO 540
  537 TERM(J,L)=CM(JN,K,I)*BFBN(2,I)+CM1(JN,K,I)*BFBN1(2,I)
  540 CONTINUE
  550 CONTINUE
  560 CONTINUE
      IF(N-1)610,565,565
  565 GO TO(570,600),NC3
  570 GO TO(1571,571),NX
  571 DO 576  J=1,6
      DO 575 K=1,6
      DET(J,K)=TERM(J,K)
  575 CONTINUE
  576 CONTINUE
      DER=-(-1.)**MB*DC
      M=XDETRMF(6,6,DET,DER)
      GO TO(580,590,595),M
  580 RESLT=DER
      GO TO(235,335),MI
  590 XN=X/PI
      WRITE OUTPUT TAPE 3,59,XN,Y,DC,DER,((TERM(L1,L2),L2=1,6),L1=1,6)
   59 FORMAT(1H0,40H OVERFLOW OR UNDERFLOW IN XDETRMF FOR X=F10.5,8H
     1  Y=F10.5,8H AND  D=E10.3,13H   D-CONTENT=E14.6//24H  DETERMINANT
     2TERMS(I,J)/(6E20.8))
  595 RESLT=0.
      GO TO(235,335),MI
  600 GO TO(1601,601),NX
  601 A=TERM(4,1)*TERM(5,2)*TERM(6,3)
      B=TERM(4,2)*TERM(5,3)*TERM(6,1)
      C=TERM(4,3)*TERM(5,1)*TERM(6,2)
      D=TERM(6,1)*TERM(5,2)*TERM(4,3)
      E=TERM(6,2)*TERM(5,3)*TERM(4,1)
      F=TERM(6,3)*TERM(5,1)*TERM(4,2)
      RESLT=((A-D)+(B-E)+(C-F))*DER
      GO TO(235,335),MI
  610 DO 615 K=1,6
      TERM(1,K)=TERM(1,K)+TERM(2,K)
  615 TERM(4,K)=TERM(4,K)+TERM(5,K)
      GO TO (620,670),NC3
  620 GO TO(1621,621),NX
  621 DO 625  K=1,5
      DET(1,K)=TERM(1,K)
      DET(2,K)=TERM(3,K)
      DET(3,K)=TERM(4,K)
      DET(4,K)=TERM(6,K)
  625 CONTINUE
      DO 630 J=1,4
      DET(J,3)=DET(J,4)
      DET(J,4)=DET(J,5)
  630 CONTINUE
      DO 636 K=1,4
      DO 635 J=1,4
      TERM(J,K)=DET(J,K)
  635 CONTINUE
  636 CONTINUE
```

```
      DER=-(-1.)**MB*DC
      M=XDETRMF(6,4,DET,DER)
      GO TO(640,650,660),M
  640 RESLT=DER
      GO TO(235,335),MI
  650 XN=X/PI
      WRITE OUTPUT TAPE 3,65,XN,Y,DC,DER,((TERM(L1,L2),L2=1,4),L1=1,4)
   65 FORMAT(1H0,40H OVERFLOW OR UNDERFLOW IN XDETRMF FOR X=F10.5,8H
     1  Y=F10.5,8H AND  D=E10.3,13H   D-CONTENT=E14.6//24H  DETERMINANT
     2TERMS(I,J)/(4E27.8))
  660 RESLT=0.
      GO TO (235,335),MI
  670 GO TO(1671,671),NX
  671 RESLT=(TERM(4,1)*TERM(6,2)-TERM(6,1)*TERM(4,2))*DER
      GO TO(235,335),MI
 1571 GO TO(1572,1590),NZM
 1572 DO 1573 K=1,6
      DET(1,K)=TERM(1,K)
      DET(2,K)=TERM(2,K)
      DET(3,K)=TERM(4,K)
      DET(4,K)=TERM(5,K)
 1573 CONTINUE
      DO 1574 J=1,4
      DET(J,2)=DET(J,3)
      DET(J,3)=DET(J,4)
      DET(J,4)=DET(J,6)
 1574 CONTINUE
      DER=DC
      M=XDETRMF(6,4,DET,DER)
      GO TO(1580,1581,1582),M
 1580 RESLT=DER
      GO TO 235
 1581 WRITE OUTPUT TAPE 3,5581,RN(IR),Y,DC,DER,((TERM(L1,L2),L2=1,6),L1=
     11,6)
 5581 FORMAT (///48H  OVERFLOW OR UNDERFLOW IN XDETRMF FOR X=0  A/H=F12.
     16,5H   Y=F10.5,8H AND  D=E10.3,13H   D-CONTENT=E14.6//24H  DETERMI
     2NANT TERMS(I,J)/(6E20.8))
 1582 RESLT=0.
      GO TO 235
 1590 RESLT=TERM(3,2)*TERM(6,5)-TERM(6,2)*TERM(3,5)
      GO TO 235
 1601 GO TO(1602,1604),NZM
 1602 RESLT=TERM(4,1)*TERM(5,3)-TERM(5,1)*TERM(4,3)
      GO TO 235
 1604 RESLT=TERM(6,2)
      GO TO 235
 1621 GO TO(1622,1590),NZM
 1622 RESLT=TERM(1,1)*TERM(4,4)-TERM(4,1)*TERM(1,4)
      GO TO 235
 1671 GO TO (1672,1604),NZM
 1672 RESLT=TERM(4,1)
      GO TO 235
 1271 GO TO(1272,1280),NZM
 1272 WRITE OUTPUT TAPE 3,5272,RN(IR),N,ANU,(ROOT(K),K=1,LRT)
 5272 FORMAT (///44H  EXTENSIONAL OR COUPLED CUT-OFF FREQUENCIES//15X,5H
     1H/R =F10.5,17X,3HN =I3,16X,4HNU =F5.3//57X,5HROOTS//(5E22.8))
```

```
C                 SAVE EXTENSIONAL ROOTS
      DO 1300 K=1,LRT
 1300 RTSV(K)=ROOT(K)
      LRT1=LRT+1
      NZM=NZM+1
      LRT=0
      GO TO 215
 1280 WRITE OUTPUT TAPE 3,5280,RN(IR),N,ANU,(ROOT(K),K=1,LRT)
 5280 FORMAT (///40H  LONGITUDINAL SHEAR CUT-OFF FREQUENCIES//15X,5HH/R
     1=F10.5,17X,3HN =I3,16X,4HNU =F5.3//57X,5HROOTS//(5E22.8))
C                 STORE      LONGIT ROOTS
      LRT2=LRT1+LRT-1
      DO 1301 K=LRT1,LRT2
       K1 = K-LRT1+1
 1301    RTSV(K) = ROOT(K1)
      DO 1302 K=1,LRT2
      DO 1302 K1=K,LRT2
       TEMP = MIN1F(RTSV(K),RTSV(K1))
        RTSV(K1) = MAX1F(RTSV(K),RTSV(K1))
       RTSV(K) = TEMP
 1302  CONTINUE
      ZERO=0.
      WRITE OUTPUT TAPE 5,6003,ZERO, (RTSV(K),K=1,LM)
      CALL RITE (15,EN,RN(IR),0.,LM,RTSV)
      GO TO 280
C     WRITE SPECIAL TERMINAL RECORD
 6000 CALL RITE(15,EN,RN(IR),-99.,1,0.)
      END FILE 15
      REWIND 15
      ENDFILE 5
       REWIND 5
      CALL EXIT
C     READ NEW X VALUES ONLY
 7000 READ INPUT TAPE 2,12,XLN,DXN,XMN,DELXN,EX1,EX2
         IF(EX1) 9001,9000,9001
 9000    EX1 = 1.E-5
 9001     IF(EX2) 9003,9002,9003
 9002    EX2 = 1.E-5
 9003    CONTINUE
      XL=PI*XLN
      DX=PI*DXN
      XM=PI*XMN
      DELX=PI*DELXN
      GO TO 140
      END(1,1,0,0,0,0,1,0,0,0,0,0,0,0,0)
```